STUDENT'S SOLUTIONS MANUAL

FINITE MATHEMATICS AND CALCULUS APPLIED TO THE REAL WORLD

Waner • Costenoble

Stefan Waner

Steven Costenoble

Hofstra University

ADDISON-WESLEY

An imprint of Addison Wesley Longman, Inc.

Reading, Massachusetts • Menlo Park, California • New York • Harlow, England
Don Mills, Ontario • Sydney • Mexico City • Madrid • Amsterdam

John and Suzanne Garlow, of Tarrant County Junior College, are the authors of the solutions for the calculus chapters 1, and 9-15.

Reproduced by Addison-Wesley Educational Publishers Inc. from camera-ready copy supplied by the authors.

ISBN   0-321-00457-4

3 4 5 6 7 8 9 10 ML 9998

CHAPTER 1
FUNCTIONS AND GRAPHS

1.1 EXERCISES

1. $P(-2, 4)$; $Q(1, 3)$; $R(-3, -4)$; $S(3, -3)$; $T(0, -2)$; $U(-9/2, 0)$; $V(5/2, 3/2)$; $W(5/2, 0)$.

3.

5.

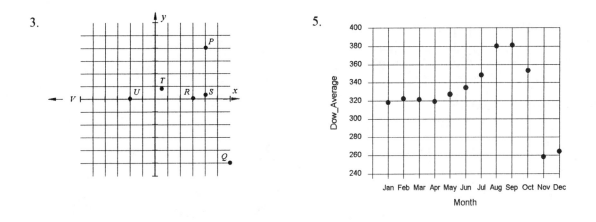

7.

Month	M	A	M	J	J	A	S	O	N	D	J	F	M
Deficit	-5.5	-7	-7.5	-6.5	-7.5	-8.5	-8.5	-7	-7.5	-6.5	-7.5	-8.0	-10.0

($Billions)

9. (a)

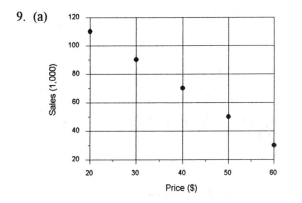

(b) A $10 increase in price has the effect of reducing sales by 20,000.

1

11.

Time	0	1	2	3	4
Height	500	480	440	350	250

13. (a)

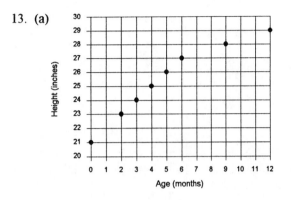

(b) Growth of 0.3 inches

15. $2x + y = 2$

$y = -2x + 2$

x	-3	0	4
y	8	2	-6

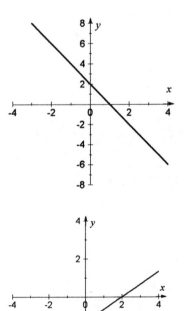

17. $2x - 3y = 4$

$y = \dfrac{2x}{3} - \dfrac{4}{3}$

x	-1	0	2
y	-2	-4/3	0

2

19. $y = -x^2 + 2$

x	-2	-1	0	1	2
y	-2	1	2	1	-2

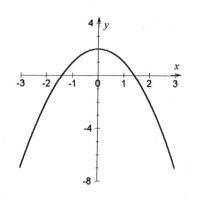

21. $y = \dfrac{1}{x^2}$

x	-2	-1	-1/2	-1/3	1/3	1/2	1	2
y	1/4	1	4	9	9	4	1	1/4

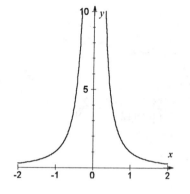

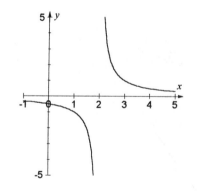

23. $y = \dfrac{1}{x-2}$

x	-1	0	1	3/2	5/2	3	4	5
y	-1/3	-1/2	-1	-2	2	1	1/2	1/3

25. $y = \dfrac{1}{(x+1)^2}$

x	-3	-2	$-3/2$	$-1/2$	0	0
y	$1/4$	1	4	4	1	$1/4$

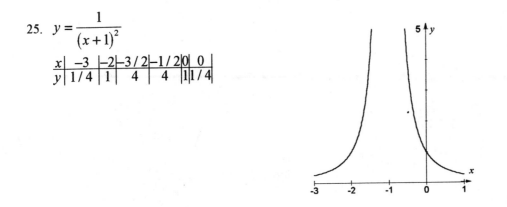

27. $y(x^2 + 2x + 1) = x$

$$y = \frac{x}{x^2 + 2x + 1}$$

x	-3	-2	$-3/2$	$-1/2$	0	1	$3/2$	2
y	$-3/4$	-2	-6	-2	0	$1/4$	$12/25$	$2/9$

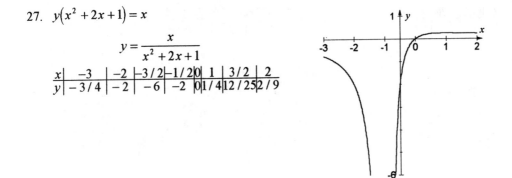

29. $y = \sqrt{x}(x-1)$

x	0	$1/4$	$1/2$	$3/4$	1	$3/2$	2	$5/2$	3	4
y	0	$-3/8$	$-\sqrt{2}/4$	$-\sqrt{3}/8$	0	$\sqrt{6}/4$	$\sqrt{2}$	$3\sqrt{10}/4$	$2\sqrt{3}$	6

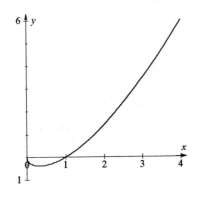

4

31. $y = x^3 - 2x - 5$

(a)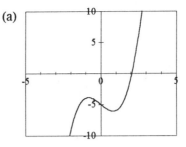

(b) Nothing is visible, since this viewing window is a square with sides of length 2 centered at the origin, and does not contain any points of the graph.

33. (a) $y = x^{1/2}(x-1)$
     (b) $y - 1 = x^{1/2}(x-1)$
     (c) $y - 2 = x^{1/2}(x-1)$
     (d) $y + 1 = x^{1/2}(x-1)$

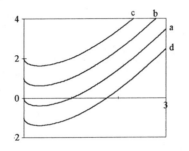

Replacing y with $(y-c)$ moves the curve up c units.

35. (a) $y = x^3 - 2x^2 + 4$
     (b) $y = (x-1)^3 - 2(x-1)^2 + 4$
     (c) $y = (x+1)^3 - 2(x+1)^2 + 4$
     (d) $y = (x-2)^3 - 2(x-2)^2 + 4$

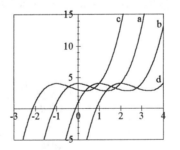

Replacing x with $(x-c)$ moves curve to the right c units.

37. $(-3, 4)$

39. $(-\infty, 0]$

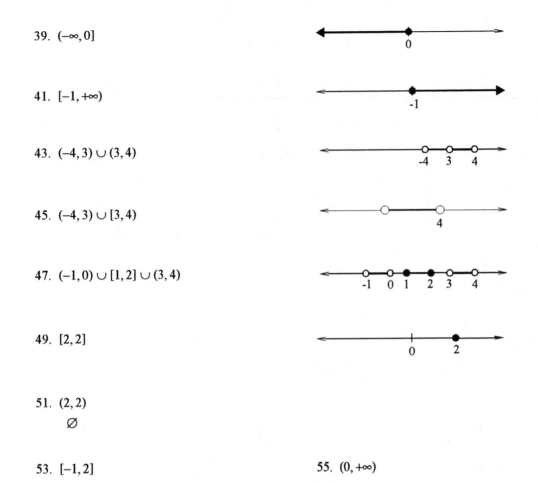

41. $[-1, +\infty)$

43. $(-4, 3) \cup (3, 4)$

45. $(-4, 3) \cup [3, 4)$

47. $(-1, 0) \cup [1, 2] \cup (3, 4)$

49. $[2, 2]$

51. $(2, 2)$
   $\varnothing$

53. $[-1, 2]$ 55. $(0, +\infty)$

57. $(-4, -2) \cup (-2, +\infty)$

1.2 EXERCISES

1. $f(x) = x^2 + 2x + 3$

   (a) $f(0) = 0^2 + 2(0) + 3 = 3$

   (b) $f(1) = 1^2 + 2(1) + 3 = 6$

   (c) $f(-1) = (-1)^2 + 2(-1) + 3 = 2$

   (d) $f(-3) = (-3)^2 + 2(-3) + 3 = 6$

   (e) $f(a) = a^2 + 2a + 3$

   (f) $f(x+h) = (x+h)^2 + 2(x+h) + 3$

   (g) $\dfrac{f(x+h) - f(x)}{h} = \dfrac{(x+h)^2 + 2(x+h) + 3 - (x^2 + 2x + 3)}{h}$

   $= \dfrac{x^2 + 2xh + h^2 + 2x + 2h + 3 - x^2 - 2x - 3}{h}$

   $= 2x + h + 2$

3. $g(s) = s^2 + \dfrac{1}{s}$

    (a) $g(1) = 1^2 + \dfrac{1}{1} = 2$

    (b) $g(-1) = (-1)^2 + \dfrac{1}{-1} = 0$

    (c) $g(4) = 4^2 + \dfrac{1}{4} = \dfrac{65}{4}$

    (d) $g(x) = x^2 + \dfrac{1}{x}$

    (e) $g(s+h) = (s+h)^2 + \dfrac{1}{(s+h)}$

    (f) $g(s+h) - g(s) = (s+h)^2 + \dfrac{1}{s+h} - \left(s^2 + \dfrac{1}{s} \right)$

5. $\phi(x) = \sqrt{x^2 + 3}$

    (a) $\phi(0) = \sqrt{0^2 + 3} = \sqrt{3}$

    (b) $\phi(-2) = \sqrt{(-2)^2 + 3} = \sqrt{7}$

    (c) $\phi(x+h) = \sqrt{(x+h)^2 + 3}$

    (d) $\phi(x) + h = \sqrt{x^2 + 3} + h$

7. $f(x) = -x^2 - 2x - 1$

$$\frac{f(x+h) - f(x)}{h} = \frac{-(x+h)^2 - 2(x+h) - 1 - (-x^2 - 2x - 1)}{h} = \frac{-x^2 - 2xh - h^2 - 2x - 2h - 1 + x^2 + 2x + 1}{h}$$

$$= -2x - h - 2$$

9. $f(x) = \dfrac{2}{x+1}$

$$\frac{f(x+h) - f(x)}{h} = \frac{\dfrac{2}{x+h+1} - \dfrac{2}{x+1}}{h} = \frac{2(x+1) - 2(x+h+1)}{h(x+h+1)(x+1)} = \frac{2x+2-2x-2h-2}{h(x+h+1)(x+1)} = \frac{-2}{(x+h+1)(x+1)}$$

11. $f(x) = x + \dfrac{1}{x}$

$$\frac{f(x+h) - f(x)}{h} = \frac{x+h+\dfrac{1}{x+h} - \left(x + \dfrac{1}{x} \right)}{h} = \frac{h + \dfrac{x-(x+h)}{x(x+h)}}{h} = \frac{h}{h} + \frac{-h}{hx(x+h)} = 1 - \frac{1}{x(x+h)}$$

13. $f(x) = \dfrac{1}{x^2}$

$$\frac{f(x+h)-f(x)}{h} = \frac{\dfrac{1}{(x+h)^2} - \dfrac{1}{x^2}}{h} = \frac{x^2 - (x+h)^2}{hx^2(x+h)^2} = \frac{x^2 - x^2 - 2xh - h^2}{hx^2(x+h)^2} = -\frac{2x+h}{x^2(x+h)^2}$$

15. $f(x) = x - \dfrac{1}{x^2}$, with domain $(0, +\infty)$

(a) $f(4) = 4 - \dfrac{1}{4^2} = \dfrac{63}{16}$, yes

(b) $f(0)$ is not defined

(c) $f(-1)$ is not defined

17. $f(x) = \sqrt{x+10}$, with domain $[-10, 0)$

(a) $f(0)$ is not defined

(b) $f(9)$ is not defined

(c) $f(-10) = \sqrt{-10+10} = 0$, yes

19. $f(x) = \sqrt{1-x}$, with domain $(-\infty, 1]$

(a) $f(0) = \sqrt{1-0} = 1$, yes

(b) $f(2)$ is not defined

(c) $f(-3) = \sqrt{1-(-3)} = 2$, yes

21. $f(x) = x^2 - 1$

$D(-\infty, +\infty)$

23. $g(x) = \sqrt{3x}$

$3x \ge 0$, $D[0, +\infty)$

25. $h(x) = \sqrt{x-1}$

$x - 1 \ge 0$

$x \ge 1$, $D[1, +\infty)$

27. $f(x) = \dfrac{1}{x}$

$x \ne 0$, $D(-\infty, 0) \cup (0, +\infty)$

29. $g(x) = 4 - \dfrac{1}{x^2}$

$x^2 \ne 0$

$x \ne 0$, $D(-\infty, 0) \cup (0, +\infty)$

31. $h(x) = \dfrac{1}{x-2}$

$x - 2 \ne 0$

$x \ne 2$, $D(-\infty, 2) \cup (2, +\infty)$

33. (a) $f(1) = 20$

(b) $f(2) = 30$

(c) $f(3) = 30$

(d) $f(5) = 22$

(e) $f(3) - f(2) = 0$

35. (a) $f(1) = 1.3$
    (b) $f(-2) = 0$
    (c) $f(0) = 2$
    (d) $f(3) = 0$
    (e) $f(3) - f(2) = -0.7$

37. (a) $f(-3) = -1$
    (b) $f(0) = 1.25$
    (c) $f(1) = 0$
    (d) $f(2) = 1$
    (e) $\dfrac{f(3) - f(2)}{3 - 2} = 0$

39. (a) $f(-3) = -0.5$
    (b) $f(-2) = 0$
    (c) $f(0) = 1$
    (d) $f(2) = 0$
    (e) $\dfrac{f(2) - f(0)}{2 - 0} = -\dfrac{1}{2}$

41. (a) I   (b) IV   (c) V   (d) VI   (e) III   (f) II

43. $f(x) = x^3$, with domain $\Re$

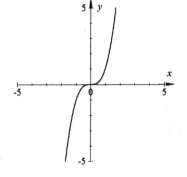

45. $f(x) = x^4$, with domain $\Re$

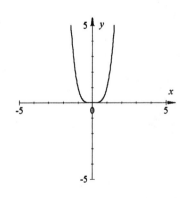

47. $f(x) = \dfrac{1}{x^2}$, with domain $(-\infty, 0) \cup (0, +\infty)$

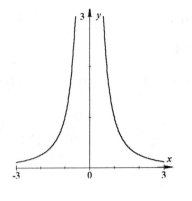

49. $f(x) = \begin{cases} x & \text{if } -4 \le x < 0 \\ 2 & \text{if } \quad 0 \le x \le 4 \end{cases}$

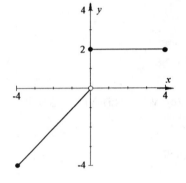

51. $f(x) = \begin{cases} x & \text{if } -1 < x \le 0 \\ x+1 & \text{if } \quad 0 < x \le 2 \\ x & \text{if } \quad 2 < x \le 4 \end{cases}$

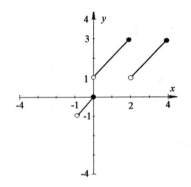

10

53. $f(x) = \begin{cases} x^2 & \text{if } -2 < x \le 0 \\ \dfrac{1}{x} & \text{if } \quad 0 < x \le 4 \end{cases}$

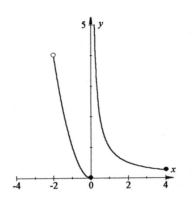

55. Greatest integer function I

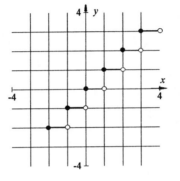

57. $f(x) = (x-1)(x-2)(x-3)(x-4)$, with domain $\Re$

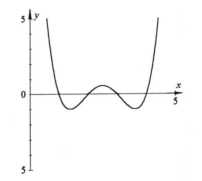

59. $f(x) = \dfrac{1}{(x-1)(x-2)(x-3)(x-4)}$

Domain: $(-\infty, 1) \cup (1, 2) \cup (2, 3) \cup (3, 4) \cup (4, +\infty)$

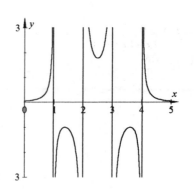

61. $f(x) = \begin{cases} \dfrac{1}{x^2+1} & \text{if } -2 \le x \le 0 \\[2mm] \dfrac{x}{x^2+1} & \text{if } \quad 0 < x \le 6 \end{cases}$

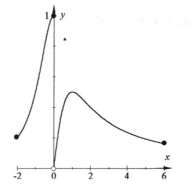

63. $f(x) = \dfrac{x^2}{(x-1)(x-2)(x-3)}, \quad -5 \le x \le 5$

Domain: $[-5, 1) \cup (1, 2) \cup (2, 3) \cup (3, 5]$

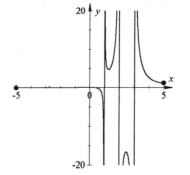

12

65. $f(x) = (x-1)(x+2)\sqrt{x} - \dfrac{1}{(x-1)(x-2)}$, $0 \le x \le 5$

  Domain: $[0, 1) \cup (1, 2) \cup (2, +\infty)$

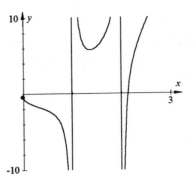

67. $f(x) = \sqrt{x}(x-1)$, with domain $(0, 5]$

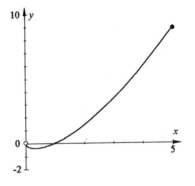

Lowest point is approximately $(0.333, -0.385)$

69. $f(x) = \dfrac{\sqrt{x}}{x-1}$

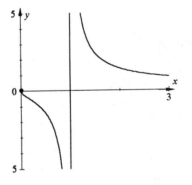

Domain: $[0, 1) \cup (1, +\infty)$
Never increasing.

13

71. $q(p) = 361{,}201 - (p+1)^2$

    (a) $q(50) = 361{,}201 - (50+1)^2 = 358{,}600$

    (b) $q(0) = 361{,}201 - (0+1)^2 = 361{,}200$

    (c) $\quad 0 = 361{,}201 - (p+1)^2$

        $(p+1)^2 = 361{,}201$

          $p+1 = 601$

            $p = 600$ cents $= \$6.00$

73. $n(t) = 5t^2 - 49t + 232$

    (a) $1994 - 1986 = 8$,   Domain: $[0, 8]$

    (b) $[0, +\infty]$ is not an appropriate domain, since it would predict investments in South Africa into the indefinite future with no basis. (It would also lead to preposterous results for large values of t).

75. $C(q) = 2000 + 100q^2$

    (a) $C(10) = 2000 + 100(10)^2 = \$12{,}000$

    (b) $N(q) = C(q) - S(q)$

          $= 2000 + 100q^2 - 500q$

    $N(20) = 2000 + 100(20)^2 - 500(20) = \$32{,}000$

77. Let $x =$ length of east and west sides

        $y =$ length of north and south sides and

        $A =$ area

    (a) If $A = 20$, $A = xy$ becomes

    $20 = xy$ or $y = \dfrac{20}{x}$, $x > 0$

    $C = 2x(4) + 2y(2)$

    $C(x) = 2x(4) + 2\left(\dfrac{20}{x}\right)(2)$

        $= 8x + \dfrac{80}{x}$ with domain $(0, +\infty)$

(b)

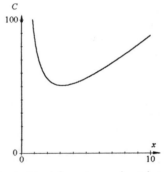

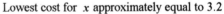

Lowest cost for x approximately equal to 3.2

79. $V = \dfrac{1}{3}\pi r^2 h$

    (a) $\dfrac{h}{r} = 3$ or $r = \dfrac{h}{3}$, $h \geq 0$

    $V(h) = \dfrac{1}{3}\pi\left(\dfrac{h}{3}\right)^2 h = \dfrac{1}{27}\pi h^3$ with domain $[0, +\infty)$

    (b) $h^3 = \dfrac{27V}{\pi}$

    $h = 3\left(\dfrac{V}{\pi}\right)^{1/3}$ with domain $[0, +\infty)$

14

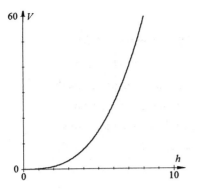

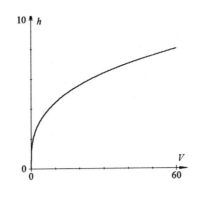

81. $R(p) = \dfrac{r}{1+kp}$

If $r = 45$ and $k = 1/8000$

$R(p) = \dfrac{45}{1 + \dfrac{p}{8000}}$

(a) Domain: $(0, +\infty)$

(b) $R(4000) = \dfrac{45}{1 + \dfrac{4000}{8000}} = 30$ per hour

(c)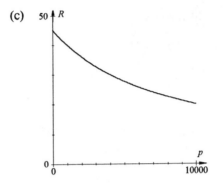

83. $\dfrac{Y - Y_p}{Y_p} = -9i$

If $i = 0.05$

$\dfrac{Y - Y_p}{Y_p} = -9(0.05)$

$Y = -0.45Y_p + Y_p$

$\quad = 0.55Y_p$ with domain $(0, +\infty)$

$Y(2) = 0.55(2) = \$1.1$ trillion

Inflation causes the actual GNP
to fall below the projected GNP.

15

85. $p(t) = 100\left(1 - \dfrac{12{,}196}{t^{4.478}}\right)$, $t \geq 8.5$

(a)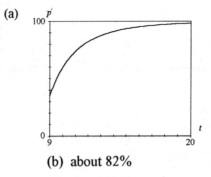

(b) about 82%

(c) about 13.7 months

87. $l(p) = l_0\sqrt{1-p^2}$, with domain $[0, 1]$

(a) $l(0.95) = 100\sqrt{1-(0.95)^2} \approx 31.22$

The rocket ship appears to be 31.22 meters in length.

(b) $\dfrac{1}{2}l_0 = l_0\sqrt{1-p^2}$

$\dfrac{1}{2} = \sqrt{1-p^2}$

$\dfrac{1}{4} = 1-p^2$

$p^2 = \dfrac{3}{4}$

$p = 0.8660$ warp, or 86.60% of the speed of light.

16

1.3 EXERCISES

1. $(0, 0)$ and $(1, 2)$
   $m = 2$

3. $(-1, -2)$ and $(0, 0)$
   $m = 2$

5. $(4, 3)$ and $(5, 1)$
   $m = -2$

7. $(1, -1)$ and $(2, -2)$
   $m = -1$

9. $(0, 1)$ and $\left(-\dfrac{1}{2}, \dfrac{3}{4}\right)$
   $m = \dfrac{1}{2}$

11. $\left(4, \sqrt{2}\right)$ and $\left(5, \sqrt{2}\right)$
    $m = 0$

13. $\left(4, \sqrt{2}\right)$ and $\left(5, 2\sqrt{2}\right)$
    $m = \sqrt{2}$

15. (a, a) and $(a, 3a)$, $a \neq 0$
    $m = \text{infinite}$

17. (a, b) and (c, d), $a \neq c$
    $m = \dfrac{d - b}{c - a}$

19. 1 and 3 are parallel

21. 5 and 7 are neither

23. 11 and 15 are perpendicular

25. 5 and 9 are perpendicular

27. Slopes of the lines are: (a) = (IV), (b) = (VII), (c) = (IX), (d) = (II), (e) = (I)
    (f) = (V), (g) = (VI), (h) = (III), (i) = (VIII).

In Exercises 29 - 43 use $m = \dfrac{y_2 - y_1}{x_2 - x_1}$ and $y - y_0 = m(x - x_0)$.

29. Through $(1, 3)$ with slope 3
    $y - 3 = 3(x - 1)$
    $y = 3x$

31. Through $\left(1, -\dfrac{3}{4}\right)$ with slope $\dfrac{1}{4}$
    $y - \left(-\dfrac{3}{4}\right) = \dfrac{1}{4}(x - 1)$
    $y = \dfrac{1}{4}x - 1$

17

33. Through $(2, -4)$ and $(1, 1)$

$$m = \frac{1 - (-4)}{1 - 2} = -5$$
$$y - 1 = -5(x - 1)$$
$$y = -5x + 6$$

35. Through $\left(1, -\frac{3}{4}\right)$ and $\left(\frac{1}{2}, \frac{3}{4}\right)$

$$m = \frac{\frac{3}{4} - \left(-\frac{3}{4}\right)}{\frac{1}{2} - 1} = -3$$
$$y - \frac{3}{4} = -3\left(x - \frac{1}{2}\right)$$
$$y = -3x + \frac{9}{4}$$

37. Through $(6, 6)$ and parallel to the line $x + y = 4$.

$l_1:\ y = -x + 4 \Rightarrow m_1 = -1$

$l_2:\ $ parallel, $m_2 = -1$

$$y - 6 = -1(x - 6)$$
$$y = -x + 12$$

39. Through $\left(\frac{1}{2}, 5\right)$ and parallel to the line $4x - 2y = 1$.

$l_1:\ -2y = -4x + 1$

$$y = 2x - \frac{1}{2} \Rightarrow m_1 = 2$$

$l_2:\ $ parallel, $m_2 = 2$

$$y - 5 = 2\left(x - \frac{1}{2}\right)$$
$$y = 2x + 4$$

41. Through $(0, 2)$ and perpendicular to the line $x + y = 4$.

$l_1:\ y = -x + 4 \Rightarrow m_1 = -1$

$l_2:\ $ perpendicular, $m_2 = -\dfrac{1}{-1} = 1$

$$y - 2 = 1(x - 0)$$
$$y = x + 2$$

43. Through $(3, -2)$ and perpendicular to the line $y = -2$.

$l_1:\ y = -2 \Rightarrow m_1 = 0$

$l_2:\ $ perpendicular, m_2 is infinite

$$x = 3$$

45. $2x + 3y = 6$

$\quad\quad y = -\dfrac{2}{3}x + 2$

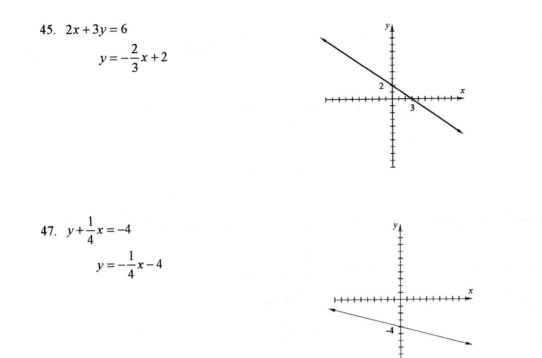

47. $y + \dfrac{1}{4}x = -4$

$\quad\quad y = -\dfrac{1}{4}x - 4$

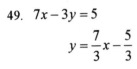

49. $7x - 3y = 5$

$\quad\quad y = \dfrac{7}{3}x - \dfrac{5}{3}$

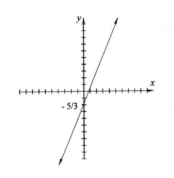

51. $3x = 8$

$x = \dfrac{8}{3}$

53. $6y = 9$

$y = \dfrac{3}{2}$

55. $2x = 3y$

$y = \dfrac{2}{3}x$

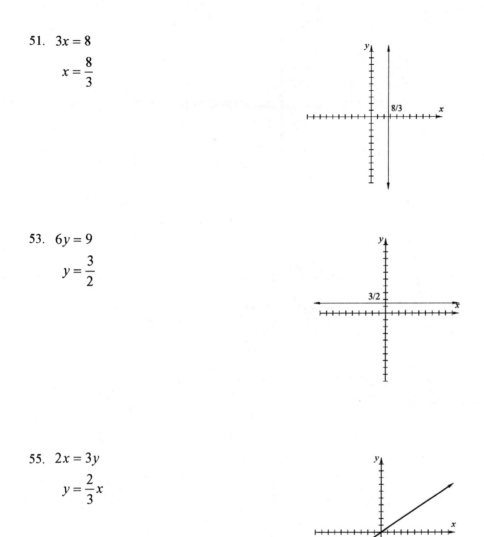

57. $y = 4.1x - 5.4$

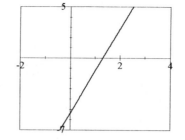

x - intercept: $x \approx 1.3$

Check: $0 = 4.1x - 5.4$

$$x = \frac{5.4}{4.1} = 1.\overline{3170}$$

59. $y = -10.4x + 10$

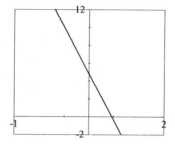

x - intercept: ≈ 1.0

Check: $0 = -10.4x + 10$

$$x = \frac{10}{10.4} \approx 0.96$$

61. $y = 10,050x + 4323$

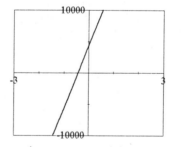
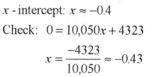

x - intercept: $x \approx -0.4$

Check: $0 = 10,050x + 4323$

$$x = \frac{-4323}{10,050} \approx -0.43$$

63. $13x - 15y = 23$

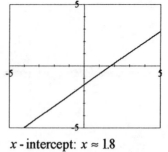

x - intercept: $x \approx 1.8$

Check: $13x = 23$

$$x = \frac{23}{13} \approx 1.77$$

65. $100x + 50y = -1020$

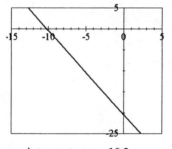

x - intercept: $x \approx -10.2$

Check: $100x = -1020$

$$x = -\frac{1020}{100} = -10.2$$

1.4 EXERCISES

1. $b = 1200$ and $m = 1500$

$$C(x) = mx + b$$
$$= 1500x + 1200$$

3. $(1, 1960)$ and $(5, 1800)$

$$m = \frac{q_2 - q_1}{p_2 - p_1} = \frac{1800 - 1960}{5 - 1} = -40$$
$$q - q_0 = m(p - p_0)$$
$$q - 1960 = -40(p - 1)$$
$$q = -40p + 2000$$

5. $b = 8.50$ and $m = 0.95$

(a) $R(t) = 0.95t + 8.50$

(b) $R(12) = 0.95(12) + 8.50 = \$19.9$ billion

(c) The model becomes unreasonable for large positive and negative values of t. It gives negative revenue for the year 1979 $(t = -9)$ and predicts revenues that rise without bound in the future.

7. $(0, 32)$ and $(100, 212)$

$$m = \frac{212 - 32}{100 - 0} = \frac{9}{5}$$
$$f - f_0 = m(c - c_0)$$
$$f - 32 = \frac{9}{5}(c - 0)$$
$$f = \frac{9}{5}c + 32$$
$$f(30) = \frac{9}{5}(30) + 32 = 86° \text{F}$$
$$f(22) = \frac{9}{5}(22) + 32 = 71.6° \text{F}$$
$$f(-10) = \frac{9}{5}(-10) + 32 = 14° \text{F}$$
$$f(-14) = \frac{9}{5}(-14) + 32 = 6.8° \text{F}$$

9. $b = 20$ and $m = 88$

$$C(x) = mx + b$$
$$C(x) = 88x + 20$$

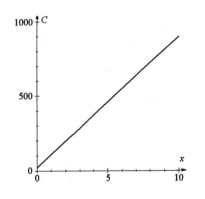

(a) $C(2) = 88(2) + 20 = \$196$

(b) $C(2) - C(1) = \$88$

(c) $C(4098) - C(4097) = \$88$

(d) Marginal cost $= m = \$88$

23

11. $b = 50,000$ and $m = 0.05$

$I = mN + b$

$I = 0.05N + 50,000$

If $I = 100,000$, then

$100,000 = 0.05N + 50,000$

$0.05N = 50,000$

$\qquad N = \$1,000,000$

Marginal income is $m = 5$ cents per dollar of net profit.

13. Let $C(x)$ be the total cost to produce x bicycles

$(100, 10,500)$ and $(120, 11,000)$

$m = \dfrac{11,000 - 10,500}{120 - 100} = 25$

$C - C_0 = m(x - x_0)$

$C - 10,500 = 25(x - 100)$

$\qquad C(x) = 25x + 8000$

Daily fixed cost $= \$8,000$

Marginal cost $= \$25$ per bicycle

15. $b = 60,000$ and $m = -0.05(b) = -3000$

(a) $v(n) = mn + b$

$\quad v(n) = -3000n + 60,000$

(b) $\quad 1000 = -3000n + 60,000$

$\quad -3000n = -59,000$

$\qquad n = 19.67$ years

(c) $0 = -3000n + 60,000$

$\qquad n = 20$ years

(d) After 20 years, the model predicts negative value.

17. Let $t(s)$ be the recovery time for s sets

$(0, 0)$ and $(3, 48)$

$m = \dfrac{48 - 0}{3 - 0} = 16$

$t(s) - t_0 = m(s - s_0)$

$t(s) - 0 = 16(s - 0)$

$\quad t(s) = 16s$

$\quad t(15) = 16(15) = 240$ hours or 10 days

This indicates that our linear model is reliable only for small numbers of sets. We need a nonlinear model to predict recovery time for large numbers of sets.

Use $P = R - C$ in Exercises 19 - 21.

19. $C = 5132,\ R = 100x$

$P(x) = 100x - 5132$ with domain $[0, 405]$

For a profit, $P(x) > 0$

$100x - 5132 > 0$

$\qquad x > 51.32$

$\qquad x \geq 52$ passengers

21. $R = 360(1.1) = 396$ in millions of dollars

$C = 0.05x$ in millions of dollars

(a) $P(x) = 396 - 0.05x$ with domain $[0,\ 1,100,000]$

(b) $\ -2000 = 396 - 0.05x$

$\quad -0.05x = -2396$

$\qquad x = 47,920$ homes damaged

Percentage of insured homes

$= \dfrac{47,920}{1,100,000} = 4.36\%$

23. $R = 2x$ with domain $[0, 5500]$

   $C = 0.4(2)x + 6000$

   $\quad = 0.8x + 6000$

   Break - even occurs when $R = C$

   $\quad 2x = 0.8x + 6000$

   $1.2x = 6000$

   $\quad x = 5000$ units sold

25. $R = SPx$

   $C = VC + FC$

   Break - even occurs when $R = C$

   $SPx = VCx + FC$

   $(SP - VC)x = FC$

   $$x = \frac{FC}{SP - VC}$$

27. $R = 50x$ and $C = 20x + 10,000$

   (a) $P(x) = 50x - (20x + 10,000)$

   $\quad = 30x - 10,000$ with domain $[0, +\infty)$

   (b) Let y be the number of new subscribers

   $C = 40y$ and $R = P(1000 + y)$

   Break - even occurs at $C = R$

   $40y = 30(1000 + y) - 10,000$

   $40y = 30,000 + 30y - 10,000$

   $10y = 20,000$

   $\quad y = 2000$ new customers

29. $(0, 55.2)$ and $(167, 15.7)$, $(1987 - 1820 = 167)$

   $$m = \frac{15.7 - 55.2}{167 - 0} = -0.2365$$

   $b(n) - b_0 = m(n - n_0)$

   $b(n) - 55.2 = -0.2365(n - 0)$

   $\quad b(n) = -0.2365n + 55.2$

   The rate drops to zero when

   $0 = -0.2365n + 55.2$

   $n = 233.4$ years or about midway through

   the year 2053.

   $(1820 + 233 = 2053)$

31. $m = 3000$ and $b = 2500$

   $p(n) = mn + b$

   $p(n) = 3000n + 2500$

   when $n = 2001 - 1992 = 9$,

   $p(9) = 3000(9) + 2500 = 29,500$ gallons

25

33. $(0, 12.2)$ and $(5, 10.8)$

$$m = \frac{10.8 - 12.2}{5 - 0} = -0.28$$

$$L(n) - L_0 = m(n - n_0)$$

$$L(n) - 12.2 = -0.28(n - 0)$$

$$L(n) = -0.28n + 12.2$$

The model makes sense for $n \geq 0$ and $L \geq 0$. If $L \geq 0$, then

$$-0.28n + 12.2 \geq 0$$

$$n \leq 43.57$$

Domain: $[0, 43.57]$

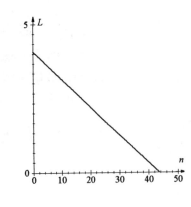

35. $P(x) = R(x) - C(x)$

$P(x) = 600x - (20{,}000 + 20x + 0.3x)$

$P(x) = 579.7x - 20{,}000$ with domain $[0, +\infty)$

Break-even

$0 = 579.7x - 20{,}000$

$x = 34.5$g per day

37. $q = 65.4 - 0.45p + 0.12b$

(a) $q(p) = 65.4 - 0.45p + 0.12(45)$

$q(p) = -0.45p + 70.8$ with domain $[0, +\infty)$

(b) $q(20) = -0.45(20) + 70.8$

$= 62$ pounds per day

(c) $0 = -0.45p + 70.8$

$p = \$1.57$ per pound

Domain: $[0, 157]$

(d) $R = qp$

$R(p) = (-0.45p + 70.8)p$

$R(p) = -0.45p^2 + 70.8p$ with domain $[0, 157]$

39. Let $n(t)$ be the number of cases per 100,000 people for t = number of years since 1920

(a) $(0, 245)$ and $(60, 30)$

$$m = \frac{30 - 245}{60 - 0} = -3.583$$

$$n(t) - n_0 = m(t - t_0)$$

$$n(t) - 245 = -3.583(t - 0)$$

$$n(t) = -3.583t + 245$$

(b)

year	1920	1930	1940	1950	1960	1970	1980	1990
t	0	10	20	30	40	50	60	70
$n(t)$	245	209	173	138	102	66	30	−5
difference	0	39	63	38	32	26	0	−55

The greatest discrepancy is for 1940.

(c) The model becomes unreliable for values of t beyond 60, predicting a negative number of cases in 1990.

41. $q = mp + b$

    (a) For $p = 2$ and $q = 3000$

    $3000 = 2m + b$

    For $p = 4$ and $q = 0$

    $0 = 4m + b \Rightarrow b = -4m$

    By substitution

    $3000 = 2m - 4m \Rightarrow m = -1500$

    and $b = -4(-1500) = 6000$

    $q(p) = -1500p + 6000$

    (b) $q(0) = -1500(0) + 6000$

    $\qquad = 6000$ per week

    (c) $R(p) = qp$

    $R(p) = (-1500p + 6000)p$

    $R(p) = -1500p^2 + 6000p$

(d)

p	1	1.5	2	2.5	2
$R(p)$	4500	5625	6000	5625	4500

Maximum $R = \$6000$ at
$p = \$2.00$ per hamburger.

43. (a) $m = 400$ and $b = 30,000$

    $C(x) = 400x + 30,000$ with domain $[0, 1000]$

    (b) $\overline{C}(x) = \dfrac{C(x)}{x} = 400 + \dfrac{30,000}{x}$ with domain $(0, 1000]$

    (c) $450 = 400 + \dfrac{30,000}{x}$

    $x = \dfrac{30,000}{50} = 600$ items per month

45. $\overline{C}(Q) = 350 + \dfrac{9000}{Q}$

    $\overline{R} = 500$

    Break - even occurs at $\overline{C} = \overline{R}$

    $500 = 350 + \dfrac{9000}{Q}$

    $Q = \dfrac{9000}{150} = 60$ units

1.5 EXERCISES

1. $f(x) = x^2 + 3x + 2$

$a = 1$, $b = 3$, $c = 2$

$a > 0$, concave up

VERTEX:

$$x = -\frac{b}{2a} = \frac{-3}{2(1)} = -\frac{3}{2}$$

$$y = f\left(-\frac{3}{2}\right) = \left(-\frac{3}{2}\right)^2 + 3\left(-\frac{3}{2}\right) + 2 = -\frac{1}{4}$$

$$V\left(-\frac{3}{2}, -\frac{1}{4}\right)$$

x-INTERCEPTS

$x^2 + 3x + 2 = 0$

$(x+2)(x+1) = 0$

$x + 2 = 0$ or $x + 1 = 0$

$x = -2$ $x = -1$

y-INTERCEPT

$y = c = 2$

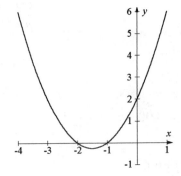

3. $f(x) = x^2 + x - 1$

$a = 1$, $b = 1$, $c = -1$

$a > 0$, concave up

VERTEX:

$$x = -\frac{b}{2a} = \frac{-(1)}{2(1)} = -\frac{1}{2}$$

$$y = f\left(-\frac{1}{2}\right) = \left(-\frac{1}{2}\right)^2 + \left(-\frac{1}{2}\right) - 1 = -\frac{5}{4}$$

$$V\left(-\frac{1}{2}, -\frac{5}{4}\right)$$

x-INTERCEPTS

$x^2 + x - 1 = 0$

$$x = \frac{-1 \pm \sqrt{5}}{2}$$

y-INTERCEPT

$y = c = -1$

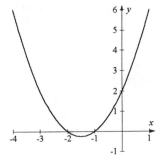

5. $f(x) = \dfrac{1}{4}x^2 + \sqrt{2}\,x - 1$

$a = \dfrac{1}{4},\ b = \sqrt{2},\ c = -1$

$a > 0$, concave up

VERTEX:

$x = -\dfrac{b}{2a} = -\dfrac{\sqrt{2}}{2(1/4)} = -2\sqrt{2}$

$y = f\left(-2\sqrt{2}\right) = \dfrac{1}{4}\left(-2\sqrt{2}\right)^2 + \sqrt{2}\left(-2\sqrt{2}\right) - 1 = -3$

$V\left(-2\sqrt{2},\, -3\right)$

x - INTERCEPTS

$\dfrac{1}{4}x^2 + \sqrt{2}\,x - 1 = 0$

$x = \dfrac{-\sqrt{2} \pm \sqrt{3}}{1/2} = 2\left(-\sqrt{2} \pm \sqrt{3}\right)$

y - INTERCEPT

$y = c = -1$

7. $f(x) = x^2 + 2x + 1$

$a = 1,\ b = 2,\ c = 1$

$a > 0$, concave up

VERTEX:

$x = -\dfrac{b}{2a} = -\dfrac{2}{2(1)} = -1$

$y = f(-1) = (-1)^2 + 2(-1) + 1 = 0$

$V(-1,\, 0)$

x - INTERCEPTS

$x^2 + 2x + 1 = 0$

$(x+1)(x+1) = 0$

$x + 1 = 0$

$x = -1$

y - INTERCEPT

$y = c = 1$

9. $f(x) = x^2$

   $a = 1,\ b = 0,\ c = 0$

   $a > 0$, concave up

   VERTEX:

   $x = -\dfrac{b}{2a} = -\dfrac{0}{2(1)} = 0$

   $y = f(0) = 0^2 = 0$

   $V(0, 0)$

   x - INTERCEPTS

   $x^2 = 0$

   $x = 0$

   y - INTERCEPT

   $y = c = 0$

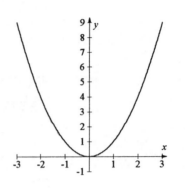

11. $f(x) = x^2 + 1$

   $a = 1,\ b = 0,\ c = 1$

   $a > 0$, concave up

   VERTEX:

   $x = -\dfrac{b}{2a} = -\dfrac{0}{2(1)} = 0$

   $y = f(0) = (0)^2 + 1 = 1$

   $V(0, 1)$

   x - INTERCEPTS

   $\Delta = b^2 - 4ac = 0^2 - 4(1)(1) = -4$

   No x - intercepts

   y - INTERCEPT

   $y = c = 1$

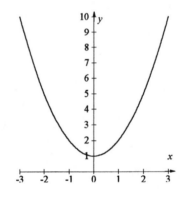

30

In Exercises 13 - 15 use p - coordinate of vertex $= -\dfrac{b}{2a}$ and R - intercept $= c$.

13. $q = -4p + 100$

$R = qp$

$R(p) = (-4p + 100)p$

$R(p) = -4p^2 + 100p$, concave down

VERTEX: $p = -\dfrac{100}{-8} = 12.5$

$R(12.5) = -4(12.5)^2 + 100(12.5) = 625$

$V(12.5, 625)$

p - INTERCEPTS

$(-4p + 100)p = 0$

$p = 0$ or $p = 25$

R - INTERCEPT: $R = 0$

Maximum revenue when $p = \$12.50$

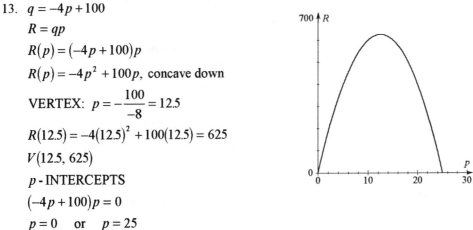

15. $q = -2p + 400$

$R = qp$

$R(p) = (-2p + 400)p$

$R(p) = -2p^2 + 400p$, concave down

VERTEX: $p = -\dfrac{400}{2(-2)} = 100$

$R(100) = -2(100)^2 + 400(100) = 20,000$

$V(100, 20,000)$

p - INTERCEPTS

$(-2p + 400)p = 0$

$p = 0$ or $p = 200$

R - INTERCEPT: $R = 0$

Maximum revenue when $p = \$100$

31

17. $(80, 100)$ and $(100, 90)$

$$m = \frac{90-100}{100-80} = -\frac{1}{2}$$

$$q - q_0 = m(p - p_0)$$

$$q - 100 = -\frac{1}{2}(p - 80)$$

$$q = -\frac{1}{2}p + 140$$

$$R = qp = \left(-\frac{1}{2}p + 140\right)p$$

$$R(p) = -\frac{1}{2}p^2 + 140p$$

VERTEX: $p = -\dfrac{140}{2(-1/2)} = 140$

$$R(140) = -\frac{1}{2}(140)^2 + 140(140) = 9800$$

Maximum revenue $= \$9,800$ when $p = \$140$

19. $(200{,}000, 40)$ and $(160{,}000, 60)$

$$m = \frac{60-40}{160{,}000 - 200{,}000} = -\frac{1}{2000}$$

$$q - q_0 = m(p - p_0)$$

$$q - 40 = -\frac{1}{2000}(p - 200{,}000)$$

$$q = -\frac{1}{2000}p + 140$$

$$R = qp = \left(-\frac{1}{2000}p + 140\right)p$$

$$R(p) = -\frac{1}{2000}p^2 + 140p$$

VERTEX: $p = -\dfrac{140}{2(-1/2000)} = 140{,}000$

$$R(140{,}000) = -\frac{1}{2000}(140{,}000)^2 + 140(140{,}000)$$

$$R(140{,}000) = 9{,}800{,}000$$

$$q = -\frac{1}{2000}(140{,}000) + 140 = 70$$

Maximum revenue $= \$9,800$ when $q = 70$ houses

21. $(25, 21.7)$ and $(14, 28.1)$

(a) $m = \dfrac{28.1 - 21.7}{14 - 25} = -0.5818$

$$q - q_0 = m(p - p_0)$$

$$q - 21.7 = -0.5818(p - 25)$$

$$q = -0.5818p + 36.2455$$

(b) $R = qp = (-0.5818p + 36.2455)p$

$$R(p) = -0.5818p^2 + 36.2455p$$

(c) VERTEX: $p = -\dfrac{36.2455}{2(-0.5818)} = 31.1484$

Largest revenue when
$p = 31$ cents per pound.

In Exercise 23 let v be the speed in miles per hour, t be the time in hours to burn 1 gallon, and R be the efficiency in miles per gallon.

23. $\left(30, \dfrac{1}{2}\right)$ and $\left(90, \dfrac{1}{6}\right)$

$m = \dfrac{\frac{1}{6} - \frac{1}{2}}{90 - 30} = -\dfrac{1}{180}$

$t - t_0 = m(v - v_0)$

$t - \dfrac{1}{2} = -\dfrac{1}{180}(v - 30)$

$t = -\dfrac{1}{180}v + \dfrac{2}{3}$

$R = tv = \left(-\dfrac{1}{180}v + \dfrac{2}{3}\right)v$

$R(v) = -\dfrac{1}{180}v^2 + \dfrac{2}{3}v$

VERTEX: $v = -\dfrac{-2/3}{2(-1/180)} = 60$

$R(60) = -\dfrac{1}{180}(60)^2 + \dfrac{2}{3}(60) = 20$

Maximum efficiency of 20 mpg at $v = 60$ mph.

25. $q = 500 - 40p$

$R = qp = (500 - 40p)p$

$\quad = 500p - 40p^2$

$C = 1000$

Break-even occurs when $R = C$

$\quad 500p - 40p^2 = 1000$

$-40p^2 + 500p - 1000 = 0$

$2p^2 - 25p + 50 = 0$

$(2p - 5)(p - 10) = 0$

$p = 2.5 \quad$ or $\quad p = 10$

The most you can charge is 10 cents per pound.

27. $h = v_0 t - 16t^2$

(a) If $v_0 = 64$,

$\quad h = 64t - 16t^2$, concave down

VERTEX: $t = -\dfrac{64}{2(-16)} = 2$

$h(2) = 64(2) - 16(2)^2 = 64$

t-INTERCEPTS

$64t - 16t^2 = 0$

$16t(4 - t) = 0$

$t = 0 \quad$ or $\quad t = 4$

h-INTERCEPT: $h = 0$

It returns to the ground when $t = 4$ seconds

(b) True; the time the ball is airborne is

given by $t = \dfrac{v_0}{16}$. Doubling v_0 results in doubling t.

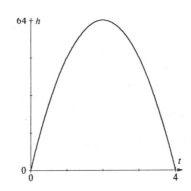

33

1.6 EXERCISES

1. $x^2 + 2x - 5$, to within ± 0.05

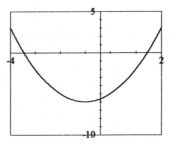

$x = -3.45 \pm 0.05$ and 1.45 ± 0.05

3. $-x^3 - 2x^2 + x - 1 = 0$, to within ± 0.01

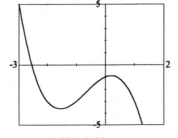

$x = -2.55 \pm 0.01$

5. $x^5 - 10x + 5 = 0$, to within ± 0.001

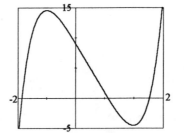

$x = -1.886 \pm 0.001$, 0.503 ± 0.001
and 1.622 ± 0.001

7. $x^7 - x^5 + x - 2 = 0$, to within ± 0.02

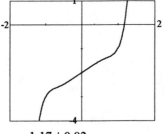

$x = 1.17 \pm 0.02$

9. $f(x) = x^2 + \frac{1}{x} - 4x$, to within ± 0.001

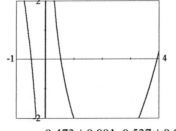

$x = -0.473 \pm 0.001$, 0.537 ± 0.001
and 3.935 ± 0.001

11. $f(x) = x^5 - x - 3$, to within ± 0.05

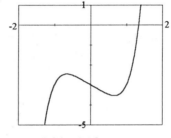

$x = 1.34 \pm 0.05$

13. $2^x = x$, to within ± 0.05

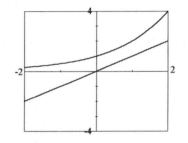

No solutions

15. $\frac{x^2+1}{x^2-1} = 2x - \sqrt{x}$, to within ± 0.05

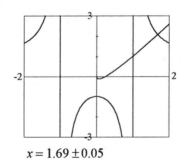

$x = 1.69 \pm 0.05$

17. $C(q) = 1000 + 100\sqrt{q}$

   $S(q) = 50q$

   They make a profit with
   32 or more employees.

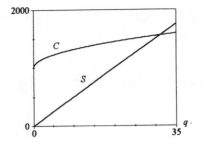

19. $A(r) = 2000\left(1 + \frac{r}{12}\right)^{120}$

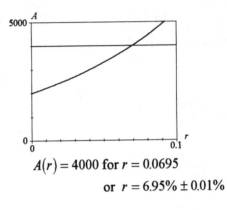

$A(r) = 4000$ for $r = 0.0695$

or $r = 6.95\% \pm 0.01\%$

CHAPTER 1   REVIEW EXERCISES

In Exercises 1 - 9 use $y - y_0 = m(x - x_0)$ and / or $m = \dfrac{y_2 - y_1}{x_2 - x_1}$.

1. Through $(0, 0)$ with $m = 3$

   $$y - 0 = 3(x - 0)$$
   $$3x - y = 0$$

3. Through $(1, -1)$ with $m = 3$

   $$y - (-1) = 3(x - 1)$$
   $$y + 1 = 3x - 3$$
   $$3x - y - 4 = 0$$

5. Through $(-3, -6)$ and $(1, -1)$

   $$m = \frac{-1 - (-6)}{1 - (-3)} = \frac{5}{4}$$
   $$y - (-1) = \frac{5}{4}(x - 1)$$
   $$4y + 4 = 5x - 5$$
   $$5x - 4y - 9 = 0$$

In Exercises 7 - 9 use $y - y_0 = m(x - x_0)$.

7. Through $(1, -2)$ and parallel to the
   line $x + 3y = 1$

   l_1: $3y = -x + 1$

   $$y = -\frac{1}{3}x + \frac{1}{3} \Rightarrow m_1 = -\frac{1}{3}$$

   l_2:   parallel, $m_2 = -\dfrac{1}{3}$

   $$y - (-2) = -\frac{1}{3}(x - 1)$$
   $$3y + 6 = -x + 1$$
   $$x + 3y + 5 = 0$$

9. Through $(1, -2)$ and perpendicular to the
   line $2x - y = 1$

   l_1: $-y = -2x + 1$

   $$y = 2x - 1 \Rightarrow m_1 = 2$$

   l_2:   perpendicular, $m_2 = -\dfrac{1}{2}$

   $$y - (-2) = -\frac{1}{2}(x - 1)$$
   $$2y + 4 = -x + 1$$
   $$x + 2y + 3 = 0$$

11. $2x + y = 6$

$\quad\quad y = -2x + 6$

x	0	2	3
y	6	2	0

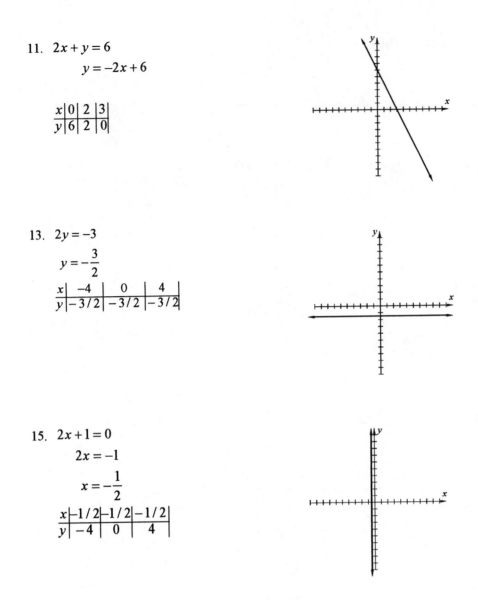

13. $2y = -3$

$\quad\quad y = -\dfrac{3}{2}$

x	-4	0	4
y	$-3/2$	$-3/2$	$-3/2$

15. $2x + 1 = 0$

$\quad\quad 2x = -1$

$\quad\quad x = -\dfrac{1}{2}$

x	$-1/2$	$-1/2$	$-1/2$
y	-4	0	4

17. $y = x^2 - 3x + 2$

x	0	1	3/2	2	3
y	2	0	$-1/4$	0	2

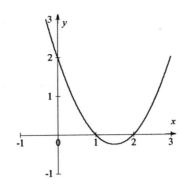

19. $y = -5x^2 + x - 2$

x	-1	0	1/10	2/10	1
y	-8	-2	$-39/20$	-2	-6

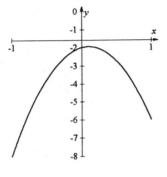

21. $y = x^2 - x - 1$

x	-1	0	1/2	1	2
y	1	-1	$-5/4$	-1	1

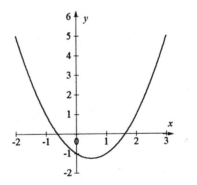

23. $f(x) = x^3$, with domain $(-\infty, 0)$

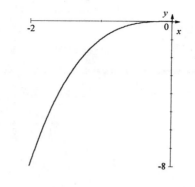

25. $f(x) = \sqrt{x}$, with domain $[0, 9]$

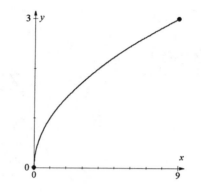

27. $f(x) = \dfrac{1}{|x|}$

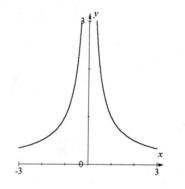

40

29. $f(x) = x^2 + x + 2$

x	-2	-1	$-1/2$	0	1
$f(x)$	4	2	$7/4$	2	4

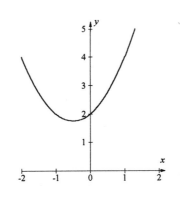

31. $f(x) = \dfrac{1}{(x+1)^2}$

x	-3	-2	$-3/2$	$-1/2$	0	1
$f(x)$	$1/4$	1	4	4	1	$1/4$

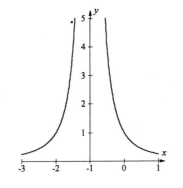

33. $f(x) = \sqrt{1 - x^2 - x}$

x	$-3/2$	-1	$-1/2$	0	$1/2$
$f(x)$	$1/2$	1	$\sqrt{5}/2$	1	$\sqrt{3}/2$

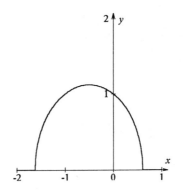

35. $x^5 - 4 = 0$

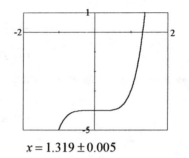

$x = 1.319 \pm 0.005$

37. $x^2 - 5\sqrt{x} = 0$

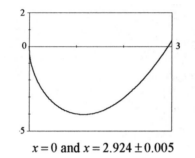

$x = 0$ and $x = 2.924 \pm 0.005$

39. $4 \doteq \sqrt{x} + \dfrac{1}{x^3}$

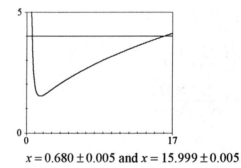

$x = 0.680 \pm 0.005$ and $x = 15.999 \pm 0.005$

41. $f(r) = \dfrac{1}{r+2}$

   (a) $f(0) = \dfrac{1}{0+2} = \dfrac{1}{2}$

   (b) $f(1) = \dfrac{1}{1+2} = \dfrac{1}{3}$

   (c) $f(-1) = \dfrac{1}{-1+2} = 1$

   (d) $f(x-2) = \dfrac{1}{x-2+2} = \dfrac{1}{x}$

   (e) $f(x^2+x) = \dfrac{1}{x^2+x+2}$

   (f) $f(x^2)+x = \dfrac{1}{x^2+2}+x$

43. $g(x) = \sqrt{x^2-1}$

   (a) $g(1) = \sqrt{(1)^2-1} = 0$

   (b) $g(-1) = \sqrt{(-1)^2-1} = 0$

   (c) $g(\sqrt{x+h}) = \sqrt{\left(\sqrt{x+h}\right)^2-1}$
   $= \sqrt{x+h-1}$

   (d) $g(\sqrt{x})+h = \sqrt{\left(\sqrt{x}\right)^2-1}+h$
   $= \sqrt{x-1}+h$

45. $f(x) = \dfrac{x^2+1}{x}$, with domain $(0,+\infty)$

   (a) $f(1) = \dfrac{(1)^2+1}{1} = 2$

   (b) $f(a^2) = \dfrac{(a^2)^2+1}{a^2} = \dfrac{a^4+1}{a^2}$

   (c) $f(x+h)-h = \dfrac{(x+h)^2+1}{x+h}-h$

   (d) $f(\sqrt{x})+h = \dfrac{(\sqrt{x})^2+1}{\sqrt{x}}+h$
   $= \dfrac{x+1}{\sqrt{x}}+h$

47. $f(x) = x^2+x-1$

   $\dfrac{f(x+h)-f(x)}{h} = \dfrac{(x+h)^2+(x+h)-1-(x^2+x-1)}{h} = \dfrac{x^2+2xh+h^2+x+h-1-x^2-x+1}{h}$
   $= 2x+h+1$

43

49. $f(x) = \dfrac{2}{2x-1}$

$$\dfrac{f(x+h)-f(x)}{h} = \dfrac{\dfrac{2}{2(x+h)-1} - \dfrac{2}{2x-1}}{h} = \dfrac{2(2x-1) - 2[2(x+h)-1]}{h[2(x+h)-1][2x-1]}$$

$$= \dfrac{4x-2-4x-4h+2}{h[2(x+h)-1][2x-1]} = \dfrac{-4}{[2(x+h)-1][2x-1]}$$

51. $(0, 96)$ and $(20, 186)$

$$m = \dfrac{186-96}{20-0} = \dfrac{9}{2}$$

$p(n) - p_0 = m(n - n_0),\ n = \text{age} - 25$

$$p(n) = 96 = \dfrac{9}{2}(n-0)$$

$$p(n) = \dfrac{9}{2}n + 96$$

$$p(65) = \dfrac{9}{2}(65) + 96 = \$388.50$$

53. $b = 45{,}000$ and $m = 55$

$s(t) = mt + b$

$s(t) = 55t + 45{,}000$

55. $(0, 0)$ and $(5, 160)$

$$m = \dfrac{160-0}{5-0} = 32$$

$V(t) - V_0 = m(t - t_0)$

$V(t) - 0 = 32(t - 0)$

$V(t) = 32t$

Acceleration = slope = 32 ft / s / s

57. Let s be annual sales and I be annual salary
$$b = 300(12) = 3600 \text{ and } m = 0.06$$
$$I(s) = ms + b$$
$$I(s) = 0.06s + 3600$$
If $I(s) = 21,600$
$$21,600 = 0.06s + 3600$$
$$18,000 = 0.06s$$
$$s = \$300,000$$

59. Let J be John's salary, D be Don's salary
and t be the time in years since 1980.
$$J(t) = 15,000 + 2450t$$
$$D(t) = 20,000 + 2000t$$
If $J(t) = D(t)$,
$$15,000 + 2450t = 20,000 + 2000t$$
$$450t = 5000$$
$$t = 11.1$$

Because John's increase each year is larger,
John's salary will be larger in the 12th year.

61. $R = qp = 120(10) = \$1200$ the first week
$(10, 120)$ and $(30, 0)$
$$m = \frac{0 - 120}{30 - 10} = -6$$
$$q(p) - q_0 = m(p - p_0)$$
$$q(p) - 0 = -6(p - 30)$$
$$q(p) = -6p + 180$$
$$R(p) = (-6p + 180)p$$
$$R(p) = -6p^2 + 180p$$
VERTEX: $p = -\dfrac{180}{2(-6)} = 15$
$$R(15) = -6(15)^2 + 180(15) = 1350$$
Maximum $R = \$1350$ when $p = \$15$

63. At break - even $R = C$ and $P = 0$
If $q = 400$ and $C = 400 + 200 = 600$

$R = qp$	Variable cost $= mq$
$600 = 400p$	$400 = m(400)$
$p = 1.5$	$m = 1$

$$P(q) = R(q) - C(q) = qp - (mq + b)$$
$$= 1.5q - (q + 200) = 0.5q - 200$$
Profit on 401st unit:
$$P(401) - P(400) = [0.5(401) - 200] - 0$$
$$= \$0.50$$

65. $q(p) = 40 - 18p - p^2$
(a) $q(1) = 40 - 18(1) - 1^2 = 21$
(b) $0 = 40 - 18p - p^2$
$$0 = (20 + p)(2 - p)$$
$$p = 2 \text{ or } p = -20 \text{ (not allowed)}$$
$$p = \$2$$

67. $(1000, 20{,}000)$ and $(2000, 30{,}000)$

(a) $m = \dfrac{30{,}000 - 20{,}000}{2000 - 1000} = 10$

$C(x) - C_0 = m(x - x_0)$

$C(x) - 20{,}000 = 10(x - 1000)$

$C(x) = 10x + 10{,}000$

(b) $\overline{C}(x) = 10 + \dfrac{10{,}000}{x}$

(c) $12.50 = 10 + \dfrac{10{,}000}{x}$

$2.50 = \dfrac{10{,}000}{x}$

$x = 4000$ fixtures

69. $l(p) = l_0\sqrt{1 - p^2}$ with domain $[0, 1]$

(a) $l(0.9) = 100\sqrt{1 - (0.9)^2}$

(b) $1 = 100\sqrt{1 - p^2}$

$0.01 = \sqrt{1 - p^2}$

$0.0001 = 1 - p^2$

$p^2 = 0.9999$

$p = 0.9999$

$= 99.995\%$ the speed of light

(c) $l(1) = l_0\sqrt{1 - 1^2} = 0$

Its apparent length would be zero.

71. Let I be the monthly earnings

(a) Commission $= 0.05\sqrt{3025x} = 2.75\sqrt{x}$

Basic wage $= 100$

$I(x) = 2.75\sqrt{x} + 100$

(b) $I(1) = 2.75\sqrt{1} + 100 = \$102.75$

$I(100) = 2.75\sqrt{100} + 100 = \$127.50$

(c) $200 = 2.75\sqrt{x} + 100$

$100 = 2.75\sqrt{x}$

$\sqrt{x} = 36.36$

$x = 1322$ sets per month

They will have difficulty hiring salespersons.

73. $(28, 30)$ and $(48, 65)$

(a) $m = \dfrac{65 - 30}{48 - 28} = \dfrac{7}{4} - 1.75$

$S(x) - S_0 = m(x - x_0)$

$S(x) - 30 = \dfrac{7}{4}(x - 28)$

$S(x) = 1.75x - 19$ with domain $[28, 48]$

(b) $S(x) = \begin{cases} 1.75x - 19 & \text{if } 28 \le x \le 48 \\ x + 17 & \text{if } 48 \le x \le 54 \\ 71 & \text{if } 54 \le x \le 58 \end{cases}$

75. $p = \dfrac{500}{q + 100}$

$q + 100 = \dfrac{500}{p}$

$q = \dfrac{500}{p} - 100$

This gives the demand corresponding to the price p.

46

Chapter 2

Section 2.1 Systems of Two Linear Equations in Two Unknowns

Pages 120–123

1. $x - y = 0$
   $x + y = 4$

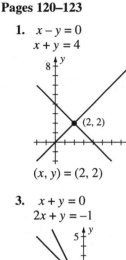

$(x, y) = (2, 2)$

3. $x + y = 0$
   $2x + y = -1$

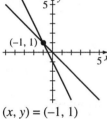

$(x, y) = (-1, 1)$

5. $x + y = 4$
   $x - y = 2$

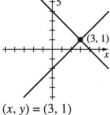

$(x, y) = (3, 1)$

7. Equation 1 is $3x - 2y = 6$. Equation 2 is $2x - 3y = -6$.
   Multiply equation 1 by -3 and equation 2 by 2, and add.
   $$\begin{array}{r} -9x + 6y = -18 \\ \underline{4x - 6y = -12} \\ -5x = -30 \\ x = 6 \end{array}$$
   Substitute $x = 6$ into equation 1.
   $3(6) - 2y = 6$
   $-2y = -12$
   $y = 6$
   $(x, y) = (6, 6)$

9. Equation 1 is $0.5x + 0.1y = 0.7$.
   Equation 2 is $0.2x - 0.2y = 0.6$.
   Multiply equation 1 by 10 to give equation 3 and equation 2 by 5 to give equation 4, and add.
   $$\begin{array}{r} 5x + y = 7 \\ \underline{x - y = 3} \\ 6x = 10 \\ x = \dfrac{5}{3} \end{array}$$
   Substitute $\dfrac{5}{3}$ for x in equation 4.
   $$\dfrac{5}{3} - y = 3$$
   $$y = \dfrac{5}{3} - 3 = -\dfrac{4}{3}$$
   $$(x, y) = \left(\dfrac{5}{3}, -\dfrac{4}{3} \right)$$

11. Equation 1 is $\dfrac{x}{3} - \dfrac{y}{2} = 1$. Equation 2 is $\dfrac{x}{4} + y = -2$.
    Multiply equation 1 by 6 and equation 2 by -8, then add.
    $$\begin{array}{r} 2x - 3y = 6 \\ \underline{-2x - 8y = 16} \\ -11y = 22 \\ y = -2 \end{array}$$
    Substitute -2 for y in equation 1.
    $$\dfrac{x}{3} - \dfrac{(-2)}{2} = 1$$
    $$\dfrac{x}{3} = 0$$
    $x = 0$
    $(x, y) = (0, -2)$

13. Equation 1 is $2x + 3y = 1$. Equation 2 is
    $-x - \dfrac{3y}{2} = -\dfrac{1}{2}$. Multiply equation 2 by 2, then add.
    $$\begin{array}{r} 2x + 3y = 1 \\ \underline{-2x - 3y = -1} \\ 0 = 0 \end{array}$$
    x is arbitrary.
    Solve for y in Equation 1.
    $$y = \dfrac{1}{3}(-2x + 1)$$
    There are an infinite number of solutions.
    $$(x, y) = \left(x, \dfrac{1}{3}(1 - 2x) \right)$$

15. Equation 1 is $2x + 3y = 2$. Equation 2 is
$-x - \dfrac{3y}{2} = -\dfrac{1}{2}$. Leave equation 1 as is and multiply
equation 2 by 2, then add.

$$\begin{array}{r} 2x + 3y = 2 \\ -2x - 3y = -1 \\ \hline 0 = 1 \end{array}$$

False, no solution

17. Equation 1 is $x = 2y$. Equation 2 is $y = x - 10$.
Substitute $x - 10$ for y into Equation 1.
$x = 2(x - 10)$
$x = 2x - 20$
$20 = x$
Substitute 20 for x in equation 2.
$y = 20 - 10 = 10$
$(x, y) = (20, 10)$

19. Equation 1 is $2x + 10 = 3y$ or, rearranged,
$2x - 3y = -10$. Equation 2 is $3x + 10 = 4y$, or,
rearranged, $3x - 4y = -10$.
Multiply equation 1 by 4 and equation 2 by -3, then
add.

$$\begin{array}{r} 8x - 12y = -40 \\ -9x + 12y = 30 \\ \hline -x = -10 \\ x = 10 \end{array}$$

Substitute 10 for x in equation 1.
$2(10) + 10 = 3y$
$10 = y$
$(x, y) = (10, 10)$

21. Let x = number of members of House voting *for* the
bill and y = number of members of House voting
against the bill. The total number of members of the
house is 435. Therefore $x + y = 435$. There were 49
more for the bill than against, so $x = y + 49$.

$$\begin{array}{r} x + y = 435 \\ x = y + 49 \\ \hline (y + 49) + y = 435 \\ 2y = 386 \\ y = 193 \end{array}$$

$x = y + 49 = 193 + 49 = 242$
$(x, y) = (242, 193)$
242 were *for* and 193 were *against*.

23. Let x = number of quarts of Creamy Vanilla and
y = number of quarts of Continental Mocha.

	x (qt)	y (qt)	Available
Eggs	2	1	500
Cream (cups)	3	3	900

eggs: $2x + y = 500$
cream: $3x + 3y = 900$
Divide the cream equation by -3 and add.

$$\begin{array}{r} 2x + y = 500 \\ -x - y = -300 \\ \hline x = 200 \end{array}$$

$2(200) + y = 500$
$y = 100$
$(x, y) = (200, 100)$
200 quarts of Creamy Vanilla and 100 quarts of
Continental Mocha.

25. Let x = number of servings of Mixed Cereal and
y = number of servings of Mango Tropical Fruit
Dessert.

Each serving:	Cereal x	Dessert y	Total to be provided
Calories	60	80	200
Carbohydrates (gm)	11	21	43

calories: $60x + 80y = 200$
carbohydrates: $11x + 21y = 43$
Divide calorie equation by 20 and multiply the result
by -11. Multiply the carbohydrate equation by 3 and
add.

$$\begin{array}{r} -33x - 44y = -110 \\ 33x + 63y = 129 \\ \hline 19y = 19 \\ y = 1 \end{array}$$

$3x + 4y = 10$
$3x + 4(1) = 10$
$3x = 6$
$x = 2$
$(x, y) = (2, 1)$
2 servings of Mixed Cereal and 1 serving of Mango
Tropical Fruit Dessert

27. Let x = amount invested in Fidelity Market Index Fund and y = amount invested in Vanguard Small Capitalization fund.

	Fidelity	Vanguard	Total
Amt invested	x	y	10,000
Growth in 1 yr	$1.1x$	$1.13y$	1.12(10,000)

Equation 1 is $x + y = 10{,}000$, and equation 2 is $1.1x + 1.13y = 11{,}200$. Multiply equation 1 by -1.13 and add.

$$-1.13x - 1.13y = -11{,}300$$
$$\underline{1.10x + 1.13y = 11{,}200}$$
$$-0.03x = -100$$
$$x = 3333.33$$

Substitute 3333.33 for x in equation 1.
$x + y = 10{,}000$
$3333.33 + y = 10{,}000$
$y = 6666.67$
$(x, y) \approx (3333.33, 6666.67)$
\$3333.33 in Fidelity and \$6666.67 in Vanguard

29. Let x = number of soccer games won and y = number of football games won.

	Soccer	Football	Total
Games won	x	y	12
Points earned	$2x$	$4y$	38

Equation 1 is $x + y = 12$, and equation 2 is $2x + 4y = 38$. Multiply equation 1 by 4 and equation 2 by -1, and add.

$$4x + 4y = 48$$
$$\underline{-2x - 4y = -38}$$
$$2x = 10$$
$$x = 5$$

$x + y = 12$
$5 + y = 12$
$y = 7$
$(x, y) = (5, 7)$
5 soccer games and 7 football games

31. Let x = number of Brand X pens purchased and y = number of Brand Y pens purchased.

	Brand X	Brand Y	total
No. purchased	x	y	12
Cost (\$)	$4x$	$2.8y$	42

Equation 1 is $x + y = 12$ and equation 2 is $4x + 2.8y = 42$. Multiply equation 1 by -28 and equation 2 by 10, and add.

$$-28x - 28y = -336$$
$$\underline{40x + 28y = 420}$$
$$12x = 84$$
$$x = 7$$

$x + y = 12$
$7 + y = 12$
$y = 5$
$(x, y) = (7, 5)$
7 Brand X pens and 5 Brand Y pens

33. Let x = number of acidified lakes in the NY Adirondacks and y = number of acidified lakes in Florida.
Total: $x + y = 664$
Florida has 302 more than the Adirondacks:
$y = x + 302$

a. $x + (x + 302) = 664$
$2x = 362$
$x = 181$

$x + y = 664$
$181 + y = 664$
$y = 483$
$(x, y) = (181, 483)$
181 acidified lakes in the Adirondacks and 483 acidified lakes in Florida

b. Adirondacks: $\dfrac{181}{1290} \approx 0.1403 \approx 14\%$

Florida: $\dfrac{483}{2098} \approx 0.2302 \approx 23\%$

Approximately 14% of the lakes in the Adirondacks are polluted, while approximately 23% of Florida lakes are polluted. Thus, Florida has the worse record.

35. Let x = number of dirty socks and y = number of dirty T-shirts.
Total: $x + y = 44$
3 times number of dirty socks as T-shirts: $x = 3y$
$(3y) + y = 44$
$4y = 44$
$y = 11$

$x + y = 44$
$x + 11 = 44$
$x = 33$
$(x, y) = (33, 11)$
33 dirty socks and 11 dirty T-shirts

37. Let x = number of trips for two to the Bahamas and y = number of VCRs.
Cost: $800x + 200y = 18{,}000$
Twice as many VCRs as trips to Bahamas: $y = 2x$
$800x + 200(2x) = 18{,}000$
$1200x = 18{,}000$
$x = 15$
$y = 2x = 2(15) = 30$
15 trips for two to the Bahamas and 30 VCRs

39. Let x = amount of raise of each part-time employee and y = amount of raise of each full-time employee.
Amount budgeted: $2x + 4y = 6000$
Raise of full-time is twice raise of part-time: $y = 2x$
Divide $2x + 4y = 6000$ by 2 to give $x + 2y = 3000$.
Substitute $2x$ for y.
$x + 2(2x) = 3000$
$5x = 3000$
$x = 600$
$y = 2x = 2(600) = 1200$
$(x, y) = (600, 1200)$
full-time: \$1200 raise
part-time: \$600 raise

41. Demand: $q = -1000p + 140{,}000$
Supply: $q = 2000p + 20{,}000$
Demand = Supply
$-1000p + 140{,}000 = 2000p + 20{,}000$
$-3000p = -120{,}000$
$p = 40$
\$40 book makes supply and demand balance.

43. Demand: $D = 85 - 5P$
Supply: $S = 25 + 5P$
Equilibrium: demand = supply
$D = S$
$85 - 5P = 25 + 5P$
$-10P = -60$
$P = 6$
$D = 85 - 5P = 85 - 5(6) = 55$
55 widgets are bought and sold when widget market is in equilibrium.

45. Let x = number of servings of beans and y = number of slices of bread

a.

	Beans, x	Bread, y	Total
Protein (g)	5	4	$\frac{1}{3}(60)$
Carbohydrates (g)	21	12	80

Equation 1 is $5x + 4y = 20$ and equation 2 is $21x + 12y = 80$. Multiply equation 1 by -3 and add.
$$
\begin{array}{r}
-15x - 12y = -60 \\
21x + 12y = 80 \\
\hline
6x = 20
\end{array}
$$
$$x = 3\frac{1}{3}$$
$$5\left(\frac{10}{3}\right) + 4y = 20$$
$$4y = 20 - \frac{50}{3} = \frac{10}{3}$$
$$y = \frac{5}{6}$$
$$(x, y) = \left(3\frac{1}{3}, \frac{5}{6}\right)$$
$3\frac{1}{3}$ servings of beans, and $\frac{5}{6}$ slice of bread

b. Equation 1 is $5x + 4y = 20$ and equation 2 is $21x + 12y = 60$. Multiply equation 1 by -1 and divide equation 2 by 3 and add the results together.
$$
\begin{array}{r}
-5x - 4y = -20 \\
7x + 4y = 20 \\
\hline
2x = 0 \\
x = 0
\end{array}
$$

$0 + 4y = 20$
$y = 5$
$(x, y) = (0, 5)$
Yes; no beans and 5 slices of bread

47. a. $q = mp + b$
5500 cans at \$4 per can: $(4, 5500)$
100 cans at \$10 per can: $(10, 100)$
$$m = \frac{100 - 5500}{10 - 4} = -\frac{5400}{6} = -900$$
Using $(10, 100)$:
$100 = -900(10) + b$
$b = 9100$
Demand: $q = -900p + 9100$

b. 1900 cans at \$2 per can: $(2, 1900)$
3300 cans at \$4 per can: $(4, 3300)$
$$m = \frac{3300 - 1900}{4 - 2} = \frac{1400}{2} = 700$$
Using $(2, 1900)$:
$1900 = 700(2) + b$
$b = 500$
Supply: $q = 700p + 500$

c. supply = demand
$700p + 500 = -900p + 9100$
$1600p = 8600$
$p = 5.375$
$p = \$5.38$ per can (to the nearest cent) to balance supply and demand.

49. The three lines intersect at a common point.

51. It is possible if the slopes are not equal and the graph of the lines intersect for a positive number of product and price.

53. Applications will vary.

Section 2.2 Using Matrices to Solve Systems with Two Unknowns
Pages 137–138

1. $\begin{bmatrix} -\frac{2}{3} & \frac{1}{3} & -3 \\ \frac{1}{3} & -\frac{2}{3} & \frac{11}{4} \end{bmatrix} \begin{matrix} 3R_1 \\ 12R_2 \end{matrix} \rightarrow \begin{bmatrix} -2 & 1 & -9 \\ 4 & -8 & 33 \end{bmatrix} 2R_1 + R_2 \rightarrow$

$\begin{bmatrix} -2 & 1 & -9 \\ 0 & -6 & 15 \end{bmatrix} 6R_1 + R_2 \rightarrow$

$\begin{bmatrix} -12 & 0 & -39 \\ 0 & -6 & 15 \end{bmatrix} \begin{matrix} -\frac{1}{12}R_1 \\ -\frac{1}{6}R_2 \end{matrix} \rightarrow \begin{bmatrix} 1 & 0 & \frac{13}{4} \\ 0 & 1 & -\frac{5}{2} \end{bmatrix}$

3. $\begin{bmatrix} -2 & 1 & -3 \\ 1 & 3 & 1 \\ -1 & 4 & -2 \end{bmatrix} \begin{matrix} \\ 2R_2 + R_1 \rightarrow \\ -2R_3 + R_1 \rightarrow \end{matrix}$

$\begin{bmatrix} -2 & 1 & -3 \\ 0 & 7 & -1 \\ 0 & -7 & 1 \end{bmatrix} \begin{matrix} -7R_1 + R_2 \rightarrow \\ \\ R_3 + R_2 \rightarrow \end{matrix} \begin{bmatrix} 14 & 0 & 20 \\ 0 & 7 & -1 \\ 0 & 0 & 0 \end{bmatrix} \begin{matrix} \frac{1}{14}R_1 \rightarrow \\ \frac{1}{7}R_2 \rightarrow \\ \end{matrix}$

$\begin{bmatrix} 1 & 0 & \frac{10}{7} \\ 0 & 1 & -\frac{1}{7} \\ 0 & 0 & 0 \end{bmatrix}$

5. $\begin{bmatrix} -2 & 1 & -3 \\ 1 & 4 & 1 \\ -1 & 4 & -2 \end{bmatrix} \begin{matrix} \\ 2R_2 + R_1 \rightarrow \\ -2R_3 + R_1 \rightarrow \end{matrix}$

$\begin{bmatrix} -2 & 1 & -3 \\ 0 & 9 & -1 \\ 0 & -7 & 1 \end{bmatrix} \begin{matrix} -9R_1 + R_2 \rightarrow \\ \\ 9R_3 + 7R_2 \rightarrow \end{matrix}$

$\begin{bmatrix} 18 & 0 & 26 \\ 0 & 9 & -1 \\ 0 & 0 & 2 \end{bmatrix} \begin{matrix} R_1 - 13R_3 \rightarrow \\ 2R_2 + R_3 \rightarrow \\ \end{matrix}$

$\begin{bmatrix} 18 & 0 & 0 \\ 0 & 18 & 0 \\ 0 & 0 & 2 \end{bmatrix} \begin{matrix} \frac{1}{18}R_1 \rightarrow \\ \frac{1}{18}R_2 \rightarrow \\ \frac{1}{2}R_3 \rightarrow \end{matrix} \begin{bmatrix} 1 & 0 & 0 \\ 0 & 1 & 0 \\ 0 & 0 & 1 \end{bmatrix}$

7. $x + y = 4$
$x - y = 2$

$\begin{bmatrix} 1 & 1 & 4 \\ 1 & -1 & 2 \end{bmatrix} \begin{matrix} \\ R_2 - R_1 \rightarrow \end{matrix} \begin{bmatrix} 1 & 1 & 4 \\ 0 & -2 & -2 \end{bmatrix} \begin{matrix} 2R_1 + R_2 \rightarrow \\ -R_2 \rightarrow \end{matrix}$

$\begin{bmatrix} 2 & 0 & 6 \\ 0 & 2 & 2 \end{bmatrix} \begin{matrix} \frac{1}{2}R_1 \rightarrow \\ \frac{1}{2}R_2 \rightarrow \end{matrix} \begin{bmatrix} 1 & 0 & 3 \\ 0 & 1 & 1 \end{bmatrix}$

Thus, $x = 3$, $y = 1$.
$(x, y) = (3, 1)$

9. $3x - 2y = 5$
$2x - 3y = -5$

$\begin{bmatrix} 3 & -2 & 5 \\ 2 & -3 & -5 \end{bmatrix} \begin{matrix} \\ 3R_2 - 2R_1 \rightarrow \end{matrix} \begin{bmatrix} 3 & -2 & 5 \\ 0 & -5 & -25 \end{bmatrix} -\frac{1}{5}R_2 \rightarrow$

$\begin{bmatrix} 3 & -2 & 5 \\ 0 & 1 & 5 \end{bmatrix} R_1 + 2R_2 \rightarrow \begin{bmatrix} 3 & 0 & 15 \\ 0 & 1 & 5 \end{bmatrix} \frac{1}{3}R_1 \rightarrow$

$\begin{bmatrix} 1 & 0 & 5 \\ 0 & 1 & 5 \end{bmatrix}$

Thus, $x = 5$, $y = 5$.
$(x, y) = (5, 5)$

11. $0.5x + 0.1y = 0.7$
$0.2x - 0.2y = 0.6$

$\begin{bmatrix} 0.5 & 0.1 & 0.7 \\ 0.2 & -0.2 & 0.6 \end{bmatrix} \begin{matrix} 10R_1 \\ 10R_2 \end{matrix} \rightarrow \begin{bmatrix} 5 & 1 & 7 \\ 2 & -2 & 6 \end{bmatrix} 5R_2 - 2R_1 \rightarrow$

$\begin{bmatrix} 5 & 1 & 7 \\ 0 & -12 & 16 \end{bmatrix} 12R_1 + R_2 \rightarrow$

$\begin{bmatrix} 60 & 0 & 100 \\ 0 & -12 & 16 \end{bmatrix} \begin{matrix} \frac{1}{60}R_1 \rightarrow \\ -\frac{1}{12}R_2 \rightarrow \end{matrix} \begin{bmatrix} 1 & 0 & \frac{5}{3} \\ 0 & 1 & -\frac{4}{3} \end{bmatrix}$

Thus, $x = \dfrac{5}{3}$, $y = -\dfrac{4}{3}$.

$(x, y) = \left(\dfrac{5}{3}, -\dfrac{4}{3} \right)$

13. $\dfrac{x}{3} - \dfrac{y}{2} = 1$

$\dfrac{x}{4} + y = -2$

$\begin{bmatrix} \frac{1}{3} & -\frac{1}{2} & 1 \\ \frac{1}{4} & 1 & -2 \end{bmatrix} \begin{matrix} 6R_1 \\ 4R_2 \end{matrix} \rightarrow \begin{bmatrix} 2 & -3 & 6 \\ 1 & 4 & -8 \end{bmatrix} 2R_2 - R_1 \rightarrow$

$\begin{bmatrix} 2 & -3 & 6 \\ 0 & 11 & -22 \end{bmatrix} 11R_1 + 3R_2 \rightarrow \begin{bmatrix} 22 & 0 & 0 \\ 0 & 11 & -22 \end{bmatrix} \begin{matrix} \frac{1}{22}R_1 \rightarrow \\ \frac{1}{11}R_2 \rightarrow \end{matrix}$

$\begin{bmatrix} 1 & 0 & 0 \\ 0 & 1 & -2 \end{bmatrix}$

Thus, $x = 0$, $y = -2$.
$(x, y) = (0, -2)$

15. $2x + 3y = 1$

$$-x - \frac{3y}{2} = -\frac{1}{2}$$

$\begin{bmatrix} 2 & 3 & 1 \\ -1 & -\frac{3}{2} & -\frac{1}{2} \end{bmatrix} 2R_2 \rightarrow \begin{bmatrix} 2 & 3 & 1 \\ -2 & -3 & -1 \end{bmatrix} R_2 + R_1 \rightarrow$

$\begin{bmatrix} 2 & 3 & 1 \\ 0 & 0 & 0 \end{bmatrix} \frac{1}{2}R_1 \rightarrow \begin{bmatrix} 1 & \frac{3}{2} & \frac{1}{2} \\ 0 & 0 & 0 \end{bmatrix}$

$x + \frac{3}{2}y = \frac{1}{2}$

$x = \frac{1}{2} - \frac{3}{2}y$

Thus, $x = \frac{1}{2}(1 - 3y)$, y is arbitrary.

$(x, y) = \left(\frac{1}{2}(1 - 3y), \ y \right)$

17. $2x + 3y = 2$

$$-x - \frac{3y}{2} = -\frac{1}{2}$$

$\begin{bmatrix} 2 & 3 & 2 \\ -1 & -\frac{3}{2} & -\frac{1}{2} \end{bmatrix} 2R_2 \rightarrow \begin{bmatrix} 2 & 3 & 2 \\ -2 & -3 & -1 \end{bmatrix} R_2 + R_1 \rightarrow$

$\begin{bmatrix} 2 & 3 & 2 \\ 0 & 0 & 1 \end{bmatrix}$

$0x + 0y = 1$

$0 = 1$

False, no solution

19. $x = 2y \quad \rightarrow x - 2y = 0$

$y = x - 10 \rightarrow -x + y = -10$

$\begin{bmatrix} 1 & -2 & 0 \\ -1 & 1 & -10 \end{bmatrix} R_2 + R_1 \rightarrow$

$\begin{bmatrix} 1 & -2 & 0 \\ 0 & -1 & -10 \end{bmatrix} R_1 - 2R_2 \rightarrow \begin{bmatrix} 1 & 0 & 20 \\ 0 & -1 & -10 \end{bmatrix} -R_2 \rightarrow$

$\begin{bmatrix} 1 & 0 & 20 \\ 0 & 1 & 10 \end{bmatrix}$

Thus, $x = 20$, $y = 10$.

$(x, y) = (20, 10)$

21. $2x + 10 = 3y \rightarrow 2x - 3y = -10$

$3x + 10 = 4y \rightarrow 3x - 4y = -10$

$\begin{bmatrix} 2 & -3 & -10 \\ 3 & -4 & -10 \end{bmatrix} 2R_2 - 3R_1 \rightarrow$

$\begin{bmatrix} 2 & -3 & -10 \\ 0 & 1 & 10 \end{bmatrix} R_1 + 3R_2 \rightarrow \begin{bmatrix} 2 & 0 & 20 \\ 0 & 1 & 10 \end{bmatrix} \frac{1}{2}R_1 \rightarrow$

$\begin{bmatrix} 1 & 0 & 10 \\ 0 & 1 & 10 \end{bmatrix}$

Thus, $x = 10$, $y = 10$.

$(x, y) = (10, 10)$

23. $x + y = 1$

$3x - y = 0$

$x - 3y = -2$

$\begin{bmatrix} 1 & 1 & 1 \\ 3 & -1 & 0 \\ 1 & -3 & -2 \end{bmatrix} \begin{matrix} \\ R_2 - 3R_1 \rightarrow \\ R_3 - R_1 \rightarrow \end{matrix} \begin{bmatrix} 1 & 1 & 1 \\ 0 & -4 & -3 \\ 0 & -4 & -3 \end{bmatrix} \begin{matrix} 4R_1 + R_2 \rightarrow \\ \\ -R_3 + R_2 \rightarrow \end{matrix}$

$\begin{bmatrix} 4 & 0 & 1 \\ 0 & -4 & -3 \\ 0 & 0 & 0 \end{bmatrix} \begin{matrix} \frac{1}{4}R_1 \rightarrow \\ -\frac{1}{4} \rightarrow \end{matrix} \begin{bmatrix} 1 & 0 & \frac{1}{4} \\ 0 & 1 & \frac{3}{4} \\ 0 & 0 & 0 \end{bmatrix}$

Thus, $x = \frac{1}{4}$, $y = \frac{3}{4}$.

$(x, y) = \left(\frac{1}{4}, \frac{3}{4} \right)$

25. $x + y = 0$

$3x - y = 1$

$x - y = -1$

$\begin{bmatrix} 1 & 1 & 0 \\ 3 & -1 & 1 \\ 1 & -1 & -1 \end{bmatrix} \begin{matrix} \\ R_2 - 3R_1 \rightarrow \\ R_3 - R_1 \rightarrow \end{matrix} \begin{bmatrix} 1 & 1 & 0 \\ 0 & -4 & 1 \\ 0 & -2 & -1 \end{bmatrix} \begin{matrix} 4R_1 + R_2 \rightarrow \\ \\ 2R_3 - R_2 \rightarrow \end{matrix}$

$\begin{bmatrix} 4 & 0 & 1 \\ 0 & -4 & 1 \\ 0 & 0 & -3 \end{bmatrix} \begin{matrix} \frac{1}{4}R_1 \rightarrow \\ -\frac{1}{4}R_2 \rightarrow \end{matrix} \begin{bmatrix} 1 & 0 & \frac{1}{4} \\ 0 & 1 & -\frac{1}{4} \\ 0 & 0 & -3 \end{bmatrix}$

$0x + 0y = -3$

$0 = -3$

False, no solution

27. $0.5x + 0.1y = 1.7$

$0.1x - 0.1y = 0.3$

$$x + y = \frac{11}{3}$$

$\begin{bmatrix} 0.5 & 0.1 & 1.7 \\ 0.1 & -0.1 & 0.3 \\ 1 & 1 & \frac{11}{3} \end{bmatrix} \begin{matrix} 10R_1 \rightarrow \\ 10R_2 \rightarrow \\ 3R_3 \rightarrow \end{matrix} \begin{bmatrix} 5 & 1 & 17 \\ 1 & -1 & 3 \\ 3 & 3 & 11 \end{bmatrix} \begin{matrix} \\ 5R_2 - R_1 \rightarrow \\ 5R_3 - 3R_1 \rightarrow \end{matrix}$

$\begin{bmatrix} 5 & 1 & 17 \\ 0 & -6 & -2 \\ 0 & 12 & 4 \end{bmatrix} \begin{matrix} 6R_1 + R_2 \rightarrow \\ \\ R_3 + 2R_2 \rightarrow \end{matrix} \begin{bmatrix} 30 & 0 & 100 \\ 0 & -6 & -2 \\ 0 & 0 & 0 \end{bmatrix} \begin{matrix} \frac{1}{30}R_1 \rightarrow \\ -\frac{1}{6}R_2 \rightarrow \end{matrix}$

$\begin{bmatrix} 1 & 0 & \frac{10}{3} \\ 0 & 1 & \frac{1}{3} \\ 0 & 0 & 0 \end{bmatrix}$

Thus, $x = \frac{10}{3}$, $y = \frac{1}{3}$.

$(x, y) = \left(\frac{10}{3}, \frac{1}{3} \right)$

29. $2.1x - 3.9y = 4.3$
$4.4x + 6.7y = 6.6$

$$\begin{bmatrix} 2.1 & -3.9 & 4.3 \\ 4.4 & 6.7 & 6.6 \end{bmatrix} 2.1R_2 - 4.4R_1 \rightarrow$$

$$\begin{bmatrix} 2.1 & -3.9 & 4.3 \\ 0 & 31.23 & -5.06 \end{bmatrix} 31.23R_1 + 3.9R_2 \rightarrow$$

$$\begin{bmatrix} 65.583 & 0 & 114.555 \\ 0 & 31.23 & -5.06 \end{bmatrix} \begin{array}{l} \frac{1}{65.583}R_1 \rightarrow \\ \frac{1}{31.23}R_2 \rightarrow \end{array}$$

$$\begin{bmatrix} 1 & 0 & 1.7467 \\ 0 & 1 & -0.1620 \end{bmatrix}$$

$x \approx 1.7467,\ y \approx -0.1620$
$(x, y) \approx (1.7467, -0.1620)$

31. $3.12x + 1.11y = 26.9$
$4.43x - 26.7y = 0$

$$\begin{bmatrix} 3.12 & 1.11 & 26.9 \\ 4.43 & -26.7 & 0 \end{bmatrix} 3.12R_2 - 4.43R_1 \rightarrow$$

$$\begin{bmatrix} 3.12 & 1.11 & 26.9 \\ 0 & -88.2213 & -119.167 \end{bmatrix} 88.2213R_1 + 1.11R_2 \rightarrow$$

$$\begin{bmatrix} 275.25046 & 0 & 2240.8776 \\ 0 & -88.2213 & -119.167 \end{bmatrix} \begin{array}{l} \frac{1}{275.25046}R_1 \rightarrow \\ -\frac{1}{88.2213}R_2 \rightarrow \end{array}$$

$$\begin{bmatrix} 1 & 0 & 8.1412 \\ 0 & 1 & 1.3508 \end{bmatrix}$$

$x \approx 8.1412,\ y \approx 1.3508$
$(x, y) \approx (8.1412, 1.3508)$

33. $\quad 31.3x - 1.71y = 1.3$
$-12.52x + 0.684y = -0.52$

$$\begin{bmatrix} 31.3 & -1.71 & 1.3 \\ -12.52 & 0.684 & -0.52 \end{bmatrix} 31.3R_2 + 12.52R_1 \rightarrow$$

$$\begin{bmatrix} 31.3 & -1.71 & 1.3 \\ 0 & 0 & 0 \end{bmatrix} \frac{1}{31.31}R_1 \rightarrow$$

$$\begin{bmatrix} 1 & -0.0546 & 0.0415 \\ 0 & 0 & 0 \end{bmatrix}$$

$x - 0.0546y = 0.0415$
$x = 0.0546y + 0.0415,\ y$ is arbitrary.
$(x, y) \approx (0.0546y + 0.0415, y)$

35. Examples will vary. It is often simpler since only the pivot row needs to be multiplied by a constant to clear an entry in the pivot column. Calculations are simpler.

37. With a calculator it simplifies the process of clearing a pivot column. One only needs to multiply the pivot by a constant and add to a row to clear an entry in the pivot column.

Section 2.3 Using Matrices to Solve Systems with Three or More Unknowns
Pages 147–150

1. $-x + 2y - z = 0$
$-x - y + 2z = 0$
$\qquad 2x - z = 4$

The augmented matrix for this system is

$$\begin{bmatrix} -1 & 2 & -1 & 0 \\ -1 & -1 & 2 & 0 \\ 2 & 0 & -1 & 4 \end{bmatrix} \begin{array}{l} \\ R_2 - R_1 \rightarrow \\ R_3 + 2R_1 \rightarrow \end{array}$$

$$\begin{bmatrix} -1 & 2 & -1 & 0 \\ 0 & -3 & 3 & 0 \\ 0 & 4 & -3 & 4 \end{bmatrix} \frac{1}{3}R_2 \rightarrow$$

$$\begin{bmatrix} -1 & 2 & -1 & 0 \\ 0 & -1 & 1 & 0 \\ 0 & 4 & -3 & 4 \end{bmatrix} \begin{array}{l} R_1 + 2R_2 \rightarrow \\ \\ R_3 + 4R_2 \rightarrow \end{array}$$

$$\begin{bmatrix} -1 & 0 & 1 & 0 \\ 0 & -1 & 1 & 0 \\ 0 & 0 & 1 & 4 \end{bmatrix} \begin{array}{l} R_1 - R_3 \rightarrow \\ R_2 - R_3 \rightarrow \end{array} \begin{bmatrix} -1 & 0 & 0 & -4 \\ 0 & -1 & 0 & -4 \\ 0 & 0 & 1 & 4 \end{bmatrix} \begin{array}{l} -R_1 \rightarrow \\ -R_2 \rightarrow \end{array}$$

$$\begin{bmatrix} 1 & 0 & 0 & 4 \\ 0 & 1 & 0 & 4 \\ 0 & 0 & 1 & 4 \end{bmatrix}$$

$x = 4,\ y = 4,\ z = 4$ or $(x, y, z) = (4, 4, 4)$

3.

$$x + y + 6z = -1$$
$$\frac{1}{3}x - \frac{1}{3}y + \frac{2}{3}z = 1$$
$$\frac{1}{2}x + z = 0$$

$$\begin{bmatrix} 1 & 1 & 6 & -1 \\ \frac{1}{3} & -\frac{1}{3} & \frac{2}{3} & 1 \\ \frac{1}{2} & 0 & 1 & 0 \end{bmatrix} \begin{matrix} \\ 3R_2 \rightarrow \\ 2R_3 \rightarrow \end{matrix} \begin{bmatrix} 1 & 1 & 6 & -1 \\ 1 & -1 & 2 & 3 \\ 1 & 0 & 2 & 0 \end{bmatrix} \begin{matrix} \\ R_2 - R_1 \rightarrow \\ R_3 - R_1 \rightarrow \end{matrix}$$

$$\begin{bmatrix} 1 & 1 & 6 & -1 \\ 0 & -2 & -4 & 4 \\ 0 & -1 & -4 & 1 \end{bmatrix} \begin{matrix} \\ \frac{1}{2}R_2 \rightarrow \\ \end{matrix}$$

$$\begin{bmatrix} 1 & 1 & 6 & -1 \\ 0 & -1 & -2 & 2 \\ 0 & -1 & -4 & 1 \end{bmatrix} \begin{matrix} R_1 + R_2 \rightarrow \\ \\ R_3 - R_2 \rightarrow \end{matrix}$$

$$\begin{bmatrix} 1 & 0 & 4 & 1 \\ 0 & -1 & -2 & 2 \\ 0 & 0 & -2 & -1 \end{bmatrix} \begin{matrix} R_1 + 2R_3 \rightarrow \\ R_2 - R_3 \rightarrow \\ \end{matrix}$$

$$\begin{bmatrix} 1 & 0 & 0 & -1 \\ 0 & -1 & 0 & 3 \\ 0 & 0 & -2 & -1 \end{bmatrix} \begin{matrix} \\ -R_2 \rightarrow \\ -\frac{1}{2}R_3 \rightarrow \end{matrix} \begin{bmatrix} 1 & 0 & 0 & -1 \\ 0 & 1 & 0 & -3 \\ 0 & 0 & 1 & \frac{1}{2} \end{bmatrix}$$

$$x = -1, \ y = -3, \ z = \frac{1}{2}$$

$$(x, y, z) = \left(-1, -3, \frac{1}{2}\right)$$

5.

$$-x + 2y - z = 0$$
$$-x - y + 2z = 0$$
$$2x - y - z = 0$$

$$\begin{bmatrix} -1 & 2 & -1 & 0 \\ -1 & -1 & 2 & 0 \\ 2 & -1 & -1 & 0 \end{bmatrix} \begin{matrix} \\ R_2 - R_1 \rightarrow \\ R_3 + 2R_1 \rightarrow \end{matrix}$$

$$\begin{bmatrix} -1 & 2 & -1 & 0 \\ 0 & -3 & 3 & 0 \\ 0 & -3 & 3 & 0 \end{bmatrix} \begin{matrix} \\ \frac{1}{3}R_2 \rightarrow \\ -\frac{1}{3}R_3 \rightarrow \end{matrix}$$

$$\begin{bmatrix} -1 & 2 & -1 & 0 \\ 0 & -1 & 1 & 0 \\ 0 & 1 & -1 & 0 \end{bmatrix} \begin{matrix} R_1 + 2R_2 \rightarrow \\ \\ R_3 + R_2 \rightarrow \end{matrix}$$

$$\begin{bmatrix} -1 & 0 & 1 & 0 \\ 0 & -1 & 1 & 0 \\ 0 & 0 & 0 & 0 \end{bmatrix} \begin{matrix} -R_1 \rightarrow \\ -R_2 \rightarrow \\ \end{matrix} \begin{bmatrix} 1 & 0 & -1 & 0 \\ 0 & 1 & -1 & 0 \\ 0 & 0 & 0 & 0 \end{bmatrix}$$

$$x - z = 0$$
$$y - z = 0$$
$$x = z, \ y = z, \ z \text{ is arbitrary.}$$
$$(x, y, z) = (z, z, z), \ z \text{ arbitrary}$$

7.

$$x + y + 2z = -1$$
$$2x + 2y + 2z = 2$$
$$\frac{3}{4}x + \frac{3}{4}y + z = \frac{1}{4}$$
$$-x - 2z = 21$$

$$\begin{bmatrix} 1 & 1 & 2 & -1 \\ 2 & 2 & 2 & 2 \\ \frac{3}{4} & \frac{3}{4} & 1 & \frac{1}{4} \\ -1 & 0 & -2 & 21 \end{bmatrix} \begin{matrix} \\ \frac{1}{2}R_2 \rightarrow \\ 4R_3 \rightarrow \\ \end{matrix}$$

$$\begin{bmatrix} 1 & 1 & 2 & -1 \\ 1 & 1 & 1 & 1 \\ 3 & 3 & 4 & 1 \\ -1 & 0 & -2 & 21 \end{bmatrix} \begin{matrix} \\ R_2 - R_1 \rightarrow \\ R_3 - 3R_1 \rightarrow \\ R_4 + R_1 \rightarrow \end{matrix}$$

$$\begin{bmatrix} 1 & 1 & 2 & -1 \\ 0 & 0 & -1 & 2 \\ 0 & 0 & -2 & 4 \\ 0 & 1 & 0 & 20 \end{bmatrix} \begin{matrix} \\ R_2 \leftrightarrow R_4 \\ \\ \end{matrix}$$

$$\begin{bmatrix} 1 & 1 & 2 & -1 \\ 0 & 1 & 0 & 20 \\ 0 & 0 & -2 & 4 \\ 0 & 0 & -1 & 2 \end{bmatrix} \begin{matrix} R_1 - R_2 \rightarrow \\ \\ \\ \end{matrix}$$

$$\begin{bmatrix} 1 & 0 & 2 & -21 \\ 0 & 1 & 0 & 20 \\ 0 & 0 & -2 & 4 \\ 0 & 0 & -1 & 2 \end{bmatrix} \begin{matrix} R_1 + R_3 \rightarrow \\ \\ \\ 2R_4 - R_3 \rightarrow \end{matrix}$$

$$\begin{bmatrix} 1 & 0 & 0 & -17 \\ 0 & 1 & 0 & 20 \\ 0 & 0 & -2 & 4 \\ 0 & 0 & 0 & 0 \end{bmatrix} \begin{matrix} \\ \\ -\frac{1}{2}R_3 \rightarrow \\ \end{matrix} \begin{bmatrix} 1 & 0 & 0 & -17 \\ 0 & 1 & 0 & 20 \\ 0 & 0 & 1 & -2 \\ 0 & 0 & 0 & 0 \end{bmatrix}$$

$$x = -17, \ y = 20, \ z = -2$$
$$(x, y, z) = (-17, 20, -2)$$

9.

$$2x - y + z = 4$$
$$3x - y + z = 5$$

$$\begin{bmatrix} 2 & -1 & 1 & 4 \\ 3 & -1 & 1 & 5 \end{bmatrix} \begin{matrix} \\ 2R_2 - 3R_1 \rightarrow \end{matrix}$$

$$\begin{bmatrix} 2 & -1 & 1 & 4 \\ 0 & 1 & -1 & -2 \end{bmatrix} \begin{matrix} R_1 + R_2 \rightarrow \\ \end{matrix}$$

$$\begin{bmatrix} 2 & 0 & 0 & 2 \\ 0 & 1 & -1 & -2 \end{bmatrix} \begin{matrix} \frac{1}{2}R_1 \rightarrow \\ \end{matrix} \begin{bmatrix} 1 & 0 & 0 & 1 \\ 0 & 1 & -1 & -2 \end{bmatrix}$$

$$x = 1$$
$$y - z = -2$$
$$y = z - 2, \ z \text{ is arbitrary.}$$
$$(x, y, z) = (1, z - 2, z), \ z \text{ arbitrary}$$

11. $0.75x - 0.75y - z = 4$
$x - y + 4z = 0$

$\begin{bmatrix} 0.75 & -0.75 & -1 & 4 \\ 1 & -1 & 4 & 0 \end{bmatrix} 4R_1 \rightarrow$

$\begin{bmatrix} 3 & -3 & -4 & 16 \\ 1 & -1 & 4 & 0 \end{bmatrix} 3R_2 - R_1 \rightarrow$

$\begin{bmatrix} 3 & -3 & -4 & 16 \\ 0 & 0 & 16 & -16 \end{bmatrix} \frac{1}{16} R_2 \rightarrow$

$\begin{bmatrix} 3 & -3 & -4 & 16 \\ 0 & 0 & 1 & -1 \end{bmatrix} R_1 + 4R_2 \rightarrow$

$\begin{bmatrix} 3 & -3 & 0 & 12 \\ 0 & 0 & 1 & -1 \end{bmatrix} \frac{1}{3} R_1 \rightarrow \begin{bmatrix} 1 & -1 & 0 & 4 \\ 0 & 0 & 1 & -1 \end{bmatrix}$

$z = -1$
$x - y = 4$
$x = y + 4$, y arbitrary.
$x = y + 4$, y is arbitrary, $z = -1$

$(x, y, z) = (y + 4, y, -1)$, y arbitrary

13. $3x + y - z = 12$

$[3 \quad 1 \quad -1 \quad 12] \frac{1}{3} R \rightarrow [1 \quad \frac{1}{3} \quad -\frac{1}{3} \quad 4]$

$x + \frac{1}{3}y - \frac{1}{3}z = 4$

$x = 4 - \frac{y}{3} + \frac{z}{3}$, y arbitrary, z arbitrary

$(x, y, z) = \left(4 - \frac{y}{3} + \frac{z}{3}, y, z\right)$, y arbitrary, z arbitrary

15. $x + y + 5z = 1$
$y + 2z + w = 1$
$x + 3y + 7y + 2w = 2$
$x + y + 5z + w = 1$

$\begin{bmatrix} 1 & 1 & 5 & 0 & 1 \\ 0 & 1 & 2 & 1 & 1 \\ 1 & 3 & 7 & 2 & 2 \\ 1 & 1 & 5 & 1 & 1 \end{bmatrix} \begin{matrix} \\ \\ R_3 - R_1 \rightarrow \\ R_4 - R_1 \rightarrow \end{matrix}$

$\begin{bmatrix} 1 & 1 & 5 & 0 & 1 \\ 0 & 1 & 2 & 1 & 1 \\ 0 & 2 & 2 & 2 & 1 \\ 0 & 0 & 0 & 1 & 0 \end{bmatrix} \begin{matrix} R_1 - R_2 \rightarrow \\ \\ R_3 - 2R_2 \rightarrow \\ \end{matrix}$

$\begin{bmatrix} 1 & 0 & 3 & -1 & 0 \\ 0 & 1 & 2 & 1 & 1 \\ 0 & 0 & -2 & 0 & -1 \\ 0 & 0 & 0 & 1 & 0 \end{bmatrix} \begin{matrix} 2R_1 + 3R_3 \rightarrow \\ R_2 + R_3 \rightarrow \\ \\ \end{matrix}$

$\begin{bmatrix} 2 & 0 & 0 & -2 & -3 \\ 0 & 1 & 0 & 1 & 0 \\ 0 & 0 & -2 & 0 & -1 \\ 0 & 0 & 0 & 1 & 0 \end{bmatrix} \begin{matrix} R_1 + 2R_4 \rightarrow \\ R_2 - R_4 \rightarrow \\ \\ \end{matrix}$

$\begin{bmatrix} 2 & 0 & 0 & 0 & -3 \\ 0 & 1 & 0 & 0 & 0 \\ 0 & 0 & -2 & 0 & -1 \\ 0 & 0 & 0 & 1 & 0 \end{bmatrix} \begin{matrix} \frac{1}{2}R_1 \rightarrow \\ \\ -\frac{1}{2}R_3 \rightarrow \\ \end{matrix} \begin{bmatrix} 1 & 0 & 0 & 0 & -\frac{3}{2} \\ 0 & 1 & 0 & 0 & 0 \\ 0 & 0 & 1 & 0 & \frac{1}{2} \\ 0 & 0 & 0 & 1 & 0 \end{bmatrix}$

$x = -\frac{3}{2}$, $y = 0$, $z = \frac{1}{2}$, $w = 0$

$(x, y, z, w) = \left(-\frac{3}{2}, 0, \frac{1}{2}, 0\right)$

17. $-\dfrac{3}{2}x + 2y - \dfrac{3}{4}z = 0$

$\qquad -x - y + 2z = 0$

$\qquad x - \dfrac{1}{3}y - z = 0$

$\begin{bmatrix} -\frac{3}{2} & 2 & -\frac{3}{4} & 0 \\ -1 & -1 & 2 & 0 \\ 1 & -\frac{1}{3} & -1 & 0 \end{bmatrix} \begin{matrix} 4R_1 \to \\ \\ 3R_3 \to \end{matrix}$

$\begin{bmatrix} -6 & 8 & -3 & 0 \\ -1 & -1 & 2 & 0 \\ 3 & -1 & -3 & 0 \end{bmatrix} \begin{matrix} \\ 6R_2 - R_1 \to \\ 2R_3 + R_1 \to \end{matrix}$

$\begin{bmatrix} -6 & 8 & -3 & 0 \\ 0 & -14 & 15 & 0 \\ 0 & 6 & -9 & 0 \end{bmatrix} \begin{matrix} \\ \\ \frac{1}{3}R_3 \to \end{matrix}$

$\begin{bmatrix} -6 & 8 & -3 & 0 \\ 0 & -14 & 15 & 0 \\ 0 & 2 & -3 & 0 \end{bmatrix} \begin{matrix} 7R_1 + 4R_2 \to \\ \\ 7R_3 + R_2 \to \end{matrix}$

$\begin{bmatrix} -42 & 0 & 39 & 0 \\ 0 & -14 & 15 & 0 \\ 0 & 0 & -6 & 0 \end{bmatrix} \begin{matrix} -\frac{1}{3}R_1 \to \\ \\ -\frac{1}{6}R_3 \to \end{matrix}$

$\begin{bmatrix} 14 & 0 & -13 & 0 \\ 0 & -14 & 15 & 0 \\ 0 & 0 & 1 & 0 \end{bmatrix} \begin{matrix} R_1 + 13R_3 \to \\ R_2 - 15R_3 \to \\ \\ \end{matrix}$

$\begin{bmatrix} 14 & 0 & 0 & 0 \\ 0 & -14 & 0 & 0 \\ 0 & 0 & 1 & 0 \end{bmatrix} \begin{matrix} \frac{1}{14}R_1 \to \\ -\frac{1}{14}R_2 \to \\ \\ \end{matrix} \begin{bmatrix} 1 & 0 & 0 & 0 \\ 0 & 1 & 0 & 0 \\ 0 & 0 & 1 & 0 \end{bmatrix}$

$x = 0, y = 0, z = 0$

$(x, y, z) = (0, 0, 0)$

19. $\quad x + y + 2z = -1$

$\quad 2x + 2y + 2z = 2$

$\quad 3x + 3y + 3z = 2$

$\begin{bmatrix} 1 & 1 & 2 & -1 \\ 2 & 2 & 2 & 2 \\ 3 & 3 & 3 & 2 \end{bmatrix} \begin{matrix} \\ \frac{1}{2}R_2 \to \\ \\ \end{matrix} \begin{bmatrix} 1 & 1 & 2 & -1 \\ 1 & 1 & 1 & 1 \\ 3 & 3 & 3 & 2 \end{bmatrix} \begin{matrix} \\ R_2 - R_1 \to \\ R_3 - 3R_1 \to \end{matrix}$

$\begin{bmatrix} 1 & 1 & 2 & -1 \\ 0 & 0 & -1 & 2 \\ 0 & 0 & -3 & 5 \end{bmatrix} \begin{matrix} 3R_1 + 2R_3 \to \\ 3R_2 - R_3 \to \\ \\ \end{matrix}$

$\begin{bmatrix} 3 & 3 & 0 & 7 \\ 0 & 0 & 0 & 1 \\ 0 & 0 & -3 & 5 \end{bmatrix} \begin{matrix} \frac{1}{3}R_1 \to \\ \\ -\frac{1}{3}R_3 \to \end{matrix} \begin{bmatrix} 1 & 1 & 0 & \frac{7}{3} \\ 0 & 0 & 0 & 1 \\ 0 & 0 & 1 & -\frac{5}{3} \end{bmatrix}$

$0x + 0y + 0z = 1$

$0 = 1$

False, no solution

21. $-0.5x + 0.5y + 0.5z = 1.5$

$\qquad 4.2x + 2.1y + 2.1z = 0$

$\qquad 0.2x + 0.2z = 0$

$\begin{bmatrix} -0.5 & 0.5 & 0.5 & 1.5 \\ 4.2 & 2.1 & 2.1 & 0 \\ 0.2 & 0 & 0.2 & 0 \end{bmatrix} \begin{matrix} 2R_1 \to \\ \frac{1}{2.1}R_2 \to \\ 5R_3 \to \end{matrix}$

$\begin{bmatrix} -1 & 1 & 1 & 3 \\ 2 & 1 & 1 & 0 \\ 1 & 0 & 1 & 0 \end{bmatrix} \begin{matrix} \\ R_2 + 2R_1 \to \\ R_3 + R_1 \to \end{matrix} \begin{bmatrix} -1 & 1 & 1 & 3 \\ 0 & 3 & 3 & 6 \\ 0 & 1 & 2 & 3 \end{bmatrix} \begin{matrix} \\ \frac{1}{3}R_2 \to \\ \\ \end{matrix}$

$\begin{bmatrix} -1 & 1 & 1 & 3 \\ 0 & 1 & 1 & 2 \\ 0 & 1 & 2 & 3 \end{bmatrix} \begin{matrix} R_1 - R_2 \to \\ \\ R_3 - R_2 \to \end{matrix}$

$\begin{bmatrix} -1 & 0 & 0 & 1 \\ 0 & 1 & 1 & 2 \\ 0 & 0 & 1 & 1 \end{bmatrix} \begin{matrix} -R_1 \to \\ R_2 - R_3 \to \\ \\ \end{matrix} \begin{bmatrix} 1 & 0 & 0 & -1 \\ 0 & 1 & 0 & 1 \\ 0 & 0 & 1 & 1 \end{bmatrix}$

$x = -1, y = 1, z = 1$

$(x, y, z) = (-1, 1, 1)$

23. $\quad 0.2x + 0.2y - z = 0$

$\qquad x - 0.5y - 0.5z = 0$

$\quad 0.4x + 0.1y - 1.1z = 0$

$\begin{bmatrix} 0.2 & 0.2 & -1 & 0 \\ 1 & -0.5 & -0.5 & 0 \\ 0.4 & 0.1 & -1.1 & 0 \end{bmatrix} \begin{matrix} 5R_1 \to \\ 2R_2 \to \\ 10R_3 \to \end{matrix}$

$\begin{bmatrix} 1 & 1 & -5 & 0 \\ 2 & -1 & -1 & 0 \\ 4 & 1 & -11 & 0 \end{bmatrix} \begin{matrix} \\ R_2 - 2R_1 \to \\ R_3 - 4R_1 \to \end{matrix}$

$\begin{bmatrix} 1 & 1 & -5 & 0 \\ 0 & -3 & 9 & 0 \\ 0 & -3 & 9 & 0 \end{bmatrix} \begin{matrix} \\ \frac{1}{3}R_2 \to \\ \frac{1}{3}R_3 \to \end{matrix} \begin{bmatrix} 1 & 1 & -5 & 0 \\ 0 & -1 & 3 & 0 \\ 0 & -1 & 3 & 0 \end{bmatrix} \begin{matrix} R_1 + R_2 \to \\ -R_2 \to \\ R_3 - R_2 \to \end{matrix}$

$\begin{bmatrix} 1 & 0 & -2 & 0 \\ 0 & 1 & -3 & 0 \\ 0 & 0 & 0 & 0 \end{bmatrix}$

$x - 2z = 0$

$x = 2z, z$ arbitrary

$y - 3z = 0$

$y = 3z, z$ arbitrary

$(x, y, z) = (2z, 3z, z), z$ arbitrary

25.
$$x + y + 5z = 1$$
$$y + 2z + w = 1$$
$$x + y + 5z + w = 1$$
$$x + 2y + 7z + 2w = 2$$

$$\begin{bmatrix} 1 & 1 & 5 & 0 & 1 \\ 0 & 1 & 2 & 1 & 1 \\ 1 & 1 & 5 & 1 & 1 \\ 1 & 2 & 7 & 2 & 2 \end{bmatrix} \begin{matrix} \\ \\ R_3 - R_1 \to \\ R_4 - R_1 \to \end{matrix} \begin{bmatrix} 1 & 1 & 5 & 0 & 1 \\ 0 & 1 & 2 & 1 & 1 \\ 0 & 0 & 0 & 1 & 0 \\ 0 & 1 & 2 & 2 & 1 \end{bmatrix} \begin{matrix} R_1 - R_2 \to \\ \\ \\ R_4 - R_2 \to \end{matrix} \begin{bmatrix} 1 & 0 & 3 & -1 & 0 \\ 0 & 1 & 2 & 1 & 1 \\ 0 & 0 & 0 & 1 & 0 \\ 0 & 0 & 0 & 1 & 0 \end{bmatrix} \begin{matrix} R_1 + R_4 \to \\ R_2 - R_4 \to \\ R_3 - R_4 \to \\ \\ \end{matrix} \begin{bmatrix} 1 & 0 & 3 & 0 & 0 \\ 0 & 1 & 2 & 0 & 1 \\ 0 & 0 & 0 & 0 & 0 \\ 0 & 0 & 0 & 1 & 0 \end{bmatrix}$$

$x + 3z = 0$; $x = -3z$, z arbitrary
$y + 2z = 1$; $y = 1 - 2z$, z arbitrary
$w = 0$
$(x, y, z, w) = (-3z, 1 - 2z, z, 0)$, z arbitrary

27. $x_1 - 2x_2 + x_3 - 4x_4 = 1$
$x_1 + 3x_2 + 7x_3 + 2x_4 = 2$
$2x_1 + x_2 + 8x_3 - 2x_4 = 3$

$$\begin{bmatrix} 1 & -2 & 1 & -4 & 1 \\ 1 & 3 & 7 & 2 & 2 \\ 2 & 1 & 8 & -2 & 3 \end{bmatrix} \begin{matrix} \\ R_2 - R_1 \to \\ R_3 - 2R_1 \to \end{matrix} \begin{bmatrix} 1 & -2 & 1 & -4 & 1 \\ 0 & 5 & 6 & 6 & 1 \\ 0 & 5 & 6 & 6 & 1 \end{bmatrix} \begin{matrix} 5R_1 + 2R_2 \to \\ \\ R_3 - R_2 \to \end{matrix} \begin{bmatrix} 5 & 0 & 17 & -8 & 7 \\ 0 & 5 & 6 & 6 & 1 \\ 0 & 0 & 0 & 0 & 0 \end{bmatrix} \begin{matrix} \frac{1}{5} R_1 \to \\ \frac{1}{5} R_2 \to \\ \\ \end{matrix} \begin{bmatrix} 1 & 0 & \frac{17}{5} & -\frac{8}{5} & \frac{7}{5} \\ 0 & 1 & \frac{6}{5} & \frac{6}{5} & \frac{1}{5} \\ 0 & 0 & 0 & 0 & 0 \end{bmatrix}$$

$$x_1 + \frac{17}{5}x_3 - \frac{8}{5}x_4 = \frac{7}{5}$$
$$x_2 + \frac{6}{5}x_3 + \frac{6}{5}x_4 = \frac{1}{5}$$

$$x_1 = \frac{7}{5} - \frac{17}{5}x_3 + \frac{8}{5}x_4, \ x_3 \text{ and } x_4 \text{ arbitrary}$$

$$x_2 = \frac{1}{5} - \frac{6}{5}x_3 - \frac{6}{5}x_4, \ x_3 \text{ and } x_4 \text{ arbitrary}$$

$$(x_1, x_2, x_3, x_4) = \left(\frac{1}{5}(7 - 17x_3 + 8x_4), \frac{1}{5}(1 - 6x_3 - 6x_4), x_3, x_4 \right), \ x_3, x_4 \text{ arbitrary}$$

29.
$$z = 4(x - y) \to \quad -4x + 4y + z = 0$$
$$z = x - 2 \qquad \qquad -x + z = -2$$
$$0.1x + 0.1y + 0.1z = 1.5 \qquad 0.1x + 0.1y + 0.1z = 1.5$$

$$\begin{bmatrix} -4 & 4 & 1 & 0 \\ -1 & 0 & 1 & -2 \\ 0.1 & 0.1 & 0.1 & 1.5 \end{bmatrix} \begin{matrix} \\ \\ 10R_3 \to \end{matrix} \begin{bmatrix} -4 & 4 & 1 & 0 \\ -1 & 0 & 1 & -2 \\ 1 & 1 & 1 & 15 \end{bmatrix} \begin{matrix} \\ 4R_2 - R_1 \to \\ 4R_3 + R_1 \to \end{matrix} \begin{bmatrix} -4 & 4 & 1 & 0 \\ 0 & -4 & 3 & -8 \\ 0 & 8 & 5 & 60 \end{bmatrix} \begin{matrix} R_1 + R_2 \to \\ \\ R_3 + 2R_2 \to \end{matrix} \begin{bmatrix} -4 & 0 & 4 & -8 \\ 0 & -4 & 3 & -8 \\ 0 & 0 & 11 & 44 \end{bmatrix} \begin{matrix} -\frac{1}{4} R_1 \to \\ \\ \frac{1}{11} R_3 \to \end{matrix}$$

$$\begin{bmatrix} 1 & 0 & -1 & 2 \\ 0 & -4 & 3 & -8 \\ 0 & 0 & 1 & 4 \end{bmatrix} \begin{matrix} R_1 + R_3 \to \\ R_2 - 3R_3 \to \\ \\ \end{matrix} \begin{bmatrix} 1 & 0 & 0 & 6 \\ 0 & -4 & 0 & -20 \\ 0 & 0 & 1 & 4 \end{bmatrix} \begin{matrix} \\ -\frac{1}{4} R_2 \to \\ \\ \end{matrix} \begin{bmatrix} 1 & 0 & 0 & 6 \\ 0 & 1 & 0 & 5 \\ 0 & 0 & 1 & 4 \end{bmatrix}$$

$x = 6$, $y = 5$, $z = 4$
$(x, y, z) = (6, 5, 4)$

31. $p = -x + y + z$ $\rightarrow$ $x - y - z + p = 0$
$y = x - z$ $-x + y + z = 0$
$y = 2x$ $-2x + y = 0$
$p = 2x - 2y + 2$ $-2x + 2y + p = 2$
(x, y, z, and p are the unknowns.)

$$\begin{bmatrix} 1 & -1 & -1 & 1 & 0 \\ -1 & 1 & 1 & 0 & 0 \\ -2 & 1 & 0 & 0 & 0 \\ -2 & 2 & 0 & 1 & 2 \end{bmatrix} \begin{array}{l} \\ R_2 + R_1 \rightarrow \\ R_3 + 2R_1 \rightarrow \\ R_4 + 2R_1 \rightarrow \end{array} \begin{bmatrix} 1 & -1 & -1 & 1 & 0 \\ 0 & 0 & 0 & 1 & 0 \\ 0 & -1 & -2 & 2 & 0 \\ 0 & 0 & -2 & 3 & 2 \end{bmatrix} \begin{array}{l} \\ R_3 \rightarrow R_2 \\ R_4 \rightarrow R_3 \\ R_2 \rightarrow R_4 \end{array} \begin{bmatrix} 1 & -1 & -1 & 1 & 0 \\ 0 & -1 & -2 & 2 & 0 \\ 0 & 0 & -2 & 3 & 2 \\ 0 & 0 & 0 & 1 & 0 \end{bmatrix} \begin{array}{l} R_1 - R_2 \rightarrow \\ \\ \\ \end{array}$$

$$\begin{bmatrix} 1 & 0 & 1 & -1 & 0 \\ 0 & -1 & -2 & 2 & 0 \\ 0 & 0 & -2 & 3 & 2 \\ 0 & 0 & 0 & 1 & 0 \end{bmatrix} \begin{array}{l} 2R_1 + R_3 \rightarrow \\ R_2 - R_3 \rightarrow \\ \\ \end{array} \begin{bmatrix} 2 & 0 & 0 & 1 & 2 \\ 0 & -1 & 0 & -1 & -2 \\ 0 & 0 & -2 & 3 & 2 \\ 0 & 0 & 0 & 1 & 0 \end{bmatrix} \begin{array}{l} R_1 - R_4 \rightarrow \\ R_2 + R_4 \rightarrow \\ R_3 - 3R_4 \rightarrow \\ \end{array}$$

$$\begin{bmatrix} 2 & 0 & 0 & 0 & 2 \\ 0 & -1 & 0 & 0 & -2 \\ 0 & 0 & -2 & 0 & 2 \\ 0 & 0 & 0 & 1 & 0 \end{bmatrix} \begin{array}{l} \frac{1}{2}R_1 \rightarrow \\ -R_2 \rightarrow \\ -\frac{1}{2}R_3 \rightarrow \\ \end{array} \begin{bmatrix} 1 & 0 & 0 & 0 & 1 \\ 0 & 1 & 0 & 0 & 2 \\ 0 & 0 & 1 & 0 & -1 \\ 0 & 0 & 0 & 1 & 0 \end{bmatrix}$$

$x = 1$, $y = 2$, $z = -1$, $p = 0$
$(x, y, z, p) = (1, 2, -1, 0)$

33. $2x + y + s = 300$ $2x + y + s = 300$
$x + 2y + t = 300$ $x + 2y + t = 300$
$\qquad\quad c = x + y$ $\rightarrow$ $-x - y + c = 0$
(The unknowns are x, y, c, s, and t.)

$$\begin{array}{ccccc} x & y & c & s & t \\ \end{array}$$
$$\begin{bmatrix} 2 & 1 & 0 & 1 & 0 & 300 \\ 1 & 2 & 0 & 0 & 1 & 300 \\ -1 & -1 & 1 & 0 & 0 & 0 \end{bmatrix} \begin{array}{l} \\ 2R_2 - R_1 \rightarrow \\ 2R_3 + R_1 \rightarrow \end{array} \begin{bmatrix} 2 & 1 & 0 & 1 & 0 & 300 \\ 0 & 3 & 0 & -1 & 2 & 300 \\ 0 & -1 & 2 & 1 & 0 & 300 \end{bmatrix} \begin{array}{l} 3R_1 - R_2 \rightarrow \\ \\ 3R_3 + R_2 \rightarrow \end{array} \begin{bmatrix} 6 & 0 & 0 & 4 & -2 & 600 \\ 0 & 3 & 0 & -1 & 2 & 300 \\ 0 & 0 & 6 & 2 & 2 & 1200 \end{bmatrix} \begin{array}{l} \frac{1}{6}R_1 \rightarrow \\ \frac{1}{3}R_2 \rightarrow \\ \frac{1}{6}R_3 \rightarrow \end{array}$$

$$\begin{bmatrix} 1 & 0 & 0 & \frac{2}{3} & -\frac{1}{3} & 100 \\ 0 & 1 & 0 & -\frac{1}{3} & \frac{2}{3} & 100 \\ 0 & 0 & 1 & \frac{1}{3} & \frac{1}{3} & 200 \end{bmatrix}$$

$x + \dfrac{2}{3}s - \dfrac{1}{3}t = 100$

$y - \dfrac{1}{3}s + \dfrac{2}{3}t = 100$

$c + \dfrac{1}{3}s + \dfrac{1}{3}t = 200$

$x = 100 - \dfrac{2}{3}s + \dfrac{1}{3}t$, s and t arbitrary

$y = 100 + \dfrac{1}{3}s - \dfrac{2}{3}t$, s and t arbitrary

$c = 200 - \dfrac{1}{3}s - \dfrac{1}{3}t$, s and t arbitrary

$(x, y, c, s, t) = \left(100 - \dfrac{2}{3}s + \dfrac{1}{3}t,\ 100 + \dfrac{1}{3}s - \dfrac{2}{3}t,\ 200 - \dfrac{1}{3}s - \dfrac{1}{3}t,\ s,\ t \right)$, s, t arbitrary

35.

$$-0.9x + 0.3z = 0$$
$$0.2x - 0.4y + 0.1z = 0$$
$$0.7x + 0.4y - 0.4z = 0$$
$$x + y + z = 1$$

$$\begin{bmatrix} -0.9 & 0 & 0.3 & 0 \\ 0.2 & -0.4 & 0.1 & 0 \\ 0.7 & 0.4 & -0.4 & 0 \\ 1 & 1 & 1 & 1 \end{bmatrix} \begin{matrix} \frac{10}{3}R_1 \rightarrow \\ 10R_2 \rightarrow \\ 10R_3 \rightarrow \\ \end{matrix}$$

$$\begin{bmatrix} -3 & 0 & 1 & 0 \\ 2 & -4 & 1 & 0 \\ 7 & 4 & -4 & 0 \\ 1 & 1 & 1 & 1 \end{bmatrix} \begin{matrix} \\ 3R_2 + 2R_1 \rightarrow \\ 3R_3 + 7R_1 \rightarrow \\ 3R_4 + R_1 \rightarrow \end{matrix}$$

$$\begin{bmatrix} -3 & 0 & 1 & 0 \\ 0 & -12 & 5 & 0 \\ 0 & 12 & -5 & 0 \\ 0 & 3 & 4 & 3 \end{bmatrix} \begin{matrix} \\ \\ R_3 + R_2 \rightarrow \\ 4R_4 + R_2 \rightarrow \end{matrix}$$

$$\begin{bmatrix} -3 & 0 & 1 & 0 \\ 0 & -12 & 5 & 0 \\ 0 & 0 & 0 & 0 \\ 0 & 0 & 21 & 12 \end{bmatrix} \begin{matrix} \\ \\ R_3 \leftrightarrow R_4 \\ \end{matrix}$$

$$\begin{bmatrix} -3 & 0 & 1 & 0 \\ 0 & -12 & 5 & 0 \\ 0 & 0 & 21 & 12 \\ 0 & 0 & 0 & 0 \end{bmatrix} \begin{matrix} \\ \\ \frac{1}{3}R_3 \rightarrow \\ \end{matrix}$$

$$\begin{bmatrix} -3 & 0 & 1 & 0 \\ 0 & -12 & 5 & 0 \\ 0 & 0 & 7 & 4 \\ 0 & 0 & 0 & 0 \end{bmatrix} \begin{matrix} 7R_1 - R_3 \rightarrow \\ 7R_2 - 5R_3 \rightarrow \\ \\ \end{matrix}$$

$$\begin{bmatrix} -21 & 0 & 0 & -4 \\ 0 & -84 & 0 & -20 \\ 0 & 0 & 7 & 4 \\ 0 & 0 & 0 & 0 \end{bmatrix} \begin{matrix} -\frac{1}{21}R_1 \rightarrow \\ -\frac{1}{84}R_2 \rightarrow \\ \frac{1}{7}R_3 \rightarrow \\ \end{matrix} \begin{bmatrix} 1 & 0 & 0 & \frac{4}{21} \\ 0 & 1 & 0 & \frac{5}{21} \\ 0 & 0 & 1 & \frac{4}{7} \\ 0 & 0 & 0 & 0 \end{bmatrix}$$

$$x = \frac{4}{21}, \ y = \frac{5}{21}, \ z = \frac{4}{7}$$

$$(x, y, z) = \left(\frac{4}{21}, \frac{5}{21}, \frac{4}{7} \right)$$

37.

$$\frac{x}{y} = -\frac{1}{2} \quad \rightarrow \quad 2x + y = 0$$

$$\frac{z-y}{x} = 4 \qquad -4x - y + z = 0$$

$$\frac{z+10}{2x+y+z} = \frac{3}{2} \qquad 6x + 3y + z = 20$$

$$\begin{bmatrix} 2 & 1 & 0 & 0 \\ -4 & -1 & 1 & 0 \\ 6 & 3 & 1 & 20 \end{bmatrix} \begin{matrix} \\ R_2 + 2R_1 \rightarrow \\ R_3 - 3R_1 \rightarrow \end{matrix}$$

$$\begin{bmatrix} 2 & 1 & 0 & 0 \\ 0 & 1 & 1 & 0 \\ 0 & 0 & 1 & 20 \end{bmatrix} \begin{matrix} R_1 - R_2 \rightarrow \\ \\ \end{matrix}$$

$$\begin{bmatrix} 2 & 0 & -1 & 0 \\ 0 & 1 & 1 & 0 \\ 0 & 0 & 1 & 20 \end{bmatrix} \begin{matrix} R_1 + R_3 \rightarrow \\ R_2 - R_3 \rightarrow \\ \end{matrix} \begin{bmatrix} 2 & 0 & 0 & 20 \\ 0 & 1 & 0 & -20 \\ 0 & 0 & 1 & 20 \end{bmatrix} \begin{matrix} \frac{1}{2}R_1 \rightarrow \\ \\ \end{matrix}$$

$$\begin{bmatrix} 1 & 0 & 0 & 10 \\ 0 & 1 & 0 & -20 \\ 0 & 0 & 1 & 20 \end{bmatrix}$$

$$x = 10, \ y = -20, \ z = 20$$
$$(x, y, z) = (10, -20, 20)$$

39. $x + y + z + u + v = 15$
$y - z + u - v = -2$
$z + u + v = 12$
$u - v = -1$
$v = 5$

Using matrices:

$$\begin{bmatrix} 1 & 1 & 1 & 1 & 1 & 15 \\ 0 & 1 & -1 & 1 & -1 & -2 \\ 0 & 0 & 1 & 1 & 1 & 12 \\ 0 & 0 & 0 & 1 & -1 & -1 \\ 0 & 0 & 0 & 0 & 1 & 5 \end{bmatrix} \begin{matrix} R_1 - R_2 \to \\ \\ \\ \\ \\ \end{matrix}$$

$$\begin{bmatrix} 1 & 0 & 2 & 0 & 2 & 17 \\ 0 & 1 & -1 & 1 & -1 & -2 \\ 0 & 0 & 1 & 1 & 1 & 12 \\ 0 & 0 & 0 & 1 & -1 & -1 \\ 0 & 0 & 0 & 0 & 1 & 5 \end{bmatrix} \begin{matrix} R_1 - 2R_3 \to \\ R_2 + R_3 \to \\ \\ \\ \\ \end{matrix}$$

$$\begin{bmatrix} 1 & 0 & 0 & -2 & 0 & -7 \\ 0 & 1 & 0 & 2 & 0 & 10 \\ 0 & 0 & 1 & 1 & 1 & 12 \\ 0 & 0 & 0 & 1 & -1 & -1 \\ 0 & 0 & 0 & 0 & 1 & 5 \end{bmatrix} \begin{matrix} R_1 + 2R_4 \to \\ R_2 - 2R_4 \to \\ R_3 - R_4 \to \\ \\ \\ \end{matrix}$$

$$\begin{bmatrix} 1 & 0 & 0 & 0 & -2 & -9 \\ 0 & 1 & 0 & 0 & 2 & 12 \\ 0 & 0 & 1 & 0 & 2 & 13 \\ 0 & 0 & 0 & 1 & -1 & -1 \\ 0 & 0 & 0 & 0 & 1 & 5 \end{bmatrix} \begin{matrix} R_1 + 2R_5 \to \\ R_2 - 2R_5 \to \\ R_3 - 2R_5 \to \\ R_4 + R_5 \to \\ \\ \end{matrix}$$

$$\begin{bmatrix} 1 & 0 & 0 & 0 & 0 & 1 \\ 0 & 1 & 0 & 0 & 0 & 2 \\ 0 & 0 & 1 & 0 & 0 & 3 \\ 0 & 0 & 0 & 1 & 0 & 4 \\ 0 & 0 & 0 & 0 & 1 & 5 \end{bmatrix}$$

$x = 1, y = 2, z = 3, u = 4, v = 5$
$(x, y, z, u, v) = (1, 2, 3, 4, 5)$
You could also easily solve this system by working backwards and substituting for each variable.
$v = 5$
$u - v = -1$: $u - 5 = -1$, $u = 4$
$z + u + v = 12$: $z + 4 + 5 = 12$, $z = 3$
$y - z + u - v = -2$: $y - 3 + 4 - 5 = -2$, $y = 2$
$x + y + z + u + v = 15$: $x + 2 + 3 + 4 + 5 = 15$, $x = 1$
$(x, y, z, u, v) = (1, 2, 3, 4, 5)$

41. $x - y + z - u + v = 0$
$y - z + u - v = -2$
$x - 2v = -2$
$2x - y + z - u - 3v = -2$
$4x - y + z - u - 7v = -6$

$$\begin{bmatrix} 1 & -1 & 1 & -1 & 1 & 0 \\ 0 & 1 & -1 & 1 & -1 & -2 \\ 1 & 0 & 0 & 0 & -2 & -2 \\ 2 & -1 & 1 & -1 & -3 & -2 \\ 4 & -1 & 1 & -1 & -7 & -6 \end{bmatrix} \begin{matrix} R_1 + R_2 \to \\ \\ \\ R_4 - R_1 \to \\ R_5 - R_1 \to \end{matrix}$$

$$\begin{bmatrix} 1 & 0 & 0 & 0 & 0 & -2 \\ 0 & 1 & -1 & 1 & -1 & -2 \\ 1 & 0 & 0 & 0 & -2 & -2 \\ 1 & 0 & 0 & 0 & -4 & -2 \\ 3 & 0 & 0 & 0 & -8 & -6 \end{bmatrix} \begin{matrix} \\ \\ R_3 - R_1 \to \\ R_4 - R_1 \to \\ R_5 - 3R_1 \to \end{matrix}$$

$$\begin{bmatrix} 1 & 0 & 0 & 0 & 0 & -2 \\ 0 & 1 & -1 & 1 & -1 & -2 \\ 0 & 0 & 0 & 0 & -2 & 0 \\ 0 & 0 & 0 & 0 & -4 & 0 \\ 0 & 0 & 0 & 0 & -8 & 0 \end{bmatrix} \begin{matrix} \\ \\ -\frac{1}{2}R_3 \to \\ -\frac{1}{4}R_4 \to \\ -\frac{1}{8}R_5 \to \end{matrix}$$

$$\begin{bmatrix} 1 & 0 & 0 & 0 & 0 & -2 \\ 0 & 1 & -1 & 1 & -1 & -2 \\ 0 & 0 & 0 & 0 & 1 & 0 \\ 0 & 0 & 0 & 0 & 1 & 0 \\ 0 & 0 & 0 & 0 & 1 & 0 \end{bmatrix} \begin{matrix} \\ R_2 + R_5 \to \\ R_3 - R_5 \to \\ R_4 - R_5 \to \\ \end{matrix}$$

$$\begin{bmatrix} 1 & 0 & 0 & 0 & 0 & -2 \\ 0 & 1 & -1 & 1 & 0 & -2 \\ 0 & 0 & 0 & 0 & 0 & 0 \\ 0 & 0 & 0 & 0 & 0 & 0 \\ 0 & 0 & 0 & 0 & 1 & 0 \end{bmatrix}$$

$x = -2$
$y - z + u = -2$: $y = z - u - 2$; z, u arbitrary
$v = 0$
$(x, y, z, u, v) = (-2, z - u - 2, z, u, 0)$, z, u arbitrary

43.–49. Use a graphing calculator or computer software to solve each system. Input each matrix.

43. $1.6x + 2.4y - 3.2z = 4.4$
$5.1x - 6.3y + 0.6z = -3.2$
$4.2x + 3.5y + 4.9z = 10.1$

$$\begin{bmatrix} 1.6 & 2.4 & -3.2 & 4.4 \\ 5.1 & -6.3 & 0.6 & -3.2 \\ 4.2 & 3.5 & 4.9 & 10.1 \end{bmatrix}$$

$$\begin{bmatrix} 1.6 & 2.4 & -3.2 & 4.4 \\ 0 & -22.32 & 17.28 & -27.56 \\ 0 & -4.48 & 21.28 & -2.32 \end{bmatrix}$$

$$\begin{bmatrix} 35.715 & 0 & -29.952 & 32.064 \\ 0 & -22.32 & 17.28 & -27.56 \\ 0 & 0 & 397.5552 & 71.6864 \end{bmatrix}$$

$$\begin{bmatrix} 14197.49 & 0 & 0 & 14894.361 \\ 0 & -8873.4321 & 0 & -12243.536 \end{bmatrix}$$

$$\begin{bmatrix} 1 & 0 & 0 & 1.0491 \\ 0 & 1 & 0 & 1.3798 \\ 0 & 0 & 1 & 0.1803 \end{bmatrix}$$

$x \approx 1.0491 \approx 1.0$
$y \approx 1.3798 \approx 1.4$
$z \approx 0.1802 \approx 0.2$
$(x, y, t) = (1.0, 1.4, 0.2)$

45. $-0.2x + 0.3y + 0.4z - t = 4.5$
$2.2x + 1.1y - 4.7z + 2t = 8.3$
$3.4x + 0.5z - 3.4t = 0.1$
$9.2y - 1.3t = 0$

$$\begin{bmatrix} -0.2 & 0.3 & 0.4 & -1 & 4.5 \\ 2.2 & 1.1 & -4.7 & 2 & 8.3 \\ 3.4 & 0 & 0.5 & -3.4 & 0.1 \\ 0 & 9.2 & 0 & -1.3 & 0 \end{bmatrix}$$

$$\begin{bmatrix} 1 & 0 & 0 & 0 & -5.5178 \\ 0 & 1 & 0 & 0 & -0.9374 \\ 0 & 0 & 1 & 0 & -7.3912 \\ 0 & 0 & 0 & 1 & -6.6342 \end{bmatrix}$$

$x \approx -5.5, y \approx -0.9, z \approx -7.4, t \approx -6.6$
$(x, y, z, t) \approx (-5.5, -0.9, -7.4, -6.6)$

47. $x + 2y - z + w = 30$
$2x - z + 2w = 30$
$x + 3y + 3z - 4w = 2$
$2x - 9y + w = 4$

$$\begin{bmatrix} 1 & 2 & -1 & 1 & 30 \\ 2 & 0 & -1 & 2 & 30 \\ 1 & 3 & 3 & -4 & 2 \\ 2 & -9 & 0 & 1 & 4 \end{bmatrix} \rightarrow$$

$$\begin{bmatrix} 1 & 2 & -1 & 1 & 30 \\ 0 & -4 & 1 & 0 & -30 \\ 0 & 1 & 4 & -5 & -28 \\ 0 & -13 & 2 & -1 & -56 \end{bmatrix} \rightarrow$$

$$\begin{bmatrix} 2 & 0 & -1 & 2 & 30 \\ 0 & -4 & 1 & 0 & -30 \\ 0 & 0 & 17 & -20 & -142 \\ 0 & 0 & -5 & -4 & 166 \end{bmatrix}$$

$$\begin{bmatrix} 34 & 0 & 0 & 14 & 368 \\ 0 & -68 & 0 & 20 & -368 \\ 0 & 0 & 17 & -20 & -142 \\ 0 & 0 & 0 & -168 & 2112 \end{bmatrix} \rightarrow$$

$$\begin{bmatrix} 17 & 0 & 0 & 7 & 184 \\ 0 & -17 & 0 & 5 & -92 \\ 0 & 0 & 17 & -20 & -142 \\ 0 & 0 & 0 & -7 & 88 \end{bmatrix}$$

$$\begin{bmatrix} 17 & 0 & 0 & 0 & 272 \\ 0 & -119 & 0 & 0 & -204 \\ 0 & 0 & -119 & 0 & 2754 \end{bmatrix} \rightarrow$$

$$\begin{bmatrix} 1 & 0 & 0 & 0 & 16 \\ 0 & 1 & 0 & 0 & \frac{12}{7} \\ 0 & 0 & 1 & 0 & -\frac{162}{7} \\ 0 & 0 & 0 & 1 & -\frac{88}{7} \end{bmatrix}$$

$(x, y, z, t) = \left(16, \dfrac{12}{17}, -\dfrac{162}{7}, -\dfrac{88}{7} \right)$

49. $x + 2y + 3z + 4w + 5t = 6$
$2x + 3y + 4z + 5w + t = 5$
$3x + 4y + 5z + w + 2t = 4$
$4x + 5y + z + 2w + 3t = 3$
$5x + y + 2z + 3w + 4t = 2$

$$\begin{bmatrix} 1 & 2 & 3 & 4 & 5 & 6 \\ 2 & 3 & 4 & 5 & 1 & 5 \\ 3 & 4 & 5 & 1 & 2 & 4 \\ 4 & 5 & 1 & 2 & 3 & 3 \\ 5 & 1 & 2 & 3 & 4 & 2 \end{bmatrix} \rightarrow \begin{bmatrix} 1 & 2 & 3 & 4 & 5 & 6 \\ 0 & 1 & 2 & 3 & 9 & 7 \\ 0 & 2 & 4 & 11 & 13 & 14 \\ 0 & 3 & 11 & 14 & 17 & 21 \\ 0 & 6 & 2 & 3 & 4 & 7 \end{bmatrix} \rightarrow \begin{bmatrix} 1 & 0 & -1 & -2 & -13 & -8 \\ 0 & 1 & 2 & 3 & 9 & 7 \\ 0 & 0 & 0 & 5 & -5 & 0 \\ 0 & 0 & 5 & 5 & -10 & 0 \\ 0 & 0 & -10 & -15 & -50 & -35 \end{bmatrix}$$

$$\begin{bmatrix} 1 & 0 & -1 & -2 & -13 & -8 \\ 0 & 1 & 2 & 3 & 9 & 7 \\ 0 & 0 & 1 & 1 & -2 & 0 \\ 0 & 0 & 0 & 1 & -1 & 0 \\ 0 & 0 & 2 & 3 & 10 & 7 \end{bmatrix} \rightarrow \begin{bmatrix} 1 & 0 & 0 & -1 & -15 & -8 \\ 0 & 1 & 0 & 1 & 13 & 7 \\ 0 & 0 & 1 & 1 & -2 & 0 \\ 0 & 0 & 0 & 1 & -1 & 0 \\ 0 & 0 & 0 & 1 & 14 & 7 \end{bmatrix} \rightarrow \begin{bmatrix} 1 & 0 & 0 & 0 & -16 & -8 \\ 0 & 1 & 0 & 0 & 14 & 7 \\ 0 & 0 & 1 & 0 & -1 & 0 \\ 0 & 0 & 0 & 1 & -1 & 0 \\ 0 & 0 & 0 & 0 & 15 & 7 \end{bmatrix}$$

$$\begin{bmatrix} 15 & 0 & 0 & 0 & 0 & -8 \\ 0 & 15 & 0 & 0 & 0 & 7 \\ 0 & 0 & 15 & 0 & 0 & 7 \\ 0 & 0 & 0 & 15 & 0 & 7 \\ 0 & 0 & 0 & 0 & 0 & 7 \end{bmatrix} \rightarrow \begin{bmatrix} 1 & 0 & 0 & 0 & 0 & -\frac{8}{15} \\ 0 & 1 & 0 & 0 & 0 & \frac{7}{15} \\ 0 & 0 & 1 & 0 & 0 & \frac{7}{15} \\ 0 & 0 & 0 & 1 & 0 & \frac{7}{15} \\ 0 & 0 & 0 & 0 & 1 & \frac{7}{15} \end{bmatrix}$$

$$(x, y, z, w, t) = \left(-\frac{8}{15}, \frac{7}{15}, \frac{7}{15}, \frac{7}{15}, \frac{7}{15} \right)$$

51. The number of pivots is equal to one less than the number of columns in the matrix. Thus, each unknown has exactly one corresponding value, so the solution is unique.

53. Reasons will vary.

55. Examples will vary.

57. Answers will vary.

Section 2.4 Applications of Systems of Linear Equations
Pages 156–163

1. Let x = number of batches of Creamy Vanilla, y = number of batches of Continental Mocha, and z = number of batches of Succulent Strawberry.

	Vanilla (x batches)	Mocha (y batches)	Strawberry (z batches)	Total Available
Eggs	2	1	1	350
Milk (cups)	1	1	2	350
Cream (cups)	2	2	1	400

$$2x + y + z = 350$$
$$x + y + 2z = 350$$
$$2x + 2y + z = 400$$

$$\begin{bmatrix} 2 & 1 & 1 & 350 \\ 1 & 1 & 2 & 350 \\ 2 & 2 & 1 & 400 \end{bmatrix} \begin{matrix} \\ 2R_2 - R_1 \to \\ R_3 - R_1 \to \end{matrix} \begin{bmatrix} 2 & 1 & 1 & 350 \\ 0 & 1 & 3 & 350 \\ 0 & 1 & 0 & 50 \end{bmatrix} \begin{matrix} R_1 - R_2 \to \\ \\ R_3 - R_2 \to \end{matrix} \begin{bmatrix} 2 & 0 & -2 & 0 \\ 0 & 1 & 3 & 350 \\ 0 & 0 & -3 & -300 \end{bmatrix} \begin{matrix} \frac{1}{2}R_1 \to \\ \\ -\frac{1}{3}R_3 \to \end{matrix}$$

$$\begin{bmatrix} 1 & 0 & -1 & 0 \\ 0 & 1 & 3 & 350 \\ 0 & 0 & 1 & 100 \end{bmatrix} \begin{matrix} R_1 + R_3 \to \\ R_2 - 3R_3 \to \\ \end{matrix} \begin{bmatrix} 1 & 0 & 0 & 100 \\ 0 & 1 & 0 & 50 \\ 0 & 0 & 1 & 100 \end{bmatrix}$$

$x = 100$, $y = 50$, $z = 100$ or $(x, y, z) = (100, 50, 100)$.
You should order 100 batches of vanilla, 50 batches of mocha, and 100 batches of strawberry.

3. Let x = number of Boeing 747s, y = number of Boeing 777s and z = number of Airbus A321s.

	Boeing 747s (x)	Boeing 777s (y)	Airbus A321s (z)	Total
Passengers	400	300	200	4500
Cost ($millions)	150	115	60	1625

Also, $x + y = 2z$ or $x + y - 2z = 0$.

$$400x + 300y + 200z = 4500$$
$$150x + 115y + 60z = 1625$$
$$x + y - 2z = 0$$

$$\begin{bmatrix} 400 & 300 & 200 & 4500 \\ 150 & 115 & 60 & 1625 \\ 1 & 1 & -2 & 0 \end{bmatrix} \begin{matrix} \frac{1}{100}R_1 \to \\ \frac{1}{5}R_2 \to \\ \end{matrix} \begin{bmatrix} 4 & 3 & 2 & 45 \\ 30 & 23 & 12 & 325 \\ 1 & 1 & -2 & 0 \end{bmatrix} \begin{matrix} \\ 2R_2 - 15R_1 \to \\ 4R_3 - R_1 \to \end{matrix}$$

$$\begin{bmatrix} 4 & 3 & 2 & 45 \\ 0 & 1 & -6 & -25 \\ 0 & 1 & -10 & -45 \end{bmatrix} \begin{matrix} R_1 - 3R_2 \to \\ \\ R_3 - R_2 \to \end{matrix} \begin{bmatrix} 4 & 0 & 20 & 120 \\ 0 & 1 & -6 & -25 \\ 0 & 0 & -4 & -20 \end{bmatrix} \begin{matrix} \frac{1}{4}R_1 \to \\ \\ -\frac{1}{4}R_3 \to \end{matrix} \begin{bmatrix} 1 & 0 & 5 & 30 \\ 0 & 1 & -6 & -25 \\ 0 & 0 & 1 & 5 \end{bmatrix} \begin{matrix} R_1 - 5R_3 \to \\ R_2 + 6R_3 \to \\ \end{matrix} \begin{bmatrix} 1 & 0 & 0 & 5 \\ 0 & 1 & 0 & 5 \\ 0 & 0 & 1 & 5 \end{bmatrix}$$

$x = 5$, $y = 5$, $z = 5$ or $(x, y, z) = (5, 5, 5)$.
You should order 5 of each.

5. Let x = number of baby sharks, y = number of piranhas and z = number of squids.

	Sharks (x)	Piranhas (y)	Squids (z)	Total
Goldfish (per day)	1	1	1	21
Angelfish (per day)	2	0	1	21
Butterflyfish (per day)	2	3	0	35

$$x + y + z = 21$$
$$2x + z = 21$$
$$2x + 3y = 35$$

$$\begin{bmatrix} 1 & 1 & 1 & 21 \\ 2 & 0 & 1 & 21 \\ 2 & 3 & 0 & 35 \end{bmatrix} \begin{matrix} R_3 \to R_1 \\ R_1 \to R_2 \\ R_2 \to R_3 \end{matrix} \begin{bmatrix} 2 & 3 & 0 & 35 \\ 1 & 1 & 1 & 21 \\ 2 & 0 & 1 & 21 \end{bmatrix} \begin{matrix} \\ 2R_2 - R_1 \to \\ R_3 - R_1 \to \end{matrix} \begin{bmatrix} 2 & 3 & 0 & 35 \\ 0 & -1 & 2 & 7 \\ 0 & -3 & 1 & -14 \end{bmatrix} \begin{matrix} R_1 + 3R_2 \to \\ \\ R_3 - 3R_2 \to \end{matrix} \begin{bmatrix} 2 & 0 & 6 & 56 \\ 0 & -1 & 2 & 7 \\ 0 & 0 & -5 & -35 \end{bmatrix} \begin{matrix} \frac{1}{2}R_1 \to \\ -R_2 \to \\ -\frac{1}{5}R_3 \to \end{matrix}$$

$$\begin{bmatrix} 1 & 0 & 3 & 28 \\ 0 & 1 & -2 & -7 \\ 0 & 0 & 1 & 7 \end{bmatrix} \begin{matrix} R_1 - 3R_3 \to \\ R_2 + 2R_3 \to \\ \end{matrix} \begin{bmatrix} 1 & 0 & 0 & 7 \\ 0 & 1 & 0 & 7 \\ 0 & 0 & 1 & 7 \end{bmatrix}$$

$x = 7$, $y = 7$, $z = 7$ or $(x, y, z) = (7, 7, 7)$.
You should have 7 of each creature.

7. Let x = number of sections of Finite Math, y = number of sections of Business Calculus, and z = number of sections of Computer Methods.

	Finite (x)	Calculus (y)	C. Methods (z)	Total
No. of students	40	40	10	210
Revenue ($/student)	1000(40)	1500(40)	2000(10)	260,000
Number of sections	1	1	1	6

$$40x + 40y + 10z = 210$$
$$40,000 + 60,000y + 20,000z = 260,000$$
$$x + y + z = 6$$

$$\begin{bmatrix} 40 & 40 & 10 & 210 \\ 40,000 & 60,000 & 20,000 & 260,000 \\ 1 & 1 & 1 & 6 \end{bmatrix} \begin{matrix} \frac{1}{10}R_1 \to \\ \frac{1}{20,000}R_2 \to \\ {} \end{matrix} \begin{bmatrix} 4 & 4 & 1 & 21 \\ 2 & 3 & 1 & 13 \\ 1 & 1 & 1 & 6 \end{bmatrix} \begin{matrix} {} \\ 2R_2 - R_1 \to \\ 4R_3 - R_1 \to \end{matrix}$$

$$\begin{bmatrix} 4 & 4 & 1 & 21 \\ 0 & 2 & 1 & 5 \\ 0 & 0 & 3 & 3 \end{bmatrix} \frac{1}{3}R_3 \to \begin{bmatrix} 4 & 4 & 1 & 21 \\ 0 & 2 & 1 & 5 \\ 0 & 0 & 1 & 1 \end{bmatrix} \begin{matrix} R_1 - 2R_2 \to \\ {} \\ {} \end{matrix} \begin{bmatrix} 4 & 0 & -1 & 11 \\ 0 & 2 & 1 & 5 \\ 0 & 0 & 1 & 1 \end{bmatrix} \begin{matrix} R_1 + R_3 \to \\ R_2 - R_3 \to \\ {} \end{matrix} \begin{bmatrix} 4 & 0 & 0 & 12 \\ 0 & 2 & 0 & 4 \\ 0 & 0 & 1 & 1 \end{bmatrix} \begin{matrix} \frac{1}{4}R_1 \to \\ \frac{1}{2}R_2 \to \\ {} \end{matrix} \begin{bmatrix} 1 & 0 & 0 & 3 \\ 0 & 1 & 0 & 2 \\ 0 & 0 & 1 & 1 \end{bmatrix}$$

$(x, y, z) = (3, 2, 1)$
The college should offer 3 sections of Finite Math, 2 sections of Business Calculus and 1 section of Computer Methods.

9. Let x = change of $value of Volvo A.B., y = change of $value of Volvo Car Corporation and z = change of $value of Volvo Truck Corporation.
1. Total value of Volvo assets ($million): $x + y + z = 2$
2. Value of Renault's Volvo holding ($million): $0.1x + 0.25y + 0.45z = 0.4$
3. Value of Volvo's A.B. and Truck ($million): $0.1x + 0.45z = 0.65$
Using equations 2 and 3, solve for y: $y = -1$
Using $y = -1$, solve for x and z using equations 1 and 3: $x = 2$, $z = 1$
$(x, y, z) = (2, -1, 1)$
Thus, Volvo A.B., up $2 million; Volvo Car Corp., down $1 million; Volvo Truck Corp., up $1 million.

11. Let x = number of pension plans scrapped in 1980, y = number of pension plans scrapped in 1985 and z = number of pension plans scrapped in 1987.
Total pension plans scrapped: $x + y + z = 187,210$
Worst year, 1985: $y = x + z - 14,932$
Least damaging year, 1980: $x = z - 45,815$
We have the following system of equations.
$$x + y + z = 187,210$$
$$x - y + z = 14,932$$
$$x - z = -45,815$$

$$\begin{bmatrix} 1 & 1 & 1 & 187,210 \\ 1 & -1 & 1 & 14,932 \\ 1 & 0 & -1 & -45,815 \end{bmatrix} \frac{1}{2}(R_1 + R_2) \to \begin{bmatrix} 1 & 0 & 1 & 101,071 \\ 1 & -1 & 1 & 14,932 \\ 1 & 0 & -1 & -45,815 \end{bmatrix} \begin{matrix} \frac{1}{2}(R_1 + R_3) \to \\ R_1 - R_2 \to \\ {} \end{matrix}$$

$$\begin{bmatrix} 1 & 0 & 0 & 27,628 \\ 0 & 1 & 0 & 86,139 \\ 1 & 0 & -1 & -45,815 \end{bmatrix} R_1 - R_3 \to \begin{bmatrix} 1 & 0 & 0 & 27,628 \\ 0 & 1 & 0 & 86,139 \\ 0 & 0 & 1 & 73,443 \end{bmatrix}$$

$(x, y, z) = (27,628, 86,139, 73,443)$
Thus, 1980: 27,628 plans junked, 1985: 86,139 plans junked, and 1987: 73,443 plans junked.

13. Let x = number of tons of cream cheese from Cheesy Cream Corp. (CCC),
y = number of tons of cream cheese from Super Smooth & Sons (SSS), and
z = number of tons of cream cheese from Bagel's Best Friend (BBF).

	CCC (x)	SSS (y)	BBF (z)	Total
Order (tons)	1	1	1	100
Cost ($ per ton)	80	50	65	5990

Also, $x = z$.
We have the following system of equations,
$$x + y + z = 100$$
$$80x + 50y + 65z = 5990$$
$$x - z = 0$$
Solving the system, we get $(x, y, z) = (22, 56, 22)$.
Thus, the bagel store orders 22 tons from CCC, 56 tons from SSS and 22 tons from BBF.

15. Let x = number of evil sorcerers, y = number of warriors and z = number of orcs.

	Sorcerers (x)	Warriors (y)	Orcs (z)	Total
Victims	1	1	1	560
Thrusts	2	2	1	620

Also, $y = 5x$.
We have the following system of equations.
$$x + y + z = 560$$
$$2x + 2y + z = 620$$
$$-5x + y = 0$$

$$\begin{bmatrix} 1 & 1 & 1 & 560 \\ 2 & 2 & 1 & 620 \\ -5 & 1 & 0 & 0 \end{bmatrix} \begin{matrix} R_2 \to R_1 \\ R_3 \to R_2 \\ R_1 \to R_3 \end{matrix} \begin{bmatrix} 2 & 2 & 1 & 620 \\ -5 & 1 & 0 & 0 \\ 1 & 1 & 1 & 560 \end{bmatrix} \begin{matrix} R_1 - R_3 \to \\ \\ \end{matrix} \begin{bmatrix} 1 & 1 & 0 & 60 \\ -5 & 1 & 0 & 0 \\ 1 & 1 & 1 & 560 \end{bmatrix} \begin{matrix} \frac{1}{6}(R_1 - R_2) \to \\ \frac{1}{6}(R_2 + 5R_1) \to \\ R_3 - R_2 \to \end{matrix} \begin{bmatrix} 1 & 0 & 0 & 10 \\ 0 & 1 & 0 & 50 \\ 0 & 0 & 1 & 500 \end{bmatrix}$$

$(x, y, z) = (10, 50, 500)$
Thus, Conan the Great has slain 10 evil sorcerers, 50 warriors, and 500 orcs.

17. Let x = number of 32-year-old males, y = number of 50-year-old males and z = number of 60-year-old males.

	32-year-old (x)	50-year-old (y)	60-year-old (z)	Total
1980 mortality rates (in millions)	0.2%	0.7%	1.5%	0.202
number (in millions)	1	1	1	26.5

Also, $x = y + 0.5$ (in millions).
We have the following system of equations.
$$0.002x + 0.007y + 0.015z = 0.202$$
$$x + y + z = 26.5$$
$$x - y = 0.5$$

$$\begin{bmatrix} 0.002 & 0.007 & 0.015 & 0.202 \\ 1 & 1 & 1 & 26.5 \\ 1 & -1 & 0 & 0.5 \end{bmatrix} \begin{matrix} 1000R_1 \to \\ 2R_2 \to \\ 2R_3 \to \end{matrix} \begin{bmatrix} 2 & 7 & 15 & 202 \\ 2 & 2 & 2 & 53 \\ 2 & -2 & 0 & 1 \end{bmatrix} \begin{matrix} \\ -R_2 + R_1 \to \\ -R_3 + R_1 \to \end{matrix} \begin{bmatrix} 2 & 7 & 15 & 202 \\ 0 & 5 & 13 & 149 \\ 0 & 9 & 15 & 201 \end{bmatrix} \begin{matrix} 5R_1 - 7R_2 \to \\ \\ 5R_3 - 9R_2 \to \end{matrix}$$

$$\begin{bmatrix} 10 & 0 & -16 & -33 \\ 0 & 5 & 13 & 149 \\ 0 & 0 & -42 & -336 \end{bmatrix} \begin{matrix} \\ \\ -\frac{1}{42}R_3 \to \end{matrix} \begin{bmatrix} 10 & 0 & -16 & -33 \\ 0 & 5 & 13 & 149 \\ 0 & 0 & 1 & 8 \end{bmatrix} \begin{matrix} R_1 + 16R_3 \to \\ R_2 - 13R_3 \to \\ \end{matrix} \begin{bmatrix} 10 & 0 & 0 & 95 \\ 0 & 5 & 0 & 45 \\ 0 & 0 & 1 & 8 \end{bmatrix} \begin{matrix} \frac{1}{10}R_1 \to \\ \frac{1}{5}R_2 \to \\ \end{matrix} \begin{bmatrix} 1 & 0 & 0 & 9.5 \\ 0 & 1 & 0 & 9 \\ 0 & 0 & 1 & 8 \end{bmatrix}$$

$(x, y, z) = (9.5, 9, 8)$
Thus, the number of males in each designated age group is 9.5 million 32-year-olds, 9 million 50-year-olds, and 8 million 60-year-olds.

19. Let x = number of Democrat representatives in favor of bill ($333 - x$ against),
y = number of Republican representatives in favor of bill ($89 - y$ against), and
z = number of representatives from other parties in favor of bill ($13 - z$ against).
31 more votes in favor than against:
$x + y + z = 31 + [(333 - x) + (89 - y) + (13 - z)]$
10 times as many Democrats as Republicans voting in favor: $x = 10y$
36 more non-Democrats voting against than in favor of:
$(89 - y) + (13 - z) = 36 + y + z$
We now have the following system of equations:
$$2x + 2y + 2z = 466$$
$$x - 10y = 0$$
$$2y + 2z = 66$$

$$\begin{bmatrix} 2 & 2 & 2 & 466 \\ 1 & -10 & 0 & 0 \\ 0 & 2 & 2 & 66 \end{bmatrix} \begin{matrix} \frac{1}{2}R_1 \to \\ \\ \frac{1}{2}R_3 \to \end{matrix} \begin{bmatrix} 1 & 1 & 1 & 233 \\ 1 & -10 & 0 & 0 \\ 0 & 1 & 1 & 33 \end{bmatrix} \begin{matrix} R_1 - R_3 \to \\ R_1 - R_2 \to \end{matrix} \begin{bmatrix} 1 & 0 & 0 & 200 \\ 0 & 11 & 1 & 233 \\ 0 & 1 & 1 & 33 \end{bmatrix} \frac{1}{10}(R_2 - R_3) \to$$

$$\begin{bmatrix} 1 & 0 & 0 & 200 \\ 0 & 1 & 0 & 20 \\ 0 & 1 & 1 & 33 \end{bmatrix} R_3 - R_2 \to \begin{bmatrix} 1 & 0 & 0 & 200 \\ 0 & 1 & 0 & 20 \\ 0 & 0 & 1 & 13 \end{bmatrix}$$

$(x, y, z) = (200, 20, 13)$
Thus, the number of representatives voting in favor of the bill was 200 Democrats, 20 Republicans and 13 of other parties.

21. Let x = number of books from Brooklyn to Manhattan, y = number of books from Brooklyn to Long Island,
z = number of books from Queens to Manhattan, and w = number of books from Queens to Long Island.

a.

	x	y	z	w	Total
Books in Brooklyn warehouse	1	1	0	0	1000
Books in Queens warehouse	0	0	1	1	2000
Books ordered by Manhattan store	1	0	1	0	1500
Books ordered by Long Island store	0	1	0	1	1500
Cost ($) to ship books	1	5	2	4	9000

We have the following system of equations in four unknowns.

$$x + y = 1000$$
$$z + w = 2000$$
$$x + z = 1500$$
$$y + w = 1500$$
$$x + 5y + 2z + 4w = 9000$$

$$\begin{bmatrix} 1 & 1 & 0 & 0 & 1000 \\ 0 & 0 & 1 & 1 & 2000 \\ 1 & 0 & 1 & 0 & 1500 \\ 0 & 1 & 0 & 1 & 1500 \\ 1 & 5 & 2 & 4 & 9000 \end{bmatrix} \text{rearrange} \rightarrow \begin{bmatrix} 1 & 1 & 0 & 0 & 1000 \\ 1 & 5 & 2 & 4 & 9000 \\ 1 & 0 & 1 & 0 & 1500 \\ 0 & 1 & 0 & 1 & 1500 \\ 0 & 0 & 1 & 1 & 2000 \end{bmatrix} \begin{matrix} \\ R_2 - R_1 \rightarrow \\ R_3 - R_1 \rightarrow \\ \\ \end{matrix} \begin{bmatrix} 1 & 1 & 0 & 0 & 1000 \\ 0 & 4 & 2 & 4 & 8000 \\ 0 & -1 & 1 & 0 & 500 \\ 0 & 1 & 0 & 1 & 1500 \\ 0 & 0 & 1 & 1 & 2000 \end{bmatrix} \begin{matrix} R_1 + R_3 \rightarrow \\ \frac{1}{2} R_2 \rightarrow \\ R_3 + R_4 \rightarrow \\ \\ \end{matrix}$$

$$\begin{bmatrix} 1 & 0 & 1 & 0 & 1500 \\ 0 & 2 & 1 & 2 & 4000 \\ 0 & 0 & 1 & 1 & 2000 \\ 0 & 1 & 0 & 1 & 1500 \\ 0 & 0 & 1 & 1 & 2000 \end{bmatrix} \begin{matrix} \\ \\ \\ R_2 - 2R_4 \rightarrow \\ \end{matrix} \begin{bmatrix} 1 & 0 & 1 & 0 & 1500 \\ 0 & 2 & 1 & 2 & 4000 \\ 0 & 0 & 1 & 1 & 2000 \\ 0 & 0 & 1 & 0 & 1000 \\ 0 & 0 & 1 & 1 & 2000 \end{bmatrix} R_3 \leftrightarrow R_4 \begin{bmatrix} 1 & 0 & 1 & 0 & 1500 \\ 0 & 2 & 1 & 2 & 4000 \\ 0 & 0 & 1 & 0 & 1000 \\ 0 & 0 & 1 & 1 & 2000 \\ 0 & 0 & 1 & 1 & 2000 \end{bmatrix} \begin{matrix} R_1 - R_3 \rightarrow \\ R_2 - R_4 \rightarrow \\ \\ R_4 - R_3 \rightarrow \\ R_5 - R_3 \rightarrow \end{matrix}$$

$$\begin{bmatrix} 1 & 0 & 0 & 0 & 500 \\ 0 & 2 & 0 & 1 & 2000 \\ 0 & 0 & 1 & 0 & 1000 \\ 0 & 0 & 0 & 1 & 1000 \\ 0 & 0 & 0 & 1 & 1000 \end{bmatrix} \frac{1}{2}(R_2 - R_4) \rightarrow \begin{bmatrix} 1 & 0 & 0 & 0 & 500 \\ 0 & 1 & 0 & 0 & 500 \\ 0 & 0 & 1 & 0 & 1000 \\ 0 & 0 & 0 & 1 & 1000 \\ 0 & 0 & 0 & 1 & 1000 \end{bmatrix}$$

$(x, y, z, w) = (500, 500, 1000, 1000)$

Brooklyn to Manhattan: 500 books; Brooklyn to Long Island: 500 books; Queens to Manhattan: 1000 books; Queens to Long Island: 1000 books

b. Yes, there is a way to ship the books for less money. The cost to ship each book from Brooklyn to Manhattan is $1, which is less than the $5/book to ship from Brooklyn to Long Island. Thus, ship all 1000 books from the Brooklyn warehouses to Manhattan. The Manhattan store can receive 500 books from the Queens warehouse, and the remaining 1500 books from the Queens warehouses can be shipped to the Long Island store.

Thus,

Brooklyn to Manhattan: 1000 books (cost: 1000 × $1 = $1000)
Brooklyn to Long Island: none (cost: $0)
Queens to Manhattan: 500 books (cost: 500 × $2 = $1000)
Queens to Long Island: 1500 books (cost: 1500 × $4 = $6000)
Total Cost: $8000

23. Let x = amount invested in Company X,
y = amount invested in Company Y,
z = amount invested in Company Z, and
w = amount invested in Company W.
Total amount of investment is $65 million:
$x + y + z + w = 65$ (in millions)
Total amount of interest: $8 million
Company X earned 15%: +0.15x
Company Y depreciated 20%: –0.20y
Company Z neither earned nor depreciated: 0z
Company W earned 20%: +0.20w
Thus, $0.15x - 0.20y + 0z + 0.20w = 8$ (in millions).
There is twice as much invested in Company X as in Company Z: $x = 2z$ or $x - 2z = 0$.
There is three times as much in Company W as in Company Z: $w = 3z$ or $w - 3z = 0$.
We now have the following system of four equations in four unknowns.
$$x + y + z + w = 65$$
$$0.15x - 0.20y + 0.20w = 8$$
$$x - 2z = 0$$
$$w - 3z = 0$$
Solving the system, we get the solution
$(x, y, z, w) = (20, 5, 10, 30)$.
Yes, Smiley has sufficient information.
The investment portfolio is:
$20 million in Company X,
$5 million in Company Y,
$10 million in Company Z,
and $30 million in Company W.

25. Let x = number of cars per day on Eastward Blvd.
y = number of cars per day on Northwest Ln., and
z = number of cars per day on Southwest Ln.

a. At the intersection of NW Ln. and SW Ln.,
$z = s$ where s is arbitrary
$y = s + 50$
At the intersecting NW Ln. and E. Blvd.,
$x = y + 150 = s + 200$
Thus, it is not possible to determine the daily flow of traffic along each of the through streets from the information given. The answer is no.
The general solution is
Eastward Blvd: $s + 200$; Northwest Lane: $s + 50$;
Southwest Lane: s, where s is arbitrary.
Thus, it would suffice to know the traffic along Southwest Lane.

b. If $s = 60$ vehicles per day, then the answer is yes. It leads to the following solutions.
Eastward Blvd: 260; Northwest Lane: 110;
Southwest: 60

27. a. Define the unknowns such that
x = the number of vehicles per hour on April,
y = the number of vehicles per hour on Broadway,
z = the number of vehicles per hour on Division,
and w = the number of vehicles per hour on Embankment. Thus we have the following equations.
$x + y = 300$, $x = z + 100$, $w = y + 100$, $z + w = 300$
From these we have the following system of four equations with four unknowns:
$$x + y = 300$$
$$x - z = 100$$
$$-y + w = 100$$
$$z + w = 300$$
Solving this system, we get many solutions, which can be written in the following way.
$$x = 400 - w$$
$$y = -100 + w$$
$$z = 300 - w$$
$100 \le w \le 300$ since x, y, z, and w are not negative.

b. You could put a counter on April, Broadway, Division, or Embankment in order to determine the flow completely.

29. Let M = money stock, C = currently in circulation, R = bank reserves, D = deposits of the public, and H = high-powered money.
$$M = C + D$$
$$C = 0.2D$$
$$R = 0.1D$$
$$H = R + C$$
If $M = 120$ ($billion), find R.
$$C + D = 120$$
$$C - 0.2D = 0$$
Solving for D gives $D = 100$.
$R = 0.1D = 0.1(100) = 10$
The bank reserves have to be $10 billion.

31. Let x = donation to the MPBF,
y = donation to the SCN, and
z = donation to the NY Jets.
There was twice as much donated to the NY Jets as to the MPBF so $z = 2x$.
Equal amounts were donated to MPBF and SCN so $x = y$
Donations in bank account:
$x + 2y + 2z = 4200$
We have the following system of equations:
$$x + 2y + 2z = 4200$$
$$2x - z = 0$$
$$x - y = 0$$
Solving the system, we get the solution
$(x, y, z) = (600, 600, 1200)$.
Thus, it donated $600 to each of the MPBF and the SCN, and $1200 to the Jets.

33. Using the X-ray absorption diagram, we have the following system of equations.

$$\begin{array}{ll} x+z+1000=3030 & x+z=2030 \\ x+y+1000=3020 \quad \text{or} & x+y=2020 \\ \quad y+x+z=3050 & x+y+z=3050 \end{array}$$

Solving the system, we have the solution for the absorption, which is $(x, y, z) = (1000, 1020, 1030)$.
From the table,
the type of absorption for $x = 1000$ is water,
the type of absorption for $y = 1020$ is grey matter,
and the type of absorption for $z = 1030$ is tumor.
Thus, $x =$ water, $y =$ grey matter, and $z =$ tumor.

35. From the diagram, we have the following system of equations.

$$\begin{array}{ll} x+u+x+u+x=3000 & 3x+2u=3000 \\ \quad y+x+u=3000 \quad \text{or} & x+y+u=3000 \\ z+u+x+y=4030 & x+y+z+u=4030 \\ \quad z+u+z=2060 & 2z+u=2060 \end{array}$$

Solving the system, we have the solution for the absorption, which is $(x, y, z, u) = (1000, 2000, 1030, 0)$.
Thus, the type of absorption is $x =$ water, $y =$ bone, $z =$ tumor, and $u =$ air.

37. Let y be the absorption of the region above site x. Let z be the absorption of the region below site x. Let w be the combined absorption of the four regions to the left of site x. Then we get the following equations:
$y + w = 6130$, $x + w = 6110$, $z + w = 6080$, and $x + y + z = 3080$.
Solving this system of equations, we get $x = 1030$, $y = 1050$, $z = 1000$, and $z = 5080$.
Thus the composition of site x is a tumor.

39. Let $x =$ number of empty seats on TWA, $y =$ number of empty seats on Northwest and $z =$ number of empty seats on American.
We have the following system of equations:

$$\begin{array}{ll} x+y+z=160 & x+y+z=160 \\ 0.092(3000)x+0.097(3000)y+0.098(3000)z=46,230 \quad \text{or} & 92x+97y+98z=15,410 \\ \quad y=3x & 3x-y=0 \end{array}$$

Solving the system, we have $(x, y, z) = (30, 90, 40)$.
Thus, the number of empty airline seats is TWA: 30; Northwest: 90; American: 40.

41. This can be represented by a linear equation.
$y = 0.3(x + y + z)$

43. This can be represented by a linear equation.
$x = 100$

45. This can be represented by a linear equation.
$x = 2(y + z)$

47. Examples will vary.

You're the Expert
Pages 166–167

A general parabola for the cost is $C = at^2 + bt + c$.

1. $t = 8$, $C = 20$ given $20 = 64a + 8b + c$.
   $t = 10$, $C = 35$ given $35 = 100a + 10b + c$.
   $t = 12$, $C = 94$ given $94 = 144a + 12b + c$.
   There are three linear equations in three unknowns.
   Solving the system gives $a = 5.5$, $b = -91.5$, $c = 400$.
   The cost equation becomes $C = 5.5t^2 - 91.5t + 400$.
   Substitute $t = 15$ to get $C = 265$.
   You can predict an annual cost to the utility industry of $265 billion. A general parabola for the jobs equation is $J = at^2 + bt + c$.
   $t = 8$, $J = 14$ (thousand) gives $14 = 64a + 8b + c$.
   $t = 10$, $J = 22$ (thousand) gives $22 = 100a + 10b + c$.
   $t = 12$, $J = 13$ (thousand) gives $13 = 144a + 12b + c$.
   Solving the system gives $a = -2.125$, $b = 42.25$, $c = -188$.
   The jobs equation becomes
   $J = -2.125t^2 + 42.25t - 188$ (in thousands).
   Substitute $t = 15$ to get $J = -32.375$ (thousand) or $J = -32,375$.
   There would be 32,375 new jobs created.
   Projection: a 15-million-ton rollback will result in an annual cost of $265 billion, but in the creations of 32,375 new jobs.

3. If the 10-million-ton-rollback data had not been available, you would be able to create linear equations to describe the situation. Describe cost by an equation of the form $C = at + b$. Substitute the given values of C and t into the equation.
   $t = 8$, $C = 20.4$ gives $20.4 = 8a + b$.
   $t = 12$, $C = 93.6$ gives $93.6 = 12a + b$.
   Solve this system to get $a = 18.3$, $b = -126$. Thus, the cost equation is $C = 18.3t - 126$. Substitute $t = 15$ into the equation to get $C = 148.5$. Describe jobs lost by an equation of the form $J = at + b$. Substitute the given values of J and t into the equation.
   $t = 8$, $J = 14,100$ gives $14,100 = 8a + b$.
   $t = 12$, $J = 13,400$ gives $13,400 = 12a + b$.
   Solve this system to get $a = -175$, $b = 15,500$. Thus the cost equation is $J = -175t + 15,500$. Substitute $t = 15$ into the equation to get $J = 12,875$. Thus you project that a 15-million-ton-rollback will result in an annual cost of $148.5 billion to the utility industry and a loss of 12,875 jobs in the coal-mining industry.

5. $(1, 2)$: $2 = a + b + c$
   $(2, 9)$: $9 = 4a + 2b + c$
   $(3, 16)$: $16 = 9a + 3b + c$
   Solving the system gives $a = 0$, $b = 7$, $c = -5$.
   $y = 7x - 5$, a linear equation
   Thus, a parabola does not pass through the points.

7. Using a graphing utility, graph the equation $C = 5.625t^2 - 94.2t + 414$. From the graph, we estimate that a rollback of about 8.375 million tons will result in the lowest annual cost to utilities of about $19.5 billion.

Chapter 2 Review Exercises
Pages 167–177

1. Equation 1 is $x + 2y = 4$ and equation 2 is $2x - y = 1$.
   Multiply equation 2 by 2 and add.
   $$\begin{array}{r} x + 2y = 4 \\ 4x - 2y = 2 \\ \hline 5x = 6 \\ x = \dfrac{6}{5} \end{array}$$
   $y = 2x - 1 = \dfrac{12}{5} - 1 = \dfrac{7}{5}$
   $(x, y) = \left(\dfrac{6}{5}, \dfrac{7}{5} \right)$

3. $3x - y = 0$
   $2x - y = 0$
   Subtracting gives $x = 0$ and $y = 2x = 0$.
   $(x, y) = (0, 0)$

5. Equation 1 is $0.2x - 0.1y = 0.3$ and equation 2 is $0.2x + 0.2y = 0.4$. Multiply equation 1 by -10 and equation 2 by 10 and add.
   $$\begin{array}{r} -2x + y = -3 \\ 2x + 2y = 4 \\ \hline 3y = 1 \\ y = \dfrac{1}{3} \end{array}$$
   $x + y = 2$ or $x = 2 - y = 2 - \dfrac{1}{3} = \dfrac{5}{3}$
   $(x, y) = \left(\dfrac{5}{3}, \dfrac{1}{3} \right)$

7. Equation 1 is $\dfrac{2x}{3} - \dfrac{3y}{2} = \dfrac{1}{3}$ and equation 2 is $\dfrac{x}{4} + 2y = 0$. Multiply equation 1 by 6 and equation 2 by -16 and add.
   $$\begin{array}{r} 4x - 9y = 2 \\ -4x - 32y = 0 \\ \hline -41y = 2 \\ y = -\dfrac{2}{41} \end{array}$$
   $x + 8y = 0$ or $x = -8y$, $x = -8\left(\dfrac{-2}{41} \right) = \dfrac{16}{41}$
   $(x, y) = \left(\dfrac{16}{41}, -\dfrac{2}{41} \right)$

9. Equation 1 is $x - 3y = 0$ and equation 2 is

$-x - \dfrac{3y}{2} = 0$. Multiply equation 2 by 2 and add.

$$\begin{aligned} x - 3y &= 0 \\ \underline{-2x - 3y} &= \underline{0} \\ -x &= 0 \\ x &= 0 \end{aligned}$$

$y = 3y = 0$

$(x, y) = (0, 0)$

11. Equation 1 is $2x + 3y = 2$ and equation 2 is

$-x - \dfrac{3y}{2} = \dfrac{1}{2}$. Multiply equation 2 by 2 and add.

$$\begin{aligned} 2x + 3y &= 2 \\ \underline{-\ 2x - 3y} &= \underline{1} \\ 0 &= 3 \end{aligned}$$

False, no solution

13. $x - 2y = 1$

$3x + 2y = \dfrac{1}{2}$

$x + 6y = -\dfrac{1}{2}$

$$\begin{bmatrix} 1 & -2 & 1 \\ 3 & 2 & \frac{1}{2} \\ 1 & 6 & -\frac{1}{2} \end{bmatrix} \begin{matrix} \\ 2R_2 \to \\ 2R_3 \to \end{matrix} \begin{bmatrix} 1 & -2 & 1 \\ 6 & 4 & 1 \\ 2 & 12 & -1 \end{bmatrix} \begin{matrix} \\ R_2 - 6R_1 \to \\ R_3 - 2R_1 \to \end{matrix}$$

$$\begin{bmatrix} 1 & -2 & 1 \\ 0 & 16 & -5 \\ 0 & 16 & -3 \end{bmatrix} \begin{matrix} 8R_1 + R_2 \to \\ \\ R_3 - R_2 \to \end{matrix} \begin{bmatrix} 8 & 0 & 3 \\ 0 & 16 & -5 \\ 0 & 0 & 2 \end{bmatrix} \begin{matrix} \frac{1}{8}R_1 \to \\ \frac{1}{16}R_2 \to \\ \end{matrix}$$

$$\begin{bmatrix} 1 & 0 & \frac{3}{8} \\ 0 & 1 & -\frac{5}{16} \\ 0 & 0 & 2 \end{bmatrix}$$

$0x + 0y = 2$

$\qquad\ 0 = 2$

False, no solution

15. $8x - 6y = 0$

$6x - 3y = \dfrac{1}{2}$

$2x - 3y = -\dfrac{1}{2}$

$$\begin{bmatrix} 8 & -6 & 0 \\ 6 & -3 & \frac{1}{2} \\ 2 & -3 & -\frac{1}{2} \end{bmatrix} \begin{matrix} \frac{1}{4}R_1 \to \\ \frac{1}{3}R_2 \to \\ \end{matrix} \begin{bmatrix} 2 & -\frac{3}{2} & 0 \\ 2 & -1 & \frac{1}{6} \\ 2 & -3 & -\frac{1}{2} \end{bmatrix} \begin{matrix} \\ R_2 - R_1 \to \\ R_3 - R_1 \to \end{matrix}$$

$$\begin{bmatrix} 2 & -\frac{3}{2} & 0 \\ 0 & \frac{1}{2} & \frac{1}{6} \\ 0 & -\frac{3}{2} & -\frac{1}{2} \end{bmatrix} \begin{matrix} 2R_1 \to \\ 6R_2 \to \\ 2R_3 \to \end{matrix} \begin{bmatrix} 4 & -3 & 0 \\ 0 & 3 & 1 \\ 0 & -3 & -1 \end{bmatrix} \begin{matrix} R_1 + R_2 \to \\ \\ R_3 + R_2 \to \end{matrix}$$

$$\begin{bmatrix} 4 & 0 & 1 \\ 0 & 3 & 1 \\ 0 & 0 & 0 \end{bmatrix} \begin{matrix} \frac{1}{4}R_1 \to \\ \frac{1}{3}R_2 \to \\ \end{matrix} \begin{bmatrix} 1 & 0 & \frac{1}{4} \\ 0 & 1 & \frac{1}{3} \\ 0 & 0 & 0 \end{bmatrix}$$

$x = \dfrac{1}{4}, \ y = \dfrac{1}{3}$

$(x, y) = \left(\dfrac{1}{4}, \dfrac{1}{3} \right)$

17. $x - 0.1y = 0$

$0.2x + 0.1y = 0$

$\quad\ 2x - y = 0$

$$\begin{bmatrix} 1 & -0.1 & 0 \\ 0.2 & 0.1 & 0 \\ 2 & -1 & 0 \end{bmatrix} \begin{matrix} 10R_1 \to \\ 10R_2 \to \\ \end{matrix} \begin{bmatrix} 10 & -1 & 0 \\ 2 & 1 & 0 \\ 2 & -1 & 0 \end{bmatrix} \begin{matrix} \\ 5R_2 - R_1 \to \\ 5R_3 - R_1 \to \end{matrix}$$

$$\begin{bmatrix} 10 & -1 & 0 \\ 0 & 6 & 0 \\ 0 & -4 & 0 \end{bmatrix} \begin{matrix} \\ \frac{1}{6}R_2 \to \\ \frac{1}{4}R_3 \to \end{matrix} \begin{bmatrix} 10 & -1 & 0 \\ 0 & 1 & 0 \\ 0 & -1 & 0 \end{bmatrix} \begin{matrix} \frac{1}{10}(R_1 + R_2) \to \\ \\ R_3 + R_2 \to \end{matrix}$$

$$\begin{bmatrix} 1 & 0 & 0 \\ 0 & 1 & 0 \\ 0 & 0 & 0 \end{bmatrix}$$

$x = 0, \ y = 0$

$(x, y) = (0, 0)$

19. $x + y = 1$

$2x + y = 0.3$

$3x + 2y = \dfrac{13}{10}$

$$\begin{bmatrix} 1 & 1 & 1 \\ 2 & 1 & 0.3 \\ 3 & 2 & 1.3 \end{bmatrix} \begin{matrix} \\ R_2 - 2R_1 \to \\ R_3 - 3R_1 \to \end{matrix} \begin{bmatrix} 1 & 1 & 1 \\ 0 & -1 & -1.7 \\ 0 & -1 & -1.7 \end{bmatrix} \begin{matrix} R_1 + R_2 \to \\ -R_2 \to \\ R_3 - R_2 \to \end{matrix}$$

$$\begin{bmatrix} 1 & 0 & -0.7 \\ 0 & 1 & 1.7 \\ 0 & 0 & 0 \end{bmatrix}$$

$x = -0.7 \text{ or } -\dfrac{7}{10}, \ y = 1.7 \text{ or } \dfrac{17}{10}$

$(x, y) = \left(-\dfrac{7}{10}, \dfrac{17}{10} \right)$

21. $-x + 2y + z = 2$
   $-x + y + 2z = 2$
         $y - z = 0$

$$\begin{bmatrix} -1 & 2 & 1 & 2 \\ -1 & 1 & 2 & 2 \\ 0 & 1 & -1 & 0 \end{bmatrix} R_2 - R_1 \rightarrow$$

$$\begin{bmatrix} -1 & 2 & 1 & 2 \\ 0 & -1 & 1 & 0 \\ 0 & 1 & -1 & 0 \end{bmatrix} \begin{matrix} -(R_1 + 2R_2) \rightarrow \\ -R_2 \rightarrow \\ R_3 + R_2 \rightarrow \end{matrix} \begin{bmatrix} 1 & 0 & -3 & -2 \\ 0 & 1 & -1 & 0 \\ 0 & 0 & 0 & 0 \end{bmatrix}$$

$\begin{matrix} x - 3z = -2 \\ y - z = 0 \end{matrix}$ or $\begin{matrix} x = 3z - 2 \\ y = z \end{matrix}$

There are an infinite number of solutions.
$(x, y, z) = (3z - 2, z, z)$, z arbitrary

23. $x + 2y + z = 40$
   $\frac{1}{3}x + \frac{1}{3}y + \frac{2}{3}z = \frac{40}{3}$
        $\frac{1}{2}x + \frac{1}{2}z = 10$

$$\begin{matrix} \\ 3R_2 \rightarrow \\ 2R_3 \rightarrow \end{matrix} \begin{bmatrix} 1 & 2 & 1 & 40 \\ 1 & 1 & 2 & 40 \\ 1 & 0 & 1 & 20 \end{bmatrix} \begin{matrix} \\ R_2 - R_1 \rightarrow \\ R_3 - R_1 \rightarrow \end{matrix}$$

$$\begin{bmatrix} 1 & 2 & 1 & 40 \\ 0 & -1 & 1 & 0 \\ 0 & -2 & 0 & -20 \end{bmatrix} \begin{matrix} R_1 + 2R_2 \rightarrow \\ -R_2 \rightarrow \\ -\frac{1}{2}(R_3 - 2R_2) \rightarrow \end{matrix}$$

$$\begin{bmatrix} 1 & 0 & 3 & 40 \\ 0 & 1 & -1 & 0 \\ 0 & 0 & 1 & 10 \end{bmatrix} \begin{matrix} R_1 - 3R_3 \rightarrow \\ R_2 + R_3 \rightarrow \\ \end{matrix} \begin{bmatrix} 1 & 0 & 0 & 10 \\ 0 & 1 & 0 & 10 \\ 0 & 0 & 1 & 10 \end{bmatrix}$$

$x = 10, y = 10, z = 10$
$(x, y, z) = (10, 10, 10)$

25. $\begin{matrix} x = -2y + z \\ y = x + 2z \\ z = -2x - y + 14 \end{matrix}$ or $\begin{matrix} x + 2y - z = 0 \\ -x + y - 2z = 0 \\ 2x + y + z = 14 \end{matrix}$

$$\begin{bmatrix} 1 & 2 & -1 & 0 \\ -1 & 1 & -2 & 0 \\ 2 & 1 & 1 & 14 \end{bmatrix} \begin{matrix} \\ \frac{1}{3}(R_2 + R_1) \rightarrow \\ R_3 - 2R_1 \rightarrow \end{matrix}$$

$$\begin{bmatrix} 1 & 2 & -1 & 0 \\ 0 & 1 & -1 & 0 \\ 0 & -3 & 3 & 14 \end{bmatrix} \begin{matrix} R_1 - 2R_2 \rightarrow \\ \\ R_3 + 3R_2 \rightarrow \end{matrix} \begin{bmatrix} 1 & 0 & 1 & 0 \\ 0 & 1 & -1 & 0 \\ 0 & 0 & 0 & 14 \end{bmatrix}$$

$0 = 14$
False, no solution

27. $x + y + 2z = 2$
   $2x - y + 2z = \frac{3}{2}$
   $\frac{1}{2}x + \frac{1}{2}y + \frac{1}{2}z = \frac{3}{4}$
        $-2x - 2z = -2$

$$\begin{matrix} \\ 2R_2 \rightarrow \\ 4R_3 \rightarrow \\ -\frac{1}{2}R_4 \rightarrow \end{matrix} \begin{bmatrix} 1 & 1 & 2 & 2 \\ 4 & -2 & 4 & 3 \\ 2 & 2 & 2 & 3 \\ 1 & 0 & 1 & 1 \end{bmatrix} \begin{matrix} \\ R_2 - 4R_1 \rightarrow \\ R_3 - 2R_1 \rightarrow \\ R_1 - R_4 \rightarrow \end{matrix}$$

$$\begin{bmatrix} 1 & 1 & 2 & 2 \\ 0 & -6 & -4 & -5 \\ 0 & 0 & -2 & -1 \\ 0 & 1 & 1 & 1 \end{bmatrix} \begin{matrix} 6R_1 + R_2 \rightarrow \\ \\ \\ 6R_4 + R_2 \rightarrow \end{matrix}$$

$$\begin{bmatrix} 6 & 0 & 8 & 7 \\ 0 & -6 & -4 & -5 \\ 0 & 0 & -2 & -1 \\ 0 & 0 & 2 & 1 \end{bmatrix} \begin{matrix} R_1 + 4R_3 \rightarrow \\ R_2 - 2R_3 \rightarrow \\ \\ R_4 + R_3 \rightarrow \end{matrix}$$

$$\begin{bmatrix} 6 & 0 & 0 & 3 \\ 0 & -6 & 0 & -3 \\ 0 & 0 & -2 & -1 \\ 0 & 0 & 0 & 0 \end{bmatrix} \begin{matrix} \frac{1}{6}R_1 \rightarrow \\ -\frac{1}{6}R_2 \rightarrow \\ -\frac{1}{2}R_3 \rightarrow \\ \end{matrix} \begin{bmatrix} 1 & 0 & 0 & \frac{1}{2} \\ 0 & 1 & 0 & \frac{1}{2} \\ 0 & 0 & 1 & \frac{1}{2} \\ 0 & 0 & 0 & 0 \end{bmatrix}$$

$x = \frac{1}{2}, y = \frac{1}{2}, z = \frac{1}{2}$

$(x, y, z) = \left(\frac{1}{2}, \frac{1}{2}, \frac{1}{2} \right)$

29.
$$x + y + 5z + w = 1$$
$$y + 2z + w = 0$$
$$x + 3y + 7z + 2w = 1$$
$$x + y + 5z + w = 1$$

$$\begin{bmatrix} 1 & 1 & 5 & 1 & 1 \\ 0 & 1 & 2 & 1 & 0 \\ 1 & 3 & 7 & 2 & 1 \\ 1 & 1 & 5 & 1 & 1 \end{bmatrix} \rightarrow \begin{bmatrix} 1 & 1 & 5 & 1 & 1 \\ 0 & 1 & 2 & 1 & 0 \\ 0 & -2 & -2 & -1 & 0 \\ 0 & 0 & 0 & 0 & 0 \end{bmatrix} \rightarrow$$

$$\begin{bmatrix} 1 & 0 & 3 & 0 & 1 \\ 0 & 1 & 2 & 1 & 0 \\ 0 & 0 & 2 & 1 & 0 \\ 0 & 0 & 0 & 0 & 0 \end{bmatrix} \rightarrow \begin{bmatrix} 2 & 0 & 0 & -3 & 2 \\ 0 & 1 & 0 & 0 & 0 \\ 0 & 0 & 2 & 1 & 0 \\ 0 & 0 & 0 & 0 & 0 \end{bmatrix} \rightarrow$$

$$\begin{bmatrix} 1 & 0 & 0 & -\frac{3}{2} & 1 \\ 0 & 1 & 0 & 0 & 0 \\ 0 & 0 & 1 & \frac{1}{2} & 0 \\ 0 & 0 & 0 & 0 & 0 \end{bmatrix}$$

$$x = \frac{3}{2}w + 1$$
$$y = 0$$
$$z = -\frac{1}{2}w$$

w is arbitrary.

$$(x, y, z, w) = \left(\frac{3}{2}w + 1, 0, -\frac{1}{2}w, w\right) \ w \text{ arbitrary.}$$

31.
$$-x + 2y - z = 0$$
$$-\frac{1}{2}x - \frac{1}{2}y + z = 0$$
$$3y - 3z = 0$$

$$\begin{matrix} -R_1 \rightarrow \\ 2R_2 \rightarrow \\ \frac{1}{3}R_3 \rightarrow \end{matrix} \begin{bmatrix} 1 & -2 & 1 & 0 \\ -1 & -1 & 2 & 0 \\ 0 & 1 & -1 & 0 \end{bmatrix} \frac{1}{3}(R_2 + R_1) \rightarrow$$

$$\begin{bmatrix} 1 & -2 & 1 & 0 \\ 0 & -1 & 1 & 0 \\ 0 & 1 & -1 & 0 \end{bmatrix} \begin{matrix} R_1 - 2R_2 \rightarrow \\ -R_2 \rightarrow \\ R_2 + R_3 \rightarrow \end{matrix} \begin{bmatrix} 1 & 0 & -1 & 0 \\ 0 & 1 & -1 & 0 \\ 0 & 0 & 0 & 0 \end{bmatrix}$$

$x - z = 0$ or $x = z$
$y - z = 0$ or $y = z$
z is arbitrary.
$(x, y, z) = (z, z, z)$, z arbitrary.

33.
$$x + y - 2z = -1$$
$$-2x - 2y + 4z = 2$$
$$\frac{3}{4}x + \frac{3}{4}y - \frac{3}{2}z = -\frac{3}{4}$$

$$\begin{matrix} \frac{1}{2}R_2 \rightarrow \\ \frac{4}{3}R_3 \rightarrow \end{matrix} \begin{bmatrix} 1 & 1 & -2 & -1 \\ -1 & -1 & 2 & 1 \\ 1 & 1 & -2 & -1 \end{bmatrix} \begin{matrix} R_2 + R_1 \rightarrow \\ R_3 - R_1 \rightarrow \end{matrix}$$

$$\begin{bmatrix} 1 & 1 & -2 & -1 \\ 0 & 0 & 0 & 0 \\ 0 & 0 & 0 & 0 \end{bmatrix}$$

$$x + y - 2z = -1$$
$$x = -y + 2z - 1$$

y, z are arbitrary.
$(-y + 2z - 1, y, z)$ where y, z are arbitrary.

35.
$$x - y + z = 1$$
$$y - z + w = 1$$
$$x + z - w = 1$$
$$2x + z = 3$$

$$\begin{bmatrix} 1 & -1 & 1 & 0 & 1 \\ 0 & 1 & -1 & 1 & 1 \\ 1 & 0 & 1 & -1 & 1 \\ 2 & 0 & 1 & 0 & 3 \end{bmatrix} \begin{matrix} \\ \\ R_3 - R_1 \rightarrow \\ R_4 - 2R_1 \rightarrow \end{matrix}$$

$$\begin{bmatrix} 1 & -1 & 1 & 0 & 1 \\ 0 & 1 & -1 & 1 & 1 \\ 0 & 1 & 0 & -1 & 0 \\ 0 & 2 & -1 & 0 & 1 \end{bmatrix} \begin{matrix} R_2 + R_1 \rightarrow \\ \\ R_3 - R_2 \rightarrow \\ R_4 - 2R_2 \rightarrow \end{matrix}$$

$$\begin{bmatrix} 1 & 0 & 0 & 1 & 2 \\ 0 & 1 & -1 & 1 & 1 \\ 0 & 0 & 1 & -2 & -1 \\ 0 & 0 & 1 & -2 & -1 \end{bmatrix} \begin{matrix} \\ R_2 + R_3 \rightarrow \\ \\ R_4 - R_3 \rightarrow \end{matrix}$$

$$\begin{bmatrix} 1 & 0 & 0 & 1 & 2 \\ 0 & 1 & 0 & -1 & 0 \\ 0 & 0 & 1 & -2 & -1 \\ 0 & 0 & 0 & 0 & 0 \end{bmatrix}$$

$x + w = 2$ or $x = -w + 2$
$y - w = 0$ $y = w$
$z - 2w = -1$ $z = 2w - 1$
w is arbitrary.
$(x, y, z, w) = (-w + 2, w, 2w - 1, w)$ where w is arbitrary.

37. $x + y + w = 100$
$x + z + w = 100$
$y + z + w = 100$

$$\begin{bmatrix} 1 & 1 & 0 & 1 & 100 \\ 1 & 0 & 1 & 1 & 100 \\ 0 & 1 & 1 & 1 & 100 \end{bmatrix} \rightarrow \begin{bmatrix} 1 & 1 & 0 & 1 & 100 \\ 0 & 1 & -1 & 0 & 0 \\ 0 & 1 & 1 & 1 & 100 \end{bmatrix} \rightarrow$$

$$\begin{bmatrix} 1 & 0 & 1 & 1 & 100 \\ 0 & 1 & -1 & 0 & 0 \\ 0 & 0 & 2 & 1 & 100 \end{bmatrix} \rightarrow \begin{bmatrix} 2 & 0 & 0 & 1 & 100 \\ 0 & 2 & 0 & 1 & 100 \\ 0 & 0 & 2 & 1 & 100 \end{bmatrix} \rightarrow$$

$$\begin{bmatrix} 1 & 0 & 0 & \frac{1}{2} & 50 \\ 0 & 1 & 0 & \frac{1}{2} & 50 \\ 0 & 0 & 1 & \frac{1}{2} & 50 \end{bmatrix}$$

$x + \dfrac{1}{2}w = 50$ or $x = -\dfrac{1}{2}w + 50$

$y + \dfrac{1}{2}w = 50$ or $y = -\dfrac{1}{2}w + 50$

$z + \dfrac{1}{2}w = 50$ or $z = -\dfrac{1}{2}w + 50$

w is arbitrary.

$(x, y, z, w) = \left(50 - \dfrac{1}{2}w, \ 50 - \dfrac{1}{2}w, \ 50 - \dfrac{1}{2}w, \ w\right)$

where w is arbitrary.

39.
$$\begin{bmatrix} \frac{2}{3} & \frac{2}{3} & -3 \\ \frac{1}{2} & -\frac{2}{3} & \frac{11}{3} \end{bmatrix} \begin{matrix} 3R_1 \rightarrow \\ 6R_2 \rightarrow \end{matrix} \begin{bmatrix} 2 & 2 & -9 \\ 3 & -4 & 22 \end{bmatrix} 2R_2 - 3R_1 \rightarrow$$

$$\begin{bmatrix} 2 & 2 & -9 \\ 0 & -14 & 71 \end{bmatrix} \begin{matrix} 7R_1 + R_2 \rightarrow \\ \end{matrix} \begin{bmatrix} 14 & 0 & 8 \\ 0 & -14 & 71 \end{bmatrix} \begin{matrix} \frac{1}{14}R_1 \rightarrow \\ -\frac{1}{14}R_2 \rightarrow \end{matrix}$$

$$\begin{bmatrix} 1 & 0 & \frac{4}{7} \\ 0 & 1 & -\frac{71}{14} \end{bmatrix}$$

41.
$$\begin{bmatrix} -2 & 1 & -3 \\ 1 & 3 & 1 \\ -3 & -2 & -5 \end{bmatrix} \begin{matrix} \\ 2R_2 + R_1 \rightarrow \\ 2R_3 - 3R_1 \rightarrow \end{matrix}$$

$$\begin{bmatrix} -2 & 1 & -3 \\ 0 & 7 & -1 \\ 0 & -7 & -1 \end{bmatrix} \begin{matrix} \frac{1}{2}(7R_1 - R_2) \rightarrow \\ \\ R_2 + R_3 \rightarrow \end{matrix}$$

$$\begin{bmatrix} -7 & 0 & -10 \\ 0 & 7 & -1 \\ 0 & 0 & -2 \end{bmatrix} \begin{matrix} R_1 - 5R_3 \rightarrow \\ 2R_2 - R_3 \rightarrow \end{matrix}$$

$$\begin{bmatrix} -7 & 0 & 0 \\ 0 & 14 & 0 \\ 0 & 0 & -2 \end{bmatrix} \begin{matrix} -\frac{1}{7}R_1 \rightarrow \\ \frac{1}{14}R_2 \rightarrow \\ -\frac{1}{2}R_3 \rightarrow \end{matrix} \begin{bmatrix} 1 & 0 & 0 \\ 0 & 1 & 0 \\ 0 & 0 & 1 \end{bmatrix}$$

43.
$$\begin{bmatrix} 3 & 0 & 0 & 4 & -1 \\ 1 & 2 & 0 & 0 & 0 \\ 0 & 2 & 0 & 0 & 0 \\ 1 & 1 & 1 & 1 & 1 \end{bmatrix} \begin{matrix} \\ 3R_2 - R_1 \rightarrow \\ \\ 3R_3 - R_1 \rightarrow \end{matrix}$$

$$\begin{bmatrix} 3 & 0 & 0 & 4 & -1 \\ 0 & 6 & 0 & -4 & 1 \\ 0 & 2 & 0 & 0 & 0 \\ 0 & 3 & 3 & -1 & 4 \end{bmatrix} \begin{matrix} \\ \\ 3R_3 - R_2 \rightarrow \\ 2R_4 - R_2 \rightarrow \end{matrix}$$

$$\begin{bmatrix} 3 & 0 & 0 & 4 & -1 \\ 0 & 6 & 0 & -4 & 1 \\ 0 & 0 & 0 & 4 & -1 \\ 0 & 0 & 6 & 2 & 7 \end{bmatrix} \begin{matrix} \\ \\ R_3 \leftrightarrow R_4 \end{matrix}$$

$$\begin{bmatrix} 3 & 0 & 0 & 4 & -1 \\ 0 & 6 & 0 & -4 & 1 \\ 0 & 0 & 6 & 2 & 7 \\ 0 & 0 & 0 & 4 & -1 \end{bmatrix} \begin{matrix} R_1 - R_4 \rightarrow \\ R_2 + R_4 \rightarrow \\ 2R_3 - R_4 \rightarrow \end{matrix}$$

$$\begin{bmatrix} 3 & 0 & 0 & 0 & 0 \\ 0 & 6 & 0 & 0 & 0 \\ 0 & 0 & 12 & 0 & 15 \\ 0 & 0 & 0 & 4 & -1 \end{bmatrix} \begin{matrix} \frac{1}{3}R_1 \rightarrow \\ \frac{1}{6}R_2 \rightarrow \\ \frac{1}{12}R_3 \rightarrow \\ \frac{1}{4}R_4 \rightarrow \end{matrix} \begin{bmatrix} 1 & 0 & 0 & 0 & 0 \\ 0 & 1 & 0 & 0 & 0 \\ 0 & 0 & 1 & 0 & \frac{5}{4} \\ 0 & 0 & 0 & 1 & -\frac{1}{4} \end{bmatrix}$$

45.
$$\begin{bmatrix} 0 & 1 & 0 & -4 \\ 1 & 2 & 0 & 0 \\ 0 & 2 & 0 & 0 \\ 1 & 1 & 1 & 0 \\ 1 & -1 & 2 & 2 \end{bmatrix}$$

Rearrange.

$$\begin{bmatrix} 1 & 2 & 0 & 0 \\ 0 & 1 & 0 & -4 \\ 1 & 1 & 1 & 0 \\ 1 & -1 & 2 & 2 \\ 0 & 2 & 0 & 0 \end{bmatrix} \begin{matrix} \\ \\ R_1 - R_3 \to \\ R_1 - R_4 \to \\ \end{matrix}$$

$$\begin{bmatrix} 1 & 2 & 0 & 0 \\ 0 & 1 & 0 & -4 \\ 0 & 1 & -1 & 0 \\ 0 & 3 & -2 & -2 \\ 0 & 2 & 0 & 0 \end{bmatrix} \begin{matrix} R_1 - 2R_2 \to \\ \\ R_3 - R_2 \to \\ R_4 - 3R_2 \to \\ R_5 - 2R_2 \to \end{matrix}$$

$$\begin{bmatrix} 1 & 0 & 0 & 8 \\ 0 & 1 & 0 & -4 \\ 0 & 0 & -1 & 4 \\ 0 & 0 & -2 & 10 \\ 0 & 0 & 0 & 8 \end{bmatrix} \begin{matrix} \\ \\ \\ -\frac{1}{2}R_4 \to \\ \frac{1}{8}R_5 \to \end{matrix}$$

$$\begin{bmatrix} 1 & 0 & 0 & 8 \\ 0 & 1 & 0 & -4 \\ 0 & 0 & -1 & 4 \\ 0 & 0 & 1 & -5 \\ 0 & 0 & 0 & 1 \end{bmatrix} \begin{matrix} \\ \\ -R_3 \to \\ R_4 + R_3 \to \\ \end{matrix}$$

$$\begin{bmatrix} 1 & 0 & 0 & 8 \\ 0 & 1 & 0 & -4 \\ 0 & 0 & 1 & -4 \\ 0 & 0 & 0 & -1 \\ 0 & 0 & 0 & 1 \end{bmatrix} \begin{matrix} R_1 + 8R_4 \to \\ R_2 - 4R_4 \to \\ R_3 - 4R_4 \to \\ -R_4 \to \\ R_5 + R_4 \to \end{matrix} \begin{bmatrix} 1 & 0 & 0 & 0 \\ 0 & 1 & 0 & 0 \\ 0 & 0 & 1 & 0 \\ 0 & 0 & 0 & 1 \\ 0 & 0 & 0 & 0 \end{bmatrix}$$

47.
$$\begin{bmatrix} \frac{1}{2} & 0 & 1 & 1 & 0 & 0 \\ 2 & -2 & 1 & 0 & 1 & 0 \\ 3 & 0 & 0 & 0 & 0 & 1 \end{bmatrix} \begin{matrix} 2R_1 \to \\ R_2 - 4R_1 \to \\ R_3 - 6R_1 \to \end{matrix}$$

$$\begin{bmatrix} 1 & 0 & 2 & 2 & 0 & 0 \\ 0 & -2 & -3 & -4 & 1 & 0 \\ 0 & 0 & -6 & -6 & 0 & 1 \end{bmatrix} \begin{matrix} 3R_1 + R_3 \to \\ 2R_2 - R_3 \to \\ \end{matrix}$$

$$\begin{bmatrix} 3 & 0 & 0 & 0 & 0 & 1 \\ 0 & -4 & 0 & -2 & 2 & -1 \\ 0 & 0 & -6 & -6 & 0 & 1 \end{bmatrix} \begin{matrix} \frac{1}{3}R_1 \to \\ -\frac{1}{4}R_2 \to \\ -\frac{1}{6}R_3 \to \end{matrix}$$

$$\begin{bmatrix} 1 & 0 & 0 & 0 & 0 & \frac{1}{3} \\ 0 & 1 & 0 & \frac{1}{2} & -\frac{1}{2} & \frac{1}{4} \\ 0 & 0 & 1 & 1 & 0 & -\frac{1}{6} \end{bmatrix}$$

49.
$$\begin{bmatrix} 0 & 1 & 0 & 0 \\ 1 & 0 & 0 & 0 \\ 0 & 0 & 1 & 0 \\ 0 & 0 & 0 & 1 \end{bmatrix} R_1 \leftrightarrow R_2 \begin{bmatrix} 1 & 0 & 0 & 0 \\ 0 & 1 & 0 & 0 \\ 0 & 0 & 1 & 0 \\ 0 & 0 & 0 & 1 \end{bmatrix}$$

51. Let x = number of Gauss Jordans,
y = number of Roeboks, and
y = number of K Scottish.

	x	y	z	Total
pair of shoes	1	1	1	120
cost($)/pair	40	50	50	5700

Also, $x = y$.
We have the following system of equations.
$$\begin{aligned} x + y + z &= 120 \\ 40x + 50y + 50z &= 5700 \\ x - y &= 0 \end{aligned}$$
Solving the system we get $(x, y, z) = (30, 30, 60)$.
Thus, the number of pairs of each jogging shoe is 30 Gauss Jordans, 30 Roebecks, and 60 K Scottish.

53. Let x = cost of hamburgers, y = cost of coleslaw, and z = cost of french fries.

x	y	z	Total
1	1	0	3.95
1	0	1	4.40

Also, $z = 2y$. We have the following system of equations.
$$\begin{aligned} x + y &= 3.95 \\ x + z &= 4.40 \\ -2y + z &= 0 \end{aligned}$$
Solving the system, we get
$(x, y, z) = (3.50, 0.45, 0.90)$.
French fries cost $0.90.

55. Let x = amount youngest child received and T = total amount.
Total amount = estate amount + previous amount
$$\begin{aligned} T &= 111,000 + (15,000 + 10,000 + 2000) \\ &= \$138,000 \end{aligned}$$

Total amount each child receives = $\frac{1}{3}$ of total
= \$46,000
Youngest child receives x(from estate) + 2000.
But,
$$\begin{aligned} x + 2000 &= 46,000 \\ x &= 44,000 \end{aligned}$$
The youngest child receives $44,000 from the estate.

57. Let x = Nielsen rating of ABC,
y = Nielsen rating of TBS and
z = Nielsen rating of ESPN.

	x	y	z	Total
Viewers/rating	607,000	600,000	611,000	6,372,000

So $607,000x + 600,000y + 611,000z = 6,372,000$ and $x = z$.
$1,218,000x + 600,000y = 6,372,000$
Also, $600,000y = 1.5 \times 10^6 = 1,500,000$ so $y = 2.5$. Then $x = z = 4$. Therefore, $(x, y, z) = (4.0, 2.5, 4.0)$.
The Nielsen ratings are ABC: 4.0, TBS: 2.5, and ESPN: 4.0.

59. Let x = number of 12-ounce servings of beer, y = number of 4-ounce servings of wine and
z = number of 3-ounce servings of sherry.

	x	y	z	Total
amount of alcohol (ounces)	at 4%, 0.48	at 10%, 0.4	at 12%, 0.36	≤1 for each person and 4 for all 4 persons
total number of drinks for 4 people	1	1	1	10

$$x + y + z = 10$$
$$0.48x + 0.40y + 0.36z = 4$$

Solving these two equations, we get $(x, y, z) = \left(\dfrac{1}{2}z, \ -\dfrac{3}{2}z + 10, \ z \right)$.

Possible combinations are $(1, 7, 2)$, $(2, 4, 4)$, and $(3, 1, 6)$.

61. Let x = number of Science credits, y = number of Fine Arts credits, z = number of Liberal Arts credits,
and w = number of Mathematics credits.
We have the following system of equations.
$$x + y + z + w = 124$$
$$x = y$$
$$w = 4x$$
$$z = w + \dfrac{1}{3}y$$
Solving the system, we get $(x, y, z, w) = (12, 12, 52, 48)$.
A student is forced to take exactly the following combination: 12 credits of Sciences, 12 credits of Fine Arts, 52 credits of Liberal Arts and 48 credits of Mathematics.

63.
$$C = 10 + 0.8Y_D$$
$$Y_D = Y - T$$
$$T = 0.25Y$$
$$Y = C + \overline{G} + \overline{I}$$
Solving the system, we get
$(C, Y_D, Y, T) = (25 + 1.5\overline{G} + 1.5\overline{I}, \ 18.75 + 1.875\overline{G} + 1.875\overline{I}, \ 25 + 2.5\overline{G} + 2.5\overline{I}, \ 6.25 + 0.625\overline{G} + 0.625\overline{I})$.
If $\overline{I} \rightarrow \overline{I} + 10$, then
$(C, Y_D, Y, T) = (40 + 1.5\overline{G} + 1.5\overline{I}, \ 37.5 + 1.875\overline{G} + 1.875\overline{I}, \ 50 + 2.5\overline{G} + 2.5\overline{I}, \ 12.5 + 1.875\overline{G} + 1.875\overline{I})$.
The equilibrium total income (Y) will increase by $50 - 25$, or 25 units.

Chapter 3

1. $A = \begin{bmatrix} 1 & -1 & 0 & 2 & \frac{1}{4} \end{bmatrix}$

   rows: 1
   columns: 5
   1×5 matrix

3. $C = \begin{bmatrix} \frac{5}{2} \\ 1 \\ -2 \\ 8 \end{bmatrix}$

   rows: 4
   columns: 1
   4×1 matrix

5. $E = \begin{bmatrix} e_{11} & e_{12} & e_{13} & \cdots & e_{1q} \\ e_{21} & e_{22} & e_{23} & \cdots & e_{2q} \\ \cdots & \cdots & \cdots & \cdots & \cdots \\ e_{p1} & e_{p2} & e_{p3} & \cdots & e_{pq} \end{bmatrix}$

   rows: p
   columns: q
   $p \times q$ matrix

7. $B = \begin{bmatrix} 1 & 3 \\ 5 & -6 \end{bmatrix}$

   rows: 2
   columns: 2
   2×2 matrix

9. $D = \begin{bmatrix} d_1 & d_2 & \cdots & d_n \end{bmatrix}$

   rows: 1
   columns: n
   $1 \times n$ matrix

11. a_{11} is the entry in the first row and first column of A, so $a_{11} = 1$

13. B_{11} is the entry in the first row and first column and the only entry in matrix B, so $B_{11} = 44$

15. C_{31} is the entry in the third row and first column of C, so $C_{31} = -2$

17. e_{13} is the entry in the first row and third column of E, so $e_{13} = e_{13}$

19. a_{11} is the entry in the first row and first column of A, so $a_{11} = 2$

21. B_{12} is the entry in the first row and second column of B, so $B_{12} = 3$

23. c_{14} is the entry in the first row and fourth column of C, so $c_{14} = e$

25. d_{1n} is the entry in the first row and n^{th} column, so $d_{1n} = d_n$

27. $\begin{bmatrix} x+y & x+z \\ y+z & w \end{bmatrix} = \begin{bmatrix} 3 & 4 \\ 5 & 4 \end{bmatrix}$

    $\begin{cases} x+y = 3 \\ y+z = 5 \\ x+z = 4 \\ \quad w = 4 \end{cases}$

    Solving the equations, we get $x = 1, y = 2, z = 3$, $w = 4$.

29. $A + B = \begin{bmatrix} 0 & -1 \\ 1 & 0 \\ -1 & 2 \\ 5 & 0 \end{bmatrix} + \begin{bmatrix} \frac{1}{4} & -1 \\ 0 & \frac{1}{3} \\ -1 & 3 \\ 5 & 0 \end{bmatrix} = \begin{bmatrix} \frac{1}{4} & -2 \\ 1 & \frac{1}{3} \\ -2 & 5 \\ 10 & 0 \end{bmatrix}$

31. $A + B - C = \begin{bmatrix} 0 & -1 \\ 1 & 0 \\ -1 & 2 \\ 5 & 0 \end{bmatrix} + \begin{bmatrix} \frac{1}{4} & -1 \\ 0 & \frac{1}{3} \\ -1 & 3 \\ 5 & 0 \end{bmatrix} - \begin{bmatrix} 1 & -1 \\ 1 & 1 \\ -1 & -1 \\ 1 & 1 \end{bmatrix}$

    $= \begin{bmatrix} -\frac{3}{4} & -1 \\ 0 & -\frac{2}{3} \\ -1 & 6 \\ 9 & -1 \end{bmatrix}$

33. $2A - C = 2\begin{bmatrix} 0 & -1 \\ 1 & 0 \\ -1 & 2 \\ 5 & 0 \end{bmatrix} - \begin{bmatrix} 1 & -1 \\ 1 & 1 \\ -1 & -1 \\ 1 & 1 \end{bmatrix}$

    $= \begin{bmatrix} 0 & -2 \\ 2 & 0 \\ -2 & 4 \\ 10 & 0 \end{bmatrix} - \begin{bmatrix} 1 & -1 \\ 1 & 1 \\ -1 & -1 \\ 1 & 1 \end{bmatrix} = \begin{bmatrix} -1 & -1 \\ 1 & -1 \\ -1 & 5 \\ 9 & -1 \end{bmatrix}$

35. $2A^T = 2\begin{bmatrix} 0 & -1 \\ 1 & 0 \\ -1 & 2 \\ 5 & 0 \end{bmatrix}^T = 2\begin{bmatrix} 0 & 1 & -1 & 5 \\ -1 & 0 & 2 & 0 \end{bmatrix} = \begin{bmatrix} 0 & 2 & -2 & 10 \\ -2 & 0 & 4 & 0 \end{bmatrix}$

37. $A + B = \begin{bmatrix} 1 & -1 & 0 \\ 0 & 2 & -1 \end{bmatrix} + \begin{bmatrix} 3 & 0 & -1 \\ 5 & -1 & 1 \end{bmatrix} = \begin{bmatrix} 4 & -1 & -1 \\ 5 & 1 & 0 \end{bmatrix}$

39. $A - B + C = \begin{bmatrix} 1 & -1 & 0 \\ 0 & 2 & -1 \end{bmatrix} - \begin{bmatrix} 3 & 0 & -1 \\ 5 & -1 & 1 \end{bmatrix} + \begin{bmatrix} x & 1 & w \\ z & r & 4 \end{bmatrix} = \begin{bmatrix} -2+x & 0 & 1+w \\ -5+z & 3+r & 2 \end{bmatrix}$

41. $2A - B = 2\begin{bmatrix} 1 & -1 & 0 \\ 0 & 2 & -1 \end{bmatrix} - \begin{bmatrix} 3 & 0 & -1 \\ 5 & -1 & 1 \end{bmatrix} = \begin{bmatrix} 2 & -2 & 0 \\ 0 & 4 & -2 \end{bmatrix} - \begin{bmatrix} 3 & 0 & -1 \\ 5 & -1 & 1 \end{bmatrix} = \begin{bmatrix} -1 & -2 & 1 \\ -5 & 5 & -3 \end{bmatrix}$

43. $3B^T = 3\begin{bmatrix} 3 & 5 \\ 0 & -1 \\ -1 & 1 \end{bmatrix} = \begin{bmatrix} 9 & 15 \\ 0 & -3 \\ -3 & 3 \end{bmatrix}$

45. $A - C = \begin{bmatrix} -8.5 & -22.35 & -24.4 \\ 54.2 & 20 & 42.2 \end{bmatrix}$

47. $1.1B = \begin{bmatrix} 1.54 & 8.58 \\ 5.94 & 0 \\ 6.16 & 7.26 \end{bmatrix}$

49. $A^T + 4.2B = \begin{bmatrix} 7.38 & 76.96 \\ 20.33 & 0 \\ 29.12 & 39.92 \end{bmatrix}$

51. $(2.1A - 2.3C)^T = \begin{bmatrix} -19.85 & 115.82 \\ -50.935 & 46 \\ -57.24 & 94.62 \end{bmatrix}$

53. a. Annual salaries in 1992 ($) = Annual salaries in 1991 ($) + Increase in 1992 ($)

$= \begin{bmatrix} 890,000 \\ 675,000 \\ 275,000 \\ 275,000 \end{bmatrix} + \begin{bmatrix} 576 \\ 411 \\ 20,822 \\ 411 \end{bmatrix} = \begin{bmatrix} 890,576 \\ 675,411 \\ 295,822 \\ 275,411 \end{bmatrix}$

b. Annual salaries in 1993 ($) = Annual salaries in 1992 ($) + Increase in 1993 ($)

$= \begin{bmatrix} 890,576 \\ 675,411 \\ 295,822 \\ 275,411 \end{bmatrix} + \begin{bmatrix} 34,424 \\ 24,589 \\ 54,370 \\ 24,781 \end{bmatrix} = \begin{bmatrix} 925,000 \\ 700,000 \\ 350,192 \\ 300,192 \end{bmatrix}$

55. Total revenue $=\begin{bmatrix}5 & 25 & 40 & 70 & 110 & 150\end{bmatrix}+\begin{bmatrix}10 & 20 & 20 & 15 & 10 & 5\end{bmatrix}=\begin{bmatrix}15 & 45 & 60 & 85 & 120 & 155\end{bmatrix}$

Expenditures = Revenue – Profit

$\begin{bmatrix}15 & 45 & 60 & 85 & 120 & 155\end{bmatrix}-\begin{bmatrix}0 & 0.5 & -5 & 2 & 0 & 4\end{bmatrix}=\begin{bmatrix}15 & 44.5 & 65 & 83 & 120 & 151\end{bmatrix}$

57.

U.S. population
(millions)

	Northeast	Midwest	South	West
1970	49.1	56.6	62.8	34.8
1980	49.1	58.9	75.4	43.2

1970 distribution $= A = \begin{bmatrix}49.1 & 56.6 & 62.8 & 34.8\end{bmatrix}$

1980 distribution $= B = \begin{bmatrix}49.1 & 58.9 & 75.4 & 43.2\end{bmatrix}$

Net change from 1970 to 1980 $= B - A = \begin{bmatrix}0 & 2.3 & 12.6 & 8.4\end{bmatrix}$ all net increases.

59. Sales $=\begin{bmatrix}\text{SF} \\ \text{LA}\end{bmatrix}=\begin{bmatrix}700 & 1300 & 2000 \\ 400 & 300 & 500\end{bmatrix}$

Inventory remaining in each store at end of January = Inventory at beginning of January – Sales

$=\begin{bmatrix}1000 & 2000 & 5000 \\ 1000 & 5000 & 2000\end{bmatrix}-\begin{bmatrix}700 & 1300 & 2000 \\ 400 & 300 & 500\end{bmatrix}=\begin{bmatrix}300 & 700 & 3000 \\ 600 & 4700 & 1500\end{bmatrix}$

61. a.

$\begin{matrix} & \text{Proc.} & \text{Mem.} & \text{Tubes} \\ \text{Pom II} \\ \text{Pom Classic}\end{matrix}\begin{bmatrix}2 & 16 & 20 \\ 1 & 4 & 40\end{bmatrix} = \text{Use}$

$\text{Inventory} =\begin{bmatrix}500 & 5000 & 10,000 \\ 200 & 2000 & 20,000\end{bmatrix}$

Inventory after two months = Inventory – 2 month $\times$ 50/month $\times$ Use = Inventory – 100 Use

$=\begin{bmatrix}500 & 5000 & 10,000 \\ 200 & 2000 & 20,000\end{bmatrix}-\begin{bmatrix}200 & 1600 & 2000 \\ 100 & 400 & 4000\end{bmatrix}=\begin{bmatrix}300 & 3400 & 8000 \\ 100 & 1600 & 16,000\end{bmatrix}$

b. Let x = number of months.

Microbucks will run out of one of the parts when some entry of Inventory – $50\times$ Use is 0.

$\begin{bmatrix}500-100x & 5000-800x & 10,000-1000x \\ 200-50x & 2000-200x & 20,000-2000x\end{bmatrix}$

Solving for x to see when the inventory of some part is brought down to zero, this will happen first to Pom Classic processor chips, after 4 months.

63. $(A+B)_{ij} = A_{ij} + B_{ij}$

The ij^{th} **entry** of the matrix $A + B$ is **equal** to the sum of the individual ij^{th} entries of the matrices A and B.

65. $A = \begin{bmatrix} a_{11} & 0 & 0 & \cdots & 0 \\ 0 & a_{22} & 0 & \cdots & 0 \\ \cdots & \cdots & \cdots & \cdots & \cdots \\ 0 & 0 & 0 & \cdots & a_{ii} \end{bmatrix}$

67. Answers will vary.

Section 3.2 Matrix Multiplication
Pages 199–203

1. $\begin{bmatrix} 1 & 3 & -1 \end{bmatrix} \begin{bmatrix} 9 \\ 1 \\ -1 \end{bmatrix} = \begin{bmatrix} 1 \cdot 9 + 3 \cdot 1 + (-1) \cdot (-1) \end{bmatrix} = \begin{bmatrix} 9 + 3 + 1 \end{bmatrix} = \begin{bmatrix} 13 \end{bmatrix}$

3. $\begin{bmatrix} -1 & \frac{1}{2} \end{bmatrix} \begin{bmatrix} -\frac{1}{3} \\ 1 \end{bmatrix} = \begin{bmatrix} -1 \cdot -\frac{1}{3} + \frac{1}{2} \cdot 1 \end{bmatrix} = \begin{bmatrix} \frac{1}{3} + \frac{1}{2} \end{bmatrix} = \begin{bmatrix} \frac{5}{6} \end{bmatrix}$

5. $\begin{bmatrix} 0 & -2 & 1 \end{bmatrix} \begin{bmatrix} x \\ y \\ z \end{bmatrix} = \begin{bmatrix} 0 \cdot x + (-2) \cdot y + 1 \cdot z \end{bmatrix} = \begin{bmatrix} -2y + z \end{bmatrix}$

7. $\begin{bmatrix} -1 & 1 \end{bmatrix} \begin{bmatrix} -3 & 1 & 4 & 3 \\ 0 & 1 & -2 & 1 \end{bmatrix} = \begin{bmatrix} -1 \cdot -3 + 1 \cdot 0 & -1 \cdot 1 + 1 \cdot 1 & -1 \cdot 4 + 1 \cdot -2 & -1 \cdot 3 + 1 \cdot 1 \end{bmatrix} = \begin{bmatrix} 3 + 0 & -1 + 1 & -4 - 2 & -3 + 1 \end{bmatrix}$

$= \begin{bmatrix} 3 & 0 & -6 & -2 \end{bmatrix}$

9. $\begin{bmatrix} 1 & -1 & 2 & 3 \end{bmatrix} \begin{bmatrix} -1 & 2 & 0 \\ 2 & -1 & 0 \\ 0 & 5 & 2 \\ -1 & 8 & 1 \end{bmatrix} = \begin{bmatrix} -1 - 2 + 0 - 3 & 2 + 1 + 10 + 24 & 0 - 0 + 4 + 3 \end{bmatrix} = \begin{bmatrix} -6 & 37 & 7 \end{bmatrix}$

11.

$\begin{bmatrix} 1 & 0 & -\frac{1}{2} & 1 \\ -1 & 1 & \frac{1}{4} & -2 \end{bmatrix} \begin{bmatrix} 0 & 1 & -1 \\ 1 & 0 & 1 \\ 4 & 8 & 0 \\ -2 & 8 & -1 \end{bmatrix}$

$= \begin{bmatrix} 1 \cdot 0 + 0 \cdot 1 + \left(-\frac{1}{2}\right) \cdot 4 + 1 \cdot -2 & 1 \cdot 1 + 0 \cdot 0 + \left(-\frac{1}{2}\right) \cdot 8 + 1 \cdot 8 & 1 \cdot -1 + 0 \cdot 1 + \left(-\frac{1}{2}\right) \cdot 0 + 1 \cdot -1 \\ -1 \cdot 0 + 1 \cdot 1 + \frac{1}{4} \cdot 4 + (-2) \cdot -2 & -1 \cdot 1 + 1 \cdot 0 + \frac{1}{4} \cdot 8 - 2 \cdot 8 & -1 \cdot -1 + 1 \cdot 1 + \frac{1}{4} \cdot 0 + (-2) \cdot -1 \end{bmatrix}$

$= \begin{bmatrix} 0 + 0 - 2 - 2 & 1 + 0 - 4 + 8 & -1 + 0 + 0 - 1 \\ 0 + 1 + 1 + 4 & -1 + 0 + 2 - 16 & 1 + 1 + 0 + 2 \end{bmatrix} = \begin{bmatrix} -4 & 5 & -2 \\ 6 & -15 & 4 \end{bmatrix}$

13. $\begin{bmatrix} 1 & 0 \\ 1 & -1 \end{bmatrix} \begin{bmatrix} 0 & 1 \\ 0 & 1 \end{bmatrix} = \begin{bmatrix} 0 + 0 & 1 + 0 \\ 0 + 0 & 1 - 1 \end{bmatrix} = \begin{bmatrix} 0 & 1 \\ 0 & 0 \end{bmatrix}$

15. $\begin{bmatrix} 1 & -1 \\ 1 & -1 \end{bmatrix} \begin{bmatrix} 2 & 3 \\ 2 & 3 \end{bmatrix} = \begin{bmatrix} 2 - 2 & 3 - 3 \\ 2 - 2 & 3 - 3 \end{bmatrix} = \begin{bmatrix} 0 & 0 \\ 0 & 0 \end{bmatrix}$

17. $\begin{bmatrix} 1 & 0 & -1 \\ 2 & -2 & 1 \\ 0 & 0 & 1 \end{bmatrix} \begin{bmatrix} 1 & -1 & 4 \\ 1 & 1 & 0 \\ 0 & 4 & 1 \end{bmatrix} = \begin{bmatrix} 1 + 0 + 0 & -1 + 0 - 4 & 4 + 0 - 1 \\ 2 - 2 + 0 & -2 - 2 + 4 & 8 - 0 + 1 \\ 0 + 0 + 0 & 0 + 0 + 4 & 0 + 0 + 1 \end{bmatrix} = \begin{bmatrix} 1 & -5 & 3 \\ 0 & 0 & 9 \\ 0 & 4 & 1 \end{bmatrix}$

19. $\begin{bmatrix} 1 & 0 & 1 & 0 \\ -1 & 1 & 0 & 1 \\ -2 & 0 & 1 & 4 \\ 0 & -1 & 0 & 1 \end{bmatrix} \begin{bmatrix} 1 \\ -3 \\ 2 \\ 0 \end{bmatrix} = \begin{bmatrix} 1 + 0 + 2 + 0 \\ -1 - 3 + 0 + 0 \\ -2 + 0 + 2 + 0 \\ 0 + 3 + 0 + 0 \end{bmatrix} = \begin{bmatrix} 3 \\ -4 \\ 0 \\ 3 \end{bmatrix}$

21. $A^2 = A \cdot A = \begin{bmatrix} 0 & 1 & 1 & 1 \\ 0 & 0 & 1 & 1 \\ 0 & 0 & 0 & 1 \\ 0 & 0 & 0 & 0 \end{bmatrix}\begin{bmatrix} 0 & 1 & 1 & 1 \\ 0 & 0 & 1 & 1 \\ 0 & 0 & 0 & 1 \\ 0 & 0 & 0 & 0 \end{bmatrix}$

$= \begin{bmatrix} 0 & 0 & 1 & 2 \\ 0 & 0 & 0 & 1 \\ 0 & 0 & 0 & 0 \\ 0 & 0 & 0 & 0 \end{bmatrix}$

$A^3 = A^2 \cdot A = \begin{bmatrix} 0 & 0 & 1 & 2 \\ 0 & 0 & 0 & 1 \\ 0 & 0 & 0 & 0 \\ 0 & 0 & 0 & 0 \end{bmatrix}\begin{bmatrix} 0 & 1 & 1 & 1 \\ 0 & 0 & 1 & 1 \\ 0 & 0 & 0 & 1 \\ 0 & 0 & 0 & 0 \end{bmatrix}$

$= \begin{bmatrix} 0 & 0 & 0 & 1 \\ 0 & 0 & 0 & 0 \\ 0 & 0 & 0 & 0 \\ 0 & 0 & 0 & 0 \end{bmatrix}$

$A^4 = A^3 \cdot A = \begin{bmatrix} 0 & 0 & 0 & 1 \\ 0 & 0 & 0 & 0 \\ 0 & 0 & 0 & 0 \\ 0 & 0 & 0 & 0 \end{bmatrix}\begin{bmatrix} 0 & 1 & 1 & 1 \\ 0 & 0 & 1 & 1 \\ 0 & 0 & 0 & 1 \\ 0 & 0 & 0 & 0 \end{bmatrix}$

$= \begin{bmatrix} 0 & 0 & 0 & 0 \\ 0 & 0 & 0 & 0 \\ 0 & 0 & 0 & 0 \\ 0 & 0 & 0 & 0 \end{bmatrix} = 0$

$A^n = 0$ for $n \geq 4$.

Thus, $A^{100} = 0$.

23. $AB = \begin{bmatrix} 0 & -1 & 0 & 1 \\ 10 & 0 & 1 & 0 \end{bmatrix}\begin{bmatrix} 0 & -1 \\ 1 & 1 \\ -1 & 3 \\ 5 & 0 \end{bmatrix}$

$= \begin{bmatrix} 0-1+0+5 & 0-1+0+0 \\ 0+0-1+0 & -10+0+3+0 \end{bmatrix} = \begin{bmatrix} 4 & -1 \\ -1 & -7 \end{bmatrix}$

25. $A(B-C) = \begin{bmatrix} 0 & -1 & 0 & 1 \\ 10 & 0 & 1 & 0 \end{bmatrix}\left(\begin{bmatrix} 0 & -1 \\ 1 & 1 \\ -1 & 3 \\ 5 & 0 \end{bmatrix} - \begin{bmatrix} 1 & -1 \\ 1 & 1 \\ 1 & 1 \\ 1 & 1 \end{bmatrix}\right)$

$= \begin{bmatrix} 0 & -1 & 0 & 1 \\ 10 & 0 & 1 & 0 \end{bmatrix}\begin{bmatrix} -1 & 0 \\ 0 & 0 \\ -2 & 2 \\ 4 & -1 \end{bmatrix}$

$= \begin{bmatrix} 0+0+0+4 & 0+0+0-1 \\ -10+0-2+0 & 0+0+2+0 \end{bmatrix} = \begin{bmatrix} 4 & -1 \\ -12 & 2 \end{bmatrix}$

27. $AB = \begin{bmatrix} 1 & -1 \\ 0 & 2 \\ 0 & -2 \end{bmatrix}\begin{bmatrix} 3 & 0 & -1 \\ 5 & -1 & 1 \end{bmatrix}$

$= \begin{bmatrix} 3-5 & 0+1 & -1-1 \\ 0+10 & 0-2 & 0+2 \\ 0-10 & 0+2 & 0-2 \end{bmatrix} = \begin{bmatrix} -2 & 1 & -2 \\ 10 & -2 & 2 \\ -10 & 2 & -2 \end{bmatrix}$

29. $A(B+C) = \begin{bmatrix} 1 & -1 \\ 0 & 2 \\ 0 & -2 \end{bmatrix}\left(\begin{bmatrix} 3 & 0 & -1 \\ 5 & -1 & 1 \end{bmatrix} + \begin{bmatrix} x & 1 & w \\ z & r & 4 \end{bmatrix}\right)$

$= \begin{bmatrix} 1 & -1 \\ 0 & 2 \\ 0 & -2 \end{bmatrix}\begin{bmatrix} 3+x & 1 & -1+w \\ 5+z & -1+r & 5 \end{bmatrix}$

$= \begin{bmatrix} -2+x-z & 2-r & -6+w \\ 10+2z & -2+2r & 10 \\ -10-2z & 2-2r & -10 \end{bmatrix}$

31. $BD = \begin{bmatrix} -1.2 & 0 & 0 \\ 0 & -1.2 & 0 \\ 0 & 0 & -1.2 \end{bmatrix}\begin{bmatrix} 350 \\ 591 \\ 911 \end{bmatrix} = \begin{bmatrix} -420 \\ -709.2 \\ -1093.2 \end{bmatrix}$

33. $AD = \begin{bmatrix} 1.1 & -2.1 & 4.5 \end{bmatrix}\begin{bmatrix} 350 \\ 591 \\ 911 \end{bmatrix} = \begin{bmatrix} 3243.4 \end{bmatrix}$

35. $BC = \begin{bmatrix} -1.2 & 0 & 0 \\ 0 & -1.2 & 0 \\ 0 & 0 & -1.2 \end{bmatrix}\begin{bmatrix} 0.01 & 1.1 \\ 1.2 & -1.1 \\ 0 & 0 \end{bmatrix}$

$= \begin{bmatrix} -0.012 & -1.32 \\ -1.44 & 1.32 \\ 0 & 0 \end{bmatrix}$

37. $A(BD) = \begin{bmatrix} 1.1 & -2.1 & 4.5 \end{bmatrix}\begin{bmatrix} -420 \\ -709.2 \\ -1093.2 \end{bmatrix} = \begin{bmatrix} -3892.08 \end{bmatrix}$

39. $B^4 = B \cdot B \cdot B \cdot B = B^2 \cdot B^2$

$= \begin{bmatrix} 1.44 & 0 & 0 \\ 0 & 1.44 & 0 \\ 0 & 0 & 1.44 \end{bmatrix}\begin{bmatrix} 1.44 & 0 & 0 \\ 0 & 1.44 & 0 \\ 0 & 0 & 1.44 \end{bmatrix}$

$= \begin{bmatrix} 2.0736 & 0 & 0 \\ 0 & 2.0736 & 0 \\ 0 & 0 & 2.0736 \end{bmatrix}$

41. $-\dfrac{1}{1.2}(BC) = -\dfrac{1}{1.2}\begin{bmatrix} -0.012 & -1.32 \\ -1.44 & 1.32 \\ 0 & 0 \end{bmatrix} = \begin{bmatrix} 0.01 & 1.1 \\ 1.2 & -1.1 \\ 0 & 0 \end{bmatrix}$

43. $P^2 = P \cdot P = \begin{bmatrix} \frac{1}{2} & \frac{1}{2} \\ \frac{1}{4} & \frac{3}{4} \end{bmatrix}\begin{bmatrix} \frac{1}{2} & \frac{1}{2} \\ \frac{1}{4} & \frac{3}{4} \end{bmatrix}$

$= \begin{bmatrix} \frac{3}{8} & \frac{5}{8} \\ \frac{5}{16} & \frac{11}{16} \end{bmatrix} = \begin{bmatrix} 0.375 & 0.625 \\ 0.3125 & 0.6875 \end{bmatrix}$

$P^4 = P^2 \cdot P^2 = \begin{bmatrix} 0.375 & 0.625 \\ 0.3125 & 0.6875 \end{bmatrix}\begin{bmatrix} 0.375 & 0.625 \\ 0.3125 & 0.6875 \end{bmatrix}$

$\approx \begin{bmatrix} 0.3359 & 0.6641 \\ 0.3320 & 0.6680 \end{bmatrix}$

$P^8 = P^4 \cdot P^4 \approx \begin{bmatrix} 0.3359 & 0.6641 \\ 0.3320 & 0.6680 \end{bmatrix}\begin{bmatrix} 0.3359 & 0.6641 \\ 0.3320 & 0.6680 \end{bmatrix}$

$\approx \begin{bmatrix} 0.3333 & 0.6667 \\ 0.3333 & 0.6667 \end{bmatrix}$

$P^{16} = P^8 \cdot P^8 \approx \begin{bmatrix} 0.3333 & 0.6667 \\ 0.3333 & 0.6667 \end{bmatrix}$

Thus, $P^{1000} \approx \begin{bmatrix} 0.3333 & 0.6667 \\ 0.3333 & 0.6667 \end{bmatrix}$

45. $P^2 = P \cdot P$

$= \begin{bmatrix} 0.25 & 0.25 & 0.50 \\ 0.25 & 0.25 & 0.50 \\ 0.25 & 0.25 & 0.50 \end{bmatrix}\begin{bmatrix} 0.25 & 0.25 & 0.50 \\ 0.25 & 0.25 & 0.50 \\ 0.25 & 0.25 & 0.50 \end{bmatrix}$

$= \begin{bmatrix} 0.25 & 0.25 & 0.50 \\ 0.25 & 0.25 & 0.50 \\ 0.25 & 0.25 & 0.50 \end{bmatrix}$

$P^4 = P^2 \cdot P^2$

$= \begin{bmatrix} 0.25 & 0.25 & 0.50 \\ 0.25 & 0.25 & 0.50 \\ 0.25 & 0.25 & 0.50 \end{bmatrix}\begin{bmatrix} 0.25 & 0.25 & 0.50 \\ 0.25 & 0.25 & 0.50 \\ 0.25 & 0.25 & 0.50 \end{bmatrix}$

$= \begin{bmatrix} 0.25 & 0.25 & 0.50 \\ 0.25 & 0.25 & 0.50 \\ 0.25 & 0.25 & 0.50 \end{bmatrix}$

$P^8 = P^4 \cdot P^4 = \begin{bmatrix} 0.25 & 0.25 & 0.50 \\ 0.25 & 0.25 & 0.50 \\ 0.25 & 0.25 & 0.50 \end{bmatrix}$

$P^{1000} = \begin{bmatrix} 0.25 & 0.25 & 0.50 \\ 0.25 & 0.25 & 0.50 \\ 0.25 & 0.25 & 0.50 \end{bmatrix}$

47. $\begin{aligned} 2x - y + 4z &= 3 \\ -4x + \tfrac{3}{4}y + \tfrac{1}{3}z &= -1 \\ -3x &= 0 \end{aligned}$

49. $\begin{aligned} x - y + w &= -1 \\ x + y + 2z + 4w &= 2 \end{aligned}$

51. $\begin{bmatrix} 1 & -1 \\ 2 & -1 \end{bmatrix}\begin{bmatrix} x \\ y \end{bmatrix} = \begin{bmatrix} 4 \\ 0 \end{bmatrix}$

53. $\begin{bmatrix} 1 & 1 & -1 \\ 2 & 1 & 1 \\ \frac{3}{4} & 0 & \frac{1}{2} \end{bmatrix}\begin{bmatrix} x \\ y \\ z \end{bmatrix} = \begin{bmatrix} 8 \\ 4 \\ 1 \end{bmatrix}$

55.
$\text{Price (\$)} = \begin{bmatrix} \overset{\text{tie-dye}}{15} & \overset{\text{Crew}}{10} & \overset{\text{Ts}}{12} \end{bmatrix}$

$\text{Quantity} = \begin{bmatrix} 50 \\ 40 \\ 30 \end{bmatrix}\begin{matrix} \text{tie-dye} \\ \text{Crew} \\ \text{Ts} \end{matrix}$

$\text{Revenue} = \text{Price} \times \text{Quantity}$

$= \begin{bmatrix} 15 & 10 & 12 \end{bmatrix}\begin{bmatrix} 50 \\ 40 \\ 30 \end{bmatrix} = [750 + 400 + 360] = [1510]$

So revenue = \$1510.

57.

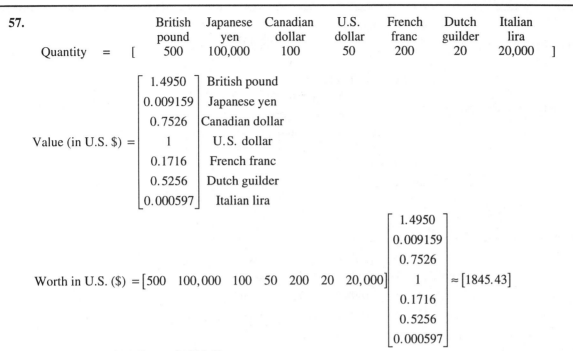

	British pound	Japanese yen	Canadian dollar	U.S. dollar	French franc	Dutch guilder	Italian lira
Quantity = [	500	100,000	100	50	200	20	20,000]

$$\text{Value (in U.S. \$)} = \begin{bmatrix} 1.4950 \\ 0.009159 \\ 0.7526 \\ 1 \\ 0.1716 \\ 0.5256 \\ 0.000597 \end{bmatrix} \begin{matrix} \text{British pound} \\ \text{Japanese yen} \\ \text{Canadian dollar} \\ \text{U.S. dollar} \\ \text{French franc} \\ \text{Dutch guilder} \\ \text{Italian lira} \end{matrix}$$

$$\text{Worth in U.S. (\$)} = \begin{bmatrix} 500 & 100,000 & 100 & 50 & 200 & 20 & 20,000 \end{bmatrix} \begin{bmatrix} 1.4950 \\ 0.009159 \\ 0.7526 \\ 1 \\ 0.1716 \\ 0.5256 \\ 0.000597 \end{bmatrix} \approx \begin{bmatrix} 1845.43 \end{bmatrix}$$

So worth in U.S. dollars = $1845.43.

59. AC^T represents the value, in each of the four listed currencies, of a wallet containing 10£, 0¥, Can $100 and U.S. $10.

61. $\dfrac{1}{4}A^2 \approx \dfrac{1}{4}\begin{bmatrix} 3.9899402 & 0.0245068 & 2.0136697 & 2.6756134 \\ 651.24319 & 4.0000375 & 328.67382 & 436.71714 \\ 7.9055939 & 0.0485577 & 3.9898743 & 5.3014456 \\ 5.9648446 & 0.036637 & 3.010378 & 3.9999647 \end{bmatrix}$

$\dfrac{1}{4}A^2 \approx \begin{bmatrix} 0.997485 & 0.0061267 & 0.5034174 & 0.6689034 \\ 162.8108 & 1.0000094 & 82.168454 & 109.17928 \\ 1.9763985 & 0.0121394 & 0.9974686 & 1.3253614 \\ 1.4912112 & 0.0091593 & 0.7525945 & 0.9999912 \end{bmatrix}$

$\dfrac{1}{4}A^2 \approx A$

The reason for this is that the ij^{th} entry of A^2 gives the value, in units of currency i, of a purse containing the equivalent of four units of currency j, so this is 4 times the value of A_{ij}.

63. $\text{price data (\$)} = P = \begin{bmatrix} 30 \\ 10 \\ 15 \end{bmatrix} \begin{matrix} \text{hard} \\ \text{soft} \\ \text{plastic} \end{matrix}$

$\text{sales data} = S = \begin{bmatrix} \text{hard} & \text{soft} & \text{plastic} \\ 700 & 1300 & 2000 \\ 400 & 300 & 500 \end{bmatrix} \begin{matrix} \\ \text{SF} \\ \text{LA} \end{matrix}$

$\text{Total revenue (\$)} = S \cdot P = \begin{bmatrix} 700 & 1300 & 2000 \\ 400 & 300 & 500 \end{bmatrix} \begin{bmatrix} 30 \\ 10 \\ 15 \end{bmatrix} = \begin{bmatrix} 64,000 \\ 22,500 \end{bmatrix} \begin{matrix} \text{SF} \\ \text{LA} \end{matrix}$

Total revenue for the San Francisco store is $64,000 and for the Los Angeles store is $22,500.

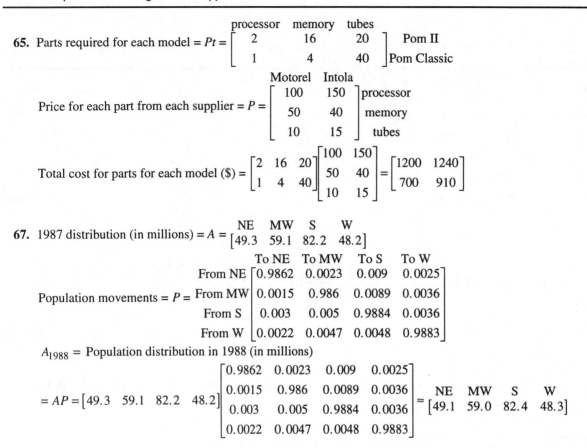

65. Parts required for each model $= Pt = $
$\begin{array}{c} \text{processor} \quad \text{memory} \quad \text{tubes} \end{array}$
$\begin{bmatrix} 2 & 16 & 20 \\ 1 & 4 & 40 \end{bmatrix} \begin{array}{l} \text{Pom II} \\ \text{Pom Classic} \end{array}$

Price for each part from each supplier $= P = $
$\begin{array}{c} \text{Motorel} \quad \text{Intola} \end{array}$
$\begin{bmatrix} 100 & 150 \\ 50 & 40 \\ 10 & 15 \end{bmatrix} \begin{array}{l} \text{processor} \\ \text{memory} \\ \text{tubes} \end{array}$

Total cost for parts for each model (\$) $= \begin{bmatrix} 2 & 16 & 20 \\ 1 & 4 & 40 \end{bmatrix} \begin{bmatrix} 100 & 150 \\ 50 & 40 \\ 10 & 15 \end{bmatrix} = \begin{bmatrix} 1200 & 1240 \\ 700 & 910 \end{bmatrix}$

67. 1987 distribution (in millions) $= A = $
$\begin{array}{cccc} \text{NE} & \text{MW} & \text{S} & \text{W} \end{array}$
$\begin{bmatrix} 49.3 & 59.1 & 82.2 & 48.2 \end{bmatrix}$

Population movements $= P = $
$\begin{array}{l} \text{From NE} \\ \text{From MW} \\ \text{From S} \\ \text{From W} \end{array}$
$\begin{array}{cccc} \text{To NE} & \text{To MW} & \text{To S} & \text{To W} \end{array}$
$\begin{bmatrix} 0.9862 & 0.0023 & 0.009 & 0.0025 \\ 0.0015 & 0.986 & 0.0089 & 0.0036 \\ 0.003 & 0.005 & 0.9884 & 0.0036 \\ 0.0022 & 0.0047 & 0.0048 & 0.9883 \end{bmatrix}$

$A_{1988} = $ Population distribution in 1988 (in millions)

$= AP = \begin{bmatrix} 49.3 & 59.1 & 82.2 & 48.2 \end{bmatrix} \begin{bmatrix} 0.9862 & 0.0023 & 0.009 & 0.0025 \\ 0.0015 & 0.986 & 0.0089 & 0.0036 \\ 0.003 & 0.005 & 0.9884 & 0.0036 \\ 0.0022 & 0.0047 & 0.0048 & 0.9883 \end{bmatrix} \approx \begin{array}{cccc} \text{NE} & \text{MW} & \text{S} & \text{W} \end{array}$
$\begin{bmatrix} 49.1 & 59.0 & 82.4 & 48.3 \end{bmatrix}$

69. The matrix equation $AX = B$ is equivalent to a system of linear equations where A is the coeffiicient matrix of the system of equations, X is the column matrix of the unknowns and B is the column matrix consisting of the right-hand sides of the equations. Thus, *every matrix equation represents a system of equations.*

71. Scenarios will vary.

Section 3.3 Matrix Inversion
Pages 213–216

1. $AB = \begin{bmatrix} 0 & 1 \\ 1 & 0 \end{bmatrix} \begin{bmatrix} 0 & 1 \\ 1 & 0 \end{bmatrix} = \begin{bmatrix} 1 & 0 \\ 0 & 1 \end{bmatrix} = I$

So, yes, A and B are an inverse pair.

3. $AB = \begin{bmatrix} 2 & 1 & 1 \\ 0 & 1 & 1 \\ 0 & 0 & 1 \end{bmatrix} \begin{bmatrix} \frac{1}{2} & -\frac{1}{2} & 0 \\ 0 & 1 & -1 \\ 0 & 0 & 1 \end{bmatrix} = \begin{bmatrix} 1 & 0 & 0 \\ 0 & 1 & 0 \\ 0 & 0 & 1 \end{bmatrix} = I$

So, yes, A and B are an inverse pair.

5. $AB = \begin{bmatrix} a & 0 & 0 \\ 0 & b & 0 \\ 0 & 0 & 0 \end{bmatrix} \begin{bmatrix} a^{-1} & 0 & 0 \\ 0 & b^{-1} & 0 \\ 0 & 0 & 0 \end{bmatrix} = \begin{bmatrix} 1 & 0 & 0 \\ 0 & 1 & 0 \\ 0 & 0 & 0 \end{bmatrix} \neq I$

So, no, A and B are an inverse pair.

7. $\begin{bmatrix} 1 & 1 & | & 1 & 0 \\ 2 & 1 & | & 0 & 1 \end{bmatrix} R_2 - 2R_1 \rightarrow \begin{bmatrix} 1 & 1 & | & 1 & 0 \\ 0 & -1 & | & -2 & 1 \end{bmatrix} \begin{matrix} R_1 + R_2 \rightarrow \\ -R_2 \rightarrow \end{matrix} \begin{bmatrix} 1 & 0 & | & -1 & 1 \\ 0 & 1 & | & 2 & -1 \end{bmatrix}$

So, $A^{-1} = \begin{bmatrix} -1 & 1 \\ 2 & -1 \end{bmatrix}$ or, $A^{-1} = \dfrac{1}{1-2}\begin{bmatrix} 1 & -1 \\ -2 & 1 \end{bmatrix} = \begin{bmatrix} -1 & 1 \\ 2 & -1 \end{bmatrix}$.

9. $\begin{bmatrix} 0 & 1 & | & 1 & 0 \\ 1 & 0 & | & 0 & 1 \end{bmatrix} R_2 \leftrightarrow R_1 \begin{bmatrix} 1 & 0 & | & 0 & 1 \\ 0 & 1 & | & 1 & 0 \end{bmatrix}$

$A^{-1} = \begin{bmatrix} 0 & 1 \\ 1 & 0 \end{bmatrix}$ or, $A^{-1} = \dfrac{1}{0-1}\begin{bmatrix} 0 & -1 \\ -1 & 0 \end{bmatrix} = \begin{bmatrix} 0 & 1 \\ 1 & 0 \end{bmatrix}$

11. $A^{-1} = A = \begin{bmatrix} 1 & 0 \\ 0 & 1 \end{bmatrix}$

13. $\begin{bmatrix} 1 & 1 & | & 1 & 0 \\ 1 & -1 & | & 0 & 1 \end{bmatrix} R_2 - R_1 \rightarrow \begin{bmatrix} 1 & 1 & | & 1 & 0 \\ 0 & -2 & | & -1 & 1 \end{bmatrix} \begin{matrix} 2R_1 + R_2 \rightarrow \\ {} \end{matrix} \begin{bmatrix} 2 & 0 & | & 1 & 1 \\ 0 & -2 & | & -1 & 1 \end{bmatrix} \begin{matrix} \frac{1}{2}R_1 \rightarrow \\ -\frac{1}{2}R_2 \rightarrow \end{matrix} \begin{bmatrix} 1 & 0 & | & \frac{1}{2} & \frac{1}{2} \\ 0 & 1 & | & \frac{1}{2} & -\frac{1}{2} \end{bmatrix}$

So, $A^{-1} = \begin{bmatrix} \frac{1}{2} & \frac{1}{2} \\ \frac{1}{2} & -\frac{1}{2} \end{bmatrix}$ or, $A^{-1} = \dfrac{1}{-1-1}\begin{bmatrix} -1 & -1 \\ -1 & 1 \end{bmatrix} = \begin{bmatrix} \frac{1}{2} & \frac{1}{2} \\ \frac{1}{2} & -\frac{1}{2} \end{bmatrix}$

15. $\begin{bmatrix} 3 & 0 & | & 1 & 0 \\ 0 & \frac{1}{2} & | & 0 & 1 \end{bmatrix} \begin{matrix} \frac{1}{3}R_1 \rightarrow \\ 2R_2 \rightarrow \end{matrix} \begin{bmatrix} 1 & 0 & | & \frac{1}{3} & 0 \\ 0 & 1 & | & 0 & 2 \end{bmatrix}$

So, $A^{-1} = \begin{bmatrix} \frac{1}{3} & 0 \\ 0 & 2 \end{bmatrix}$ or, $A^{-1} = \dfrac{1}{\frac{3}{2}}\begin{bmatrix} \frac{1}{2} & 0 \\ 0 & 3 \end{bmatrix} = \begin{bmatrix} \frac{1}{3} & 0 \\ 0 & 2 \end{bmatrix}$.

17. $\begin{bmatrix} 1 & 1 & | & 1 & 0 \\ 6 & 6 & | & 0 & 1 \end{bmatrix} R_2 - 6R_1 \rightarrow \begin{bmatrix} 1 & 1 & | & 1 & 0 \\ 0 & 0 & | & -6 & 1 \end{bmatrix} \leftarrow \text{A row of zeros}$

No inverse exists.

$\begin{bmatrix} 1 & 1 \\ 6 & 6 \end{bmatrix}$ is a singular matrix.

19. $\begin{bmatrix} \frac{1}{6} & -\frac{1}{6} & | & 1 & 0 \\ 0 & \frac{1}{6} & | & 0 & 1 \end{bmatrix} R_1 + R_2 \rightarrow \begin{bmatrix} \frac{1}{6} & 0 & | & 1 & 1 \\ 0 & \frac{1}{6} & | & 0 & 1 \end{bmatrix} \begin{matrix} 6R_1 \rightarrow \\ 6R_2 \rightarrow \end{matrix} \begin{bmatrix} 1 & 0 & | & 6 & 6 \\ 0 & 1 & | & 0 & 6 \end{bmatrix}$

So, $A^{-1} = \begin{bmatrix} 6 & 6 \\ 0 & 6 \end{bmatrix}$ or, $A^{-1} \dfrac{1}{\frac{1}{36}}\begin{bmatrix} \frac{1}{6} & \frac{1}{6} \\ 0 & \frac{1}{6} \end{bmatrix} = \begin{bmatrix} 6 & 6 \\ 0 & 6 \end{bmatrix}$.

21. $\begin{bmatrix} 1 & 0 & | & 1 & 0 \\ \frac{3}{4} & 0 & | & 0 & 1 \end{bmatrix} 4R_2 - 3R_1 \rightarrow \begin{bmatrix} 1 & 0 & | & 1 & 0 \\ 0 & 0 & | & -3 & 4 \end{bmatrix} \leftarrow \text{A row of zeros}$

No inverse exists.

$\begin{bmatrix} 1 & 0 \\ \frac{3}{4} & 0 \end{bmatrix}$ is a singular matrix.

23. $\begin{bmatrix} 1 & 1 & 1 & | & 1 & 0 & 0 \\ 0 & 1 & 1 & | & 0 & 1 & 0 \\ 0 & 0 & 1 & | & 0 & 0 & 1 \end{bmatrix} \begin{matrix} R_1 - R_2 \rightarrow \\ R_2 - R_3 \rightarrow \\ \end{matrix} \begin{bmatrix} 1 & 0 & 0 & | & 1 & -1 & 0 \\ 0 & 1 & 0 & | & 0 & 1 & -1 \\ 0 & 0 & 1 & | & 0 & 0 & 1 \end{bmatrix}$

$A^{-1} = \begin{bmatrix} 1 & -1 & 0 \\ 0 & 1 & -1 \\ 0 & 0 & 1 \end{bmatrix}$

25. $\begin{bmatrix} 1 & 1 & 1 & | & 1 & 0 & 0 \\ 1 & 0 & 2 & | & 0 & 1 & 0 \\ 1 & -1 & 1 & | & 0 & 0 & 1 \end{bmatrix} \begin{matrix} \\ R_2 - R_1 \rightarrow \\ R_3 - R_1 \rightarrow \end{matrix} \begin{bmatrix} 1 & 1 & 1 & | & 1 & 0 & 0 \\ 0 & -1 & 1 & | & -1 & 1 & 0 \\ 0 & -2 & 0 & | & -1 & 0 & 1 \end{bmatrix} \begin{matrix} R_1 + R_2 \rightarrow \\ -R_2 \rightarrow \\ R_3 - 2R_2 \rightarrow \end{matrix}$

$\begin{bmatrix} 1 & 0 & 2 & | & 0 & 1 & 0 \\ 0 & 1 & -1 & | & 1 & -1 & 0 \\ 0 & 0 & -2 & | & 1 & -2 & 1 \end{bmatrix} \begin{matrix} R_1 + R_3 \rightarrow \\ 2R_2 - R_3 \rightarrow \\ -\frac{1}{2}R_3 \rightarrow \end{matrix} \begin{bmatrix} 1 & 0 & 0 & | & 1 & -1 & 1 \\ 0 & 2 & 0 & | & 1 & 0 & -1 \\ 0 & 0 & 1 & | & -\frac{1}{2} & 1 & -\frac{1}{2} \end{bmatrix} \begin{matrix} \\ \frac{1}{2}R_2 \rightarrow \\ \end{matrix}$

$\begin{bmatrix} 1 & 0 & 0 & | & 1 & -1 & 1 \\ 0 & 1 & 0 & | & \frac{1}{2} & 0 & -\frac{1}{2} \\ 0 & 0 & 1 & | & -\frac{1}{2} & 1 & -\frac{1}{2} \end{bmatrix}$ So, $A^{-1} = \begin{bmatrix} 1 & -1 & 1 \\ \frac{1}{2} & 0 & -\frac{1}{2} \\ -\frac{1}{2} & 1 & -\frac{1}{2} \end{bmatrix}$

27. $\begin{bmatrix} 1 & 1 & 1 & | & 1 & 0 & 0 \\ 1 & -1 & 0 & | & 0 & 1 & 0 \\ 1 & 2 & 3 & | & 0 & 0 & 1 \end{bmatrix} \begin{matrix} \\ R_2 - R_1 \rightarrow \\ R_3 - R_1 \rightarrow \end{matrix} \begin{bmatrix} 1 & 1 & 1 & | & 1 & 0 & 0 \\ 0 & -2 & -1 & | & -1 & 1 & 0 \\ 0 & 1 & 2 & | & -1 & 0 & 1 \end{bmatrix} \begin{matrix} 2R_1 + R_2 \rightarrow \\ -R_2 \rightarrow \\ 2R_3 + R_2 \rightarrow \end{matrix}$

$\begin{bmatrix} 2 & 0 & 1 & | & 1 & 1 & 0 \\ 0 & 2 & 1 & | & 1 & -1 & 0 \\ 0 & 0 & 3 & | & -3 & 1 & 2 \end{bmatrix} \begin{matrix} 3R_1 - R_3 \rightarrow \\ 3R_2 - R_3 \rightarrow \\ \end{matrix} \begin{bmatrix} 6 & 0 & 0 & | & 6 & 2 & -2 \\ 0 & 6 & 0 & | & 6 & -4 & -2 \\ 0 & 0 & 3 & | & -3 & 1 & 2 \end{bmatrix} \begin{matrix} \frac{1}{6}R_1 \rightarrow \\ \frac{1}{6}R_2 \rightarrow \\ \frac{1}{3}R_3 \rightarrow \end{matrix} \begin{bmatrix} 1 & 0 & 0 & | & 1 & \frac{1}{3} & -\frac{1}{3} \\ 0 & 1 & 0 & | & 1 & -\frac{2}{3} & -\frac{1}{3} \\ 0 & 0 & 1 & | & -1 & \frac{1}{3} & \frac{2}{3} \end{bmatrix}$

So, $A^{-1} = \begin{bmatrix} 1 & \frac{1}{3} & -\frac{1}{3} \\ 1 & -\frac{2}{3} & -\frac{1}{3} \\ -1 & \frac{1}{3} & \frac{2}{3} \end{bmatrix}$

29. $\begin{bmatrix} 1 & 1 & 1 & | & 1 & 0 & 0 \\ 1 & 0 & 1 & | & 0 & 1 & 0 \\ 1 & -1 & 1 & | & 0 & 0 & 1 \end{bmatrix} \begin{matrix} \\ R_2 - R_1 \rightarrow \\ R_3 - R_1 \rightarrow \end{matrix} \begin{bmatrix} 1 & 1 & 1 & | & 1 & 0 & 0 \\ 0 & -1 & 0 & | & -1 & 1 & 0 \\ 0 & -2 & 0 & | & -1 & 0 & 1 \end{bmatrix} \begin{matrix} R_1 + R_2 \rightarrow \\ -R_2 \rightarrow \\ R_3 - 2R_2 \rightarrow \end{matrix} \begin{bmatrix} 1 & 0 & 1 & | & 0 & 1 & 0 \\ 0 & 1 & 0 & | & 1 & -1 & 0 \\ 0 & 0 & 0 & | & 1 & -2 & 1 \end{bmatrix} \leftarrow$ A row of zeros

No inverse exists.

$\begin{bmatrix} 1 & 1 & 1 \\ 1 & 0 & 1 \\ 1 & -1 & 1 \end{bmatrix}$ is a singular matrix.

31.
$$\left[\begin{array}{cccc|cccc} 1 & 0 & 1 & 0 & 1 & 0 & 0 & 0 \\ -1 & 1 & 0 & 1 & 0 & 1 & 0 & 0 \\ -1 & 0 & 0 & 1 & 0 & 0 & 1 & 0 \\ 0 & -1 & 0 & 1 & 0 & 0 & 0 & 1 \end{array}\right] \begin{array}{l} \\ R_2+R_1 \rightarrow \\ R_3+R_1 \rightarrow \\ \\ \end{array} \left[\begin{array}{cccc|cccc} 1 & 0 & 1 & 0 & 1 & 0 & 0 & 0 \\ 0 & 1 & 1 & 1 & 1 & 1 & 0 & 0 \\ 0 & 0 & 1 & 1 & 1 & 0 & 1 & 0 \\ 0 & -1 & 0 & 1 & 0 & 0 & 0 & 1 \end{array}\right] \begin{array}{l} \\ \\ \\ R_4+R_2 \rightarrow \end{array}$$

$$\left[\begin{array}{cccc|cccc} 1 & 0 & 1 & 0 & 1 & 0 & 0 & 0 \\ 0 & 1 & 1 & 1 & 1 & 1 & 0 & 0 \\ 0 & 0 & 1 & 1 & 1 & 0 & 1 & 0 \\ 0 & 0 & 1 & 2 & 1 & 1 & 0 & 1 \end{array}\right] \begin{array}{l} R_1-R_3 \rightarrow \\ R_2-R_3 \rightarrow \\ \\ R_4-R_3 \rightarrow \end{array} \left[\begin{array}{cccc|cccc} 1 & 0 & 0 & -1 & 0 & 0 & -1 & 0 \\ 0 & 1 & 0 & 0 & 0 & 1 & -1 & 0 \\ 0 & 0 & 1 & 1 & 1 & 0 & 1 & 0 \\ 0 & 0 & 0 & 1 & 0 & 1 & -1 & 1 \end{array}\right] \begin{array}{l} R_1+R_4 \rightarrow \\ \\ R_3-R_4 \rightarrow \\ \end{array}$$

$$\left[\begin{array}{cccc|cccc} 1 & 0 & 0 & 0 & 0 & 1 & -2 & 1 \\ 0 & 1 & 0 & 0 & 0 & 1 & -1 & 0 \\ 0 & 0 & 1 & 0 & 1 & -1 & 2 & -1 \\ 0 & 0 & 0 & 1 & 0 & 1 & -1 & 1 \end{array}\right] \text{So, } A^{-1} = \left[\begin{array}{cccc} 0 & 1 & -2 & 1 \\ 0 & 1 & -1 & 0 \\ 1 & -1 & 2 & -1 \\ 0 & 1 & -1 & 1 \end{array}\right]$$

33.
$$\left[\begin{array}{cccc|cccc} 1 & 2 & 3 & 4 & 1 & 0 & 0 & 0 \\ 0 & 1 & 2 & 3 & 0 & 1 & 0 & 0 \\ 0 & 0 & 1 & 2 & 0 & 0 & 1 & 0 \\ 0 & 0 & 0 & 1 & 0 & 0 & 0 & 1 \end{array}\right] R_1-2R_2 \rightarrow \left[\begin{array}{cccc|cccc} 1 & 0 & -1 & -2 & 1 & -2 & 0 & 0 \\ 0 & 1 & 2 & 3 & 0 & 1 & 0 & 0 \\ 0 & 0 & 1 & 2 & 0 & 0 & 1 & 0 \\ 0 & 0 & 0 & 1 & 0 & 0 & 0 & 1 \end{array}\right] \begin{array}{l} R_1+R_3 \rightarrow \\ R_2-2R_3 \rightarrow \\ \\ \end{array}$$

$$\left[\begin{array}{cccc|cccc} 1 & 0 & 0 & 0 & 1 & -2 & 1 & 0 \\ 0 & 1 & 0 & -1 & 0 & 1 & -2 & 0 \\ 0 & 0 & 1 & 2 & 0 & 0 & 1 & 0 \\ 0 & 0 & 0 & 1 & 0 & 0 & 0 & 1 \end{array}\right] \begin{array}{l} \\ R_2+R_4 \rightarrow \\ R_3-2R_4 \rightarrow \\ \end{array} \left[\begin{array}{cccc|cccc} 1 & 0 & 0 & 0 & 1 & -2 & 1 & 0 \\ 0 & 1 & 0 & 0 & 0 & 1 & -2 & 1 \\ 0 & 0 & 1 & 0 & 0 & 0 & 1 & -2 \\ 0 & 0 & 0 & 1 & 0 & 0 & 0 & 1 \end{array}\right]$$

$$\text{So, } A^{-1} \left[\begin{array}{cccc} 1 & -2 & 1 & 0 \\ 0 & 1 & -2 & 1 \\ 0 & 0 & 1 & -2 \\ 0 & 0 & 0 & 1 \end{array}\right]$$

35. $A^{-1} \approx \left[\begin{array}{cc} 0.38 & 0.45 \\ 0.49 & -0.41 \end{array}\right]$

37. $A^{-1} = \left[\begin{array}{cc} 0.00 & -0.99 \\ 0.81 & 2.87 \end{array}\right]$

39. A^{-1} does not exist. A is a singular matrix.

41. $A^{-1} = \left[\begin{array}{cccc} 91.35 & -8.65 & 0 & -71.30 \\ -0.07 & -0.07 & 0 & 2.49 \\ 2.60 & 2.60 & -4.35 & 1.37 \\ 2.69 & 2.69 & 0 & -2.10 \end{array}\right]$

43. $A = \begin{bmatrix} 1 & 1 \\ 1 & -1 \end{bmatrix}$, $X = \begin{bmatrix} x \\ y \end{bmatrix}$, $B = \begin{bmatrix} 4 \\ 2 \end{bmatrix}$

$AX = B$

$A^{-1} = \dfrac{1}{-1-1} \begin{bmatrix} -1 & -1 \\ -1 & 1 \end{bmatrix} = \begin{bmatrix} \frac{1}{2} & \frac{1}{2} \\ \frac{1}{2} & -\frac{1}{2} \end{bmatrix}$

$X = A^{-1}B = \begin{bmatrix} \frac{1}{2} & \frac{1}{2} \\ \frac{1}{2} & -\frac{1}{2} \end{bmatrix} \begin{bmatrix} 4 \\ 2 \end{bmatrix} = \begin{bmatrix} 3 \\ 1 \end{bmatrix}$

So, $x = 3$ and $y = 1$ is the solution to the system.

45. $A = \begin{bmatrix} \frac{1}{3} & -\frac{1}{2} \\ \frac{1}{4} & 1 \end{bmatrix}$, $B = \begin{bmatrix} 1 \\ -2 \end{bmatrix}$

$A^{-1} = \dfrac{1}{\frac{1}{3} + \frac{1}{8}} \begin{bmatrix} 1 & \frac{1}{2} \\ -\frac{1}{4} & \frac{1}{3} \end{bmatrix} = \begin{bmatrix} \frac{24}{11} & \frac{12}{11} \\ -\frac{6}{11} & \frac{8}{11} \end{bmatrix}$

$X = A^{-1}B = \begin{bmatrix} \frac{24}{11} & \frac{12}{11} \\ -\frac{6}{11} & \frac{8}{11} \end{bmatrix} \begin{bmatrix} 1 \\ -2 \end{bmatrix} = \begin{bmatrix} 0 \\ -2 \end{bmatrix}$

So, $x = 0$, $y = -2$ is the solution to the system.

47. $A = \begin{bmatrix} -1 & 2 & -1 \\ -1 & -1 & 2 \\ 2 & 0 & -1 \end{bmatrix}$, $B = \begin{bmatrix} 0 \\ 0 \\ 6 \end{bmatrix}$

$A^{-1} = \begin{bmatrix} \frac{1}{3} & \frac{2}{3} & 1 \\ 1 & 1 & 1 \\ \frac{2}{3} & \frac{4}{3} & 1 \end{bmatrix}$

$X = A^{-1}B = \begin{bmatrix} \frac{1}{3} & \frac{2}{3} & 1 \\ 1 & 1 & 1 \\ \frac{2}{3} & \frac{4}{3} & 1 \end{bmatrix} \begin{bmatrix} 0 \\ 0 \\ 6 \end{bmatrix} = \begin{bmatrix} 6 \\ 6 \\ 6 \end{bmatrix}$

So, $(x, y, z) = (6, 6, 6)$.

49. $A = \begin{bmatrix} 1 & 1 & 6 \\ \frac{1}{3} & -\frac{1}{3} & \frac{2}{3} \\ \frac{1}{2} & 0 & 1 \end{bmatrix}$, $B = \begin{bmatrix} -1 \\ 1 \\ 0 \end{bmatrix}$

$A^{-1} = \begin{bmatrix} -\frac{1}{2} & -\frac{3}{2} & 4 \\ 0 & -3 & 2 \\ \frac{1}{4} & \frac{3}{4} & -1 \end{bmatrix}$

$X = A^{-1}B = \begin{bmatrix} -\frac{1}{2} & -\frac{3}{2} & 4 \\ 0 & -3 & 2 \\ \frac{1}{4} & \frac{3}{4} & -1 \end{bmatrix} \begin{bmatrix} -1 \\ 1 \\ 0 \end{bmatrix} = \begin{bmatrix} -1 \\ -3 \\ \frac{1}{2} \end{bmatrix}$

So, $(x, y, z) = \left(-1, -3, \dfrac{1}{2}\right)$.

51. $A = \begin{bmatrix} 1 & 1 & 5 & 0 \\ 0 & 1 & 2 & 1 \\ 1 & 3 & 7 & 2 \\ 1 & 1 & 5 & 1 \end{bmatrix}$, $B = \begin{bmatrix} 1 \\ 1 \\ 2 \\ 1 \end{bmatrix}$

$A^{-1} = \begin{bmatrix} -\frac{3}{2} & -4 & \frac{3}{2} & 1 \\ 0 & -1 & 1 & -1 \\ \frac{1}{2} & 1 & -\frac{1}{2} & 0 \\ -1 & 0 & 0 & 1 \end{bmatrix}$

$X = A^{-1}B = \begin{bmatrix} -\frac{3}{2} & -4 & \frac{3}{2} & 1 \\ 0 & -1 & 1 & -1 \\ \frac{1}{2} & 1 & -\frac{1}{2} & 0 \\ -1 & 0 & 0 & 1 \end{bmatrix} \begin{bmatrix} 1 \\ 1 \\ 2 \\ 1 \end{bmatrix} = \begin{bmatrix} -\frac{3}{2} \\ 0 \\ \frac{1}{2} \\ 0 \end{bmatrix}$

So, $(x, y, z, w) = \left(-\dfrac{3}{2}, 0, \dfrac{1}{2}, 0\right)$.

53. Let x = number of servings of Pork & Beans and y = number of servings of rye bread.

a.

	P & B	Rye bread	Total
Protein	5	4	20
Carbohydrates	21	12	80

$5x + 4y = 20$

$21x + 12y = 80$

$A = \begin{bmatrix} 5 & 4 \\ 21 & 12 \end{bmatrix}$, $B = \begin{bmatrix} 20 \\ 80 \end{bmatrix}$

$A^{-1} = \begin{bmatrix} -\frac{1}{2} & \frac{1}{6} \\ \frac{7}{8} & -\frac{5}{24} \end{bmatrix}$

$X = A^{-1}B = \begin{bmatrix} -\frac{1}{2} & \frac{1}{6} \\ \frac{7}{8} & -\frac{5}{24} \end{bmatrix} \begin{bmatrix} 20 \\ 80 \end{bmatrix} = \begin{bmatrix} \frac{10}{3} \\ \frac{5}{6} \end{bmatrix}$

So, you should prepare $3\frac{1}{3}$ servings of Pork & Beans and $\frac{5}{6}$ slices of rye bread.

b. Using the information from part a:

Let A = the total grams of protein and B = the total grams of carbohydrates.

Then, $\begin{bmatrix} -\frac{1}{2} & \frac{1}{6} \\ \frac{7}{8} & -\frac{5}{24} \end{bmatrix} \begin{bmatrix} A \\ B \end{bmatrix} = \begin{bmatrix} -\frac{1}{2}A + \frac{1}{6}B \\ \frac{7}{8}A - \frac{5}{24}B \end{bmatrix}$

or

number of servings of Pork & Beans $= -\dfrac{1}{2}A + \dfrac{1}{6}B$

number of slices of rye bread $= \dfrac{7}{8}A - \dfrac{5}{24}B$

55. Let x = gallons of PineOrange, y = gallons of PineKiwi, and z = gallons of OrangeKiwi.

a.

	PineOrange	PineKiwi	OrangeKiwi	Total
Pineapple (qt)	2	3	0	800
Orange (qt)	2	0	3	650
Kiwi (qt)	0	1	1	350

$2x + 3y = 800$
$2x + 3z = 650$
$y + z = 350$

$$A = \begin{bmatrix} 2 & 3 & 0 \\ 2 & 0 & 3 \\ 0 & 1 & 1 \end{bmatrix}, \quad B = \begin{bmatrix} 800 \\ 650 \\ 350 \end{bmatrix}$$

$$A^{-1} = \begin{bmatrix} \frac{1}{4} & \frac{1}{4} & -\frac{3}{4} \\ \frac{1}{6} & -\frac{1}{6} & \frac{1}{2} \\ -\frac{1}{6} & \frac{1}{6} & \frac{1}{2} \end{bmatrix}$$

$$X = A^{-1}B = \begin{bmatrix} \frac{1}{4} & \frac{1}{4} & -\frac{3}{4} \\ \frac{1}{6} & -\frac{1}{6} & \frac{1}{2} \\ -\frac{1}{6} & \frac{1}{6} & \frac{1}{2} \end{bmatrix} \begin{bmatrix} 800 \\ 650 \\ 350 \end{bmatrix} = \begin{bmatrix} 100 \\ 200 \\ 150 \end{bmatrix}$$

So, $(x,\ y,\ z) = (100,\ 200,\ 150)$.

100 gallons of PineOrange, 200 gallons of PineKiwi, and 150 gallons of OrangeKiwi can be made.

b. With 850 qt of pineapple juice, 600 qt of orange juice, and 400 qt of kiwi juice:

$$B = \begin{bmatrix} 850 \\ 600 \\ 400 \end{bmatrix}$$

$$X = A^{-1}B = \begin{bmatrix} \frac{1}{4} & \frac{1}{4} & -\frac{3}{4} \\ \frac{1}{6} & -\frac{1}{6} & \frac{1}{2} \\ -\frac{1}{6} & \frac{1}{6} & \frac{1}{2} \end{bmatrix} \begin{bmatrix} 850 \\ 600 \\ 400 \end{bmatrix} = \begin{bmatrix} 62.5 \\ 241.\overline{6} \\ 158.\overline{3} \end{bmatrix}$$

So, $(x,\ y,\ z) = \left(62\frac{1}{2},\ 241\frac{2}{3},\ 158\frac{1}{3}\right)$

$62\frac{1}{2}$ gallons of PineOrange, $241\frac{2}{3}$ gallons of PineKiwi, and $158\frac{1}{3}$ gallons of OrangeKiwi can be made.

c. With A qt of pineapple juice, B qt of orange juice, and C qt of kiwi juice:

$$B = \begin{bmatrix} A \\ B \\ C \end{bmatrix}$$

$$X = A^{-1}B = \begin{bmatrix} \frac{1}{4} & \frac{1}{4} & -\frac{3}{4} \\ \frac{1}{6} & -\frac{1}{6} & \frac{1}{2} \\ -\frac{1}{6} & \frac{1}{6} & \frac{1}{2} \end{bmatrix} \begin{bmatrix} A \\ B \\ C \end{bmatrix}$$

Gallons of PineOrange $= \frac{1}{4}A + \frac{1}{4}B - \frac{3}{4}C$, gallons of PineKiwi $= \frac{1}{6}A - \frac{1}{6}B + \frac{1}{2}C$,

and gallons of OrangeKiwi $= -\frac{1}{6}A + \frac{1}{6}B + \frac{1}{2}C$.

57. 1987 distribution (in millions) = A_{1987}=

	NE	MW	S	W
	49.3	59.1	82.2	48.2

Population movements = P =

	To NE	To MW	To S	To W
From NE	0.9862	0.0023	0.009	0.0025
From MW	0.0015	0.986	0.0089	0.0036
From S	0.003	0.005	0.9884	0.0036
From W	0.0022	0.0047	0.0048	0.9883

If P also describes the population movements from 1986 to 1987, then $A_{1986}P = A_{1987}$ or $A_{1986} = A_{1987}P^{-1}$

$$P^{-1} \approx \begin{bmatrix} 1.0140 & -0.0023 & -0.0092 & -0.0025 \\ -0.0015 & 1.0143 & -0.0091 & -0.0037 \\ -0.0031 & -0.0051 & 1.0118 & -0.0037 \\ -0.0022 & -0.0048 & -0.0049 & 1.0119 \end{bmatrix}$$

Population Distribution in 1986 (in millions) = $A_{1986} = A_{1987}P^{-1} \approx$

	NE	MW	S	W
	49.5	59.2	81.9	48.1

59. $A = \begin{bmatrix} a & b \\ a & b \end{bmatrix}$

$$\begin{bmatrix} a & b & | & 1 & 0 \\ a & b & | & 0 & 1 \end{bmatrix} R_2 - R_1 \rightarrow \begin{bmatrix} a & b & | & 1 & 0 \\ 0 & 0 & | & -1 & 1 \end{bmatrix} \leftarrow \text{A row of zeros}$$

No inverse exists. So, A is a singular matrix.

61. $A = \begin{bmatrix} a & b \\ c & d \end{bmatrix}$

$$A^{-1} = \begin{bmatrix} a & b \\ c & d \end{bmatrix}^{-1} = \frac{1}{ad - bc} \begin{bmatrix} d & -b \\ -c & a \end{bmatrix} (ad - bc \neq 0)$$

$$AA^{-1} = \begin{bmatrix} a & b \\ c & d \end{bmatrix} \begin{bmatrix} \frac{d}{ab-bc} & \frac{-b}{ad-bc} \\ \frac{-c}{ab-bc} & \frac{a}{ad-bc} \end{bmatrix} = \begin{bmatrix} \frac{ad}{ad-bc} - \frac{bc}{ad-bc} & -\frac{ab}{ad-bc} + \frac{ab}{ad-bc} \\ \frac{cd}{ad-bc} - \frac{cd}{ad-bc} & -\frac{bc}{ad-bc} + \frac{ad}{ad-bc} \end{bmatrix} = \begin{bmatrix} 1 & 0 \\ 0 & 1 \end{bmatrix} = I$$

63. A square matrix which reduces to a matrix with a row of zeros does not have an inverse so the matrix is singular and *not* invertible.
Answer: No

65. Suppose A is invertible. Then $A^{-1}(AB) = (A^{-1}A)B = IB = B$, but $A^{-1}(AB) = A^{-1}0 = 0$, so $B = 0$. Since neither A nor B is the zero matrix, A cannot be invertible.
Suppose B is invertible. Then $(AB)B^{-1} = A(BB^{-1}) = AI = A$, but $(AB)B^{-1} = 0A^{-1} = 0$. Thus, B cannot be invertible and neither A nor B is invertible.

67. A and B are invertible.
$$(AB)(B^{-1}A^{-1}) = A(BB^{-1})A^{-1} = A(I)A^{-1} = I$$
So $B^{-1}A^{-1}$ is the inverse of AB.

Section 3.4 Input-Output Models
Pages 227–231

1. $I - A = \begin{bmatrix} 1 & 0 \\ 0 & 1 \end{bmatrix} - \begin{bmatrix} 0.5 & 0.4 \\ 0 & 0.5 \end{bmatrix} = \begin{bmatrix} 0.5 & -0.4 \\ 0 & 0.5 \end{bmatrix}$

$(I - A)^{-1} = \begin{bmatrix} 2 & 1.6 \\ 0 & 2 \end{bmatrix}$

$X = (I = A)^{-1} D$

$X = \begin{bmatrix} 2 & 1.6 \\ 0 & 2 \end{bmatrix}\begin{bmatrix} 10,000 \\ 20,000 \end{bmatrix}$

$= \begin{bmatrix} 52,000 \\ 40,000 \end{bmatrix}$

3. $I - A = \begin{bmatrix} 0.9 & -0.4 \\ -0.2 & 0.5 \end{bmatrix}$

$(I - A)^{-1} = \begin{bmatrix} \frac{50}{37} & \frac{40}{37} \\ \frac{20}{37} & \frac{90}{37} \end{bmatrix}$

$X = (I - A)^{-1} D = \begin{bmatrix} \frac{50}{37} & \frac{40}{37} \\ \frac{20}{37} & \frac{90}{37} \end{bmatrix}\begin{bmatrix} 25,000 \\ 15,000 \end{bmatrix}$

$= \begin{bmatrix} 50,000 \\ 50,000 \end{bmatrix}$

5. $I - A = \begin{bmatrix} 0.5 & -0.1 & 0 \\ 0 & 0.5 & -0.1 \\ 0 & 0 & 0.5 \end{bmatrix}$

$(I - A)^{-1} = \begin{bmatrix} 2 & 0.4 & 0.08 \\ 0 & 2 & 0.4 \\ 0 & 0 & 2 \end{bmatrix}$

$X = (I - A)^{-1} D = \begin{bmatrix} 2 & 0.4 & 0.08 \\ 0 & 2 & 0.4 \\ 0 & 0 & 2 \end{bmatrix}\begin{bmatrix} 1000 \\ 1000 \\ 2000 \end{bmatrix}$

$= \begin{bmatrix} 2560 \\ 2800 \\ 4000 \end{bmatrix}$

7. $I - A = \begin{bmatrix} 0.8 & -0.2 & 0 \\ -0.2 & 0.6 & -0.2 \\ 0 & -0.2 & 0.8 \end{bmatrix}$

$(I - A)^{-1} = \begin{bmatrix} \frac{11}{8} & \frac{1}{2} & \frac{1}{8} \\ \frac{1}{2} & 2 & \frac{1}{2} \\ \frac{1}{8} & \frac{1}{2} & \frac{11}{8} \end{bmatrix}$

$X = (I - A)^{-1} D = \begin{bmatrix} 27,000 \\ 28,000 \\ 17,000 \end{bmatrix}$

9. $I - A = \begin{bmatrix} 0.6 & -0.1 & 0 & 0 \\ -0.1 & 0.6 & 0 & 0 \\ 0 & 0 & 0.7 & -0.2 \\ 0 & 0 & -0.2 & 0.7 \end{bmatrix}$

$(I - A)^{-1} = \begin{bmatrix} \frac{12}{7} & \frac{2}{7} & 0 & 0 \\ \frac{2}{7} & \frac{12}{7} & 0 & 0 \\ 0 & 0 & \frac{14}{9} & \frac{4}{9} \\ 0 & 0 & \frac{4}{9} & \frac{14}{9} \end{bmatrix}$

$X = (I - A)^{-1} D = \begin{bmatrix} 138,000 \\ 128,000 \\ 168,000 \\ 138,000 \end{bmatrix}$

In Exercises 11–20, read off the answers from the columns of the $(I - A)^{-1}$ matrices from Exercises 1-10.
Column 1 corresponds to the changes in production resulting from a unit increase in demand for the products of the first sector, column 2 corresponds to the second sector, and so on.

11. $(I - A)^{-1} = \begin{bmatrix} 2 & 1.6 \\ 0 & 2 \end{bmatrix}$

13. $(I - A)^{-1} = \begin{bmatrix} \frac{50}{37} & \frac{40}{37} \\ \frac{20}{37} & \frac{90}{37} \end{bmatrix} \approx \begin{bmatrix} 1.3514 & 1.0811 \\ 0.5405 & 2.4324 \end{bmatrix}$

15. $(I - A)^{-1} = \begin{bmatrix} 2 & 0.4 & 0.08 \\ 0 & 2 & 0.4 \\ 0 & 0 & 2 \end{bmatrix}$

17. $(I - A)^{-1} = \begin{bmatrix} \frac{11}{8} & \frac{1}{2} & \frac{1}{8} \\ \frac{1}{2} & 2 & \frac{1}{2} \\ \frac{1}{8} & \frac{1}{2} & \frac{11}{8} \end{bmatrix} = \begin{bmatrix} 1.375 & 0.5 & 0.125 \\ 0.5 & 2 & 0.5 \\ 0.125 & 0.5 & 1.375 \end{bmatrix}$

19. $(I - A)^{-1} = \begin{bmatrix} \frac{12}{7} & \frac{2}{7} & 0 & 0 \\ \frac{2}{7} & \frac{12}{7} & 0 & 0 \\ 0 & 0 & \frac{14}{9} & \frac{4}{9} \\ 0 & 0 & \frac{4}{9} & \frac{14}{9} \end{bmatrix}$

$\approx \begin{bmatrix} 1.7143 & 0.2857 & 0 & 0 \\ 0.2857 & 1.7143 & 0 & 0 \\ 0 & 0 & 1.5556 & 0.4444 \\ 0 & 0 & 0.4444 & 1.5556 \end{bmatrix}$

21. We first need to convert the input-output matrix into a technology matrix. We do this by dividing each column by the total output of that sector.

$$A = \begin{bmatrix} \frac{1653}{28,333} & \frac{2}{14,790} \\ \frac{4742}{28,333} & \frac{2123}{14,790} \end{bmatrix} \approx \begin{bmatrix} 0.058342 & 0.000135 \\ 0.167367 & 0.143543 \end{bmatrix}$$

To see how each section will react to rising external demands, we calculate $(I-A)^{-1}$:

$$I - A = \begin{bmatrix} \frac{26,680}{28,333} & -\frac{2}{14,790} \\ -\frac{4742}{28,333} & \frac{12,667}{14,790} \end{bmatrix} \approx \begin{bmatrix} 0.941658 & -0.000135 \\ -0.167367 & 0.856457 \end{bmatrix}$$

$$(I-A)^{-1} \approx \begin{bmatrix} 1.062 & 0.0001677 \\ 0.2075 & 1.168 \end{bmatrix}$$

For a \$1,000 million increase in demand for Sector 1, $D = \begin{bmatrix} 1000 \\ 0 \end{bmatrix}$ and $(I-A)^{-1}D \approx \begin{bmatrix} 1062 \\ 207.5 \end{bmatrix}$. For a \$1000 million increase

in demand for Sector 2, $D = \begin{bmatrix} 0 \\ 1000 \end{bmatrix}$ and $(I-A)^{-1}D \approx \begin{bmatrix} 0.1677 \\ 1168 \end{bmatrix}$.

23. $A = \begin{bmatrix} \frac{8905}{48,603} & \frac{1336}{63,176} & 0 & \frac{213}{8293} \\ \frac{13,769}{48,603} & \frac{2498}{63,176} & 0 & \frac{162}{8293} \\ 0 & 0 & \frac{32}{5170} & \frac{16}{8293} \\ \frac{1900}{48,603} & \frac{2520}{63,176} & \frac{457}{5170} & \frac{60}{8293} \end{bmatrix}$

$I - A = \begin{bmatrix} \frac{39,698}{48,603} & -\frac{1336}{63,176} & 0 & -\frac{213}{8293} \\ -\frac{13,769}{48,603} & \frac{60,678}{63,176} & 0 & -\frac{162}{8293} \\ 0 & 0 & \frac{5138}{5170} & -\frac{16}{8293} \\ -\frac{1900}{48,603} & -\frac{2520}{63,176} & -\frac{457}{5170} & \frac{8233}{8293} \end{bmatrix}$

$(I-A)^{-1} \approx \begin{bmatrix} 1.236 & 0.02856 & 0.002894 & 0.03254 \\ 0.3658 & 1.0505 & 0.002681 & 0.03014 \\ 0.0001230 & 0.00008414 & 1.00640 & 0.001961 \\ 0.06337 & 0.04334 & 0.08983 & 1.009955 \end{bmatrix}$

For a \$1000 million increase in demand for Sector 1, $D = \begin{bmatrix} 1000 \\ 0 \\ 0 \\ 0 \end{bmatrix}$ and $(I-A)^{-1}D \approx \begin{bmatrix} 1236 \\ 365.8 \\ 0.1230 \\ 63.37 \end{bmatrix}$.

For a \$1000 million increase in demand for Sector 2, $D = \begin{bmatrix} 0 \\ 1000 \\ 0 \\ 0 \end{bmatrix}$ and $(I-A)^{-1}D \approx \begin{bmatrix} 28.56 \\ 1050.5 \\ 0.08414 \\ 43.34 \end{bmatrix}$.

For a \$1000 million increase in demand for Sector 3, $D = \begin{bmatrix} 0 \\ 0 \\ 1000 \\ 0 \end{bmatrix}$ and $(I-A)^{-1}D \approx \begin{bmatrix} 2.894 \\ 2.681 \\ 1006.40 \\ 89.83 \end{bmatrix}$.

For a \$1000 million increase in demand for Sector 4, $D = \begin{bmatrix} 0 \\ 0 \\ 0 \\ 1000 \end{bmatrix}$ and $(I-A)^{-1}D \approx \begin{bmatrix} 32.54 \\ 30.14 \\ 1.961 \\ 1009.955 \end{bmatrix}$.

25. $A = \begin{bmatrix} \frac{364.9}{1971.1} & \frac{282.5}{2226.0} \\ \frac{7.6}{1971.1} & \frac{318.0}{2226.0} \end{bmatrix}$

$I - A = \begin{bmatrix} \frac{1606.2}{1971.1} & -\frac{282.5}{2226} \\ -\frac{7.6}{1971.1} & \frac{1908}{2226} \end{bmatrix}$

$(I - A)^{-1} \approx \begin{bmatrix} 1.2280 & 0.1818 \\ 0.005524 & 1.1675 \end{bmatrix}$

For a \$100 million increase in demand for Sector 1, $(I - A)^{-1} D \approx \begin{bmatrix} 122.80 \\ 0.5524 \end{bmatrix}$.

For a \$100 million increase in demand for Sector 2, $(I - A)^{-1} D \approx \begin{bmatrix} 18.18 \\ 116.75 \end{bmatrix}$.

27. $A = \begin{bmatrix} \frac{678.4}{9401.3} & \frac{3.7}{685.8} & \frac{3341.5}{6997.3} & \frac{1023.5}{4818.3} \\ \frac{15.5}{9401.3} & \frac{6.9}{685.8} & \frac{17.1}{6997.3} & \frac{124.5}{4818.3} \\ \frac{47.3}{9401.3} & \frac{4.3}{685.8} & \frac{893.1}{6997.3} & \frac{145.8}{4818.3} \\ \frac{312.5}{9401.3} & \frac{22.1}{685.8} & \frac{83.2}{6997.3} & \frac{693.5}{4818.3} \end{bmatrix}$

$I - A = \begin{bmatrix} \frac{8722.9}{9401.3} & -\frac{3.7}{685.8} & -\frac{3341.5}{6997.3} & -\frac{1023.5}{4818.3} \\ -\frac{15.5}{9401.3} & \frac{678.9}{685.8} & -\frac{17.1}{6997.3} & -\frac{124.5}{4818.3} \\ -\frac{47.3}{9401.3} & -\frac{4.3}{685.8} & \frac{6104.2}{6997.3} & -\frac{145.8}{4818.3} \\ -\frac{312.5}{9401.3} & -\frac{22.1}{685.8} & -\frac{83.2}{6997.3} & \frac{4124.8}{4818.3} \end{bmatrix}$

$(I - A)^{-1} \approx \begin{bmatrix} 1.0916 & 0.019287 & 0.60157 & 0.292697 \\ 0.0029492 & 1.01123 & 0.00487558 & 0.031426 \\ 0.0077943 & 0.00872996 & 1.1512 & 0.042889 \\ 0.042603 & 0.038936 & 0.039531 & 1.1813 \end{bmatrix}$

For a \$100 million increase in demand for Sector 1, $(I - A)^{-1} D \approx \begin{bmatrix} 109.16 \\ 0.29492 \\ 0.77943 \\ 7.2603 \end{bmatrix}$.

For a \$100 million increase in demand for Sector 2, $(I - A)^{-1} D \approx \begin{bmatrix} 1.9287 \\ 101.123 \\ 0.872996 \\ 3.8936 \end{bmatrix}$.

For a \$100 million increase in demand for Sector 3, $(I - A)^{-1} D \approx \begin{bmatrix} 60.157 \\ 0.487558 \\ 115.12 \\ 3.9531 \end{bmatrix}$.

For a \$100 million increase in demand for Sector 4, $(I - A)^{-1} D \approx \begin{bmatrix} 29.2697 \\ 3.1426 \\ 4.2889 \\ 118.13 \end{bmatrix}$.

29. $A = \begin{bmatrix} \frac{1462.9}{2,927,501.9} & \frac{127,914.0}{523,447.3} \\ \frac{16,776.5}{2,927,501.9} & \frac{73,232.9}{523,447.3} \end{bmatrix}$

$I - A = \begin{bmatrix} \frac{2,926,039}{2,927,501.9} & -\frac{127,914.0}{523,447.3} \\ -\frac{16,776.5}{2,927,501.9} & \frac{450,214.4}{523,447.3} \end{bmatrix}$

$(I - A)^{-1} \approx \begin{bmatrix} 1.002132429 & 0.2847238282 \\ 0.006677023 & 1.16455934 \end{bmatrix}$

Using an increase in demand of 100,000 million rupiahs for each sector, we have:

Sector 1, $(I - A)^{-1} D \approx \begin{bmatrix} 100,213.2429 \\ 667.7023 \end{bmatrix}$; Sector 2, $(I - A)^{-1} D \approx \begin{bmatrix} 28,472.38282 \\ 116,455.934 \end{bmatrix}$.

31. $A = \begin{bmatrix} \frac{61,432.8}{3,436,221.0} & \frac{1,035,167.7}{4,471,902.0} & 0 & \frac{5940.4}{1,929,474.8} \\ 0 & \frac{133,184.3}{4,471,902.0} & \frac{1316.7}{3,643,267.5} & \frac{9448.0}{1,929,474.8} \\ 0 & \frac{420.7}{4,471,902.0} & \frac{589,918.8}{3,643,267.5} & \frac{10,186.3}{1,929,474.8} \\ \frac{235.0}{3,436,221.0} & \frac{1886.6}{4,471,902.0} & \frac{8809.7}{3,643,267.5} & \frac{1013.3}{1,929,474.8} \end{bmatrix}$

$I - A = \begin{bmatrix} \frac{3,374,788.2}{3,436,221.0} & -\frac{1,035,167.7}{4,471,902.0} & 0 & -\frac{5940.4}{1,929,474.8} \\ 0 & \frac{4,338,717.7}{4,471,902.0} & -\frac{1316.7}{3,643,267.5} & -\frac{9448.0}{1,929,474.8} \\ 0 & -\frac{420.7}{4,471,902.0} & \frac{3,053,348.7}{3,643,267.5} & -\frac{10,186.3}{1,929,474.8} \\ -\frac{235}{3,436,221.0} & -\frac{1886.6}{4,471,902.0} & -\frac{8809.7}{3,643,267.5} & \frac{1,928,461.5}{1,929,474.8} \end{bmatrix}$

$(I - A)^{-1} \approx \begin{bmatrix} 1.018203755 & 0.2429334944 & 0.0001172458 & 0.0043272669 \\ 0.0000003518 & 1.030699022 & 0.0004590464 & 0.00505207 \\ 0.0000004389 & 0.0001185453 & 1.193222113 & 0.0063032843 \\ 0.0000696718 & 0.0004519679 & 0.0028870204 & 1.000543123 \end{bmatrix}$

Using an increase in demand of 100,000 million rupiahs for each sector, we have:

Sector 1, $(I - A)^{-1} D \approx \begin{bmatrix} 101,820.3755 \\ 0.03518 \\ 0.04389 \\ 6.96718 \end{bmatrix}$; Sector 2, $(I - A)^{-1} D \approx \begin{bmatrix} 24,293.34944 \\ 103,069.9022 \\ 11.85453 \\ 45.19679 \end{bmatrix}$; Sector 3, $(I - A)^{-1} D \approx \begin{bmatrix} 11.72458 \\ 45.90464 \\ 119,322.2113 \\ 288.70204 \end{bmatrix}$;

Sector 4, $(I - A)^{-1} D \approx \begin{bmatrix} 432.72669 \\ 505.207 \\ 630.32843 \\ 100,054.3123 \end{bmatrix}$.

33. There is no surplus or deficit in the number of units produced in that sector for the other sectors. In other words, internal demand equals production.

35. If the ijth entry in the matrix $(I - A)^{-1}$ is zero, it means that an increase in demand for Sector j does not affect the production in Sector i.

37. We expect the off-diagonal entries to be less than 1 since a one unit increase in demand for a sector should not require more than one unit of increase in production for a different sector.

You're The Expert
Page 234

1. The (2, 11) entry in the matrix A, which is 0, tells us that no units for Sector 2 (mining) are used in production for Sector 11 (office supplies).

3. None of the entries are negative since an increase in demand for a sector would not require a decrease of production for any other sector.

5. An increase in demand for products in Sector 7 (real estate rent) would have the least impact on the mining sector since the (2, 7) entry is the smallest in the 2nd row.

7. Sum up each row of the input-output matrix and divide by the total output for that sector to calculate the percentage of output used for internal consumption. The smallest percentage is in the public services sector, $\dfrac{2448}{132,752} \approx 1.8\%$. Thus it has the greatest percentage available for external consumption.

Chapter 3 Review Exercises
Pages 235–237

1. $A = \begin{bmatrix} 1 & 2 & 3 \\ 6 & 5 & 4 \end{bmatrix}$

dimensions: 2×3

$A^T = \begin{bmatrix} 1 & 6 \\ 2 & 5 \\ 3 & 4 \end{bmatrix}$

3. $C = \begin{bmatrix} 1 & 3 \\ 5 & 7 \\ 9 & 11 \end{bmatrix}$

dimensions: 3×2

$C^T = \begin{bmatrix} 1 & 5 & 9 \\ 3 & 7 & 11 \end{bmatrix}$

5. $E = \begin{bmatrix} 1 & -1 \\ -1 & 2 \end{bmatrix}$

dimensions: 2×2

$E^T = \begin{bmatrix} 1 & -1 \\ -1 & 2 \end{bmatrix}$

7. $G = \begin{bmatrix} 1 & -1 & 2 & -2 \\ -1 & 1 & -2 & 2 \end{bmatrix}$

dimensions: 2×4

$G^T = \begin{bmatrix} 1 & -1 \\ -1 & 1 \\ 2 & -2 \\ -2 & 2 \end{bmatrix}$

9. $A + B$ is undefined because A and B have different dimensions. A is a 2×3 matrix and B is a 2×2 matrix.

11. $A - D = \begin{bmatrix} 1 & 2 & 3 \\ 4 & 5 & 6 \end{bmatrix} - \begin{bmatrix} -3 & -2 & -1 \\ 1 & 2 & 3 \end{bmatrix} = \begin{bmatrix} 4 & 4 & 4 \\ 3 & 3 & 3 \end{bmatrix}$

13. $2A^T + C = 2\begin{bmatrix} 1 & 4 \\ 2 & 5 \\ 3 & 6 \end{bmatrix} + \begin{bmatrix} -1 & 0 \\ 1 & 1 \\ 0 & 1 \end{bmatrix} = \begin{bmatrix} 1 & 8 \\ 5 & 11 \\ 6 & 13 \end{bmatrix}$

15. AB is undefined because the column dimension of A does not equal the row dimension of B. A is a 2×3 matrix and B is a 2×2 matrix.

$(2 \times 3)(2 \times 2)$
$\quad\uparrow\quad\uparrow$
do not match

17. $A^T B = \begin{bmatrix} 1 & 4 \\ 2 & 5 \\ 3 & 6 \end{bmatrix} \begin{bmatrix} 1 & -1 \\ 0 & 1 \end{bmatrix} = \begin{bmatrix} 1+0 & -1+4 \\ 2+0 & -2+5 \\ 3+0 & -3+6 \end{bmatrix} = \begin{bmatrix} 1 & 3 \\ 2 & 3 \\ 3 & 3 \end{bmatrix}$

19. BC is undefined because the column dimension of B (2) does not match the row dimension of C (3).
B: 2×2
C: 3×2

21. $BC^T = \begin{bmatrix} 1 & -1 \\ 0 & 1 \end{bmatrix} \begin{bmatrix} -1 & 1 & 0 \\ 0 & 1 & 1 \end{bmatrix} = \begin{bmatrix} -1 & 0 & -1 \\ 0 & 1 & 1 \end{bmatrix}$

23. $A^2 = A \cdot A$ is undefined because the column dimension of A (3) does not match the row dimension of A (2).
A: 2×3

25. $AA^T = \begin{bmatrix} 1 & 2 & 3 \\ 4 & 5 & 6 \end{bmatrix} \begin{bmatrix} 1 & 4 \\ 2 & 5 \\ 3 & 6 \end{bmatrix}$

$= \begin{bmatrix} 1+4+9 & 4+10+18 \\ 4+10+18 & 16+25+36 \end{bmatrix} = \begin{bmatrix} 14 & 32 \\ 32 & 77 \end{bmatrix}$

27. $BB^T = \begin{bmatrix} 1 & -1 \\ 0 & 1 \end{bmatrix} \begin{bmatrix} 1 & 0 \\ -1 & 1 \end{bmatrix} = \begin{bmatrix} 2 & -1 \\ -1 & 1 \end{bmatrix}$

29. $A = \begin{bmatrix} 1 & -1 \\ 0 & 1 \end{bmatrix}$

$\begin{bmatrix} 1 & -1 & | & 1 & 0 \\ 0 & 1 & | & 0 & 1 \end{bmatrix} \begin{matrix} R_1 + R_2 \rightarrow \\ \, \end{matrix} \begin{bmatrix} 1 & 0 & | & 1 & 1 \\ 0 & 1 & | & 0 & 1 \end{bmatrix}$

$A^{-1} = \begin{bmatrix} 1 & 1 \\ 0 & 1 \end{bmatrix}$ or $A^{-1} = \dfrac{1}{1+0}\begin{bmatrix} 1 & 1 \\ 0 & 1 \end{bmatrix} = \begin{bmatrix} 1 & 1 \\ 0 & 1 \end{bmatrix}$

31. $\begin{bmatrix} 1 & 2 \\ 0 & 0 \end{bmatrix} \leftarrow$ A row of zeros.

No inverse exists.

$\begin{bmatrix} 1 & 2 \\ 0 & 0 \end{bmatrix}$ is a singular matrix.

33. $A = \begin{bmatrix} 1 & 2 & 3 \\ 0 & 4 & 1 \\ 0 & 0 & 1 \end{bmatrix}$

$\left[\begin{array}{ccc|ccc} 1 & 2 & 3 & 1 & 0 & 0 \\ 0 & 4 & 1 & 0 & 1 & 0 \\ 0 & 0 & 1 & 0 & 0 & 1 \end{array}\right] \begin{array}{l} 2R_1 - R_2 \rightarrow \end{array} \left[\begin{array}{ccc|ccc} 2 & 0 & 5 & 2 & -1 & 0 \\ 0 & 4 & 1 & 0 & 1 & 0 \\ 0 & 0 & 1 & 0 & 0 & 1 \end{array}\right] \begin{array}{l} R_1 - 5R_3 \rightarrow \\ R_2 - R_3 \rightarrow \end{array}$

$\left[\begin{array}{ccc|ccc} 2 & 0 & 0 & 2 & -1 & -5 \\ 0 & 4 & 0 & 0 & 1 & -1 \\ 0 & 0 & 1 & 0 & 0 & 1 \end{array}\right] \begin{array}{l} \frac{1}{2}R_1 \rightarrow \\ \frac{1}{4}R_2 \rightarrow \end{array} \left[\begin{array}{ccc|ccc} 1 & 0 & 0 & 1 & -\frac{1}{2} & -\frac{5}{2} \\ 0 & 1 & 0 & 0 & \frac{1}{4} & -\frac{1}{4} \\ 0 & 0 & 1 & 0 & 0 & 1 \end{array}\right]$

So, $A^{-1} = \begin{bmatrix} 1 & -\frac{1}{2} & -\frac{5}{2} \\ 0 & \frac{1}{4} & -\frac{1}{4} \\ 0 & 0 & 1 \end{bmatrix}$

35. $A = \begin{bmatrix} 1 & 2 & 3 \\ 1 & 3 & 2 \\ 1 & 1 & 4 \end{bmatrix}$

$\left[\begin{array}{ccc|ccc} 1 & 2 & 3 & 1 & 0 & 0 \\ 1 & 3 & 2 & 0 & 1 & 0 \\ 1 & 1 & 4 & 0 & 0 & 1 \end{array}\right] \begin{array}{l} R_2 - R_1 \rightarrow \\ R_3 - R_1 \rightarrow \end{array} \left[\begin{array}{ccc|ccc} 1 & 2 & 3 & 1 & 0 & 0 \\ 0 & 1 & -1 & -1 & 1 & 0 \\ 0 & -1 & 1 & -1 & 0 & 1 \end{array}\right] \begin{array}{l} R_1 - 2R_2 \rightarrow \\ R_3 + R_2 \rightarrow \end{array}$

$\left[\begin{array}{ccc|ccc} 1 & 0 & 5 & 3 & -2 & 0 \\ 0 & 1 & -1 & -1 & 1 & 0 \\ 0 & 0 & 0 & -2 & 1 & 1 \end{array}\right] \leftarrow$ A row of zeros

No inverse exists.

$\begin{bmatrix} 1 & 2 & 3 \\ 1 & 3 & 2 \\ 1 & 1 & 4 \end{bmatrix}$ is a singular matrix.

37. $A = \begin{bmatrix} 1 & 2 & 3 & 4 \\ 0 & 1 & 2 & 3 \\ 0 & 0 & 1 & 2 \\ 0 & 0 & 0 & 1 \end{bmatrix}$

$\begin{bmatrix} 1 & 2 & 3 & 4 & | & 1 & 0 & 0 & 0 \\ 0 & 1 & 2 & 3 & | & 0 & 1 & 0 & 0 \\ 0 & 0 & 1 & 2 & | & 0 & 0 & 1 & 0 \\ 0 & 0 & 0 & 1 & | & 0 & 0 & 0 & 1 \end{bmatrix} \begin{matrix} R_1 - 2R_2 \to \\ \\ \\ \\ \end{matrix} \begin{bmatrix} 1 & 0 & -1 & -2 & | & 1 & -2 & 0 & 0 \\ 0 & 1 & 2 & 3 & | & 0 & 1 & 0 & 0 \\ 0 & 0 & 1 & 2 & | & 0 & 0 & 1 & 0 \\ 0 & 0 & 0 & 1 & | & 0 & 0 & 0 & 1 \end{bmatrix} \begin{matrix} R_1 + R_3 \to \\ R_2 - 2R_3 \to \\ \\ \\ \end{matrix}$

$\begin{bmatrix} 1 & 0 & 0 & 0 & | & 1 & -2 & 1 & 0 \\ 0 & 1 & 0 & -1 & | & 0 & 1 & -2 & 0 \\ 0 & 0 & 1 & 2 & | & 0 & 0 & 1 & 0 \\ 0 & 0 & 0 & 1 & | & 0 & 0 & 0 & 1 \end{bmatrix} \begin{matrix} \\ R_2 + R_4 \to \\ R_3 - 2R_4 \to \\ \\ \end{matrix} \begin{bmatrix} 1 & 0 & 0 & 0 & | & 1 & -2 & 1 & 0 \\ 0 & 1 & 0 & 0 & | & 0 & 1 & -2 & 1 \\ 0 & 0 & 1 & 0 & | & 0 & 0 & 1 & -2 \\ 0 & 0 & 0 & 1 & | & 0 & 0 & 0 & 1 \end{bmatrix}$

So, $A^{-1} = \begin{bmatrix} 1 & -2 & 1 & 0 \\ 0 & 1 & -2 & 1 \\ 0 & 0 & 1 & -2 \\ 0 & 0 & 0 & 1 \end{bmatrix}$

39. $A = \begin{bmatrix} 1 & 2 & 3 & 4 \\ 2 & 3 & 3 & 3 \\ 0 & 1 & 2 & 3 \\ 0 & 0 & 1 & 2 \end{bmatrix}$

$\begin{bmatrix} 1 & 2 & 3 & 4 & | & 1 & 0 & 0 & 0 \\ 2 & 3 & 3 & 3 & | & 0 & 1 & 0 & 0 \\ 0 & 1 & 2 & 3 & | & 0 & 0 & 1 & 0 \\ 0 & 0 & 1 & 2 & | & 0 & 0 & 0 & 1 \end{bmatrix} \begin{matrix} \\ R_2 - 2R_1 \to \\ \\ \\ \end{matrix} \begin{bmatrix} 1 & 2 & 3 & 4 & | & 1 & 0 & 0 & 0 \\ 0 & -1 & -3 & -5 & | & -2 & 1 & 0 & 0 \\ 0 & 1 & 2 & 3 & | & 0 & 0 & 1 & 0 \\ 0 & 0 & 1 & 2 & | & 0 & 0 & 0 & 1 \end{bmatrix} \begin{matrix} R_1 + 2R_2 \to \\ \\ R_3 + R_2 \to \\ \\ \end{matrix}$

$\begin{bmatrix} 1 & 0 & -3 & -6 & | & -3 & 2 & 0 & 0 \\ 0 & -1 & -3 & -5 & | & -2 & 1 & 0 & 0 \\ 0 & 0 & -1 & -2 & | & -2 & 1 & 1 & 0 \\ 0 & 0 & 1 & 2 & | & 0 & 0 & 0 & 1 \end{bmatrix} \begin{matrix} R_1 - 3R_3 \to \\ R_2 - 3R_3 \to \\ \\ R_4 + R_3 \to \end{matrix} \begin{bmatrix} 1 & 0 & 0 & 0 & | & 3 & -1 & -3 & 0 \\ 0 & -1 & 0 & 1 & | & 4 & -2 & -3 & 0 \\ 0 & 0 & -1 & -2 & | & -2 & 1 & 1 & 0 \\ 0 & 0 & 0 & 0 & | & -2 & 1 & 1 & 1 \end{bmatrix} \leftarrow \text{A row of zeros}$

No inverse exists.

$\begin{bmatrix} 1 & 2 & 3 & 4 \\ 2 & 3 & 3 & 3 \\ 0 & 1 & 2 & 3 \\ 0 & 0 & 1 & 2 \end{bmatrix}$ is a singular matrix.

41. $x + 2y = 0$
$3x + 4y = 2$

$$A = \begin{bmatrix} 1 & 2 \\ 3 & 4 \end{bmatrix}, \quad X = \begin{bmatrix} x \\ y \end{bmatrix}, \quad B = \begin{bmatrix} 0 \\ 2 \end{bmatrix}$$

$AX = B$

$$\begin{bmatrix} 1 & 2 \\ 3 & 4 \end{bmatrix} \begin{bmatrix} x \\ y \end{bmatrix} = \begin{bmatrix} 0 \\ 2 \end{bmatrix}$$

$$A^{-1} = \begin{bmatrix} -2 & 1 \\ \frac{3}{2} & -\frac{1}{2} \end{bmatrix}$$

$$X = A^{-1}B = \begin{bmatrix} -2 & 1 \\ \frac{3}{2} & -\frac{1}{2} \end{bmatrix} \begin{bmatrix} 0 \\ 2 \end{bmatrix} = \begin{bmatrix} 2 \\ -1 \end{bmatrix} = \begin{bmatrix} x \\ y \end{bmatrix}$$

So $x = 2$, $y = -1$ is the solution.

43. $x + y + z = 3$
$y + 2z = 4$
$y - z = 1$

$$\begin{bmatrix} 1 & 1 & 1 \\ 0 & 1 & 2 \\ 0 & 1 & -1 \end{bmatrix} \begin{bmatrix} x \\ y \\ z \end{bmatrix} = \begin{bmatrix} 3 \\ 4 \\ 1 \end{bmatrix}$$

$$A^{-1} = \begin{bmatrix} 1 & -\frac{2}{3} & -\frac{1}{3} \\ 0 & \frac{1}{3} & \frac{2}{3} \\ 0 & \frac{1}{3} & -\frac{1}{3} \end{bmatrix}$$

$$X = A^{-1}B = \begin{bmatrix} 1 & -\frac{2}{3} & -\frac{1}{3} \\ 0 & \frac{1}{3} & \frac{2}{3} \\ 0 & \frac{1}{3} & -\frac{1}{3} \end{bmatrix} \begin{bmatrix} 3 \\ 4 \\ 1 \end{bmatrix} = \begin{bmatrix} 0 \\ 2 \\ 1 \end{bmatrix} = \begin{bmatrix} x \\ y \\ z \end{bmatrix}$$

So $x = 0$, $y = 2$, $z = 1$ is the solution.

45. $x + y + z = 2$
$x + 2y + z = 3$
$x + y + 2z = 1$

$$\begin{bmatrix} 1 & 1 & 1 \\ 1 & 2 & 1 \\ 1 & 1 & 2 \end{bmatrix} \begin{bmatrix} x \\ y \\ z \end{bmatrix} = \begin{bmatrix} 2 \\ 3 \\ 1 \end{bmatrix}$$

$$A^{-1} = \begin{bmatrix} 3 & -1 & -1 \\ -1 & 1 & 0 \\ -1 & 0 & 1 \end{bmatrix}$$

$$X = A^{-1}B = \begin{bmatrix} 3 & -1 & -1 \\ -1 & 1 & 0 \\ -1 & 0 & 1 \end{bmatrix} \begin{bmatrix} 2 \\ 3 \\ 1 \end{bmatrix} = \begin{bmatrix} 2 \\ 1 \\ -1 \end{bmatrix} = \begin{bmatrix} x \\ y \\ z \end{bmatrix}$$

So $x = 2$, $y = 1$, $z = -1$ is the solution.

47. $x + y = 0$
$y + z = 1$
$z + w = 0$
$x - w = 3$

$$\begin{bmatrix} 1 & 1 & 0 & 0 \\ 0 & 1 & 1 & 0 \\ 0 & 0 & 1 & 1 \\ 1 & 0 & 0 & -1 \end{bmatrix} \begin{bmatrix} x \\ y \\ z \\ w \end{bmatrix} = \begin{bmatrix} 0 \\ 1 \\ 0 \\ 3 \end{bmatrix}$$

$$A^{-1} = \begin{bmatrix} \frac{1}{2} & -\frac{1}{2} & \frac{1}{2} & \frac{1}{2} \\ \frac{1}{2} & \frac{1}{2} & -\frac{1}{2} & -\frac{1}{2} \\ -\frac{1}{2} & \frac{1}{2} & \frac{1}{2} & \frac{1}{2} \\ \frac{1}{2} & -\frac{1}{2} & \frac{1}{2} & -\frac{1}{2} \end{bmatrix}$$

$$X = A^{-1}B = \begin{bmatrix} \frac{1}{2} & -\frac{1}{2} & \frac{1}{2} & \frac{1}{2} \\ \frac{1}{2} & \frac{1}{2} & -\frac{1}{2} & -\frac{1}{2} \\ -\frac{1}{2} & \frac{1}{2} & \frac{1}{2} & \frac{1}{2} \\ \frac{1}{2} & -\frac{1}{2} & \frac{1}{2} & -\frac{1}{2} \end{bmatrix} \begin{bmatrix} 0 \\ 1 \\ 0 \\ 3 \end{bmatrix} = \begin{bmatrix} 1 \\ -1 \\ 2 \\ -2 \end{bmatrix} = \begin{bmatrix} x \\ y \\ z \\ w \end{bmatrix}$$

So $x = 1$, $y = -1$, $z = 2$, $w = -2$ is the solution.

49. $x + 2y + 3z + 4w = 0$
$x + 3y + 4z + 2w = 3$
$y + 2z + 3w = -1$
$z + 2w = -1$

$$\begin{bmatrix} 1 & 2 & 3 & 4 \\ 1 & 3 & 4 & 2 \\ 0 & 1 & 2 & 3 \\ 0 & 0 & 1 & 2 \end{bmatrix} \begin{bmatrix} x \\ y \\ z \\ w \end{bmatrix} = \begin{bmatrix} 0 \\ 3 \\ -1 \\ -1 \end{bmatrix}$$

$$A^{-1} = \begin{bmatrix} 1 & 0 & -2 & 1 \\ \frac{1}{3} & -\frac{1}{3} & \frac{4}{3} & -\frac{7}{3} \\ -\frac{2}{3} & \frac{2}{3} & -\frac{2}{3} & \frac{5}{3} \\ \frac{1}{3} & -\frac{1}{3} & \frac{1}{3} & -\frac{1}{3} \end{bmatrix}$$

$$X = A^{-1}B = \begin{bmatrix} 1 & 0 & -2 & 1 \\ \frac{1}{3} & -\frac{1}{3} & \frac{4}{3} & -\frac{7}{3} \\ -\frac{2}{3} & \frac{2}{3} & -\frac{2}{3} & \frac{5}{3} \\ \frac{1}{3} & -\frac{1}{3} & \frac{1}{3} & -\frac{1}{3} \end{bmatrix} \begin{bmatrix} 0 \\ 3 \\ -1 \\ -1 \end{bmatrix} = \begin{bmatrix} 1 \\ 0 \\ 1 \\ -1 \end{bmatrix} = \begin{bmatrix} x \\ y \\ z \\ w \end{bmatrix}$$

So $x = 1$, $y = 0$, $z = 1$, $w = -1$ is the solution.

51. $A = \begin{bmatrix} \frac{63,080}{112,120} & \frac{4956}{143,955} \\ \frac{73}{112,120} & \frac{24,475}{143,955} \end{bmatrix}$

$I - A = \begin{bmatrix} \frac{49,040}{112,120} & -\frac{4956}{143,955} \\ -\frac{73}{112,120} & \frac{119,480}{143,955} \end{bmatrix}$

$(I - A)^{-1} \approx \begin{bmatrix} 2.286438079 & 0.0948408698 \\ 0.0017936213 & 1.204920398 \end{bmatrix}$

Using an increase in demand of DM 10,000 million for each sector, we have:

Sector 1, $(I - A)^{-1} D \approx \begin{bmatrix} 22,864.38079 \\ 17.936213 \end{bmatrix}$;

Sector 2, $(I - A)^{-1} D \approx \begin{bmatrix} 948.408698 \\ 12,049.20398 \end{bmatrix}$.

53. $A = \begin{bmatrix} \frac{1377}{11,172} & \frac{58}{112,120} & \frac{886}{143,955} & \frac{3}{13,569} \\ \frac{2}{11,172} & \frac{63,080}{112,120} & \frac{4956}{143,955} & \frac{77}{13,569} \\ \frac{72}{11,172} & \frac{73}{112,120} & \frac{24,475}{143,955} & \frac{3}{13,569} \\ 0 & 0 & 0 & \frac{1604}{13,569} \end{bmatrix}$

$I - A = \begin{bmatrix} \frac{9795}{11,172} & -\frac{58}{112,120} & -\frac{886}{143,955} & -\frac{3}{13,569} \\ -\frac{2}{11,172} & \frac{49,040}{112,120} & -\frac{4956}{143,955} & -\frac{77}{13,569} \\ -\frac{72}{11,172} & -\frac{73}{112,120} & \frac{119,480}{143,955} & -\frac{3}{13,569} \\ 0 & 0 & 0 & \frac{11,965}{13,569} \end{bmatrix}$

$(I - A)^{-1} \approx \begin{bmatrix} 1.140644798 & 0.0013617248 & 0.0085148979 & 0.0002968936 \\ 0.0011640681 & 2.286439468 & 0.0948495595 & 0.0147383101 \\ 0.0088578497 & 0.001804196 & 1.204986522 & 0.0003159596 \\ 0 & 0 & 0 & 1.134057668 \end{bmatrix}$

Using an increase in demand of DM 10,000 million for each sector, we have:

Sector 1, $(I - A)^{-1} D \approx \begin{bmatrix} 11,406.44798 \\ 11.640681 \\ 88.578497 \\ 0 \end{bmatrix}$; Sector 2, $(I - A)^{-1} D \approx \begin{bmatrix} 13.617248 \\ 22,864.39468 \\ 18.04196 \\ 0 \end{bmatrix}$; Sector 3, $(I - A)^{-1} D \approx \begin{bmatrix} 85.148979 \\ 948.495595 \\ 12,049.86522 \\ 0 \end{bmatrix}$;

Sector 4, $(I - A)^{-1} D \approx \begin{bmatrix} 2.968936 \\ 147.383101 \\ 3.159596 \\ 11,340.57668 \end{bmatrix}$.

Chapter 4

1. $2x + y \le 10$

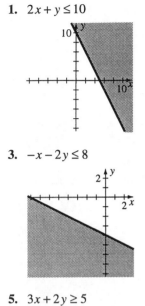

3. $-x - 2y \le 8$

5. $3x + 2y \ge 5$

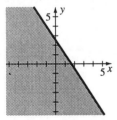

7. $x \le 3y$

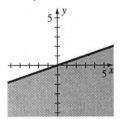

9. $\frac{3}{4}x - \frac{1}{4}y \le 1$

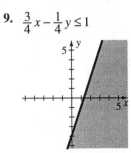

11. $x \ge -5$

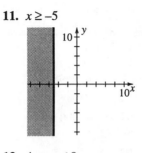

13. $4x - y \le 8$
   $x + 2y \le 2$

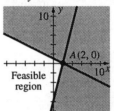

Point	Lines Through Point	Coordinates
A	$4x - y = 8$ $x + 2y = 2$	$(2, 0)$

Corner point: $(2, 0)$

15. $3x + 2y \ge 6$
   $3x - 2y \le 6$
   $x \ge 0$

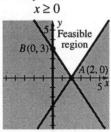

Point	Lines Through Point	Coordinates
A	$3x + 2y = 6$ $3x - 2y = 6$	$(2, 0)$
B	$3x + 2y = 6$ $x = 0$	$(0, 3)$

Corner points: $(2, 0)$, $(0, 3)$

17. $x + y \geq 5$
     $x \leq 10$
     $y \leq 8$
     $x \geq 0, \quad y \geq 0$

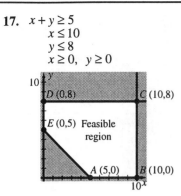

Point	Lines Through Point	Coordinates
A	$x + y = 5, y = 0$	$(5, 0)$
B	$x = 10, y = 0$	$(10, 0)$
C	$x = 10, y = 8$	$(10, 8)$
D	$x = 0, y = 8$	$(0, 8)$
E	$x + y = 5, x = 0$	$(0, 5)$

Corner points: $(5, 0)$, $(10, 0)$, $(10, 8)$, $(0, 8)$, $(0, 5)$

19. $20x + 10y \leq 100$
     $10x + 20y \leq 100$
     $10x + 10y \leq 60$
     $x \geq 0, \quad y \geq 0$

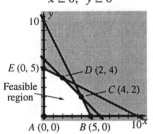

Point	Lines Through Point	Coordinates
A	$x = 0, y = 0$	$(0, 0)$
B	$20x + 10y = 100$ $y = 0$	$(5, 0)$
C	$20x + 10y = 100$ $10x + 10y = 60$	$(4, 2)$
D	$10x + 20y = 100$ $10x + 10y = 60$	$(2, 4)$
E	$10x + 20y = 100$ $x = 0$	$(0, 5)$

Corner points: $(0, 0)$, $(5, 0)$, $(4, 2)$, $(2, 4)$, $(0, 5)$

21. $20x + 10y \geq 100$
     $10x + 20y \geq 100$
     $10x + 10y \geq 80$
     $x \geq 0, \quad y \geq 0$

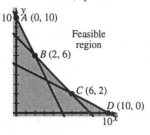

Point	Lines Through Point	Coordinates
A	$20x + 10y = 100$ $x = 0$	$(0, 10)$
B	$20x + 10y = 100$ $10x + 10y = 80$	$(2, 6)$
C	$10x + 20y = 100$ $10x + 10y = 80$	$(6, 2)$
D	$10x + 20y = 100$ $y = 0$	$(10, 0)$

Corner points: $(0, 10)$, $(2, 6)$, $(6, 2)$, $(10, 0)$

23. $-3x + 2y \leq 5$
     $3x - 2y \leq 6$
     $x \leq 2y$
     $x \geq 0, \quad y \geq 0$

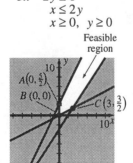

Point	Lines Through Point	Coordinates
A	$-3x + 2y = 5$ $x = 0$	$\left(0, \frac{5}{2}\right)$
B	$x = 2y$ $x = 0$	$(0, 0)$
C	$3x - 2y = 6$ $x = 2y$	$\left(3, \frac{3}{2}\right)$

Corner points: $\left(0, \frac{5}{2}\right)$, $(0, 0)$, $\left(3, \frac{3}{2}\right)$

25. $2x - y \geq 0$
$x - 3y \leq 0$
$x \geq 0, \quad y \geq 0$

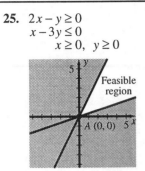

Point	Lines Through Point	Coordinates
A	$2x - y = 0$ $x - 3y = 0$	$(0, 0)$

27. $2.1x - 4.3y \geq 9.7$

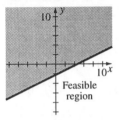

29. $-0.2x + 0.7y \geq 3.3$
$1.1x + 3.4y \geq 0$

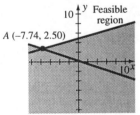

Point	Lines Through Point	Coordinates
A	$-0.2x + 0.7y = 3.3$ $1.1x + 3.4y = 0$	$(-7.74, 2.50)$

Corner point: $(-7.74, 2.50)$

31. $4.1x - 4.3y \leq 4.4$
$7.5x - 4.4y \leq 5.7$
$4.3x + 8.5y \leq 10$

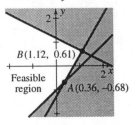

Point	Lines Through Point	Coordinates
A	$4.1x - 4.3y = 4.4$ $7.5x - 4.4y = 5.7$	$(0.36, -0.68)$
B	$7.5x - 4.4y = 5.7$ $4.3x + 8.5y = 10$	$(1.12, 0.61)$

Corner points: $(0.36, -0.68)$, $(1.12, 0.61)$

33. Let $x =$ number of quarts of Creamy Vanilla
and $y =$ number of quarts of Continental Mocha

	Vanilla	Mocha	total
eggs	2	1	≤ 500
cream (cups)	3	3	≤ 900

Restrictions: $2x + y \leq 500$
$3x + 3y \leq 900$
$x \geq 0, \quad y \geq 0$

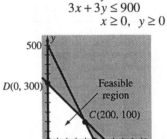

Point	Lines Through Point	Coordinates
A	$x = 0, y = 0$	$(0, 0)$
B	$y = 0$ $2x + y = 500$	$(250, 0)$
C	$2x + y = 500$ $x + y = 300$	$(200, 100)$
D	$x = 0$ $x + y = 300$	$(0, 300)$

Corner points: $(0, 0)$, $(250, 0)$, $(200, 100)$, $(0, 300)$

35. Let x = number of ounces of chicken
and y = number of ounces of grain

	chicken	grain	total
protein (grams/ounce)	10	2	≥ 200
fat (grams/ounce)	5	2	≥ 150

Restrictions: $10x + 2y \geq 200$
$5x + 2y \geq 150$
$x \geq 0, \ y \geq 0$

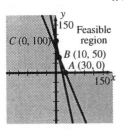

Point	Lines Through Point	Coordinates
A	$y = 0$ $5x + 2y = 150$	$(30, 0)$
B	$10x + 2y = 200$ $5x + 2y = 150$	$(10, 50)$
C	$x = 0$ $10x + 2y = 200$	$(0, 100)$

Corner points: $(30, 0)$, $(10, 50)$, $(0, 100)$

37. Let x = number of servings of Mixed Cereal for Baby
and y = number of servings of Mango Tropical Fruit
Dessert

	Mixed Cereal	Mango	total
calories	60	80	≥ 140
carbohydrates (grams)	11	21	≥ 32

Restrictions: $60x + 80y \geq 140$
$11x + 21y \geq 32$
$x \geq 0, \ y \geq 0$

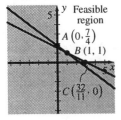

Point	Lines Through Point	Coordinates
A	$x = 0$ $60x + 80y = 140$	$\left(0, \frac{7}{4}\right)$
B	$60x + 80y = 140$ $11x + 21y = 32$	$(1, 1)$
C	$y = 0$ $11x + 21y = 32$	$\left(\frac{32}{11}, \ 0\right)$

Corner points: $\left(0, \frac{7}{4}\right)$, $(1, \ 1)$, $\left(\frac{32}{11}, \ 0\right)$

39. Let x = number of units of Muniyield Fund
and y = number of units of Municipal Bond Fund
Class B.

	Muniyield	Class B	total
cost ($)	15	12	$\leq 42,000$
yield ($)	15(0.06)	12(0.05)	≥ 2400

$15x + 12y \leq 42,000$
Restrictions: $0.9x + 0.6y \geq 2400$
$x \geq 0, \ y \geq 0$

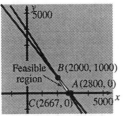

Point	Lines Through Point	Coordinates
A	$y = 0$ $15x + 12y = 42,000$	$(2800, 0)$
B	$15x + 12y = 42,000$ $0.9x + 0.6y = 2400$	$(2000, 1000)$
C	$0.9x + 0.6y = 2400$ $y = 0$	$(2667, 0)$

Corner points: $(2800, 0)$, $(2000, 1000)$, $(2667, 0)$

41. Systems will vary. For example: $x \geq 0, \ y \geq 0$. (The solution set extends infinitely for increasing positive x and y.)

43. Scenarios will vary.

45. First write equations describing the edges.

$x = 0$ describes the line containing $(0, 0)$ and $(0, 1)$.

$y = 0$ describes the line containing $(0, 0)$, and $(2, 0)$.

$x + 2y = 2$ describes the line containing $(0, 1)$ and $(2, 0)$.

The triangle can be described by:

$x \geq 0$

$y \geq 0$

$x + 2y \leq 2$

47. There is no number of first-class and business-class tickets that the airline can sell that satisfies the constraints.

Section 4.2 Solving Linear Programming Problems Graphically

Pages 270–275

1. Maximize $p = x + y$

subject to $x + 2y \leq 9$

$2x + y \leq 9$

$x \geq 0, \ y \geq 0$

We begin by drawing the feasible region for the problem, which is the set of points representing solutions to the constraints.

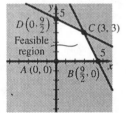

We find the coordinates of each corner point and compute the value of the objective function at each corner.

Point	Lines Through Point	Coordinates	$p = x + y$
A	$x = 0, y = 0$	$(0, 0)$	0
B	$2x + y = 9$ $y = 0$	$\left(\frac{9}{2}, 0\right)$	$\frac{9}{2}$
C	$x + 2y = 9$ $2x + y = 9$	$(3, 3)$	6
D	$x + 2y = 9$ $x = 0$	$\left(0, \frac{9}{2}\right)$	$\frac{9}{2}$

The maximum value is $p = 6$ at $x = 3, y = 3$.

3. Minimize $c = x + y$

subject to $x + 2y \geq 6$

$2x + y \geq 6$

$x \geq 0, \ y \geq 0$

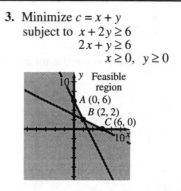

Point	Lines Through Point	Coordinates	$c = x + y$
A	$x = 0$ $2x + y = 6$	$(0, 6)$	6
B	$x + 2y = 6$ $2x + y = 6$	$(2, 2)$	4
C	$y = 0$ $x + 2y = 6$	$(6, 0)$	6

The minimum value is $c = 4$ at $x = 2, y = 2$.

5. Maximize $p = 3x + y$

subject to $3x - 7y \leq 0$

$7x - 3y \geq 0$

$x + y \leq 10$

$x \geq 0, \ y \geq 0$

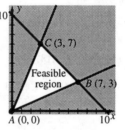

Point	Lines Through Point	Coordinates	$p = 3x + y$
A	$3x - 7y = 0$ $7x - 3y = 0$	$(0, 0)$	0
B	$3x - 7y = 0$ $x + y = 10$	$(7, 3)$	24
C	$7x - 3y = 0$ $x + y = 10$	$(3, 7)$	16

The maximum value is $p = 24$ at $x = 7, y = 3$.

7. Maximize $p = 3x + 2y$
  subject to $20x + 10y \leq 100$
  $\qquad\qquad 10x + 20y \leq 100$
  $\qquad\qquad 10x + 10y \leq 60$
  $\qquad\qquad x \geq 0, \ y \geq 0$

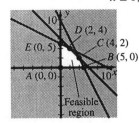

Point	Lines Through Point	Coordinates	$p = 3x + 2y$
A	$x = 0, y = 0$	$(0, 0)$	0
B	$y = 0$ $20x + 10y = 100$	$(5, 0)$	15
C	$20x + 10y = 100$ $10x + 10y = 60$	$(4, 2)$	16
D	$10x + 10y = 60$ $10x + 20y = 100$	$(2, 4)$	14
E	$x = 0$ $10x + 20y = 100$	$(0, 5)$	10

The maximum value is $p = 16$ at $x = 4$, $y = 2$.

9. Minimize $c = 2x + 3y$
  subject to $20x + 10y \geq 100$
  $\qquad\qquad 10x + 20y \geq 100$
  $\qquad\qquad 10x + 10y \geq 80$
  $\qquad\qquad x \geq 0, \ y \geq 0$

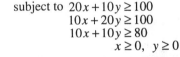
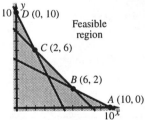

Point	Lines Through Point	Coordinates	$c = 2x + 3y$
A	$y = 0$ $10x + 20y = 100$	$(10, 0)$	20
B	$10x + 20y = 100$ $10x + 10y = 80$	$(6, 2)$	18
C	$20x + 10y = 100$ $10x + 10y = 80$	$(2, 6)$	22
D	$x = 0$ $20x + 10y = 100$	$(0, 10)$	30

The minimum value is $c = 18$ at $x = 6$, $y = 2$.

11. Maximize and minimize $p = x + 2y$
  subject to $x + y \geq 2$
  $\qquad\qquad x + y \leq 10$
  $\qquad\qquad x - y \leq 2$
  $\qquad\qquad x - y \geq -2$

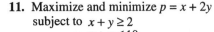
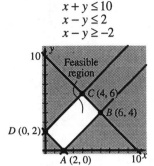

Point	Lines Through Point	Coordinates	$p = x + 2y$
A	$x + y = 2$ $x - y = 2$	$(2, 0)$	2
B	$x + y = 10$ $x - y = 2$	$(6, 4)$	14
C	$x + y = 10$ $x - y = -2$	$(4, 6)$	16
D	$x + y = 2$ $x - y = -2$	$(0, 2)$	4

The maximum value is $p = 16$ at $x = 4$, $y = 6$.
The minimum value is $p = 2$ at $x = 2$, $y = 0$.

13. Maximize $p = 2x + 3y$
  subject to $x + 2y \geq 10$
  $\qquad\qquad 2x + y \geq 10$
  $\qquad\qquad x \geq 0, \ y \geq 0$

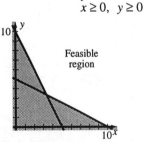

The feasible region is unbounded and p increases as x and y increase. Therefore, there is no solution.

15. Minimize $c = 2x + 4y$
  subject to $x + y \geq 10$
  $\qquad\qquad x + 2y \geq 14$
  $\qquad\qquad x \geq 0, \ y \geq 0$

Point	Lines Through Point	Coordinates	$c = 2x + 4y$
A	$y = 0$ $x + 2y = 14$	$(14, 0)$	28
B	$x + y = 10$ $x + 2y = 14$	$(6, 4)$	28
C	$x + y = 10$ $x = 0$	$(0, 10)$	40

The minimum value is $c = 28$ at $x = 14$, $y = 0$ and $x = 6$, $y = 4$ and the line connecting them.

17. Minimize $c = 3x - 3y$

subject to
$$\frac{x}{4} \le y$$
$$y \le \frac{2x}{3}$$
$$x + y \ge 5$$
$$x + 2y \le 10$$
$$x \ge 0, \quad y \ge 0$$

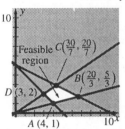

Point	Lines Through Point	Coordinates	$c = 3x - 3y$
A	$y = \dfrac{x}{4}$ $x + y = 5$	$(4, 1)$	9
B	$y = \dfrac{x}{4}$ $x + 2y = 10$	$\left(\dfrac{20}{3}, \dfrac{5}{3}\right)$	15
C	$y = \dfrac{2x}{3}$ $x + 2y = 10$	$\left(\dfrac{30}{7}, \dfrac{20}{7}\right)$	$\dfrac{30}{7}$
D	$y = \dfrac{2x}{3}$ $x + y = 5$	$(3, 2)$	3

The minimum value is $c = 3$ at $x = 3$, $y = 2$.

19. Maximize $p = x + y$

subject to
$$x + 2y \ge 10$$
$$2x + 2y \le 10$$
$$2x + y \ge 10$$
$$x \ge 0, \quad y \ge 0$$

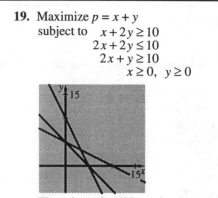

There is no feasible region bounded by the restrictions. Therefore, there is no solution.

21. Let x = number of quarts of Creamy Vanilla and y = number of quarts of Continental Mocha

	Vanilla x	Mocha y	total
eggs	2	1	≤ 500
cream (cups)	3	3	≤ 900

Produce largest profit:
Maximize $P = 3x + 2y$
subject to $2x + y \le 500$
$3x + 3y \le 900$
$x \ge 0, \quad y \ge 0$

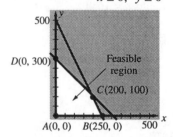

Point	Lines Through Point	Coordinates	$P = 3x + 2y$
A	$x = 0, y = 0$	$(0, 0)$	0
B	$y = 0$ $2x + y = 500$	$(250, 0)$	750
C	$2x + y = 500$ $x + y = 300$	$(200, 100)$	800
D	$x = 0$ $x + y = 300$	$(0, 300)$	600

Maximum: $P = 800$, $x = 200$, $y = 100$
You should make 200 quarts of Creamy Vanilla and 100 quarts of Continental Mocha.

23. Let x = number of ounces of chicken
and y = number of ounces of grain

	chicken x	grain y	total
protein (grams/ounce)	10	2	≥ 200
fat (grams/ounce)	5	2	≥ 150

Minimize cost (cents per ounce):
Minimize $C = 10x + y$
subject to $10x + 2y \geq 200$
$5x + 2y \geq 150$
$x \geq 0, \ y \geq 0$

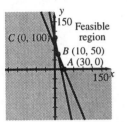

Point	Lines Through Point	Coordinates	$C = 10x + y$
A	$y = 0$ $5x + 2y = 150$	$(30, 0)$	300
B	$10x + 2y = 200$ $5x + 2y = 150$	$(10, 50)$	150
C	$x = 0$ $10x + 2y = 200$	$(0, 100)$	100

Minimum: $C = 100$, $x = 0$, $y = 100$. Ruff, Inc. should use 100 ounces of grain and no chicken.

25. Let x = number of servings of Mixed Cereal for Baby
and y = number of servings of Mango Tropical Fruit Dessert

	Mixed Cereal	Mango	total
calories	60	80	≥ 140
carbohydrates (grams)	11	21	≥ 32

Minimize cost $C = 30x + 50y$
subject to $60x + 80y \geq 140$
$11x + 21y \geq 32$
$x \geq 0, \ y \geq 0$

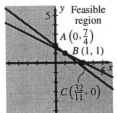

Point	Lines Through Point	Coordinates	$C = 30x + 50y$
A	$x = 0$ $60x + 80y = 140$	$\left(0, \frac{7}{4}\right)$	87.5
B	$60x + 80y = 140$ $11x + 21y = 32$	$(1, 1)$	80
C	$y = 0$ $11x + 21y = 32$	$\left(\frac{32}{11}, 0\right)$	87.3

Minimum: $C = 80$, $x = 1$, $y = 1$. Feed your child 1 serving of cereal and 1 serving of dessert.

27. Let $x =$ number of units of Muniyield Fund
and $y =$ number of units of Municipal Bond Fund Class B.

	Muniyield	Class B	total
cost ($)	15	12	≤ 42000
yield ($)	15(0.06)	12(0.05)	≥ 2400

Minimize $C = 2x + 1.5y$
subject to $15x + 12y \leq 42,000$
$0.9x + 0.6y \geq 2400$
$x \geq 0, \quad y \geq 0$

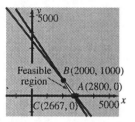

Point	Lines Through Point	Coordinates	$C = 2x + 1.5y$
A	$y = 0$ $15x + 12y = 42,000$	$(2800, 0)$	5600
B	$15x + 12y = 42,000$ $0.9x + 0.6y = 2400$	$(2000, 1000)$	5500
C	$0.9x + 0.6y = 2400$ $y = 0$	$(2666.7, 0)$	5333.4

Minimum: $C = 5333.4$, $x = 2666.7$, $y = 0$. Purchase 2666.7 units of Muniyield and no units of Municipal Bond Fund Class B.

29. Let $x =$ number of Dracula Salami
and $y =$ number of Frankenstein Sausage

	Dracula x	Frankenstein y	total
pork (lb)	1	2	≤ 1000
beef (lb)	3	2	≤ 2400

Maximize $P = x + 3y$
subject to $x + 2y \leq 1000$
$\qquad\qquad 3x + 2y \leq 2400$
$\qquad\qquad\qquad y \leq 2x$
$\qquad\qquad\quad x \geq 0, \ y \geq 0$

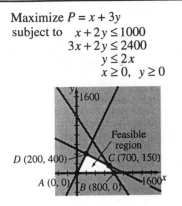

Point	Lines Through Point	Coordinates	$P = x + 3y$
A	$y = 2x$ $x = 0$ $y = 0$	$(0, 0)$	0
B	$y = 0$ $3x + 2y = 2400$	$(800, 0)$	800
C	$x + 2y = 1000$ $3x + 2y = 2400$	$(700, 150)$	1150
D	$x + 2y = 1000$ $y = 2x$	$(200, 400)$	1400

Maximum: $P = 1400$, $x = 200$, $y = 400$.
You should make 200 Dracula Salami and 400 Frankenstein Sausage.

31. Let $x =$ number of spots on wrestling shows
and $y =$ number of spots on baseball shows

	wrestling x	baseball y	total
number of spots	1	1	≥ 30
price per spot	1000	1500	$\leq 40,000$

Also $x \geq \frac{3}{4}(x + y)$

Maximize $V = 2x + 1.5y$ (in millions of viewers)
subject to $x + y \geq 30$
$\qquad\qquad x + 1.5y \leq 40$
$\qquad\qquad -x + 3y \leq 0$
$\qquad\qquad\quad x \geq 0, \ y \geq 0$

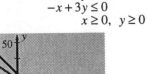
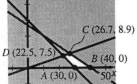

Point	Lines Through Point	Coordinates	$V = 2x + 1.5y$
A	$x + y = 30$ $y = 0$	(30, 0)	60
B	$x + 1.5y = 40$ $y = 0$	(40, 0)	80
C	$x + 1.5y = 40$ $y = \frac{1}{3}x$	(26.7, 8.9)	66.75
D	$x + y = 30$ $y = \frac{1}{3}x$	(22.5, 7.5)	56.25

Maximum: $V = 80$, $x = 40$, $y = 0$. You should purchase 40 spots on wrestling shows and no spots on baseball shows.

33. a. New York:

average annual gasoline sales per gas station $= \dfrac{5.6 \text{ billion gallons}}{7000} = 800,000$ gallons

average annual revenue of a gas station $= 800,000$ gallons ($1.20/gallon) $= \$960,000$

Connecticut:

average annual gasoline sales per gas station $= \dfrac{2 \text{ billion gallons}}{2000} = 1,000,000$ gallons

average annual revenue of a gas station $= 1,000,000$ gallons ($1.20/gallon) $= \$1,200,000$

b. Let $x =$ number of gas stations in New York

and $y =$ number of gas stations in Connecticut

Minimize $T = 0.2(800,000)x + 0.3(1,000,000)y$

or $T = 160,000x + 300,000y$

subject to: $x + y \le 20$

$960,000x + 1,200,000y \ge 20,400,000$

$x \ge 0, \ y \ge 0$

or $x + y \le 20$

$4x + 5y \ge 85$

$x \ge 0, \ y \ge 0$

Point	Lines Through Point	Coordinates	$T = 160,000x + 300,000y$
A	$x = 0$ $4x + 5y = 85$	(0, 17)	5,100,000
B	$4x + 5y = 85$ $x + y = 20$	(15, 5)	3,900,000
C	$x + y = 20$ $x = 0$	(0, 20)	6,000,000

Minimum $T = 3.9$ million, $x = 15$, $y = 5$

Solution: 15 gas stations in New York and 5 in Connecticut

35. a. Each dobbie mill produces $(4.63)(24)(30) = 3333.6$ yards of fabric in one month. Thus the Scottsville Textile Mill produces $8(3333.6) = 26,668.8$ yards of fabric in one month. Since the monthly demand is $16,000 + 12,000 = 28,000$ yards of fabric, it will not be possible to satisfy total demand.

b. Let x = the number of yards of Fabric 1 produced at the Scottsville Textile Mill
and y = the number of yards of Fabric 2 produced at the Scottsville Textile Mill.
$16,000 - x$ is the number of yards of Fabric 1 purchased for resale and $12,000 - y$ is the number of yards of Fabric 2 purchased for resale. Thus the profit is $P = 0.33x + 0.33y + 0.20(16,000 - x) + 0.16(12,000 - y) = 0.13x + 0.17y + 5120$.
Hence we want to maximize $P = 0.13x + 0.17y + 5120$ subject to $x + y \le 26,668.8$
$$x \le 16,000, \quad y \le 12,000$$
$$x \ge 0, \quad y \ge 0.$$

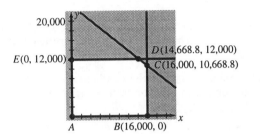

Point	Lines Through Point	Coordinates	$P = 0.13x + 0.17y + 5120$
A	$x = 0, \; y = 0$	$(0, 0)$	5120
B	$x = 16,000$ $y = 0$	$(16,000, 0)$	7200
C	$x + y = 26,668.8$ $x = 16,000$	$(16,000,$ $10,668.8)$	9013.70
D	$x + y = 26,668.8$ $y = 12,000$	$(14,668.8,$ $12,000)$	9066.94
E	$x = 0$ $y = 12,000$	$(0, 12,000)$	7160

The mill should produce 14,668.8 yards of Fabric 1 and 12,000 yards of Fabric 2.

37. Let x = hours of battle instruction per week
and y = hours of diplomacy instruction per week.
Brutus' income will be $P = 50x + 40y$. Since he can spend no more than 50 hours per week, $x + y \le 50$. Since he has decided to spend no more than one-third of the time in diplomatic instruction, $y \le \frac{1}{3}(x + y)$ or $2y \le x$. Since he has decided to spend at least 10 hours per week instructing in diplomacy, $y \ge 10$. Due to the matter of Scarlet Brew, $10x + 5y \ge 400$. Maximize $P = 50x + 40y$ subject to $x + y \le 50$, $2y \le x$, $y \ge 10$, and $10x + 5y \ge 400$.

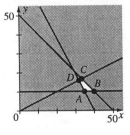

Point	Lines Through Point	Coordinates	$P = 50x + 40y$
A	$10x + 5y = 400$ $y = 10$	$(35, 10)$	2150
B	$x + y = 50$ $y = 10$	$(40, 10)$	2400
C	$x = 2y$ $x + y = 50$	$\left(33\frac{1}{3}, 16\frac{2}{3}\right)$	$2333\frac{1}{3}$
D	$x = 2y$ $10x + 5y = 400$	$(32, 16)$	2240

Brutus should spend 40 hours per week in battle instruction and 10 hours per week in diplomacy instruction.

39. Let x = number of sleep spells and y = number of shock spells. Gillian's expenditure of aural energy is $C = 500x + 750y$. There are $(24)(60) = 1440$ minutes in one day, so $2x + 3y \geq 1440$. In order to immobilize the 1200 trolls, $3x + 2y \geq 1200$. By the By-Laws,

$y \geq \frac{1}{4}(x + y)$ and $y \leq \frac{1}{3}(x + y)$ or $3y \geq x$ and $2y \leq x$.

Minimize $C = 500x + 750y$ subject to $2x + 3y \geq 1440$, $3x + 2y \geq 1200$, $3y \geq x$, and $2y \leq x$.

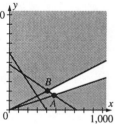

Point	Lines Through Point	Coordinates	$C = 500x + 750y$
A	$x = 3y$ $2x + 3y = 1440$	$(480, 160)$	360,000
B	$x = 2y$ $2x + 3y = 1440$	$\left(\dfrac{2880}{7}, \dfrac{1440}{7}\right)$	360,000

Gillian should use 480 sleep spells and 160 shock spells, costing 360,000 pico-shirleys of energy. (She can also use any combination of x sleep spells and y shock spells that satisfy the equation $2x + 3y = 1440$ and the inequalities $3y \geq x$ and $2y \leq x$. This will also cost her 360,000 pico-shirleys of energy.)

41. Answers will vary. For example, maximize $p = x + y$ subject to $x \geq 0$, $y \geq 0$.

43. Scenarios will vary.

45. Examples will vary.

47. There is more than one configuration of first-class and business-class tickets that can be sold for maximum profit.

Section 4.3 Solving Standard Linear Programming Problems Using the Simplex Method
Pages 292–299

1. Introduce slack variables and rewrite the constraints and objective function in standard form.

$$
\begin{array}{rcrcrcrcrcr}
x & + & 2y & + & s & & & & & = & 6 \\
-x & + & y & & & + & t & & & = & 4 \\
x & + & y & & & & & + & u & = & 4 \\
-2x & - & y & & & & & & + p & = & 0
\end{array}
$$

Set up the initial tableau. Select the x-column as the pivot column. The test ratios are $\frac{6}{1}$ and $\frac{4}{1}$, so select the pivot in the u-row. Set up instructions to clear the column.

	x	y	s	t	u	p		
s	1	2	1	0	0	0	6	$R_1 - R_3$
t	-1	1	0	1	0	0	4	$R_2 + R_3$
u	1	1	0	0	1	0	4	
p	-2	-1	0	0	0	1	0	$R_4 + 2R_3$

This gives the second tableau.

	x	y	s	t	u	p	
s	0	1	1	0	-1	0	2
t	0	2	0	1	1	0	8
x	1	1	0	0	1	0	4
p	0	1	0	0	2	1	8

Since no entry in the bottom row is negative, we are done. p has a maximum value of $\frac{8}{1} = 8$ when $x = \frac{4}{1} = 4$ and $y = 0$. (The slack variables are $s = 2$, $t = 8$, $u = 0$.)

3. Introduce slack variables and rewrite the constraints and objective function in standard form.

$$
\begin{aligned}
5x &- 5y + s &&&&= 20 \\
2x &- 10y &+ t &&&= 40 \\
-x &+ y &&+ p &&= 0
\end{aligned}
$$

Set up the initial tableau. Select the x-column as the pivot column. The test ratios are $\frac{20}{5}$ and $\frac{40}{2}$, so select the pivot in the s-row. Set up instructions to clear the column.

	x	y	s	t	p		
s	5	−5	1	0	0	20	
t	2	−10	0	1	0	40	$5R_2 - 2R_1$
p	−1	1	0	0	1	0	$5R_3 + R_1$

This gives the second tableau.

	x	y	s	t	p	
x	5	−5	1	0	0	20
t	0	−40	−2	5	0	160
p	0	0	1	0	5	20

Since no entry in the bottom row is negative, we are done. p has a maximum value of $\frac{20}{5} = 4$ when $x = \frac{20}{5} = 4$ and $y = 0$.

(The slack variables are $s = 0$, $t = \frac{160}{5} = 32$.)

5. Introduce slack variables and rewrite the constraints and objective function in standard form.

$$
\begin{aligned}
5x_1 &- x_2 + x_3 + s &&&&&= 1500 \\
2x_1 &+ 2x_2 + x_3 &+ t &&&= 2500 \\
4x_1 &+ 2x_2 + x_3 &&+ u &&= 2000 \\
-3x_1 &- 7x_2 - 8x_3 &&&+ p &= 0
\end{aligned}
$$

Set up the initial tableau. Select the x_3-column as the pivot column. The test ratios are $\frac{1500}{1}$, $\frac{2500}{1}$ and $\frac{2000}{1}$, so select the pivot in the s-row. Set up instructions to clear the column.

	x_1	x_2	x_3	s	t	u	p		
s	5	−1	1	1	0	0	0	1500	
t	2	2	1	0	1	0	0	2500	$R_2 - R_1$
u	4	2	1	0	0	1	0	2000	$R_3 - R_1$
p	−3	−7	−8	0	0	0	1	0	$R_4 + 8R_1$

This gives the second tableau. Select the x_2-column as the pivot column. The test ratios are $\frac{1000}{3}$ and $\frac{500}{3}$, so select the pivot in the u-row. Set up instructions to clear the column.

	x_1	x_2	x_3	s	t	u	p		
x_3	5	−1	1	1	0	0	0	1500	$3R_1 + R_3$
t	−3	3	0	−1	1	0	0	1000	$R_2 - R_3$
u	−1	3	0	−1	0	1	0	500	
p	37	−15	0	8	0	0	1	12,000	$R_4 + 5R_3$

This gives the third tableau.

	x_1	x_2	x_3	s	t	u	p	
x_3	14	0	3	2	0	1	0	5000
t	−2	0	0	0	1	−1	0	500
x_2	−1	3	0	−1	0	1	0	500
p	32	0	0	3	0	5	1	14,500

Since no entry in the bottom row is negative, we are done. p has a maximum value of

$\frac{14,500}{1} = 14,500$ when $x_1 = 0$, $x_2 = \frac{500}{3}$ and $x_3 = \frac{5000}{3}$. (The slack variables are $s = 0$, $t = \frac{500}{1} = 500$, $u = 0$.)

7. Introduce slack variables and rewrite the constraints and objective function in standard form.

$$
\begin{array}{rrrrrrr}
5x & & + & 5z & + & s & & & = & 100 \\
& 5y & - & 5z & & & + & t & & = & 50 \\
5x & - & 5y & & & & & + & u & = & 50 \\
-5x & + & 4y & - & 3z & & & & + & p & = & 0
\end{array}
$$

Set up the initial tableau. Select the x-column as the pivot column. The test ratios are $\dfrac{100}{5}$ and $\dfrac{50}{5}$, so select the pivot in the u-row. Set up instructions to clear the column.

	x	y	z	s	t	u	p		
s	5	0	5	1	0	0	0	100	$R_1 - R_3$
t	0	5	-5	0	1	0	0	50	
u	5	-5	0	0	0	1	0	50	
p	-5	4	-3	0	0	0	1	0	$R_4 + R_3$

This gives the second tableau. Select the z-column as the pivot column. The only test ratio is $\dfrac{50}{5}$, so select the pivot in the s-row. Set up instructions to clear the column.

	x	y	z	s	t	u	p		
s	0	5	5	1	0	-1	0	50	
t	0	5	-5	0	1	0	0	50	$R_2 + R_1$
x	5	-5	0	0	0	1	0	50	
p	0	-1	-3	0	0	1	1	50	$5R_4 + 3R_1$

This gives the third tableau.

	x	y	z	s	t	u	p	
z	0	5	5	1	0	-1	0	50
t	0	10	0	1	1	-1	0	100
x	5	-5	0	0	0	1	0	50
p	0	10	0	3	0	2	5	400

Since no entry in the bottom row is negative, we are done. p has a maximum value of

$$\frac{400}{5} = 80 \text{ when } x = \frac{50}{5} = 10, \ y = 0 \text{ and } z = \frac{50}{5} = 10. \ (\text{The slack variables are } s = 0, \ t = \frac{100}{1} = 100, \ u = 0.)$$

9. Introduce slack variables and rewrite the constraints and objective function in standard form.

$$
\begin{array}{rrrrrrr}
x & + & y & - & z & + & s & & & = & 3 \\
x & + & 2y & + & z & & & + & t & & = & 8 \\
x & + & y & & & & & & + & u & = & 5 \\
-7x & - & 5y & - & 6z & & & & & + & p & = & 0
\end{array}
$$

Set up the initial tableau. Select the x-column as the pivot column. The test ratios are $\dfrac{3}{1}$, $\dfrac{8}{1}$, and $\dfrac{5}{1}$, so select the pivot in the s-row. Set up instructions to clear the column.

	x	y	z	s	t	u	p		
s	1	1	-1	1	0	0	0	3	
t	1	2	1	0	1	0	0	8	$R_2 - R_1$
u	1	1	0	0	0	1	0	5	$R_3 - R_1$
p	-7	-5	-6	0	0	0	1	0	$R_4 + 7R_1$

This gives the second tableau. Select the z-column as the pivot column. The test ratios are $\frac{5}{2}$ and $\frac{2}{1}$, so select the pivot in the u-row. Set up instructions to clear the column.

	x	y	z	s	t	u	p		
x	1	1	−1	1	0	0	0	3	$R_1 + R_3$
t	0	1	2	−1	1	0	0	5	$R_2 - 2R_3$
u	0	0	1	−1	0	1	0	2	
p	0	2	−13	7	0	0	1	21	$R_4 + 13R_3$

This gives the third tableau. Select the s-column as the pivot column. The only test ratio is $\frac{1}{1}$, so select the pivot in the t-row. Set up instructions to clear the column.

	x	y	z	s	t	u	p		
x	1	1	0	0	0	1	0	5	
t	0	1	0	1	1	−2	0	1	
z	0	0	1	−1	0	1	0	2	$R_3 + R_2$
p	0	2	0	−6	0	13	1	47	$R_4 + 6R_2$

This gives the fourth tableau.

	x	y	z	s	t	u	p	
x	1	1	0	0	0	1	0	5
s	0	1	0	1	1	−2	0	1
z	0	1	1	0	1	−1	0	3
p	0	8	0	0	6	1	1	53

Since no entry in the bottom row is negative, we are done. p has a maximum value of $\frac{53}{1} = 53$ when $x = \frac{5}{1} = 5$, $y = 0$ and $z = \frac{3}{1} = 3$. (The slack variables are $s = \frac{1}{1} = 1$, $t = 0$, $u = 0$.)

11. Introduce slack variables and rewrite the constraints and objective function in standard form.

$$
\begin{aligned}
x + y + z \quad\quad\quad + s \quad\quad\quad\quad\quad\quad &= 3 \\
y + z + w \quad\quad\quad + t \quad\quad\quad\quad &= 4 \\
x \quad\quad + z + w \quad\quad\quad\quad + u \quad\quad &= 5 \\
x + y \quad\quad + w \quad\quad\quad\quad\quad + v \quad &= 6 \\
-x - y - z - w \quad\quad\quad\quad\quad\quad + p &= 0
\end{aligned}
$$

Set up the initial tableau. Select the x-, y-, z-, or w-column as the pivot column. We select the x-column. The test ratios are $\frac{3}{1}$, $\frac{5}{1}$ and $\frac{6}{1}$, so select the pivot in the s-row. Set up instructions to clear the column.

	x	y	z	w	s	t	u	v	p		
s	1	1	1	0	1	0	0	0	0	3	
t	0	1	1	1	0	1	0	0	0	4	
u	1	0	1	1	0	0	1	0	0	5	$R_3 - R_1$
v	1	1	0	1	0	0	0	1	0	6	$R_4 - R_1$
p	−1	−1	−1	−1	0	0	0	0	1	0	$R_5 + R_1$

This gives the second tableau. Select the w-column as the pivot column. The test ratios are $\frac{4}{1}$, $\frac{2}{1}$ and $\frac{3}{1}$, so select the pivot in the u-row. Set up instructions to clear the column.

	x	y	z	w	s	t	u	v	p		
x	1	1	1	0	1	0	0	0	0	3	
t	0	1	1	1	0	1	0	0	0	4	$R_2 - R_3$
u	0	−1	0	1	−1	0	1	0	0	2	
v	0	0	−1	1	−1	0	0	1	0	3	$R_4 - R_3$
p	0	0	0	−1	1	0	0	0	1	3	$R_5 + R_3$

This gives the third tableau. Select the y-column as the pivot column. The test ratios are $\frac{3}{1}$, $\frac{2}{2}$ and $\frac{1}{1}$, so select the pivot in either the t-row or v-row. We select the v-row. Set up instructions to clear the column.

	x	y	z	w	s	t	u	v	p		
x	1	1	1	0	1	0	0	0	0	3	$R_1 - R_4$
t	0	2	1	0	1	1	-1	0	0	2	$R_2 - 2R_4$
w	0	-1	0	1	-1	0	1	0	0	2	$R_3 + R_4$
v	0	1	-1	0	0	0	-1	1	0	1	
p	0	-1	0	0	0	0	1	0	1	5	$R_5 + R_4$

This gives the fourth tableau. Select the z-column as the pivot column. The test ratios are $\frac{2}{2}$ and $\frac{0}{3}$, so select the pivot in the t-row. Set up instructions to clear the column.

	x	y	z	w	s	t	u	v	p		
x	1	0	2	0	1	0	1	-1	0	2	$3R_1 - 2R_2$
t	0	0	3	0	1	1	1	-2	0	0	
w	0	0	-1	1	-1	0	0	1	0	3	$3R_3 + R_2$
y	0	1	-1	0	0	0	-1	1	0	1	$3R_4 + R_2$
p	0	0	-1	0	0	0	0	1	1	6	$3R_5 + R_2$

This gives the fifth tableau.

	x	y	z	w	s	t	u	v	p	
x	3	0	0	0	1	-2	1	1	0	6
z	0	0	3	0	1	1	1	0	0	0
w	0	0	0	3	-2	1	1	1	0	9
y	0	3	0	0	1	1	-2	1	0	3
p	0	0	0	0	1	1	1	3	3	18

Since no entries in the bottom row are negative, we are done. p has a maximum value of $\frac{18}{3} = 6$ when $x = \frac{6}{3} = 2$, $y = \frac{3}{3} = 1$, $z = \frac{0}{3} = 0$, and $w = \frac{9}{3} = 3$. (The slack variables are all zero.)

13. Introduce slack variables and rewrite the constraints and objective function in standard form.

$$
\begin{array}{rcccccccccl}
x & + & y & & & & & + & s & & & & & & & = & 1 \\
 & & y & + & z & & & & & + & t & & & & & = & 2 \\
 & & & & z & + & w & & & & & + & u & & & = & 3 \\
 & & & & & & w & + & v & & & & & + & r & = & 4 \\
-x & - & y & - & z & - & w & - & v & & & & & & + & p & = & 0
\end{array}
$$

Set up the initial tableau. Select the x-, y-, z-, w-, or v-column as the pivot column. We select the x-column. The only test ratio is $\frac{1}{1}$ so select the pivot in the s-row. Set up instructions to clear the column.

	x	y	z	w	v	s	t	u	r	p		
s	1	1	0	0	0	1	0	0	0	0	1	
t	0	1	1	0	0	0	1	0	0	0	2	
u	0	0	1	1	0	0	0	1	0	0	3	
r	0	0	0	1	1	0	0	0	1	0	4	
p	-1	-1	-1	-1	-1	0	0	0	0	1	0	$R_5 + R_1$

This gives the second tableau. Select the z-, w-, or v-column as the pivot column. We select the v-column. The only test ratio is $\frac{4}{1}$ so select the pivot in the r-row. Set up instructions to clear the column.

	x	y	z	w	v	s	t	u	r	p		
x	1	1	0	0	0	1	0	0	0	0	1	
t	0	1	1	0	0	0	1	0	0	0	2	
u	0	0	1	1	0	0	0	1	0	0	3	
r	0	0	0	1	1	0	0	0	1	0	4	
p	0	0	-1	-1	-1	1	0	0	0	1	1	$R_5 + R_4$

This gives the third tableau. Select the z-column as the pivot column. The test ratios are $\frac{2}{1}$ and $\frac{3}{1}$ so select the pivot in the t-row. Set up instructions to clear the column.

	x	y	z	w	v	s	t	u	r	p		
x	1	1	0	0	0	1	0	0	0	0	1	
t	0	1	1	0	0	0	1	0	0	0	2	
u	0	0	1	1	0	0	0	1	0	0	3	$R_3 - R_2$
v	0	0	0	1	1	0	0	0	1	0	4	
p	0	0	-1	0	0	1	0	0	1	1	5	$R_5 + R_2$

This gives the fourth tableau.

	x	y	z	w	v	s	t	u	r	p	
x	1	1	0	0	0	1	0	0	0	0	1
z	0	1	1	0	0	0	1	0	0	0	2
u	0	-1	0	1	0	0	-1	1	0	0	1
v	0	0	0	1	1	0	0	0	1	0	4
p	0	1	0	0	0	1	1	0	1	1	7

Since no entry in the bottom row is negative, we are done. p has a maximum value of

$\frac{7}{1} = 7$ when $x = \frac{1}{1} = 1$, $y = 0$, $z = \frac{2}{1} = 2$, $w = 0$, and $v = \frac{4}{1} = 4$. (The slack variables are $s = 0$, $t = 0$, $u = 1$, $r = 0$.)

Choosing different pivots may result in different values of the variables.

15. Introduce slack variables and set up the initial tableau as a matrix.

0.1	1	-2.2	1	0	0	0	4.5
2.1	1	1	0	1	0	0	8
1	2.2	0	0	0	1	0	5
-2.5	-4.2	-2	0	0	0	1	0

The pivot entry is in position 3, 2. Turn the pivot into a 1.

$R_3 \to \frac{1}{2.2} R_3$

0.1	1	-2.2	1	0	0	0	4.5
2.1	1	1	0	1	0	0	8
0.454545	1	0	0	0	0.454545	0	2.27273
-2.5	-4.2	-2	0	0	0	1	0

Clear the pivot column.

$R_1 \to R_1 - R_3$
$R_2 \to R_2 - R_3$
$R_4 \to R_4 + 4.2R_3$

−0.354545	0	−2.2	1	0	−0.454545	0	2.22727
1.64545	0	1	0	1	−0.454545	0	5.72727
0.454545	1	0	0	0	0.454545	0	2.27273
−0.590909	0	−2	0	0	1.90909	1	9.54545

The pivot entry is in position 2, 3. The pivot is a 1, so clear the pivot column.

$R_1 \to R_1 + 2.2R_2$
$R_4 \to R_4 + 2R_2$

3.26545	0	0	1	2.2	−1.454545	0	14.8273
1.64545	0	1	0	1	−0.454545	0	5.72727
0.454545	1	0	0	0	0.454545	0	2.27273
2.7	0	0	0	2		1	1	21

Thus, p has a maximum value of 21 when $x = 0$, $y = 2.27$, and $z = 5.73$.

17. Introduce slack variables and set up the initial tableau as a matrix.

1	2	3	0	1	0	0	0	0	3
0	1	1	2.2	0	1	0	0	0	4
1	0	1	2.2	0	0	1	0	0	5
1	1	0	2.2	0	0	0	1	0	6
−1	−2	−3	−1	0	0	0	0	1	0

The pivot entry is in position 1, 3. Turn the pivot into a 1.

$R_1 \to \frac{1}{3}R_1$

0.333333	0.666667	1	0	0.333333	0	0	0	0	1
0	1	1	2.2		0	1	0	0	0	4
1	0	1	2.2		0	0	1	0	0	5
1	1	0	2.2		0	0	0	1	0	6
−1	−2	−3	−1		0	0	0	0	1	0

Clear the pivot column.

$R_2 \to R_2 - R_1$
$R_3 \to R_3 - R_1$
$R_5 \to R_5 + 3R_1$

0.333333	0.666667	1	0	0.333333	0	0	0	0	1
−0.333333	0.333333	0	2.2	−0.333333	1	0	0	0	3
0.666667	−0.666667	0	2.2	−0.333333	0	1	0	0	4
1	1	0	2.2		0	0	0	1	0	6
0	0	0	−1		1	0	0	0	1	3

The pivot entry is in position 2, 4. Turn the pivot into a 1.

$R_2 \rightarrow \dfrac{1}{2.2} R_2$

0.333333	0.666667	1	0	0.333333		0	0	0	0	1
−0.151515	0.151515	0	1	−0.151515	0.454545	0	0	0	1.36364	
0.666667	−0.666667	0	2.2	−0.333333		0	1	0	0	4
1		1	0	2.2	0		0	0	1	0	6
0		0	0	−1		1		0	0	0	1	3

Clear the pivot column.

$R_3 \rightarrow R_3 - 2.2R_2$
$R_4 \rightarrow R_4 - 2.2R_2$
$R_5 \rightarrow R_5 + R_2$

0.333333	0.666667	1	0	0.333333		0	0	0	0	1
−0.151515	0.151515	0	1	−0.151515	0.454545	0	0	0	1.36364	
1		−1	0	0	0		−1	1	0	0	1
1.33333	0.666667	0	0	0.333333		−1	0	1	0	3
−0.151515	0.151515	0	0	0.848485	0.454545	0	0	1	4.36364	

The pivot entry is in position 3, 1. The pivot is a 1 so clear the pivot column.

$R_1 \rightarrow R_1 - 0.33333R_3$
$R_2 \rightarrow R_2 + 0.151515R_3$
$R_4 \rightarrow R_4 - 1.33333R_3$
$R_5 \rightarrow R_5 + 0.151515R_3$

0	1	1	0	0.333333	0.333333	−0.333333	0	0	0.666667
0	0	0	1	−0.151515	0.30303	0.151515	0	0	1.51515
1	−1	0	0	0	−1	1	0	0	1
0	2	0	0	0.333333	0.333333	−1.333333	1	0	1.66667
0	0	0	0	0.848485	0.30303	0.151515	0	1	4.51515

Thus, p has a maximum value of 4.52 when $x = 1$, $y = 0$, $z = 0.67$, and $w = 1.52$.

19. Introduce slack variables and set up the initial tableau as a matrix.

1	1	0	0	0	1	0	0	0	0	1.1
0	1	1	0	0	0	1	0	0	0	2.2
0	0	1	1	0	0	0	1	0	0	3.3
0	0	0	1	1	0	0	0	1	0	4.4
−1	1	−1	1	−1	0	0	0	0	1	0

We use the entry in position 1, 1 for the pivot entry. The pivot is a 1 so clear the column.

$R_5 \rightarrow R_5 + R_1$

1	1	0	0	0	1	0	0	0	0	1.1
0	1	1	0	0	0	1	0	0	0	2.2
0	0	1	1	0	0	0	1	0	0	3.3
0	0	0	1	1	0	0	0	1	0	4.4
0	2	−1	1	−1	1	0	0	0	1	1.1

We use the entry in position 2, 3 for the pivot entry. The pivot is a 1 so clear the column.
$R_3 \rightarrow R_3 - R_2$
$R_5 \rightarrow R_5 + R_2$

$$\begin{bmatrix} 1 & 1 & 0 & 0 & 0 & 1 & 0 & 0 & 0 & 0 & 1.1 \\ 0 & 1 & 1 & 0 & 0 & 0 & 1 & 0 & 0 & 0 & 2.2 \\ 0 & -1 & 0 & 1 & 0 & 0 & -1 & 1 & 0 & 0 & 1.1 \\ 0 & 0 & 0 & 1 & 1 & 0 & 0 & 0 & 1 & 0 & 4.4 \\ 0 & 3 & 0 & 1 & -1 & 1 & 1 & 0 & 0 & 1 & 3.3 \end{bmatrix}$$

The pivot entry is in position 4, 5. The pivot is a 1 so clear the column.
$R_5 \rightarrow R_5 + R_4$

$$\begin{bmatrix} 1 & 1 & 0 & 0 & 0 & 1 & 0 & 0 & 0 & 0 & 1.1 \\ 0 & 1 & 1 & 0 & 0 & 0 & 1 & 0 & 0 & 0 & 2.2 \\ 0 & -1 & 0 & 1 & 0 & 0 & -1 & 1 & 0 & 0 & 1.1 \\ 0 & 0 & 0 & 1 & 1 & 0 & 0 & 0 & 1 & 0 & 4.4 \\ 0 & 3 & 0 & 2 & 0 & 1 & 1 & 0 & 1 & 1 & 7.7 \end{bmatrix}$$

Thus, p has a maximum value of 7.7 when $x = 1.1$, $y = 0$, $z = 2.2$, $w = 0$, and $v = 4.4$.

21. Let x = number of calculus texts. Let y = number of history texts. Let z = number of marketing texts.
The objective function is the profit $p = 10x + 4y + 8z$. The constraints are $x + y + z \le 650$, $2x + y + 3z \le 1000$, $x \ge 0$, $y \ge 0$, $z \ge 0$.
Introduce slack variables.

$$
\begin{aligned}
x + y + z + s &= 650 \\
2x + y + 3z \quad\quad + t &= 1000 \\
-10x - 4y - 8z \quad\quad\quad\quad + p &= 0
\end{aligned}
$$

Set up the initial tableau and go through the procedure of the simplex method.

	x	y	z	s	t	p		
s	1	1	1	1	0	0	650	$2R_1 - R_2$
t	2	1	3	0	1	0	1000	
p	-10	-4	-8	0	0	1	0	$R_3 + 5R_2$

	x	y	z	s	t	p	
s	0	1	-1	2	-1	0	300
x	2	1	3	0	1	0	1000
p	0	1	7	0	5	1	5000

Since no entry in the bottom row is negative, we are done.

p has a maximum value of $\frac{5000}{1} = 5000$ when $x = \frac{1000}{2} = 500$, $y = 0$, and $z = 0$.

Thus you should purchase 500 calculus texts, no history texts, and no marketing texts. The maximum profit is $5000 per semester.

23. Let x = number of gallons of PineOrange. Let y = number of gallons of PineKiwi. Let z = number of gallons of OrangeKiwi.
The objective function is the profit $p = x + 2y + z$. The constraints are $2x + 3y \le 800$, $2x + 3z \le 650$, $y + z \le 350$, $x \ge 0$, $y \ge 0$, $z \ge 0$.
Introduce slack variables.

$$
\begin{aligned}
2x + 3y \quad\quad + s &= 800 \\
2x \quad\quad + 3z \quad\quad + t &= 650 \\
y + z \quad\quad\quad\quad + u &= 350 \\
-x - 2y - z \quad\quad\quad\quad\quad\quad + p &= 0
\end{aligned}
$$

Set up the initial tableau and go through the procedure of the simplex method.

	x	y	z	s	t	u	p	
s	2	3	0	1	0	0	0	800
t	2	0	3	0	1	0	0	650
u	0	1	1	0	0	1	0	350
p	-1	-2	-1	0	0	0	1	0

$3R_3 - R_1$
$3R_4 + 2R_1$

	x	y	z	s	t	u	p	
y	2	3	0	1	0	0	0	800
t	2	0	3	0	1	0	0	650
u	-2	0	3	-1	0	3	0	250
p	1	0	-3	2	0	0	3	1600

$R_2 - R_3$
$R_4 + R_3$

	x	y	z	s	t	u	p	
y	2	3	0	1	0	0	0	800
t	4	0	0	1	1	-3	0	400
z	-2	0	3	-1	0	3	0	250
p	-1	0	0	1	0	3	3	1850

$2R_1 - R_2$
$2R_3 + R_2$
$4R_4 + R_2$

	x	y	z	s	t	u	p	
y	0	6	0	1	-1	3	0	1200
x	4	0	0	1	1	-3	0	400
z	0	0	6	-1	1	3	0	900
p	0	0	0	5	1	9	12	7800

Since no entry in the bottom row is negative, we are done.

p has a maximum value of $\frac{7800}{12} = 650$ when $x = \frac{400}{4} = 100$, $y = \frac{1200}{6} = 200$ and $z = \frac{900}{6} = 150$.

Thus the Arctic Juice Company should make 100 gallons of PineOrange, 200 gallons of PineKiwi, and 150 gallons of OrangeKiwi. The maximum profit is $650.

25. Let x = number of sections of Ancient History. Let y = number of sections of Medieval History.
Let z = number of sections of Modern History.
The objective function is the profit $p = 10{,}000x + 20{,}000y + 30{,}000z$. Divide the coefficients by 10,000 to rewrite this as $p = x + 2y + 3z$. The constraints are $x + y + z \le 45$, $100x + 50y + 200z \le 5000$ (or $2x + y + 4z \le 100$), $x + y + 2z \le 60$, $x \ge 0$, $y \ge 0$, $z \ge 0$.
Introduce slack variables.

$$x + y + z + s \qquad\qquad\qquad = 45$$
$$2x + y + 4z \qquad + t \qquad\qquad = 100$$
$$x + y + 2z \qquad\qquad + u \qquad = 60$$
$$-x - 2y - 3z \qquad\qquad\qquad + p = 0$$

Set up the initial tableau and go through the procedure of the simplex method.

	x	y	z	s	t	u	p	
s	1	1	1	1	0	0	0	45
t	2	1	4	0	1	0	0	100
u	1	1	2	0	0	1	0	60
p	-1	-2	-3	0	0	0	1	0

$4R_1 - R_2$
$2R_3 - R_2$
$4R_4 + 3R_2$

	x	y	z	s	t	u	p	
s	2	3	0	4	-1	0	0	80
z	2	1	4	0	1	0	0	100
u	0	1	0	0	-1	2	0	20
p	2	-5	0	0	3	0	4	300

$R_1 - 3R_3$
$R_2 - R_3$
$R_4 + 5R_3$

	x	y	z	s	t	u	p	
s	2	0	0	4	2	−6	0	20
z	2	0	4	0	2	−2	0	80
y	0	1	0	0	−1	2	0	20
p	2	0	0	0	−2	10	4	400

$R_2 - R_1$
$2R_3 + R_1$
$R_4 + R_1$

	x	y	z	s	t	u	p	
t	2	0	0	4	2	−6	0	20
z	0	0	4	−4	0	4	0	60
y	2	2	0	4	0	−2	0	60
p	4	0	0	4	0	4	4	420

Since no entry in the bottom row is negative, we are done.

p has a maximum value of $\dfrac{420}{4} = 105$ when $x = 0$, $y = \dfrac{60}{2} = 30$ and $z = \dfrac{60}{4} = 15$.

Thus the department should offer no sections of Ancient History, 30 sections of Medieval History, and 15 sections of Modern History. The largest possible profit is $105 \cdot 10,000 = \$1,050,000$. There will be $30 + 15 = 45$ total sections, $50(30) + 200(15) = 4500$ students, and $30 + 2(15) = 60$ professors. 500 students will not get into a class.

a. Thus you should make 80 batches of Creamy Vanilla, no batches of Continental Mocha, and 20 batches of Succulent Strawberry.

b. In part a., you have used $2(80) + 20 = 180$ eggs, $80 + 2(20) = 120$ cups of milk, and $2(80) + 2(20) = 200$ cups of cream. You have not used all your ingredients.

c. The maximum profit decreases. Adding the constraint $z \leq 10$ results in a maximum profit of \$310 when $x = 90$, $y = 0$, and $z = 10$ (90 batches of Creamy Vanilla, no batches of Continental Mocha, and 10 batches of Succulent Strawberry).

27. Let $x =$ acres of tomatoes. Let $y =$ acres of lettuce. Let $z =$ acres of carrots. The objective function is the profit $p = 2000x + 1500y + 500z$. Divide the coefficients by 500 to rewrite this as $p = 4x + 3y + z$. The constraints are $x + y + z \leq 100$, $5x + 4y + 2z \leq 400$, $4x + 2y + 2z \leq 500$, $x \geq 0$, $y \geq 0$, $z \geq 0$.
Introduce slack variables.

$$
\begin{aligned}
x + y + z + s &= 100 \\
5x + 4y + 2z + t &= 400 \\
4x + 2y + 2z + u &= 500 \\
-4x - 3y - z + p &= 0
\end{aligned}
$$

Set up the initial tableau and go through the procedure of the simplex method.

	x	y	z	s	t	u	p	
s	1	1	1	1	0	0	0	100
t	5	4	2	0	1	0	0	400
u	4	2	2	0	0	1	0	500
p	−4	−3	−1	0	0	0	1	0

$5R_1 - R_2$
$5R_3 - 4R_2$
$5R_4 + 4R_2$

	x	y	z	s	t	u	p	
s	0	1	3	5	−1	0	0	100
x	5	4	2	0	1	0	0	400
u	0	−6	2	0	−4	5	0	900
p	0	1	3	0	4	0	5	1600

Since no entry in the bottom row is negative, we are done.

p has a maximum value of $\dfrac{1600}{5} = 320$ when $x = \dfrac{400}{5} = 80$, $y = 0$, and $z = 0$.

Thus you should plant 80 acres of tomatoes for a profit of $500(320) = \$160,000$. There will be 20 unused acres.

29. Let x = servings of granola. Let y = servings of nutty granola. Let z = servings of nuttiest granola. The objective function is the profit $p = 6x + 8y + 3z$. The constraints are $x + y + 5z \leq 1500$, $4x + 8y + 8z \leq 10{,}000$, $2x + 4y + 8z \leq 4000$, $x \geq 0$, $y \geq 0$, $z \geq 0$.

Introduce slack variables.

$$
\begin{aligned}
x + y + 5z + s &= 1500 \\
4x + 8y + 8z + t &= 10{,}000 \\
2x + 4y + 8z + u &= 4000 \\
-6x - 8y - 3z + p &= 0
\end{aligned}
$$

Set up the initial tableau and go through the procedure of the simplex method.

	x	y	z	s	t	u	p		
s	1	1	5	1	0	0	0	1500	$4R_1 - R_3$
t	4	8	8	0	1	0	0	10,000	$R_2 - 2R_3$
u	2	4	8	0	0	1	0	4000	
p	-6	-8	-3	0	0	0	1	0	$R_4 + 2R_3$

	x	y	z	s	t	u	p		
s	2	0	12	4	0	-1	0	2000	
t	0	0	-8	0	1	-2	0	2000	
y	2	4	8	0	0	1	0	4000	$R_3 - R_1$
p	-2	0	13	0	0	2	1	8000	$R_4 + R_1$

	x	y	z	s	t	u	p	
x	2	0	12	4	0	-1	0	2000
t	0	0	-8	0	1	-2	0	2000
y	0	4	-4	-4	0	2	0	2000
p	0	0	25	4	0	1	1	10,000

Since no entry in the bottom row is negative, we are done.

p has a maximum of $\dfrac{10{,}000}{1} = 10{,}000$ when $x = \dfrac{2000}{2} = 1000$, $y = \dfrac{2000}{4} = 500$ and $z = 0$.

The Choral Society should make 1000 servings of granola treats, 500 servings of nutty granola treats, and no servings of nuttiest granola treats. They will use $1000 + 500 = 1500$ ounces of toasted oats, $4(1000) + 8(500) = 8000$ ounces of almonds, and $2(1000) + 4(500) = 4000$ ounces of raisins. Thus they will have 2000 ounces of almonds left over.

31. Let x = millions of gallons to process A. Let y = millions of gallons to process B. Let z = millions of gallons to process C. The objective function is the revenue (in millions of dollars) $p = 4(0.6x + 0.55y + 0.5z) = 2.4x + 2.2y + 2z$. The constraints are $x + y + z \leq 50$, $150{,}000x + 100{,}000y + 50{,}000z \leq 3{,}000{,}000$ (or $15x + 10y + 5z \leq 300$), $x \geq 0$, $y \geq 0$, $z \geq 0$.

Introduce slack variables.

$$
\begin{aligned}
x + y + z + s &= 50 \\
15x + 10y + 5z + t &= 300 \\
-2.4x - 2.2y - 2z + p &= 0
\end{aligned}
$$

Set up the initial tableau and go through the procedure of the simplex method.

	x	y	z	s	t	p		
s	1	1	1	1	0	0	50	$15R_1 - R_2$
t	15	10	5	0	1	0	300	
p	-2.4	-2.2	-2	0	0	1	0	$25R_3 + 4R_2$

	x	y	z	s	t	p		
s	0	5	10	15	-1	0	450	
x	15	10	5	0	1	0	300	$2R_2 - R_1$
p	0	-15	-30	0	4	25	1200	$R_3 + 3R_1$

	x	y	z	s	t	p	
z	0	5	10	15	−1	0	450
x	30	15	0	−15	3	0	150
p	0	0	0	45	1	25	2550

Since no entry in the bottom row is negative, we are done.

p has a maximum value of $\frac{2550}{25} = 102$ when $x = \frac{150}{30} = 5$, $y = 0$ and $z = \frac{450}{10} = 45$.

Thus, allocate 5 million gallons to process A and 45 million gallons to process C.

33. Let x = amount of automobile loans (in hundreds of thousands of dollars), y = amount of furniture loans (in hundreds of thousands of dollars), z = amount of signature loans (in hundreds of thousands of dollars), and w = amount of other secured loans (in hundreds of thousands of dollars). The objective function is the revenue $p = 0.08x + 0.10y + 0.12z + 0.10w$. Multiply the coefficients by 100 to rewrite this as $p = 8x + 10y + 12z + 10w$. Since ESU has \$5,000,000 available, $x + y + z + w \leq 50$. From restriction (a), $z \leq 5$. From restriction (b), $y + w \leq x$ or $-x + y + w \leq 0$. From restriction (c), $w \leq 2x$ or $-2x + w \leq 0$.
Introduce slack variables.

$$
\begin{aligned}
x + y + z + w + s &= 50 \\
z + t &= 5 \\
-x + y + w + u &= 0 \\
-2x + w + v &= 0 \\
-8x - 10y - 12z - 12w + p &= 0
\end{aligned}
$$

Set up the initial tableau and go through the procedure of the simplex method.

	x	y	z	w	s	t	u	v	p		
s	1	1	1	1	1	0	0	0	0	50	$R_1 - R_2$
t	0	0	1	0	0	1	0	0	0	5	
u	−1	1	0	1	0	0	1	0	0	0	
v	−2	0	0	1	0	0	0	1	0	0	
p	−8	−10	−12	−10	0	0	0	0	1	0	$R_5 + 12R_2$

	x	y	z	w	s	t	u	v	p		
s	1	1	0	1	1	−1	0	0	0	45	$R_1 - R_3$
z	0	0	1	0	0	1	0	0	0	5	
u	−1	1	0	1	0	0	1	0	0	0	
v	−2	0	0	1	0	0	0	1	0	0	
p	−8	−10	0	−10	0	12	0	0	1	60	$R_5 + 10R_3$

	x	y	z	w	s	t	u	v	p		
s	2	0	0	0	1	−1	−1	0	0	45	
z	0	0	1	0	0	1	0	0	0	5	
y	−1	1	0	1	0	0	1	0	0	0	$2R_3 + R_1$
v	−2	0	0	1	0	0	0	1	0	0	$R_4 + R_1$
p	−18	0	0	0	0	12	10	0	1	60	$R_5 + 9R_1$

	x	y	z	w	s	t	u	v	p	
x	2	0	0	0	1	−1	−1	0	0	45
z	0	0	1	0	0	1	0	0	0	5
y	0	2	0	2	1	−1	1	0	0	45
v	0	0	0	1	1	−1	−1	1	0	45
p	0	0	0	0	9	3	1	0	1	465

Since no entry in the bottom row is negative, we are done.

p has a maximum value of $\frac{465}{1} = 465$ when $x = \frac{45}{2} = 22.5$, $y = \frac{45}{2} = 22.5$, $z = \frac{5}{1} = 5$, and $w = 0$.

Thus allocate \$2,250,000 to automobile loans, \$2,250,000 to furniture loans, \$500,000 to signature loans, and none to other secured loans.

35. Let x = amount invested in *Warner Music*, y = amount invested in *Sony*, z = amount invested in *Polygram*, and w = amount invested in *EMI*. The objective function is the brand name exposure $p = 8x + 2y + 7z + 8w$. The constraints are $x + y + z + w \le 100{,}000$, $0.22x + 0.15y + 0.12z + 0.11w \le 15{,}000$ or $22x + 15y + 12z + 11w \le 1{,}500{,}000$, $0.25(x + y + z + w) \le z$ or $x + y - 3z + w \le 0$, $x \ge 0$, $y \ge 0$, $z \ge 0$, $w \ge 0$.

Introduce slack variables.

$$
\begin{aligned}
x + y + z + w + s &= 100{,}000 \\
22x + 15y + 12z + 11w + t &= 1{,}500{,}000 \\
x + y - 3z + w + u &= 0 \\
-8x - 2y - 7z - 8w + p &= 0
\end{aligned}
$$

Set up the initial tableau and go through the procedure of the simplex method.

	x	y	z	w	s	t	u	p		
s	1	1	1	1	1	0	0	0	100,000	$R_1 - R_3$
t	22	15	12	11	0	1	0	0	1,500,000	$R_2 - 11R_3$
u	1	1	−3	1	0	0	1	0	0	
p	−8	−2	−7	−8	0	0	0	1	0	$R_4 + 8R_3$

	x	y	z	w	s	t	u	p		
s	0	0	4	0	1	0	−1	0	100,000	
t	11	4	45	0	0	1	−11	0	1,500,000	$4R_2 - 45R_1$
w	1	1	−3	1	0	0	1	0	0	$4R_3 + 3R_1$
p	0	6	−31	0	0	0	8	1	0	$4R_4 + 31R_1$

	x	y	z	w	s	t	u	p		
z	0	0	4	0	1	0	−1	0	100,000	
t	44	16	0	0	−45	4	1	0	1,500,000	
w	4	4	0	4	3	0	1	0	300,000	
p	0	24	0	0	31	0	1	4	3,100,000	

Since no entry in the bottom row is negative, we are done.

p has a maximum when $x = 0$, $y = 0$, $z = \dfrac{100{,}000}{4} = 25{,}000$, and $w = \dfrac{300{,}000}{4} = 75{,}000$.

Thus invest \$25,000 in *Polygram* and \$75,000 in *EMI*.

37. Let x = number of boards shipped from Tucson to Honolulu, y = number of boards shipped from Tucson to Venice Beach, z = number of boards shipped from Toronto to Honolulu, and w = number of boards shipped from Toronto to Venice Beach. The objective function is the total number shipped $p = x + y + z + w$. The constraints are $10x + 5y + 20z + 10w \leq 6550$, $x + y \leq 620$, $z + w \leq 410$, $x + z \leq 500$, $y + w \leq 530$, $x \geq 0$, $y \geq 0$, $z \geq 0$, $w \geq 0$.
Introduce slack variables.

$$
\begin{array}{rcrcrcrcrcrcrcl}
10x & + & 5y & + & 20z & + & 10w & + & q & & & & & = & 6550 \\
x & + & y & & & & & + & r & & & & & = & 620 \\
& & & & z & + & w & & & + & s & & & = & 410 \\
x & & & + & z & & & & & & & + & t & = & 500 \\
& & y & & & + & w & & & & & & + u & = & 530 \\
-x & - & y & - & z & - & w & & & & & & & + p = & 0
\end{array}
$$

Set up the initial tableau and go through the procedure of the simplex method.

	x	y	z	w	q	r	s	t	u	p		
q	10	5	20	10	1	0	0	0	0	0	6550	$R_1 - 10R_4$
r	1	1	0	0	0	1	0	0	0	0	620	$R_2 - R_4$
s	0	0	1	1	0	0	1	0	0	0	410	
t	1	0	1	0	0	0	0	1	0	0	500	
u	0	1	0	1	0	0	0	0	1	0	530	
p	−1	−1	−1	−1	0	0	0	0	0	1	0	$R_6 + R_4$

	x	y	z	w	q	r	s	t	u	p		
q	0	5	10	10	1	0	0	−10	0	0	1550	$R_1 - 5R_2$
r	0	1	−1	0	0	1	0	−1	0	0	120	
s	0	0	1	1	0	0	1	0	0	0	410	
x	1	0	1	0	0	0	0	1	0	0	500	
u	0	1	0	1	0	0	0	0	1	0	530	$R_5 - R_2$
p	0	−1	0	−1	0	0	0	1	0	1	500	$R_6 + R_2$

	x	y	z	w	q	r	s	t	u	p		
q	0	0	15	10	1	−5	0	−5	0	0	950	
y	0	1	−1	0	0	1	0	−1	0	0	120	
s	0	0	1	1	0	0	1	0	0	0	410	$10R_3 - R_1$
x	1	0	1	0	0	0	0	1	0	0	500	
u	0	0	1	1	0	−1	0	1	1	0	410	$10R_5 - R_1$
p	0	0	−1	−1	0	1	0	0	0	1	620	$10R_6 + R_1$

	x	y	z	w	q	r	s	t	u	p		
w	0	0	15	10	1	−5	0	−5	0	0	950	$3R_1 + R_5$
y	0	1	−1	0	0	1	0	−1	0	0	120	$15R_2 + R_5$
s	0	0	−5	0	−1	5	10	5	0	0	3150	$3R_3 - R_5$
x	1	0	1	0	0	0	0	1	0	0	500	$15R_4 - R_5$
u	0	0	−5	0	−1	−5	0	15	10	0	3150	
p	0	0	5	0	1	5	0	−5	0	10	7150	$3R_6 + R_5$

	x	y	z	w	q	r	s	t	u	p	
w	0	0	40	30	2	−20	0	0	10	0	6000
y	0	15	−20	0	−1	10	0	0	10	0	4950
s	0	0	−10	0	−2	20	30	0	−10	0	6300
x	15	0	20	0	1	5	0	0	−10	0	4350
t	0	0	−5	0	−1	−5	0	15	10	0	3150
p	0	0	10	0	2	10	0	0	10	30	24,600

Since no entry in the bottom row is negative, we are done.

p has a maximum value of $\dfrac{24,600}{30} = 820$ when $x = \dfrac{4350}{15} = 290$, $y = \dfrac{4950}{15} = 330$, $z = 0$, and $w = \dfrac{6000}{30} = 200$.

Thus ship 290 boards from Tucson to Honolulu, 330 boards from Tucson to Venice Beach, 0 boards from Toronto to Honolulu, and 200 boards from Toronto to Venice Beach. This gives a total of 820 boards shipped.

39. Let x = number of salespeople from Chicago to Los Angeles, y = number of salespeople from Chicago to New York, z = number of salespeople from Denver to Los Angeles, and w = number of salespeople from Denver to New York. The objective function is the total number of salespeople on the road, $p = x + y + z + w$. The constraints are $428x + 305y + 338z + 493w \le 8895$, $x + y \le 10$, $z + w \le 20$, $x + z \le 10$, $y + w \le 15$, $x \ge 0, y \ge 0, z \ge 0, w \ge 0$. Introduce slack variables.

$$
\begin{aligned}
428x + 305y + 338z + 493w + q &= 8895 \\
x + y + r &= 10 \\
z + w + s &= 20 \\
x + z + t &= 10 \\
y + w + u &= 15 \\
-x - y - z - w + p &= 0
\end{aligned}
$$

Set up the initial tableau and go through the procedure of the simplex method.

	x	y	z	w	q	r	s	t	u	p		
q	428	305	338	493	1	0	0	0	0	0	8895	$R_1 - 305R_2$
r	1	1	0	0	0	1	0	0	0	0	10	
s	0	0	1	1	0	0	1	0	0	0	20	
t	1	0	1	0	0	0	0	1	0	0	10	
u	0	1	0	1	0	0	0	0	1	0	15	$R_5 - R_2$
p	−1	−1	−1	−1	0	0	0	0	0	1	0	$R_6 + R_2$

	x	y	z	w	q	r	s	t	u	p		
q	123	0	338	493	1	−305	0	0	0	0	5845	$R_1 - 338R_4$
y	1	1	0	0	0	1	0	0	0	0	10	
s	0	0	1	1	0	0	1	0	0	0	20	$R_3 - R_4$
t	1	0	1	0	0	0	0	1	0	0	10	
u	−1	0	0	1	0	−1	0	0	1	0	5	
p	0	0	−1	−1	0	1	0	0	0	1	10	$R_6 + R_4$

	x	y	z	w	q	r	s	t	u	p		
q	−215	0	0	493	1	−305	0	−338	0	0	2465	$R_1 - 493R_5$
y	1	1	0	0	0	1	0	0	0	0	10	
s	−1	0	0	1	0	0	1	−1	0	0	10	$R_3 - R_5$
z	1	0	1	0	0	0	0	1	0	0	10	
u	−1	0	0	1	0	−1	0	0	1	0	5	
p	1	0	0	−1	0	1	0	1	0	1	20	$R_6 + R_5$

	x	y	z	w	q	r	s	t	u	p	
q	278	0	0	0	1	188	0	−338	−493	0	0
y	1	1	0	0	0	0	1	0	0	0	10
s	0	0	0	0	0	1	1	−1	−1	0	5
z	1	0	1	0	0	0	0	1	0	0	10
w	−1	0	0	1	0	−1	0	0	1	0	5
p	0	0	0	0	0	0	0	1	1	1	25

Since no entry in the bottom row is negative, we are done.

p has a maximum value of $\frac{25}{1} = 25$ when $x = 0$, $y = \frac{10}{1} = 10$, $z = \frac{10}{1} = 10$, and $w = \frac{5}{1} = 5$.

You should fly no salespeople from Chicago to Los Angeles, 10 salespeople from Chicago to New York, 10 salespeople from Denver to Los Angeles, and 5 salespeople from Denver to New York.

41. Answers will vary.

43. The simplex method is useful when there are more than two variables. We cannot use the graphical method in such cases.

45. The basic solution is obtained from setting the inactive variables to zero.

47. No; the entries of the last column of a tableau are always nonnegative and when passing from one tableau to the next, a multiple of a higher row is always added to the last row. Thus the value of the objective can either increase or stay the same.

Section 4.4 Solving Nonstandard Linear Programming Problems
Pages 315–322

1. Introduce surplus and slack variables and rewrite the system of equations:

$$
\begin{array}{rcrcrcrcrcr}
.x & + & 2y & - & s & & & & & = & 6 \\
-x & + & y & & & + & t & & & = & 4 \\
2x & + & y & & & & & + & u & = & 8 \\
-x & - & y & & & & & & & + p = & 0
\end{array}
$$

Set up the initial tableau. To remove the star from the first row select the y-column as the pivot column. Then apply the standard simplex method.

	x	y	s	t	u	p		
*s	1	2	−1	0	0	0	6	
t	−1	1	0	1	0	0	4	$2R_2 - R_1$
u	2	1	0	0	1	0	8	$2R_3 - R_1$
p	−1	−1	0	0	0	1	0	$2R_4 + R_1$

	x	y	s	t	u	p		
y	1	2	−1	0	0	0	6	$3R_1 - R_3$
t	−3	0	1	2	0	0	2	$R_2 + R_3$
u	3	0	1	0	2	0	10	
p	−1	0	−1	0	0	2	6	$3R_4 + R_3$

	x	y	s	t	u	p		
y	0	6	−4	0	−2	0	8	$R_1 + 2R_2$
t	0	0	2	2	2	0	12	
x	3	0	1	0	2	0	10	$2R_3 - R_2$
p	0	0	−2	0	2	6	28	$R_4 + R_2$

	x	y	s	t	u	p	
y	0	6	0	4	2	0	32
s	0	0	2	2	2	0	12
x	6	0	0	-2	2	0	8
p	0	0	0	2	4	6	40

Since no entry in the bottom row is negative, we are done.

p has a maximum value of $\frac{40}{6} = \frac{20}{3}$ when $x = \frac{8}{6} = \frac{4}{3}$ and $y = \frac{32}{6} = \frac{16}{3}$.

3. Introduce surplus and slack variables and rewrite the constraints and objective function in standard form.

$$
\begin{array}{rcrcrcrcrcl}
x & + & y & + & s & & & & & = & 25 \\
x & & & & & - & t & & & = & 10 \\
-x & + & 2y & & & & & - & u & = & 0 \\
-12x & - & 10y & & & & & & + p & = & 0
\end{array}
$$

Set up the initial tableau. To remove the star from the third row select the y-column as the pivot column.

	x	y	s	t	u	p		
s	1	1	1	0	0	0	25	$2R_1 - R_3$
$*t$	1	0	0	-1	0	0	10	
$*u$	-1	2	0	0	-1	0	0	
p	-12	-10	0	0	0	1	0	$R_4 + 5R_3$

To remove the star from the second row select the
x-column as the pivot column. Then apply the standard simplex method.

	x	y	s	t	u	p		
s	3	0	2	0	1	0	50	$R_1 - 3R_2$
$*t$	1	0	0	-1	0	0	10	
y	-1	2	0	0	-1	0	0	$R_3 + R_2$
p	-17	0	0	0	-5	1	0	$R_4 + 17R_2$

	x	y	s	t	u	p		
s	0	0	2	3	1	0	20	
x	1	0	0	-1	0	0	10	$3R_2 + R_1$
y	0	2	0	-1	-1	0	10	$3R_3 + R_1$
p	0	0	0	-17	-5	1	170	$3R_4 + 17R_1$

	x	y	s	t	u	p	
t	0	0	2	3	1	0	20
x	3	0	2	0	1	0	50
y	0	6	2	0	-2	0	50
p	0	0	34	0	2	3	850

Since no entry in the bottom row is negative, we are done.

p has a maximum value of $\frac{850}{3}$ when $x = \frac{50}{3}$ and $y = \frac{50}{6} = \frac{25}{3}$.

5. Introduce surplus and slack variables and rewrite the constraints and objective function in standard form.

$$x + y + z + s \qquad\qquad = 150$$
$$x + y + z \qquad - t \qquad = 100$$
$$-2x - 5y - 3z \qquad\qquad + p = 0$$

Set up the initial tableau. To remove the star from the second row select the y-column as the pivot column. Then apply the standard simplex method.

	x	y	z	s	t	p		
s	1	1	1	1	0	0	150	$R_1 - R_2$
*t	1	1	1	0	-1	0	100	
p	-2	-5	-3	0	0	1	0	$R_3 + 5R_2$

	x	y	z	s	t	p		
s	0	0	0	1	1	0	50	
y	1	1	1	0	-1	0	100	$R_2 + R_1$
p	3	0	2	0	-5	1	500	$R_4 + 5R_1$

	x	y	z	s	t	p	
t	0	0	0	1	1	0	50
y	1	1	1	1	0	0	150
p	3	0	2	5	0	1	750

Since no entry in the bottom row is negative, we are done.

p has a maximum value of $\frac{750}{1} = 750$ when $x = 0$, $y = \frac{150}{1} = 150$, and $z = 0$.

7. Introduce surplus and slack variables and rewrite the constraints and objective function in standard form.

$$x + y + z + w + s \qquad\qquad = 40$$
$$2x + y - z - w \qquad - t \qquad = 10$$
$$x + y + z + w \qquad\qquad - u \qquad = 10$$
$$-2x - 3y - z - 4w \qquad\qquad + p = 0$$

Set up the initial tableau. To remove the star from the third row select the w-column as the pivot column.

	x	y	z	w	s	t	u	p		
s	1	1	1	1	1	0	0	0	40	$R_1 - R_3$
*t	2	1	-1	-1	0	-1	0	0	10	$R_2 + R_3$
*u	1	1	1	1	0	0	-1	0	10	
p	-2	-3	-1	-4	0	0	0	1	0	$R_4 + 4R_3$

To remove the star from the second row select the x-column as the pivot column. Then apply the standard simplex method.

	x	y	z	w	s	t	u	p		
s	0	0	0	0	1	0	1	0	30	
*t	3	2	0	0	0	-1	-1	0	20	
w	1	1	1	1	0	0	-1	0	10	$3R_3 - R_2$
p	2	1	3	0	0	0	-4	1	40	$3R_4 - 2R_2$

	x	y	z	w	s	t	u	p		
s	0	0	0	0	1	0	1	0	30	
x	3	2	0	0	0	-1	-1	0	20	$R_2 + R_1$
w	0	1	3	3	0	1	-2	0	10	$R_3 + 2R_1$
p	0	-1	9	0	0	2	-10	3	80	$R_4 + 10R_1$

	x	y	z	w	s	t	u	p		
u	0	0	0	0	1	0	1	0	30	
x	3	2	0	0	1	−1	0	0	50	
w	0	1	3	3	2	1	0	0	70	$2R_3 - R_2$
p	0	−1	9	0	10	2	0	3	380	$2R_4 + R_2$

	x	y	z	w	s	t	u	p	
u	0	0	0	0	1	0	1	0	30
y	3	2	0	0	1	−1	0	0	50
w	−3	0	6	6	3	3	0	0	90
p	3	0	18	0	21	3	0	6	810

Since no entry in the bottom row is negative, we are done.

p has a maximum value of $\frac{810}{6} = 135$ when $x = 0$, $y = \frac{50}{2} = 25$, $z = 0$, and $w = \frac{90}{6} = 15$.

9. Translate into a maximization problem.

Maximize $p = -6x - 6y$ subject to $x + 2y \geq 20$, $2x + y \geq 20$, $x \geq 0$, $y \geq 0$. Introduce surplus variables and rewrite the constraints and objective function in standard form.

$$
\begin{aligned}
x + 2y - s &= 20 \\
2x + y \quad\ - t &= 20 \\
6x + 6y \quad\quad + p &= 0
\end{aligned}
$$

Set up the initial tableau. Select a pivot with the largest test ratio in the x-column.

	x	y	s	t	p		
$*s$	1	2	−1	0	0	20	
$*t$	2	1	0	−1	0	20	$R_2 - 2R_1$
p	6	6	0	0	1	0	$R_3 - 6R_1$

	x	y	s	t	p	
x	1	2	−1	0	0	20
$*t$	0	−3	2	−1	0	−20
p	0	−6	6	0	1	−120

Multiply row 2 by −1.

	x	y	s	t	p		
x	1	2	−1	0	0	20	$3R_1 - 2R_2$
t	0	3	−2	1	0	20	
p	0	−6	6	0	1	−120	$R_3 + 2R_2$

	x	y	s	t	p	
x	3	0	1	−2	0	20
y	0	3	−2	1	0	20
p	0	0	2	2	1	−80

p has a maximum value of $-\frac{80}{1} = -80$ so c has a minimum value of 80 when $x = \frac{20}{3}$ and $y = \frac{20}{3}$.

11. Translate into a maximization problem.
Maximize $p = -2x - y - 3z$ subject to $x + y + z \geq 100$, $2x + y \geq 50$, $y + z \geq 50$, $x \geq 0$, $y \geq 0$, $z \geq 0$.
Introduce surplus variables and rewrite the constraints and objective function in standard form.

$$
\begin{array}{rcrcrcrcr}
x &+& y &+& z &-& s & & &=& 100 \\
2x &+& y & & & & &-& t &=& 50 \\
& & y &+& z & & & & -u &=& 50 \\
2x &+& y &+& 3z & & & & +p &=& 0
\end{array}
$$

Set up the initial tableau. Select a pivot with the largest test ratio in the y-column.

	x	y	z	s	t	u	p		
$*s$	1	1	1	-1	0	0	0	100	
$*t$	2	1	0	0	-1	0	0	50	$R_2 - R_1$
$*u$	0	1	1	0	0	-1	0	50	$R_3 - R_1$
p	2	1	3	0	0	0	1	0	$R_4 - R_1$

	x	y	z	s	t	u	p	
y	1	1	1	-1	0	0	0	100
$*t$	1	0	-1	1	-1	0	0	-50
$*u$	-1	0	0	1	0	-1	0	-50
p	1	0	2	1	0	0	1	-100

Multiply rows 2 and 3 by –1.

	x	y	z	s	t	u	p	
y	1	1	1	-1	0	0	0	100
t	-1	0	1	-1	1	0	0	50
u	1	0	0	-1	0	1	0	50
p	1	0	2	1	0	0	1	-100

p has a maximum value of -100 so c has a minimum value of 100 when $x = 0$, $y = 100$, and $z = 0$.

13. Translate into a maximization problem.
Maximize $p = -50x - 50y - 11z$ subject to $2x + z \geq 3$, $2x + y - z \geq 2$, $3x + y - z \leq 3$, $x \geq 0$, $y \geq 0$, $z \geq 0$.
Introduce surplus and slack variables and rewrite the constraints and objective function in standard form.

$$
\begin{array}{rcrcrcrcrcr}
2x & & &+& z &-& s & & &=& 3 \\
2x &+& y &-& z & & &-& t & &=& 2 \\
3x &+& y &-& z & & & & +u &=& 3 \\
50x &+& 50y &+& 11z & & & & +p &=& 0
\end{array}
$$

Set up the initial tableau. (Note that this is not a standard minimization problem.) To remove the star from the second row select the y-column as the pivot column.

	x	y	z	s	t	u	p		
$*s$	2	0	1	-1	0	0	0	3	
$*t$	2	1	-1	0	-1	0	0	2	
u	3	1	-1	0	0	1	0	3	$R_3 - R_2$
p	50	50	11	0	0	0	1	0	$R_4 - 50R_2$

To remove the star from the first row select the z-column as the pivot column Then apply the standard simplex method.

	x	y	z	s	t	u	p		
$*s$	2	0	1	-1	0	0	0	3	
y	2	1	-1	0	-1	0	0	2	$R_2 + R_1$
u	1	0	0	0	1	1	0	1	
p	-50	0	61	0	50	0	1	-100	$R_4 - 61R_1$

	x	y	z	s	t	u	p		
z	2	0	1	−1	0	0	0	3	$R_1 - 2R_3$
y	4	1	0	−1	−1	0	0	5	$R_2 - 4R_3$
u	1	0	0	0	1	1	0	1	
p	−172	0	0	61	50	0	1	−283	$R_4 + 172R_3$

	x	y	z	s	t	u	p	
z	0	0	1	−1	−2	−2	0	1
y	0	1	0	−1	−5	−4	0	1
x	1	0	0	0	1	1	0	1
p	0	0	0	61	222	172	1	−111

p has a maximum value of −111 so c has a minimum value of 111 when $x = 1, y = 1$, and $z = 1$.

15. Translate into a maximization problem.
   Maximize $p = -x - y - z - w$ subject to $5x - y + w \geq 1000$, $z + w \leq 2000$, $x + y \leq 500$, $x \geq 0, y \geq 0, z \geq 0, w \geq 0$.
   Introduce surplus and slack variables and rewrite the constraints and objective function in standard form.

$$
\begin{aligned}
5x \;-\; y \quad\quad\quad +\; w \;-\; s \quad\quad\quad\quad\quad\quad &=\; 1000 \\
z \;+\; w \quad\quad\quad +\; t \quad\quad\quad\quad &=\; 2000 \\
x \;+\; y \quad\quad\quad\quad\quad\quad\quad\quad +\; u \quad\quad &=\; 500 \\
x \;+\; y \;+\; z \;+\; w \quad\quad\quad\quad\quad +\; p \;&=\; 0
\end{aligned}
$$

Set up the initial tableau. (Note that this is not a standard minimization problem.) To remove the star from the first row select the x-column as the pivot column.

	x	y	z	w	s	t	u	p		
*s	5	−1	0	1	−1	0	0	0	1000	
t	0	0	1	1	0	1	0	0	2000	
u	1	1	0	0	0	0	1	0	500	$5R_3 - R_1$
p	1	1	1	1	0	0	0	1	0	$5R_4 - R_1$

	x	y	z	w	s	t	u	p	
x	5	−1	0	1	−1	0	0	0	1000
t	0	0	1	1	0	1	0	0	2000
u	0	6	0	−1	1	0	5	0	1500
p	0	6	5	4	1	0	0	5	−1000

p has a maximum value of $-\dfrac{1000}{5} = -200$ so c has a minimum value of 200 when $x = 200, y = 0, z = 0$, and $w = 0$.

17. Introduce surplus and slack variables and set up the initial tableau as a matrix.

$$
\begin{bmatrix}
1.2 & 1 & 1 & 1 & 1 & 0 & 0 & 0 & 40.5 \\
2.2 & 1 & -1 & -1 & 0 & -1 & 0 & 0 & 10 \\
1.2 & 1 & 1 & 1.2 & 0 & 0 & -1 & 0 & 10.5 \\
-2 & -3 & -1.1 & -4 & 0 & 0 & 0 & 1 & 0
\end{bmatrix}
$$

Row 2 is a starred row. Select the pivot entry in position 2, 2. Clear the pivot column.
$R_1 \rightarrow R_1 - R_2$
$R_3 \rightarrow R_3 - R_2$
$R_4 \rightarrow R_4 + 3R_2$

$$
\begin{bmatrix}
-1 & 0 & 2 & 2 & 1 & 1 & 0 & 0 & 30.5 \\
2.2 & 1 & -1 & -1 & 0 & -1 & 0 & 0 & 10 \\
-1 & 0 & 2 & 2.2 & 0 & 1 & -1 & 0 & 0.5 \\
4.6 & 0 & -4.1 & -7 & 0 & -3 & 0 & 1 & 30
\end{bmatrix}
$$

Row 3 is a starred row. Select the pivot entry in position 3, 3. Turn the pivot into a 1.

$R_3 \to \frac{1}{2}R_3$

−1	0	2	2	1	1	0	0	30.5
2.2	1	−1	−1	0	−1	0	0	10
−0.5	0	1	1.1	0	0.5	−0.5	0	0.25
4.6	0	−4.1	−7	0	−3	0	1	30

Clear the pivot column.

$R_1 \to R_1 - 2R_3$
$R_2 \to R_2 + R_3$
$R_4 \to R_4 + 4.1R_3$

0	0	0	−0.2	1	0	1	0	30
1.7	1	0	0.1	0	−0.5	−0.5	0	10.25
−0.5	0	1	1.1	0	0.5	−0.5	0	0.25
2.55	0	0	−2.49	0	−0.95	−2.05	1	31.025

There are no more starred rows. The pivot entry is in position 3, 4. Turn the pivot into a 1.

$R_3 \to \frac{1}{1.1}R_3$

0	0	0	−0.2	1	0	1	0	30
1.7	1	0	0.1	0	−0.5	−0.5	0	10.25
−0.454545	0	0.909091	1	0	0.454545	−0.454545	0	0.227273
2.55	0	0	−2.49	0	−0.95	−2.05	1	31.025

Clear the pivot column.

$R_1 \to R_1 + 0.2R_3$
$R_2 \to R_2 - 0.1R_3$
$R_4 \to R_4 + 2.49R_3$

−0.0909091	0	0.181818	0	1	0.0909091	0.909091	0	30.0455
1.74545	1	−0.0909091	0	0	−0.545455	−0.454545	0	10.2273
−0.454545	0	0.909091	1	0	0.454545	−0.454545	0	0.227273
1.41818	0	2.26364	0	0	0.181818	−3.18182	1	31.5909

The pivot entry is in position 1, 7. Turn the pivot into a 1.

$R_1 \to \frac{1}{0.909091}R_1$

−0.1	0	0.2	0	1.1	0.1	1	0	33.05
1.74545	1	−0.0909091	0	0	−0.545455	−0.454545	0	10.2273
−0.454545	0	0.909091	1	0	0.454545	−0.454545	0	0.227273
1.41818	0	2.265364	0	0	0.181818	−3.18182	1	31.5909

Clear the pivot column.

$R_2 \to R_2 + 0.454545R_1$
$R_3 \to R_3 + 0.454545R_1$
$R_4 \to R_4 + 3.18182R_1$

−0.1	0	0.2	0	1.1	0.1	1	0	33.05
1.7	1	0	0	0.5	−0.5	0	0	25.25
−0.5	0	1	1	0.5	0.5	0	0	15.25
1.1	0	2.9	0	3.5	0.5	0	1	136.75

Thus, p has a maximum value of 136.75 when $x = 0$, $y = 25.25$, $z = 0$, and $w = 15.25$.

19. Translate into a maximization problem. Then introduce surplus variables and set up the initial tableau as a matrix.

1	1.5	1.2	−1	0	0	0	100
2	1.5	0	0	−1	0	0	50
0	1.5	1.1	0	0	−1	0	50
2.2	1	3.3	0	0	0	1	0

Select a pivot with the largest test ratio in the y-column. The pivot entry is in position 1, 2. Turn the pivot into a 1.

$$R_1 \to \frac{1}{1.5} R_1$$

0.666667	1	0.8	−0.666667	0	0	0	66.6667
2	1.5	0		0	−1	0	0	50
0	1.5	1.1		0	0	−1	0	50
2.2	1	3.3		0	0	0	1	0

Clear the column.
$$R_2 \to R_2 - 1.5R_1$$
$$R_3 \to R_3 - 1.5R_1$$
$$R_4 \to R_4 - R_1$$

0.666667	1	0.8	−0.666667	0	0	0	66.6667
1	0	−1.2		1	−1	0	0	−50
−1	0	−0.1		1	0	−1	0	−50
1.53333	0	2.5	0.666667	0	0	1	−66.6667

Multiply rows 2 and 3 by −1.
$$R_2 \to -R_2$$
$$R_3 \to -R_3$$

0.666667	1	0.8	−0.666667	0	0	0	66.6667
−1	0	1.2		−1	1	0	0	50
1	0	0.1		−1	0	1	0	50
1.53333	0	2.5	0.666667	0	0	1	−66.6667

c has a minimum value of 66.67 when $x = 0$, $y = 66.67$, and $z = 0$.

21. Translate into a maximization problem. Then introduce surplus and slack variables and set up the initial tableau as a matrix.

5.12	−1	0	1	1	0	0	0	1000
0	0	1	1	0	−1	0	0	2000
1.22	1	0	0	0	0	1	0	500
1.1	1	1.5	−1	0	0	0	1	0

Row 2 is a starred row. Select the pivot entry in position 2, 3. The pivot entry is a 1 so clear the column.
$$R_4 \to R_4 - 1.5R_2$$

5.12	−1	0	1	1	0	0	0	1000
0	0	1	1	0	−1	0	0	2000
1.22	1	0	0	0	0	1	0	500
1.1	1	0	−2.5	0	1.5	0	1	−3000

Select the pivot entry in position 1, 4. The pivot entry is a 1 so clear the pivot column.

$R_2 \rightarrow R_2 - R_1$
$R_4 \rightarrow R_4 + 2.5R_1$

5.12	−1	0	1	1	0	0	0	1000
−5.12	1	1	0	−1	−1	0	0	1000
1.22	1	0	0	0	0	1	0	500
13.9	−1.5	0	0	2.5	1.5	0	1	−500

Select the pivot entry in position 3, 2. The pivot entry is a 1 so clear the pivot column.

$R_1 \rightarrow R_1 + R_3$
$R_2 \rightarrow R_2 - R_3$
$R_4 \rightarrow R_4 + 1.5R_3$

6.34	0	0	1	1	0	1	0	1500
−6.34	0	1	0	−1	−1	−1	0	500
1.22	1	0	0	0	0	1	0	500
15.73	0	0	0	2.5	1.5	1.5	1	250

c has a minimum value of −250 when $x = 0$, $y = 500$, $z = 500$, and $w = 1500$.

23. Let x = quarts of orange juice. Let y = quarts of orange concentrate. The objective function is the cost $c = 0.5x + 2y$. The constraints are $x \geq 10{,}000$, $y \geq 1000$, $10x + 50y \geq 200{,}000$. Translate into a maximization problem by maximizing $p = -0.5x - 2y$.

Introduce surplus variables.

$$
\begin{array}{rcrcrcrcl}
x & & & - & s & & & = & 10{,}000 \\
& & y & & & - & t & = & 1000 \\
10x & + & 50y & & & - & u & = & 200{,}000 \\
0.5x & + & 2y & & & & + p & = & 0
\end{array}
$$

Set up the initial tableau and go through the procedure of the simplex method.

.	x	y	s	t	u	p	
*s	1	0	−1	0	0	0	10,000
*t	0	1	0	−1	0	0	1000
*u	10	50	0	0	−1	0	200,000
p	0.5	2	0	0	0	1	0

$10R_1 - R_3$

$20R_4 - R_3$

	x	y	s	t	u	p	
*s	0	−50	−10	0	1	0	−100,000
*t	0	1	0	−1	0	0	1000
x	10	50	0	0	−1	0	200,000
p	0	−10	0	0	1	20	−200,000

Multiply row 1 by −1.

	x	y	s	t	u	p	
s	0	50	10	0	−1	0	100,000
*t	0	1	0	−1	0	0	1000
x	10	50	0	0	−1	0	200,000
p	0	−10	0	0	1	20	−200,000

$R_1 - 50R_2$

$R_3 - 50R_2$

$R_4 + 10R_2$

	x	y	s	t	u	p	
s	0	0	10	50	−1	0	50,000
y	0	1	0	−1	0	0	1000
x	10	0	0	50	−1	0	150,000
p	0	0	0	−10	1	20	−190,000

$50R_2 + R_1$

$R_3 - R_1$

$5R_4 + R_1$

	x	y	s	t	u	p	
t	0	0	10	50	-1	0	50,000
y	0	50	10	0	-1	0	100,000
x	10	0	-10	0	0	0	100,000
p	0	0	10	0	4	100	-900,000

p has a maximum value of $-\dfrac{900,000}{100} = -9000$ so c has a minimum value of 9000 when

$x = \dfrac{100,000}{10} = 10,000$ and $y = \dfrac{100,000}{50} = 2000$.

Thus the company should produce 10,000 quarts of orange juice and 2000 quarts of orange concentrate to meet the demand and minimize total costs.

25. Let $x =$ servings of *Gerber* Mixed Cereal. Let $y =$ servings of *Gerber* Mango Tropical Dessert.
Let $z =$ servings of *Gerber* Apple Banana Juice.
The objective function is the cost $c = 10x + 53y + 27z$. The constraints are $60x + 80y + 60z \geq 120$, $45y + 120z \geq 120$, $x \geq 0$, $y \geq 0$, $z \geq 0$. Translate into a maximization problem by maximizing $p = -10x - 53y - 27z$.
Introduce surplus variables.

$$
\begin{array}{rcrcrcrcrcr}
60x & + & 80y & + & 60z & - & s & & & = & 120 \\
& & 45y & + & 120z & & & - & t & = & 120 \\
10x & + & 53y & + & 27z & & & & + p & = & 0
\end{array}
$$

Set up the initial tableau and go through the procedure of the simplex method.

	x	y	z	s	t	p		
*s	60	80	60	-1	0	0	120	
*t	0	45	120	0	-1	0	120	$R_2 - 2R_1$
p	10	53	27	0	0	1	0	$20R_3 - 9R_1$

	x	y	z	s	t	p	
z	60	80	60	-1	0	0	120
*t	-120	-115	0	2	-1	0	-120
p	-340	340	0	9	0	20	-1080

Multiply row 2 by -1.

	x	y	z	s	t	p		
z	60	80	60	-1	0	0	120	$2R_1 - R_2$
t	120	115	0	-2	1	0	120	
p	-340	340	0	9	0	20	-1080	$6R_3 + 17R_2$

	x	y	z	s	t	p	
z	0	45	120	0	-1	0	120
x	120	115	0	-2	1	0	120
p	0	3995	0	20	17	120	-4440

p has a maximum value of $-\dfrac{4440}{120} = -37$ so c has a minimum value of 37 when $x = \dfrac{120}{120} = 1$, $y = 0$, and $z = \dfrac{120}{120} = 1$.

Thus you should provide one serving of cereal, no servings of dessert, and one serving of juice.

27. Let x = number of mailings to the East Coast, y = number of mailings to the Midwest, and
z = number of mailings to the West Coast.
The objective function is the cost $c = 40x + 60y + 50z$. The constraints are $100x + 100y + 50z \geq 1500$,
$50x + 100y + 100z \geq 1500$, $x \geq 0$, $y \geq 0$, $z \geq 0$. Translate into a maximization problem by maximizing $p = -40x - 60y - 50z$.
Introduce surplus variables.

$$100x + 100y + 50z - s \qquad\qquad = 1500$$
$$50x + 100y + 100z \qquad - t \qquad = 1500$$
$$40x + 60y + 50z \qquad\qquad + p = 0$$

Set up the initial tableau and go through the procedure of the simplex method.

	x	y	z	s	t	p		
*s	100	100	50	−1	0	0	1500	
*t	50	100	100	0	−1	0	1500	$R_2 - 2R_1$
p	40	60	50	0	0	1	0	$R_3 - R_1$

	x	y	z	s	t	p	
z	100	100	50	−1	0	0	1500
*t	−150	−100	0	2	−1	0	−1500
p	−60	−40	0	1	0	1	−1500

Multiply row 2 by −1.

	x	y	z	s	t	p		
z	100	100	50	−1	0	0	1500	$3R_1 - 2R_2$
t	150	100	0	−2	1	0	1500	
p	−60	−40	0	1	0	1	−1500	$5R_3 + 2R_2$

	x	y	z	s	t	p	
z	0	100	150	1	−2	0	1500
x	150	100	0	−2	1	0	1500
p	0	0	0	1	2	5	−4500

p has a maximum value of $-\dfrac{4500}{5} = -900$ so c has a minimum value of 900 when

$$x = \frac{1500}{150} = 10, \quad y = 0, \quad \text{and } z = \frac{1500}{150} = 10.$$

Thus Canter should send 10 mailings to the East Coast, no mailings to the Midwest, and 10 mailings to the West Coast. The cost will be $900.

29. Let x = number of bundles from Nadir, y = number of bundles from Blunt, and z = number of bundles from Sonny.
The objective function is the cost $c = 3000x + 4000y + 5000z$. The constraints are $5x + 10y + 15z \geq 150$,
$10x + 10y + 10z \geq 200$, $15x + 10y + 10z \geq 150$, $x \geq 0$, $y \geq 0$, $z \geq 0$. Translate into a maximization problem by maximizing
$p = -3000x - 4000y - 5000z$.
Introduce surplus variables.

$$5x + 10y + 15z - s \qquad\qquad\qquad = 150$$
$$10x + 10y + 10z \qquad - t \qquad\qquad = 200$$
$$15x + 10y + 10z \qquad\qquad - u \qquad = 150$$
$$3000x + 4000y + 5000z \qquad\qquad\qquad + p = 0$$

Set up the initial tableau and go through the procedure of the simplex method.

	x	y	z	s	t	u	p		
*s	5	10	15	−1	0	0	0	150	
*t	10	10	10	0	−1	0	0	200	$R_2 - 2R_1$
*u	15	10	10	0	0	−1	0	150	$R_3 - 3R_1$
p	3000	4000	5000	0	0	0	1	0	$R_4 - 600R_1$

	x	y	z	s	t	u	p	
x	5	10	15	−1	0	0	0	150
*t	0	−10	−20	2	−1	0	0	−100
*u	0	−20	−35	3	0	−1	0	−300
p	0	−2000	−4000	600	0	0	1	−90,000

Multiply rows 2 and 3 by −1.

	x	y	z	s	t	u	p		
x	5	10	15	−1	0	0	0	150	$4R_1 - 3R_2$
t	0	10	20	−2	1	0	0	100	
u	0	20	35	−3	0	1	0	300	$4R_3 - 7R_2$
p	0	−2000	−4000	600	0	0	1	−90,000	$R_4 + 200R_2$

	x	y	z	s	t	u	p	
x	20	10	0	2	−3	0	0	300
z	0	10	20	−2	1	0	0	100
u	0	10	0	2	−7	4	0	500
p	0	0	0	200	200	0	1	−70,000

p has a maximum value of $-\dfrac{70,000}{1} = -70,000$ so c has a minimum value of 70,000 when

$x = \dfrac{300}{20} = 15$, $y = 0$, and $z = \dfrac{100}{20} = 5$.

Thus Cheapskate Electronics Store can update its inventory by buying 15 bundles from Nadir, no bundles from Blunt, and 5 bundles from Sonny. The least possible cost is $70,000.

31. Let x = number of convention-style hotels, y = number of vacation-style hotels, and z = number of small motels. The objective function is the cost $c = 100x + 20y + 4z$. The constraints are $x \geq 1$, $z \leq 2$, $500x + 200y + 50z \geq 1400$, $x \geq 0$, $y \geq 0$, $z \geq 0$. Translate into a maximization problem by maximizing $p = -100x - 20y - 4z$. Introduce surplus and slack variables.

$$
\begin{array}{rcrcrcrcr}
x & & & & & - & s & & & = & 1 \\
& & z & & & + & t & & & = & 2 \\
500x & + & 200y & + & 50z & & & - & u & = & 1400 \\
100x & + & 20y & + & 4z & & & & + p & = & 0
\end{array}
$$

Set up the initial tableau and go through the procedure of the simplex method.

	x	y	z	s	t	u	p		
*s	1	0	0	−1	0	0	0	1	
t	0	0	1	0	1	0	0	2	
*u	500	200	50	0	0	−1	0	1400	$R_3 - 500R_1$
p	100	20	4	0	0	0	1	0	$R_4 - 100R_1$

	x	y	z	s	t	u	p		
x	1	0	0	−1	0	0	0	1	
t	0	0	1	0	1	0	0	2	
*u	0	200	50	500	0	−1	0	900	
p	0	20	4	100	0	0	1	−100	$10R_4 - R_3$

	x	y	z	s	t	u	p		
x	1	0	0	−1	0	0	0	1	
t	0	0	1	0	1	0	0	2	
y	0	200	50	500	0	−1	0	900	$R_3 - 50R_2$
p	0	0	−10	500	0	1	10	−1900	$R_4 + 10R_2$

	x	y	z	s	t	u	p	
x	1	0	0	−1	0	0	0	1
z	0	0	1	0	1	0	0	2
y	0	200	0	500	−50	−1	0	800
p	0	0	0	500	10	1	10	−1880

p has a maximum value of $-\frac{1880}{10} = -188$ so c has a minimum value of 188 when $x = \frac{1}{1} = 1$, $y = \frac{800}{200} = 4$, and $z = \frac{2}{1} = 2$.

Thus you should build 1 convention-style hotel, 4 vacation-style hotels, and 2 small motels. Your total cost is $188 million, so 20% of the total cost is $37.6 million. The city's subsidy will be sufficient.

33. Let x = number of cardiologists hired, y = number of rehabilitation specialists hired, and z = number of infectious disease specialists hired.

The objective function is the revenue per week brought in by the hired specialists $p = 120,000x + 190,000y + 140,000z$. Divide the coefficients by 10,000 to rewrite this as $p = 12x + 19y + 14z$. Since there are already 3 cardiologists, the hired specialists can admit up to 170 patients per week and should bring in 27 grants per year. The constraints are $10x + 10y + 10z \leq 170$ or $x + y + z \leq 17, x + y + 3z \geq 27, x \geq 0, y \geq 0, z \geq 0$.

Introduce surplus and slack variables.

$$
\begin{aligned}
x + y + z + s &= 17 \\
x + y + 3z - t &= 27 \\
-12x - 19y - 14z + p &= 0
\end{aligned}
$$

Set up the initial tableau and go through the procedure of the simplex method.

	x	y	z	s	t	p		
s	1	1	1	1	0	0	17	$3R_1 - R_2$
*t	1	1	3	0	−1	0	27	
p	−12	−19	−14	0	0	1	0	$3R_3 + 14R_2$

	x	y	z	s	t	p		
s	2	2	0	3	1	0	24	
z	1	1	3	0	−1	0	27	$2R_2 - R_1$
p	−22	−43	0	0	−14	3	378	$2R_3 + 43R_1$

	x	y	z	s	t	p	
y	2	2	0	3	1	0	24
z	0	0	6	−3	−3	0	30
p	42	0	0	129	15	6	1788

p has a maximum value of $\frac{1788}{6} = 298$ when $x = 0$, $y = \frac{24}{2} = 12$, and $z = \frac{30}{6} = 5$.

Thus no more cardiologists, 12 rehabilitation specialists, and 5 infectious disease specialists should be hired.

35. Let x = ounces of hamburger, y = cups of powdered milk, and z = number of eggs. The objective function is the cost $c = 10x + 15y + 5z$. The constraints are $10x + 20y + 10z \geq 100, 35y \leq 350, 10x + 1020y + 30z \geq 3180$, $50x + 100y + 80z \leq 1120, x \geq 0, y \geq 0, z \geq 0$. Translate into a maximization problem by maximizing $p = -10x - 15y - 5z$. Introduce surplus and slack variables.

$$
\begin{aligned}
10x + 20y + 10z - s &= 100 \\
35y + t &= 350 \\
10x + 1020y + 30z - u &= 3180 \\
50x + 100y + 80z + v &= 1120 \\
10x + 15y + 5z + p &= 0
\end{aligned}
$$

Set up the initial tableau and go through the procedure of the simplex method.

	x	y	z	s	t	u	v	p		
*s	10	20	10	−1	0	0	0	0	100	
t	0	35	0	0	1	0	0	0	350	
*u	10	1020	30	0	0	−1	0	0	3180	$R_3 - R_1$
v	50	100	80	0	0	0	1	0	1120	$R_4 - 5R_1$
p	10	15	5	0	0	0	0	1	0	$R_5 - R_1$

	x	y	z	s	t	u	v	p		
x	10	20	10	−1	0	0	0	0	100	$50R_1 - R_3$
t	0	35	0	0	1	0	0	0	350	$200R_2 - 7R_3$
*u	0	1000	20	1	0	−1	0	0	3080	
v	0	0	30	5	0	0	1	0	620	
p	0	−5	−5	1	0	0	0	1	−100	$200R_5 + R_1$

	x	y	z	s	t	u	v	p		
x	500	0	480	−51	0	1	0	0	1920	
t	0	0	−140	−7	200	7	0	0	48,440	$24R_2 + 7R_1$
y	0	1000	20	1	0	−1	0	0	3080	$24R_3 - R_1$
v	0	0	30	5	0	0	1	0	620	$16R_4 - R_1$
p	0	0	−980	201	0	−1	0	200	−16,920	$24R_5 + 49R_1$

	x	y	z	s	t	u	v	p	
z	500	0	480	−51	0	1	0	0	1920
t	3500	0	0	−525	4800	175	0	0	1,176,000
y	−500	24,000	0	75	0	−25	0	0	72,000
v	−500	0	0	131	0	−1	16	0	8000
p	24,500	0	0	2325	0	25	0	4800	−312,000

p has a maximum value of $-\dfrac{312,000}{4800} = -65$ when $x = 0$, $y = \dfrac{72,000}{24,000} = 3$, and $z = \dfrac{1920}{480} = 4$. No hamburger, 3 cups of powdered milk, and 4 eggs should be combined. The hamburgers don't contain any ground beef!

37. Let x = amount for luxury condominiums in millions of dollars, y = amount for low-income housing in millions of dollars, and z = amount for urban development in millions of dollars.

The objective function is the return $p = 0.12x + 0.10y + 0.05z$ or $p = 12x + 10y + 5z$. The constraints are $x + y + z \le 25$, $x \ge 10$, $y \ge \frac{1}{3}(x+y+z)$ or $-x + 2y - z \ge 0$, $x \ge 0$, $y \ge 0$, $z \ge 0$.

Introduce surplus and slack variables.

$$
\begin{array}{rcrcrcrcrcrcr}
x &+& y &+& z &+& s & & & & & &= 25 \\
x & & & & & & &-& t & & & &= 10 \\
-x &+& 2y &-& z & & & & &-& u & &= 0 \\
-12x &-& 10y &-& 5z & & & & & &+& p &= 0
\end{array}
$$

Set up the initial tableau and go through the procedure of the simplex method.

	x	y	z	s	t	u	p		
s	1	1	1	1	0	0	0	25	$R_1 - R_2$
*t	1	0	0	0	-1	0	0	10	
*u	-1	2	-1	0	0	-1	0	0	$R_3 + R_2$
p	-12	-10	-5	0	0	0	1	0	$R_4 + 12R_2$

	x	y	z	s	t	u	p		
s	0	1	1	1	1	0	0	15	$2R_1 - R_3$
x	1	0	0	0	-1	0	0	10	
*u	0	2	-1	0	-1	-1	0	10	
p	0	-10	-5	0	-12	0	1	120	$R_4 + 5R_3$

	x	y	z	s	t	u	p		
s	0	0	3	2	3	1	0	20	
x	1	0	0	0	-1	0	0	10	$3R_2 + R_1$
y	0	2	-1	0	-1	-1	0	10	$3R_3 + R_1$
p	0	0	-10	0	-17	-5	1	170	$3R_4 + 17R_1$

	x	y	z	s	t	u	p	
t	0	0	3	2	3	1	0	20
x	3	0	3	2	0	1	0	50
y	0	6	0	2	0	-2	0	50
p	0	0	21	34	0	2	3	850

p has a maximum value of $\dfrac{850}{300} = 2\dfrac{5}{6}$ when $x = \dfrac{50}{3} = 16\dfrac{2}{3}$, $y = \dfrac{50}{6} = 8\dfrac{1}{3}$, and $z = 0$. It should allocate \$$16\dfrac{2}{3}$ million for luxury condominiums, \$$8\dfrac{1}{3}$ million for low-income housing, and none for urban development.

39. Let x = number of salespeople from Chicago to Los Angeles, y = number of salespeople from Chicago to New York, z = number of salespeople from Denver to Los Angeles, and w = number of salespeople from Denver to New York. The objective function is the cost $c = 428x + 305y + 338z + 493w$. Translate into a maximization problem by maximizing $p = -428x - 305y - 338z - 493w$. The constraints are $x + y \le 10$, $z + w \le 20$, $x + z \ge 10$, $y + w \ge 15$, $x \ge 0$, $y \ge 0$, $z \ge 0$, $w \ge 0$. Introduce surplus and slack variables.

$$
\begin{array}{rcrcrcrcrcrcrcrcrcr}
x & + & y & & & & & + & s & & & & & & & & & = & 10 \\
& & & & z & + & w & & & + & t & & & & & & & = & 20 \\
x & & & + & z & & & & & & & - & u & & & & & = & 10 \\
& & y & & & + & w & & & & & & & - & v & & & = & 15 \\
428x & + & 305y & + & 338z & + & 493w & & & & & & & & & + & p & = & 0
\end{array}
$$

Set up the initial tableau and go through the procedure of the simplex method.

	x	y	z	w	s	t	u	v	p		
s	1	1	0	0	1	0	0	0	0	10	
t	0	0	1	1	0	1	0	0	0	20	$R_2 - R_3$
*u	1	0	1	0	0	0	-1	0	0	10	
*v	0	1	0	1	0	0	0	-1	0	15	
p	428	305	338	493	0	0	0	0	1	0	$R_5 - 338R_3$

	x	y	z	w	s	t	u	v	p		
s	1	1	0	0	1	0	0	0	0	10	$R_1 - R_4$
t	-1	0	0	1	0	1	1	0	0	10	
z	1	0	1	0	0	0	-1	0	0	10	
*v	0	1	0	1	0	0	0	-1	0	15	
p	90	305	0	493	0	0	338	0	1	-3380	$R_5 - 305R_4$

	x	y	z	w	s	t	u	v	p	
*s	1	0	0	-1	1	0	0	1	0	-5
t	-1	0	0	1	0	1	1	0	0	10
z	1	0	1	0	0	0	-1	0	0	10
y	0	1	0	1	0	0	0	-1	0	15
p	90	0	0	188	0	0	338	305	1	-7955

Multiply row 1 by -1.

	x	y	z	w	s	t	u	v	p		
*s	-1	0	0	1	-1	0	0	-1	0	5	
t	-1	0	0	1	0	1	1	0	0	10	$R_2 - R_1$
z	1	0	1	0	0	0	-1	0	0	10	
y	0	1	0	1	0	0	0	-1	0	15	$R_4 - R_1$
p	90	0	0	188	0	0	338	305	1	-7955	$R_5 - 188R_1$

	x	y	z	w	s	t	u	v	p	
w	-1	0	0	1	-1	0	0	-1	0	5
t	0	0	0	0	1	1	1	1	0	5
z	1	0	1	0	0	0	-1	0	0	10
y	1	1	0	0	1	0	0	0	0	10
p	278	0	0	0	188	0	338	493	1	-8895

p has a maximum value of $-\dfrac{8895}{1} = -8895$ so c has a minimum value of 8895 when

$x = 0$, $y = \dfrac{10}{1} = 10$, $z = \dfrac{10}{1} = 10$, and $w = \dfrac{5}{1} = 5$. You should fly no salespeople from Chicago to Los Angeles, 10 salespeople from Chicago to New York, 10 salespeople from Denver to Los Angeles, and 5 salespeople from Denver to New York for a total cost of \$8895.

41. Let x = number of boards shipped from Tucson to Honolulu, y = number of boards shipped from Tucson to Venice Beach, z = number of boards shipped from Toronto to Honolulu, and w = number of boards shipped from Toronto to Venice Beach. The objective function is the cost $c = 10x + 5y + 20z + 10w$. Translate into a maximization problem by maximizing $p = -10x - 5y - 20z - 10w$. The constraints are $x + y \leq 620$, $z + w \leq 410$, $x + z \geq 500$, $y + w \geq 530$, $x \geq 0$, $y \geq 0$, $z \geq 0$, $w \geq 0$. Introduce slack variables.

$$
\begin{aligned}
x + y + s &= 620 \\
z + w + t &= 410 \\
x + z - u &= 500 \\
y + w - v &= 530 \\
10x + 5y + 20z + 10w + p &= 0
\end{aligned}
$$

Set up the initial tableau and go through the procedure of the simplex method.

	x	y	z	w	s	t	u	v	p		
s	1	1	0	0	1	0	0	0	0	620	$R_1 - R_3$
t	0	0	1	1	0	1	0	0	0	410	
*u	1	0	1	0	0	0	-1	0	0	500	
*v	0	1	0	1	0	0	0	-1	0	530	
p	10	5	20	10	0	0	0	0	1	0	$R_5 - 10R_3$

	x	y	z	w	s	t	u	v	p		
s	0	1	-1	0	1	0	1	0	0	120	$R_1 - R_4$
t	0	0	1	1	0	1	0	0	0	410	
x	1	0	1	0	0	0	-1	0	0	500	
*v	0	1	0	1	0	0	0	-1	0	530	
p	0	5	10	10	0	0	10	0	1	-5000	$R_5 - 5R_4$

	x	y	z	w	s	t	u	v	p		
*s	0	0	-1	-1	1	0	1	1	0	-410	
·t	0	0	1	1	0	1	0	0	0	410	
x	1	0	1	0	0	0	-1	0	0	500	
y	0	1	0	1	0	0	0	-1	0	530	
p	0	0	10	5	0	0	10	5	1	-7650	

Multiply row 1 by -1.

	x	y	z	w	s	t	u	v	p		
*s	0	0	1	1	-1	0	-1	-1	0	410	
t	0	0	1	1	0	1	0	0	0	410	$R_2 - R_1$
x	1	0	1	0	0	0	-1	0	0	500	
y	0	1	0	1	0	0	0	-1	0	530	$R_4 - R_1$
p	0	0	10	5	0	0	10	5	1	-7650	$R_5 - 5R_1$

	x	y	z	w	s	t	u	v	p		
w	0	0	1	1	-1	0	-1	-1	0	410	
t	0	0	0	0	1	1	1	1	0	0	
x	1	0	1	0	0	0	-1	0	0	500	
y	0	1	-1	0	1	0	1	0	0	120	
p	0	0	5	0	5	0	15	10	1	-9700	

p has a maximum value of $-\dfrac{9700}{1} = -9700$ so c has a minimum value of 9700 when

$x = \dfrac{500}{1} = 500$, $y = \dfrac{120}{1} = 120$, $z = 0$, and $w = \dfrac{410}{1} = 410$. Thus ship 500 boards from Tucson to Honolulu, 120 boards from Tucson to Venice Beach, 0 boards from Toronto to Honolulu, and 410 boards from Toronto to Venice Beach for a total cost of $9700.

43. Answers will vary.

45. Answers will vary.

47. We need to eliminate starred rows so that corresponding variables of the basic solution are positive.

49. We want do not want an active variable to take on negative value.

You're the Expert

1. Suppose you send a plane from Chicago to New York. This costs $10,000, which saves $10,000 since it costs $20,000 to fly a plane to New York from Atlanta. On the other hand, a plane that you would have sent to either L.A. or Boston must be sent from Atlanta. Since flying a plane from Chicago to L.A. costs $50,000 and flying a plane from Atlanta to L.A. costs $70,000, you would lose $20,000 in flying a plane to L.A. Thus you would lose $10,000 overall. Since flying a plane from Chicago to Boston costs $20,000 and flying a plane from Atlanta to Boston costs $50,000, you would lose $30,000 in flying a plane to Boston. Thus you would lose $20,000 overall. Therefore, each plane you send to New York from Chicago will cost you an additional $10,000 or $20,000.

3. If more planes are flown from Chicago to L.A., then fewer planes are flown from Chicago to Boston and additional planes must be flown from Atlanta to Boston. Flying a plane from Chicago to L.A. costs $30,000 more than flying a plane from Chicago to Boston, while flying a plane from Atlanta to Boston costs $50,000, so flying the additional planes to L.A. from Chicago results in an increase in cost of $80,000 per plane.
If the additional planes are flown from Atlanta to L.A., the increase in cost is $70,000 per plane.
Your best option is to fly 5 planes from Chicago to L.A., none from Chicago to N.Y., 10 from Chicago to Boston, 10 from Atlanta to L.A., 20 from Atlanta to N.Y., and none from Atlanta to Boston. The total cost would be $1,550,000.

5. Suppose you send a plane from Chicago to Boston rather than L.A. This saves $30,000 since it costs $50,000 to fly from Chicago to L.A. but only $20,000 to fly from Chicago to Boston. However, this means that a plane from Atlanta must be sent to L.A., which costs $70,000. Thus, the increase in cost is $40,000. This is less than the $50,000 that it costs to send a plane from Atlanta to Boston.
Therefore, your best option is to fly no planes from Chicago to L.A., none from Chicago to N.Y., 15 from Chicago to Boston, 10 from Atlanta to L.A., 20 from Atlanta to N.Y., and none from Atlanta to Boston. The total cost would be $1,400,000.

7. Having more planes in Atlanta does not change the optimal solution. Note that it costs more to fly a plane from Atlanta to L.A., N.Y., or Boston than it does to fly a plane from Chicago to the same city. Also, there were planes remaining in Atlanta in the optimal solution to the original problem, so having more planes in Atlanta does not change the best solution.

9. The N.Y. airport being closed does not affect any of the planes being flown from Chicago. Since it costs more to fly planes to L.A. and Boston from Atlanta than from Chicago, the planes that would be flown to N.Y. should remain in Atlanta. The best option is to fly 5 planes from Chicago to L.A., 10 from Chicago to Boston, 5 from Atlanta to L.A., and none from Atlanta to Boston. The total cost would be $800,000.

Chapter 4 Review Exercises
Pages 325–329

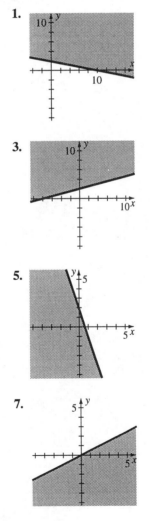

1.

3.

5.

7.

9.

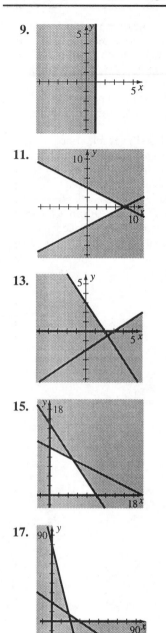

11.

13.

15.

17.

19. Sketch the feasible region.

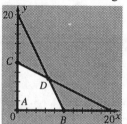

Point	Lines Through Point	Coordinates	$P = x + y$
A	$x = 0, y = 0$	$(0, 0)$	0
B	$y = 0$ $2x + y = 20$	$(10, 0)$	10
C	$x = 0$ $x + 2y = 20$	$(0, 10)$	10
D	$2x + y = 20$ $x + 2y = 20$	$\left(\dfrac{20}{3}, \dfrac{20}{3}\right)$	$\dfrac{40}{3}$

Maximum $P = \dfrac{40}{3}$ when $x = \dfrac{20}{3}$, $y = \dfrac{20}{3}$.

21. Sketch the feasible region.

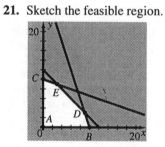

Point	Lines Through Point	Coordinates	$P = 2x + y$
A	$x = 0, y = 0$	$(0, 0)$	0
B	$y = 0$ $3x + y = 30$	$(10, 0)$	20
C	$x = 0$ $x + 3y = 30$	$(0, 10)$	10
D	$3x + y = 30$ $x + y = 12$	$(9, 3)$	21
E	$x + y = 12$ $x + 3y = 30$	$(3, 9)$	15

Maximum $P = 21$ when $x = 9, y = 3$.

23. Sketch the feasible region.

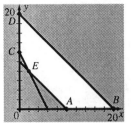

Point	Lines Through Point	Coordinates	$P = 2x + 3y$
A	$y = 0$ $x + y = 10$	$(10, 0)$	20
B	$y = 0$ $x + y = 20$	$(20, 0)$	40
C	$x = 0$ $2x + y = 12$	$(0, 12)$	36
D	$x = 0$ $x + y = 20$	$(0, 20)$	60
E	$2x + y = 12$ $x + y = 10$	$(2, 8)$	28

Maximum $P = 60$ when $x = 0$, $y = 20$.

25. Sketch the feasible region.

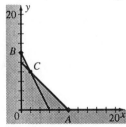

The feasible region is unbounded. There is no optimal solution.

27. Sketch the feasible region.

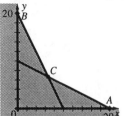

Point	Lines Through Point	Coordinates	$C = x + y$
A	$y = 0$ $x + 2y = 20$	$(20, 0)$	20
B	$x = 0$ $2x + y = 20$	$(0, 20)$	20
C	$2x + y = 20$ $x + 2y = 20$	$\left(\dfrac{20}{3}, \dfrac{20}{3}\right)$	$\dfrac{40}{3}$

Minimum $C = \dfrac{40}{3}$ when $x = \dfrac{20}{3}$, $y = \dfrac{20}{3}$.

29. Sketch the feasible region.

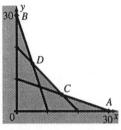

Point	Lines through Point	Coordinates	$C = 2x + y$
A	$y = 0$ $x + 3y = 30$	$(30, 0)$	60
B	$x = 0$ $3x + y = 30$	$(0, 30)$	30
C	$x + 3y = 30$ $x + y = 20$	$(15, 5)$	35
D	$3x + y = 30$ $x + y = 20$	$(5, 15)$	25

Minimum $C = 25$ when $x = 5$, $y = 15$.

31. Sketch the feasible region.

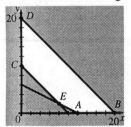

Point	Lines Through Point	Coordinates	$C = 2x + 3y$
A	$y = 0$ $x + 2y = 12$	$(12, 0)$	24
B	$y = 0$ $x + y = 20$	$(20, 0)$	40
C	$x = 0$ $x + y = 10$	$(0, 10)$	30
D	$x = 0$ $x + y = 20$	$(0, 20)$	60
E	$x + 2y = 12$ $x + y = 10$	$(8, 2)$	22

Minimum $C = 22$ when $x = 8$, $y = 2$

33. Sketch the feasible region.

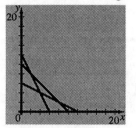

There is no feasible region. Thus there is no optimal solution.

35. Introduce slack variables.

$$
\begin{aligned}
3x &+ 2y + z + s &&= 60 \\
2x &+ y + 3z &+ t &= 60 \\
-x &- 2y - 3z &+ P &= 0
\end{aligned}
$$

Set up the initial tableau and go through the procedure of the simplex method.

	x	y	z	s	t	P		
s	3	2	1	1	0	0	60	$3R_1 - R_2$
t	2	1	3	0	1	0	60	
P	-1	-2	-3	0	0	1	0	$R_3 + R_2$

	x	y	z	s	t	P		
s	7	5	0	3	-1	0	120	
z	2	1	3	0	1	0	60	$5R_2 - R_1$
P	1	-1	0	0	1	1	60	$5R_3 + R_1$

	x	y	z	s	t	P	
y	7	5	0	3	-1	0	120
z	3	0	15	-3	6	0	180
P	12	0	0	3	4	5	420

P has a maximum value of $\dfrac{420}{5} = 84$ when $x = 0$,

$y = \dfrac{120}{5} = 24$ and $z = \dfrac{180}{15} = 12$.

37. Introduce slack variables.

$$
\begin{aligned}
3x + 2y + z + s &&&& = 60 \\
2x + y + 3z &&+ t && = 60 \\
x + y + z &&&+ u & = 22 \\
-x - 2y - 3z &&&& + P = 0
\end{aligned}
$$

Set up the initial tableau and go through the procedure of the simplex method.

	x	y	z	s	t	u	P	
s	3	2	1	1	0	0	0	60
t	2	1	3	0	1	0	0	60
u	1	1	1	0	0	1	0	22
P	−1	−2	−3	0	0	0	1	0

$3R_1 - R_2$

$3R_3 - R_2$

$R_4 + R_2$

	x	y	z	s	t	u	P	
s	7	5	0	3	−1	0	0	120
z	2	1	3	0	1	0	0	60
u	1	2	0	0	−1	3	0	6
P	1	−1	0	0	1	0	1	60

$2R_1 - 5R_3$

$2R_2 - R_3$

$2R_4 + R_3$

	x	y	z	s	t	u	P	
s	9	0	0	6	3	−15	0	210
z	3	0	6	0	3	−3	0	114
y	1	2	0	0	−1	3	0	6
P	3	0	0	0	1	3	2	126

P has a maximum value of $\dfrac{126}{2} = 63$ when $x = 0$, $y = \dfrac{6}{2} = 3$ and $z = \dfrac{114}{6} = 19$.

39. Introduce surplus and slack variables.

$$
\begin{aligned}
x + y + z - s &&&& = 100 \\
y + z &&+ t && = 80 \\
x &+ z &&&+ u & = 80 \\
-x - 2y - 3z &&&& + P = 0
\end{aligned}
$$

Set up the initial tableau and go through the procedure of the simplex method.

	x	y	z	s	t	u	P	
*s	1	1	1	−1	0	0	0	100
t	0	1	1	0	1	0	0	80
u	1	0	1	0	0	1	0	80
P	−1	−2	−3	0	0	0	1	0

$R_3 - R_1$

$R_4 + R_1$

	x	y	z	s	t	u	P	
x	1	1	1	−1	0	0	0	100
t	0	1	1	0	1	0	0	80
*u	0	−1	0	1	0	1	0	−20
P	0	−1	−2	−1	0	0	1	100

	x	y	z	s	t	u	P	
x	1	1	1	−1	0	0	0	100
t	0	1	1	0	1	0	0	80
*u	0	1	0	−1	0	−1	0	20
P	0	−1	−2	−1	0	0	1	100

$R_1 - R_3$

$R_2 - R_3$

$R_4 + R_3$

	x	y	z	s	t	u	P		
x	1	0	1	0	0	1	0	80	$R_1 - R_2$
t	0	0	1	1	1	1	0	60	
y	0	1	0	-1	0	-1	0	20	
P	0	0	-2	-2	0	-1	1	120	$R_4 + 2R_2$

	x	y	z	s	t	u	P	
x	1	0	0	-1	-1	0	0	20
z	0	0	1	1	1	1	0	60
y	0	1	0	-1	0	-1	0	20
P	0	0	0	0	2	1	1	240

P has a maximum value of 240 when $x = 20$, $y = 20$ and $z = 60$.

41. Translate into a maximization problem by maximizing $p = -x - 2y - 3z$. Introduce surplus variables.

$$3x + 2y + z - s = 60$$
$$2x + y + 3z - t = 60$$
$$x + 2y + 3z + p = 0$$

Set up the initial tableau and go through the procedure of the simplex method.

	x	y	z	s	t	p		
*s	3	2	1	-1	0	0	60	$2R_1 - 3R_2$
*t	2	1	3	0	-1	0	60	
p	1	2	3	0	0	1	0	$2R_3 - R_2$

	x	y	z	s	t	p	
*s	0	1	-7	-2	3	0	-60
x	2	1	3	0	-1	0	60
p	0	3	3	0	1	2	-60

	x	y	z	s	t	p	
s	0	-1	7	2	-3	0	60
x	2	1	3	0	-1	0	60
p	0	3	3	0	1	2	-60

p has a maximum value of $-\dfrac{60}{2} = -30$ so C has a minimum value of 30 when $x = \dfrac{60}{2} = 30$, $y = 0$ and $z = 0$.

43. Translate into a maximization problem by maximizing $p = -x - y - z$. Introduce surplus variables.

$$3x + 2y + z - s = 60$$
$$2x + y + 3z - t = 60$$
$$x + 3y + 2z - u = 60$$
$$x + y + z + p = 0$$

Set up the initial tableau and go through the procedure of the simplex method.

	x	y	z	s	t	u	p		
*s	3	2	1	-1	0	0	0	60	$R_1 - 3R_3$
*t	2	1	3	0	-1	0	0	60	$R_2 - 2R_3$
*u	1	3	2	0	0	-1	0	60	
p	1	1	1	0	0	0	1	0	$R_4 - R_3$

	x	y	z	s	t	u	p	
*s	0	-7	-5	-1	0	3	0	-120
*t	0	-5	-1	0	-1	2	0	-60
x	1	3	2	0	0	-1	0	60
p	0	-2	-1	0	0	1	1	-60

	x	y	z	s	t	u	p		
s	0	7	5	1	0	-3	0	120	$5R_1 - 7R_2$
t	0	5	1	0	1	-2	0	60	
x	1	3	2	0	0	-1	0	60	$5R_3 - 3R_2$
p	0	-2	-1	0	0	1	1	-60	$5R_4 + 2R_2$

	x	y	z	s	t	u	p		
s	0	0	18	5	-7	-1	0	180	
y	0	5	1	0	1	-2	0	60	$18R_2 - R_2$
x	5	0	7	0	-3	1	0	120	$18R_3 - 7R_1$
p	0	0	-3	0	2	1	5	-180	$6R_4 + R_1$

	x	y	z	s	t	u	p	
z	0	0	18	5	-7	-1	0	180
y	0	90	0	-5	25	-35	0	900
x	90	0	0	-35	-5	25	0	900
p	0	0	0	5	5	5	30	-900

p has a maximum value of $-\dfrac{900}{30} = -30$ so C has a minimum value of 30 when $x = \dfrac{900}{90} = 10$, $y = \dfrac{900}{90} = 10$ and

$z = \dfrac{180}{18} = 10.$

45. Since there are no points that satisfy all of the given conditions, there is no feasible region. Thus (a) is true.

47. The problem is equivalent to maximizing $p = x_1 + 4x_2 + 2x_3$. Introduce slack variables.

$$\begin{aligned}
4x_1 &+ x_2 + x_3 + s &&= 45 \\
-x_1 &+ x_2 + 2x_3 &&+ t &&= 0 \\
-x_1 &- 4x_2 - 2x_3 &&&&+ p = 0
\end{aligned}$$

Set up the initial tableau and go through the procedure of the simplex method.

	x_1	x_2	x_3	s	t	p		
s	4	1	1	1	0	0	45	$R_1 - R_2$
t	-1	1	2	0	1	0	0	
p	-1	-4	-2	0	0	1	0	$R_3 + 4R_2$

	x_1	x_2	x_3	s	t	p		
s	5	0	-1	1	-1	0	45	
x_2	-1	1	2	0	1	0	0	$5R_2 + R_1$
p	-5	0	6	0	4	1	0	$R_3 + R_1$

	x_1	x_2	x_3	s	t	p	
x_1	5	0	-1	1	-1	0	45
x_2	0	5	9	1	4	0	45
p	0	0	5	1	3	1	45

p has a maximum value of $\dfrac{45}{1} = 45$ when $x_1 = \dfrac{45}{5} = 9,$ $x_2 = \dfrac{45}{5} = 9$ and $x_3 = 0.$

Thus, the optimal value of Z is 35.

49. Let x = quarts of Continental Mocha. Let y = quarts of Succulent Strawberry.
The objective function is the profit $p = 3x + y$. The constraints are $2x + 3y \le 800$, $2x + y \le 400$, $x \ge 0$,
$y \ge 0$. The corner points of the feasible region are $(0, 0)$, $(200,0)$, $(100, 200)$, and $\left(0, 266\frac{2}{3}\right)$. At $(0, 0)$,
$p = 0$. At $(200,0)$, $p = 600$. At $(100, 200)$, $p = 500$. At $\left(0, 266\frac{2}{3}\right)$, $p = 266\frac{2}{3}$. Thus you should make 200 quarts of
Continental Mocha and no quarts of Succulent Strawberry in order to maximize profit.

51. Let x = ounces of fish. Let y = ounces of cornmeal.
The objective function is the cost $c = 5x + 5y$. The constraints are $8x + 4y \ge 48$, $4x + 8y \ge 48$, $x \ge 0$, $y \ge 0$.
The corner points of the feasible region are $(12, 0)$, $(4, 4)$, and $(0, 12)$. At $(12, 0)$, $c = 60$. At $(4, 4)$, $c = 40$. At $(0, 12)$, $c = 60$.
Thus they should use 4 ounces of fish and 4 ounces of cornmeal in each can of cat food in order to minimize costs.

53. Let x = number of the Sprinkle model, y = number of the Storm model, and z = number of the Hurricane model.
Introduce slack variables.

$$
\begin{array}{rcrcrcrcrcrcrcr}
x & + & 2y & + & 2z & + & s & & & & & & & = & 600 \\
2x & + & y & + & 3z & & & + & t & & & & & = & 600 \\
x & + & 3y & + & 6z & & & & & + & u & & & = & 600 \\
-x & - & y & - & 2z & & & & & & & + & p & = & 0
\end{array}
$$

Set up the initial tableau and go through the procedure of the simplex method.

	x	y	z	s	t	u	p		
s	1	2	2	1	0	0	0	600	$3R_1 - R_3$
t	2	1	3	0	1	0	0	600	$2R_2 - R_3$
u	1	3	6	0	0	1	0	600	
p	-1	-1	-2	0	0	0	1	0	$3R_4 + R_3$

	x	y	z	s	t	u	p		
s	2	3	0	3	0	-1	0	1200	$3R_1 - 2R_2$
t	3	-1	0	0	2	-1	0	600	
z	1	3	6	0	0	1	0	600	$3R_3 - R_2$
p	-2	0	0	0	0	1	3	600	$3R_4 + 2R_2$

	x	y	z	s	t	u	p		
s	0	11	0	9	-4	-1	0	2400	$10R_1 - 11R_3$
x	3	-1	0	0	2	-1	0	600	$10R_2 + R_3$
z	0	10	18	0	-2	4	0	1200	
p	0	-2	0	0	4	1	9	3000	$5R_4 + R_3$

	x	y	z	s	t	u	p	
s	0	0	-198	90	-18	-54	0	10,800
x	30	0	18	0	18	-6	0	7200
y	0	10	18	0	-2	4	0	1200
p	0	0	18	0	18	9	45	16,200

p has a maximum value of $\dfrac{16,200}{45} = 360$ when $x = \dfrac{7200}{30} = 240$, $y = \dfrac{1200}{10} = 120$ and $z = 0$. Thus you should build 240 of
the Sprinkle model, 120 of the Storm model and none of the Hurricane model in order to maximize profits.

55. Let x = amount invested in TIAA, y = amount invested in the stock fund, z = amount invested in the money market fund, v = amount invested in the bond fund, and w = amount invested in the social choice fund.
The objective function is the risk $c = 8y + 5z + 6v + 8w$. The constraints are $x + y + z + v + w \le 20{,}000$,
$0.0899x + 0.1445y + 0.0701z + 0.1132v + 0.1403w \ge 2000$, $x \ge 0$, $y \ge 0$, $z \ge 0$, $v \ge 0$, $w \ge 0$. Translate into a maximization problem by maximizing $p = -8y - 5z - 6v - 8w$.
The initial matrix is as follows.

1	1	1	1	1	1	0	0	20,000
0.0899	0.1445	0.0701	0.1132	0.1403	0	−1	0	2000
0	8	5	6	8	0	0	1	0

$$R_2 \to \frac{1}{0.1445} R_2$$

1	1	1	1	1	1	0	0	20,000
0.622145	1	0.485121	0.783391	0.970934	0	−6.92042	0	13,840.8
0	8	5	6	8	0	0	1	0

$$R_1 \to R_1 - R_2$$
$$R_3 \to R_3 - 8R_2$$

0.377855	0	0.514879	0.216609	0.029066	1	6.92042	0	6159.17
0.622145	1	0.485121	0.783391	0.970934	0	−6.92042	0	13,840.8
−4.97716	0	1.119030	−0.267128	0.232526	0	55.3633	1	−110,727

$$R_1 \to \frac{1}{0.377855} R_1$$

1	0	1.36264	0.573260	0.076923	2.64652	18.315	0	16,300.4
0.622145	1	0.485121	0.783391	0.970934	0	−6.92042	0	13,840.8
−4.97716	0	1.119030	−0.267128	0.232526	0	55.3633	1	−110,727

$$R_2 \to R_2 - 0.622145 R_1$$
$$R_3 \to R_3 + 4.97716 R_1$$

1	0	1.36264	0.573260	0.076923	2.64652	18.315	0	16,300.4
0	1	−0.362637	0.426740	0.923077	−1.64652	−18.315	0	3699.63
0	0	7.9011	2.58608	0.615385	13.1722	146.52	1	−29,597.1

Thus, you should allocate \$16,300 to TIAA and \$3700 to the stock fund.

1	1	1	1	1	1	0	0	20,000
0.0899	0.1445	0.0701	0.1132	0.1403	0	−1	0	2000
4	8	5	5	8	0	0	1	0

$$R_2 \to \frac{1}{0.1445} R_2$$

1	1	1	1	1	1	0	0	20,000
0.622145	1	0.485121	0.783391	0.970934	0	−6.92042	0	13,840.8
4	8	5	5	8	0	0	1	0

$R_1 \rightarrow R_1 - R_2$
$R_3 \rightarrow R_3 - 8R_2$

0.377855	0	0.514879	0.216609	0.029066	1	6.92042	0	6159.17
0.622145	1	0.485121	0.783391	0.970934	0	−6.92042	0	13,840.8
−0.977163	0	1.119030	−1.26713	0.232526	0	55.3633	1	−110,727

$R_2 \rightarrow \dfrac{1}{0.783391} R_2$

0.377855	0	0.514879	0.216609	0.029066	1	6.92042	0	6159.17
0.79417	1.2765	0.619258	1	1.2394	0	−8.83392	0	17,667.8
−0.977163	0	1.119030	−1.26713	0.232526	0	55.3633	1	−110,727

$R_1 \rightarrow R_1 - 0.216609 R_2$
$R_3 \rightarrow R_3 + 1.26713 R_2$

0.20583	−0.276502	0.380743	0	−0.239399	1	8.83392	0	2332.16
0.79417	1.2765	0.619258	1	1.2394	0	−8.83392	0	17,667.8
0.029152	1.61749	1.90371	0	1.803	0	44.1696	1	−88,339.2

Thus, you should allocate $17,668 in the bond fund.

57. Let x = number of cars from Northside to Eastside, y = number of cars from Northside to Westside, z = number of cars from Airport to Eastside, and w = number of cars from Airport to Westside.
The objective function is the cost $c = x + 2y + 2z + 4w$. The constraints are $x + y \leq 10$, $z + w \leq 15$, $x + z \geq 12$, $y + w \geq 7$, $x \geq 0$, $y \geq 0$, $z \geq 0$, $w \geq 0$.
Translate into a maximization problem by maximizing $p = -x - 2y - 2z - 4w$.

$$
\begin{array}{rcrcrcrcrcrcrcrcr}
x &+& y &&&&&&+& s &&&&&&&=& 10 \\
&& && z &+& w &&&&+& t &&&&&=& 15 \\
x &&&&+& z &&&&&&&-& u &&&=& 12 \\
&& y &&&&+& w &&&&&&&-& v &=& 7 \\
x &+& 2y &+& 2z &+& 4w &&&&&&&&&&+& p &=& 0
\end{array}
$$

Set up the initial tableau and go through the procedure of the simplex method.

	x	y	z	w	s	t	u	v	p		
s	1	1	0	0	1	0	0	0	0	10	$R_1 - R_4$
t	0	0	1	1	0	1	0	0	0	15	
*u	1	0	1	0	0	0	−1	0	0	12	
*v	0	1	0	1	0	0	0	−1	0	7	
p	1	2	2	4	0	0	0	0	1	0	$R_5 - 2R_4$

	x	y	z	w	s	t	u	v	p		
s	1	0	0	−1	1	0	0	1	0	3	
t	0	0	1	1	0	1	0	0	0	15	$R_2 - R_3$
*u	1	0	1	0	0	0	−1	0	0	12	
y	0	1	0	1	0	0	0	−1	0	7	
p	1	0	2	2	0	0	0	2	1	−14	$R_5 - 2R_3$

	x	y	z	w	s	t	u	v	p		
s	1	0	0	-1	1	0	0	1	0	3	
t	-1	0	0	1	0	1	1	0	0	3	$R_2 + R_1$
z	1	0	1	0	0	0	-1	0	0	12	$R_3 - R_1$
y	0	1	0	1	0	0	0	-1	0	7	
p	-1	0	0	2	0	0	2	2	1	-38	$R_5 + R_1$

	x	y	z	w	s	t	u	v	p	
x	1	0	0	-1	1	0	0	1	0	3
t	0	0	0	0	1	1	1	1	0	6
z	0	0	1	1	-1	0	-1	-1	0	9
y	0	1	0	1	0	0	0	-1	0	7
p	0	0	0	1	1	0	2	3	1	-35

Thus Federal should drive 3 cars from Northside to Eastside, 7 cars from Northside to Westside, 9 cars from Airport to Eastside, and no cars from Airport to Westside.

59. Let x = the amount invested in the stock fund run by Integrity Investments, y = the amount invested in the bond fund sold by Citizen's Bank, and z = the amount invested in the guaranteed annuity run by Publisher's Insurance.

The objective function is the amount invested in the stock fund $p = x$. The constraints are $x + y + z \le 10,000$,

$x + y \le \dfrac{1}{4}(x + y + z)$ or $3x + 3y - z \le 0$, $x \ge 0$, $y \ge 0$, $z \ge 0$.

Introduce slack variables.

$$
\begin{aligned}
x &+ y + z + s &&= 10,000 \\
3x &+ 3y - z &+ t &= 0 \\
-x & &+ p &= 0
\end{aligned}
$$

Set up the initial tableau and go through the procedure of the simplex method.

	x	y	z	s	t	p		
s	1	1	1	1	0	0	10,000	$3R_1 - R_2$
t	3	3	-1	0	1	0	0	
p	-1	0	0	0	0	1	0	$3R_3 + R_2$

	x	y	z	s	t	p		
s	0	0	4	3	-1	0	30,000	
x	3	3	-1	0	1	0	0	$4R_2 + R_1$
p	0	3	-1	0	1	3	0	$4R_3 + R_1$

	x	y	z	s	t	p	
z	0	0	4	3	-1	0	30,000
x	12	12	0	3	3	0	30,000
p	0	12	0	3	3	12	30,000

JoAnn should invest $\dfrac{30,000}{12} = \$2500$ in the stock fund, $\$0$ in the bond fund, and $\dfrac{30,000}{4} = \$7500$ in the annuity.

Chapter 5

1. $F = \{$spring, summer, fall, winter$\}$

3. $I = \{1, 2, 3, 4, 5, 6\}$

5. $A = \{1, 2, 3\}$

7. $B = \{2, 4, 6, 8\}$

9. $S = \{$HH, HT, TH, TT$\}$

11. $S = \{$H1, H2, H3, H4, H5, H6, T1, T2, T3, T4, T5, T6$\}$

13. $S = \{(5,1), (4,2), (3,3), (2,4), (1,5)\}$

15. $S = \varnothing$

17. $A \cup B = \{$June, Janet, Jill, Justin, Jeffrey, Jello$\} \cup \{$Janet, Jello, Justin$\} = \{$June, Janet, Jill, Justin, Jeffrey, Jello$\} = A$

19. $A \cup \varnothing = A$

21. $A \cup (B \cup C)$
    $= \{$June, Janet, Jill, Justin, Jeffrey, Jello$\} \cup (\{$Janet, Jello, Justin$\} \cup \{$Sally, Solly, Molly, Jolly, Jello$\})$
    $= \{$June, Janet, Jill, Justin, Jeffrey, Jello$\} \cup \{$Janet, Jello, Justin, Sally, Solly, Molly, Jolly$\}$
    $= \{$June, Janet, Jill, Justin, Jeffrey, Jello, Sally, Solly, Molly, Jolly$\}$

23. $C \cap B = \{$Sally, Solly, Molly, Jolly, Jello$\} \cap \{$Janet, Jello, Justin$\} = \{$Jello$\}$

25. $\dot{A} \cap \varnothing = \varnothing$

27. $(A \cap B) \cap C = \{$Janet, Jello, Justin$\} \cap \{$Sally, Solly, Molly, Jolly, Jello$\} = \{$Jello$\}$

29. $(A \cap B) \cup C = \{$Janet, Jello, Justin$\} \cup \{$Sally, Solly, Molly, Jolly, Jello$\} = \{$Janet, Jello, Justin, Sally, Solly, Molly, Jolly$\}$

31. (owe money) and (at least \$10,000 worth of business) $= A \cap B = \{$Acme, Crafts$\}$

33. (at least \$10,000 worth of business) or (employed her last year)
    $B \cup C = \{$Acme, Brothers, Crafts, DeTour, Effigy, Global, Hilbert$\}$

35. (do not owe money) and (employed her last year)
    $= A' \cap C = \{$Brothers, DeTour, Friends, Hilbert$\} \cap \{$Acme, Crafts, DeTour, Effigy, Global, Hilbert$\} = \{$DeTour, Hilbert$\}$

37. (owe money) $\cap$ {not at least \$10,000 worth of business} $\cap$ (not employed her last year) $= A \cap B' \cap C'$
    $= \{$Acme, Crafts, Effigy, Gerbal$\} \cap \{$Effigy, Friends, Global, Hilbert$\} \cap \{$Brothers, Friends$\}$
    $= \{$Effigy, Global$\} \cap \{$Brothers, Friends$\}$
    $= \varnothing$

39–43. $E = \{$at least one die shown even when two distinguishable dice are thrown$\}$
    $= \{(1, 2), (1, 4), (1, 6), (2, 1), (2, 2), (2, 3), (2, 4), (2, 5), (2, 6), (3, 2), (3, 4), (3, 6), (4, 1), (4, 2), (4, 3), (4, 4), (4, 5),$
    $(4, 6), (5, 2), (5, 4), (5, 6), (6, 1), (6, 2), (6, 3), (6, 4), (6, 5), (6, 6)\}$
    $F = \{$at least one die shows odd when two distinguishable dice are thrown$\}$
    $= \{(1, 1), (1, 2), (1, 3), (1, 4), (1, 5), (1, 6), (2, 1), (2, 3), (2, 5), (3, 1), (3, 2), (3, 3), (3, 4), (3, 5), (3, 6), (4, 1), (4, 3),$
    $(4, 5), (5, 1), (5, 2), (5, 3), (5, 4), (5, 5), (5, 6), (6, 1), (6, 3), (6, 5)\}$

39. $E' = \{(1, 1), (1, 3), (1, 5), (3, 1), (3, 3), (3, 5), (5, 1), (5, 3), (5, 5)\}$

41. $(E \cup F)' = S' = \varnothing$

43. $E' \cup F'$
={(1, 1), (1, 3), (1, 5), (3, 1), (3, 3), (3, 5), (5, 1), (5, 3), (5, 5), (2, 2), (2, 4), (2, 6), (4, 2), (4, 4), (4, 6), (6, 2), (6, 4), (6, 6)}

45. $(A \cup B)'$ is the region outside of both A and B shaded in the figure below.

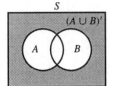

$A' \cap B'$ is the intersection of the region outside of A and the region outside of B. This is the darker shaded region in the figure below.

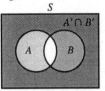

We see from the two figures that $(A \cup B)' = A' \cap B'$.

47. First shade $A \cap B$ in one tone and then shade C in another. $(A \cap B) \cap C$ is the region in both. This is the darker shaded region in the figure below.

First shade A in one tone and then shade $B \cap C$ in another. $A \cap (B \cap C)$ is the region in both. This is the darker shaded region in the figure below.

We see from the two figures that $(A \cap B) \cap C = A \cap (B \cap C)$.

49. First shade A in one tone and then shade $B \cap C$ in another. $A \cup (B \cap C)$ is all of the shaded regions in the figure below.

First shade $A \cup B$ in one tone and then shade $A \cup C$ in another. $(A \cup B) \cap (A \cup C)$ is the region in both. This is the darker shaded region in the figure below.

We see from the two figures that $A \cup (B \cap C) = (A \cup B) \cap (A \cup C)$.

51. The shaded region in the figure below is S. S' would be the unshaded region, but since there is none, $S' = \varnothing$.

S

53. $\{(1, 1), (1, 2), (1, 3), (1, 4), (1, 5), (1, 6), (2, 2), (2, 3), (2, 4), (2, 5), (2, 6), (3, 3), (3, 4), (3, 5), (3, 6), (4, 4), (4, 5), (4, 6),$
$(5, 5), (5, 6), (6, 6)\}$

55. Let A = soccer, B = rugby, and C = cricket
It is not evident if this statement implies
$A \cup (B \cap C)$ or $(A \cup B) \cap C$
These two expressions are *not* equivalent.

57. Answers will vary.

59. Let A = movies that are violent
$\qquad B$ = movies that are shorter than two hours
$\qquad C$ = movies that have a tragic ending
$\qquad D$ = movies that have an unexpected ending
She prefers

movies that are not violent, $(A'$	$\cap$	are shorter than two hours $B)$	and $\cap$	have neither a tragic ending $(C'$	nor $\cup$	an unexpected ending $D')$

Section 5.2 The Number of Elements in a Set
Pages 350–352

1. $n(A) = 4, n(B) = 5$
$n(A) + n(B) = 4 + 5 = 9$

3. $n(A \cup B) = n(A) + n(B) - n(A \cap B)$
$A \cap B = \{\text{Frans, Sarie}\}, \ n(A \cap B) = 2$
$n(A \cup B) = 4 + 5 - 2 = 7$

5. $n(A \cup (B \cap C)) = n(A) + n(B \cap C) - n(A \cap (B \cap C))$
$B \cap C = \{\text{Frans}\}, \ n(B \cap C) = 1$
$A \cap (B \cap C) = \{\text{Frans}\}, \ n(A \cap (B \cap C)) = 1$
$n(A \cup (B \cap C)) = 4 + 1 - 1 = 4$

7. Verify $n(A \cup B) = n(A) + n(B) - n(A \cap B)$
$A \cup B = \{\text{Dirk, Johan, Frans, Sarie, Tina, Klaas, Henrika}\}$
$n(A \cup B) = 7$
$n(A) = 4, n(B) = 5$
$A \cap B = \{\text{Frans, Sarie}\}, \ n(A \cap B) = 2$
$n(A) + n(B) - n(A \cap B) = 4 + 5 - 2 = 7$
So $n(A \cup B) = n(A) + n(B) - n(A \cap B)$

9. $n(A) = 43, n(B) = 20, \ n(A \cap B) = 3$
$n(A \cup B) = n(A) + n(B) - n(A \cap B) = 43 + 20 - 3 = 60$

11. $n(A \cup B) = 100, \ n(A) = n(B) = 60$
$n(A \cup B) = n(A) + n(B) - n(A \cap B)$
$n(A \cap B) = n(A) + n(B) - n(A \cup B) = 60 + 60 - 100 = 20$

13. $A' = \{$Barnsley, Manchester United, Sheffield United, Witbank Aces, Dundee United, Lyon$\}$
$n(A') = 6$

15. $A \cap B = \{$Southend$\}$, $n(A \cap B) = 1$
$n((A \cap B)') = n(S) - n(A \cap B) = 10 - 1 = 9$

17. From Exercise 13, $A' = \{$Barnsley, Manchester United, Sheffield United, Witbank Aces, Dundee United, Lyon$\}$.
$B' = \{$Sheffield United, Liverpool, Marolea Swallows, Witbank Aces, Royal Tigers, Dundee United, Lyon$\}$
$A' \cap B' = \{$Sheffield United, Witbank Aces, Dundee United, Lyon$\}$
$n(A' \cap B') = 4$

19. Verify $n((A \cap B)') = n(A') + n(B') - n((A \cup B)')$
$n((A \cap B)') = n(S) - n(A \cap B) = 10 - 1 = 9$
$n((A \cup B)') = n(S) - n(A \cup B) = 10 - 6 = 4$
$n(A') + n(B') - n((A \cup B)') = 6 + 7 - 4 = 9$
So, $n((A \cap B)') = n(A') + n(B') - n((A \cup B)')$

21.

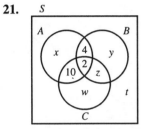

Given: $n(A) = 20, n(B) = 20, n(C) = 28$, $n(B \cap C) = 8$, $n(S) = 50$
$n(A) = x + 4 + 2 + 10 = 20$
$x + 16 = 20$
$x = 4$
$n(B \cap C) = 2 + z = 8$
$2 + z = 8$
$z = 6$
$n(B) = 4 + y + 2 + z = 20$
$y + z = 14$
$y + 6 = 14$
$y = 8$
$n(C) = w + 10 + 2 + z = 28$
$w + z = 16$
$w + 6 = 16$
$w = 10$
$n(S) = x + 4 + 2 + 10 + y + z + w + t = 50$
$x + y + z + w + t = 34$
$4 + 8 + 6 + 10 + t = 34$
$t = 6$

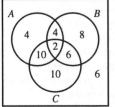

23.

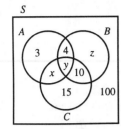

Given: $n(A) = 10$, $n(B) = 19$, $n(S) = 140$
$n(A) = x + y + 3 + 4 = 10$
$n(B) = y + z + 4 + 10 = 19$
$n(S) = 3 + 4 + x + y + z + 10 + 15 + 100 = 140$
Thus, we get the following system of equations.
$x + y = 3$
$y + z = 5$
$x + y + z = 8$
Solving, we get $(x, y, z) = (3, 0, 5)$.

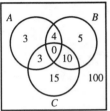

25. Let B = have black hair
W = whole row of seats to themselves
Given: $n(B \cup W) = 37$, $n(B) = 33$, $n(W) = 6$
$B \cap W$ = black-haired people who had whole rows to themselves
$n(B \cap W) = n(B) + n(W) - n(B \cup W)$
$\qquad = 33 + 6 - 37 = 2$

27. Let S = subjects in a psychological study,
H = subjects highly motivated, and
P = subjects attended preschool.
$H \cap P$ = subjects highly motivated who attended
$\qquad\qquad$ preschool
$n(S) = 400$, $n(H) = 180$, $n(P) = 220$,
$n(H \cap P) = 84$

a. $n(H \cap P') = n(H) - n(H \cap P) = 180 - 84 = 96$

b. $n(H' \cap P') = n(S) - n(H \cup P)$
$\qquad = n(S) - [n(H) + n(P) - n(H \cap P)]$
$\qquad = n(S) - n(H) - n(P) + n(H \cap P)$
$\qquad = 400 - 180 - 220 + 84$
$\qquad = 84$

29. Let S = categories of South African mutual funds,
M = mutual funds offered by Old Mutual, Inc.,
N = mutual funds offered by Sanlam, Inc., and
F = mutual funds offered by Syfrets, Inc.
$M \cap N$ = mutual funds offered by Old Mutual and Sanlam
$N \cap F$ = mutual funds offered by Sanlam and Syfrets
$F \cap M$ = mutual funds offered by Syfrets and Old Mutual
$M \cap N \cap F$ = mutual funds offered by all three
$n(S) = 30$, $n(M) = 10$, $n(N) = 6$, $n(F) = 6$,
$n(M \cap N) = 4$, $n(N \cap F) = 2$, $n(F \cap M) = 2$,
$n(M \cap N \cap F) = 2$

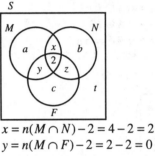

$x = n(M \cap N) - 2 = 4 - 2 = 2$
$y = n(M \cap F) - 2 = 2 - 2 = 0$
$z = n(N \cap F) - 2 = 2 - 2 = 0$
$a = n(M) - x - y - 2 = 10 - 2 - 0 - 2 = 6$
$b = n(N) - x - z - 2 = 6 - 2 - 0 - 2 = 2$
$c = n(F) - y - z - 2 = 6 - 0 - 2 = 4$
a = number of categories offered by Old Mutual and none of the other two categories = 6

31. Let S = Enormous State University students,
C = students who enjoy classical music,
R = students who enjoy popular rock music,
and H = students who enjoy heavy metal music.
$C \cap R$ = students who enjoy both classical and popular rock
$n(S) = 100$, $n(C) = 21$, $n(R) = 22$, $n(H) = 27$,
$n(C \cap R) = 5$

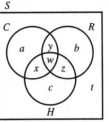

$b + z$ = number of students who enjoyed popular rock but did not enjoy classical music
$b + z = n(R) - n(C \cap R) = 22 - 5 = 17$

33. Let S = people who attend movies,
F = people who attend science fiction movies,
A = people who attend adventure movies, and
H = people who attend horror movies.
$n(S) = 100$, $n(F) = 40$, $n(A) = 55$, $n(H) = 35$,
$n(F \cap A) = b + w = 25$, $n(A \cap H) = w + c = 5$,
$n(F \cap H) = a + w = 15$, $n(F \cap A \cap H) = w = 5$

a.

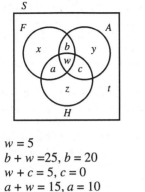

$w = 5$
$b + w = 25, b = 20$
$w + c = 5, c = 0$
$a + w = 15, a = 10$
$n(F) = x + b + w + a, x = 5$
$n(A) = y + b + w + c, y = 30$
$n(H) = z + a + w + c, z = 20$
$t = n(S) - (x + y + z + a + b + c + w), t = 10$

b. $n(F \cap H) = 15$
$n(F) = 40$
$\dfrac{\text{science fiction and horror}}{\text{science fiction}} = \dfrac{15}{40} = .375 = 37.5\%$
The survey suggests that 37.5% of science fiction movie fans are also horror movie fans.

35. Since $n(A \cup B) = n(A) + n(B) - n(A \cap B)$, $n(A \cup B) \neq n(A) + n(B)$ when $A \cap B \neq \varnothing$.

37. Answers will vary.

Section 5.3 The Multiplication Principle; Permutations
Pages 361–366

1. $6! = 6 \cdot 5 \cdot 4 \cdot 3 \cdot 2 \cdot 1 = 720$

3. $\dfrac{8!}{6!} = \dfrac{8 \cdot 7 \cdot 6!}{6!} = 56$

5. $P(6,4) = \dfrac{6!}{2!} = \dfrac{6 \cdot 5 \cdot 4 \cdot 3 \cdot 2!}{2!} = 360$

7. $\dfrac{P(6,4)}{4!} = \dfrac{6!}{4!2!} = \dfrac{6 \cdot 5 \cdot 4!}{4! \cdot 2 \cdot 1} = 15$

9. Step 1: Choose the outcome from the first die: 6 choices
Step 2: Choose the outcome from the second die: 6 choices
Thus, the total number of choices is $6 \times 6 = 36$.

11. money: 5 distinguishable letters
By the multiplication principle there is a total of $5 \cdot 4 \cdot 3 \cdot 2 \cdot 1 = 5! = 120$ possible 5-letter sequences.

13. a, e, a, a, u, k

Step 1: Choose a place for k: 6 choices
Step 2: Choose a place fo u: 5 choices
Step 3: Choose a place for e: 4 choices
Step 4: Choose a place for $3a$'s: 1 choice
possible sequences: $6 \times 5 \times 4 \times 1 = 120$

15. possible outcomes for T/F quarters $\times$ possible outcomes for M/C
(10 T/F with 2 possible answers/questions) $\times$ (2 M/C with 5 answers/questions)
$= 2^{10} \times 5^2 = 1024 \times 25 = 25,600$ possible answer sheets

17. $2 \times 2 \times 2 = 8$ possible ways
$\uparrow \quad \uparrow \quad \uparrow$
(2nd slots) (3rd slot)

19. a. Each digit has 10 possible outcomes.
$10 \times 10 \times 10 \times 10 \times 10 \times 10 \times 10 = 10^7 = 10,000,000$ possible numbers

b. 463 □ □ □ □
↓ ↓ ↓ ↓
$10 \times 10 \times 10 \times 10 = 10^4 = 10,000$ possible numbers

c. 10 choices for the first number, but 9 choices for the rest
↓
$10 \times 9 \times 9 \times 9 \times 9 \times 9 \times 9 = 10 \times 9^6 = 5,314,410$

21.

first step	second step	third step	fourth step
↓	↓	↓	↓
4	$\times$ 3	$\times$ 2	$\times$ 1

$= 4! = 24$ possible itineraries

23.

dessert or appetizer	$\times$	soups	$\times$	main courses
↓		↓		↓
$(5 + 5)$	$\times$	2	$\times$	4

$= 80$ possible three course meals

25. nine possible *wrong* digits for each digit
So, $9 + 9 + 9 + 9 = 4 \times 9 = 36$ incorrect codes for a programmed padlock.

27. a. $4 \times 4 \times 4 = 4^3 = 64$ because bases can be repeated

b. n-element DNA chains: 4^n chains

c. 2.1×10^{10} elements: $4^{2.1 \times 10^{10}}$ DNA chains possible

29. Each place in the four-digit sequence has 16 choices, so there are $16^4 = 65,536$ possible different encodings.

31. 10 persons and each person shakes hands exactly once with each of the others:
total number of handshakes = $9 + 8 + 7 + 6 + 5 + 4 + 3 + 2 + 1 = 45$
(d) 45 is the correct answer.

33. Step 1: Choose a day of the week for January 1: 7 choices
Step 2: Decide whether or not it is a leap year: 2 choices
Total: $7 \times 2 = 14$ possible calendars

35. a. First cylinder: 6 choices;
Second cylinder: 3 choices;
Third cylinder: 2 choices;
Fourth cylinder: 2 choices;
Fifth cylinder: 1 choice;
Sixth cylinder: 1 choice
Total number of good firing sequences:
$6 \times 3 \times 2 \times 2 \times 1 \times 1 = 72$

b. $3 \times 3 \times 2 \times 2 \times 1 \times 1 = 36$

37. License plate: sequence of three letters followed by three digits

a. $26^3 \times 10^3 = 17,576,000$

b. $26^2 \times 23 + 10^3 = 15,548,000$

c. $\left(26^2 \times 23 - 3\right) \times 10^3 = 15,545,000$

39. $2 \times 2 \times 2 \times 2 = 2^4 = 16$

41. a. $P(4, 1) = 4$

b. $P(4, 1) = 4$ (same)

c. Allowing for left and/or up moves could lead to infinitely many different routes.

43. 10 male and 8 female actors available for 7 male roles and 4 female roles:
$P(10, 7) \cdot P(8, 4)$
$= \dfrac{10!}{3!} \cdot \dfrac{8!}{4!} = (10 \times 9 \times 8 \times 7 \times 6 \times 5 \times 4) \times (8 \times 7 \times 6 \times 5) = 1,016,064,000$ possible casts

45. ten vertical colored lines:
Every second line starting with the first: either green or red
has 2^5 choices
while the remaining five lines: either all yellow, all pink, or all purple
has 3 choices
Thus, the total is $2 \times 2 \times 2 \times 2 \times 2 \times 3 = 2^5 \times 3 = 96$ paintings

47. 23 cities

a. each city exactly once: 23!

b. The first five stops have already been determined: $(23 - 5)! = 18!$

c. Itinerary must include the sequence Anchorage, Fairbanks, Seatlle, Chicago, and Detroit:
$$19 \quad \times \quad 18! = 19!$$
$$\uparrow \qquad\qquad \uparrow$$
possible places   5 stops already
for the 5 stops   determined
$(23 - 5 = 18$ cities left)

49. Step 1: Choose a row: m choices
Step 2: Choose a column: n choices
Thus, there are $m \cdot n$ possible outcomes.

51. each bit is either a 0 or a 1
Thus, a computer consisting of eight "bits" has $2^8 = 256$ different character representations.

53. first loop: 10 times
second loop: 19 times
third loop: 10 times
Total: $10 \times 19 \times 10 = 1900$ times

55. The first choice has n choices.
The second choice has $(n-1)$ choices.

.(etc)

The second to last choice has 2 choices.
The last choice has only 1 choice.
Thus, by the multiplication principle,
total $= n(n-1)(n-2)...(2)(1) = n!$

57. Student-specific: Answers may vary.

Section 5.4 The Addition Principle; Combinations
Pages 379–382

1. $C(3,2) = \frac{3!}{2!1!} = 3$

3. $C(10,8) = \frac{10!}{8!2!} = 45$

5. $C(20,1) = \frac{20!}{1!19!} = 20$

7. $C(100,98) = \frac{100!}{98!2!} = 4950$

9. alternative 1: 3.5"
choose density: 2 choices
choose color: 3 choices
total: $2 \times 3 = 6$ choices
alternative 2: 5.25"
choose density: 2 choices
choose color: 1 choice
total: $2 \times 1 = 2$ choices
total possible choices = 6 +2 =8

11. Part A (8 T/F): 2^8 choices
Part B (5 one correct out of five): 5^5 choices
Part C (match 5 Q with 5 A): $5!$ choices
Total: $2^8 + 5^5 + 5! = 3501$ different completed answer sheets

13. 7 contestants with 4 semifinalists:
number of possible selections which contain *neither* Ben *nor* Ann = $C(5, 4) = 5$
a is the correct answer

15. Alternative 1: letter, 26 choices
Alternative 2: letters and digits
choose letter: 26 choices
Choose digit: 10 choices
total $= 26 + 26 \times 10 = 286$ possible different variables

17. Possible groups of four marbles:
$C(10, 4) = 210$

19. Possible groups of four marbles which include all the red ones
$C(3,3) \times C(7,1) = 7$

21. Possible groups of four marbles which include none of the red ones:
$C(3,0) \times C(7,4) = 35$

23. Possible groups of four marbles which include one of each color other than lavender:
$C(3,1) \times C(2,1) \times C(2,1) \times C(2,1) = 3 \times 2 \times 2 \times 2 = 24$

25. Possible groups of five marbles which include at least two red ones:
Alternative 1: 2 red ones
$C(3,2) \times C(7,3)$
Alternative 2: 3 red ones
$C(3,3) \times C(7,2)$
Total: $C(3, 2) \cdot C(7, 3) + C(3, 3) \cdot C(7, 2) = 105 + 21 = 126$

27. Possible groups of five marbles which include at most one of the yellow ones:
$C(2, 1) \cdot C(8, 4) + C(2, 0) \cdot C(8, 5) = 196$

29. Possible groups of five marbles which include either the lavender one or a yellow one, but not both:
$C(1, 1) \cdot C(7, 4) + C(2, 1) \cdot C(7, 4) = 105$

31. exactly 5 ones: $C(30,5) \times 5^{25}$
fraction of these sequences which have exactly 5 ones
$$C(30,5)\left(\frac{1}{6}\right)^5 \left(\frac{5}{6}\right)^{25} = \frac{C(30,5) \times 5^{25}}{6^{30}} = .192$$

33. exactly 15 even numbers: 3^{15}(even) $\cdot 3^{15}$(odd)
$$C(30,15) \cdot 3^{15} \cdot C(15,15) \cdot 3^{15} = C(30,15) \cdot C(15,15) \cdot 3^{15} \cdot 3^{15}$$
$\qquad \uparrow \qquad \uparrow \qquad\qquad \uparrow$
choose   exactly   exactly
15   15 even   15 odd
$$\frac{C(30, 15) \cdot C(15, 15) \cdot 3^{15} \cdot 3^{15}}{6^{30}} = 0.144$$

35. $2^1 + 2^2 + \cdots + 2^x \geq 26$, so $2 + 4 + 8 + 16 + \cdots + 2^x \geq 26$.
$x = 4$ is the smallest number to satisfy this inequality.
The longest sequence should be 4 to allow for every possible letter of the alphabet.

37. **a.** $C(40, 3) = 9880$

b. $C(40, 2) \times 2 = 1560$ more possiblities
$9880 + 1560 + 40 = 11,480$

39. Two pairs:
$C(13, 2) \cdot C(4, 2) \cdot C(4, 2) \cdot 44 = 123,552$

41. Two of a kind:
$C(13, 1) \cdot C(4, 2) \cdot C(12, 3) \cdot C(4, 1)^3 = 1,098,240$

43. Straight:
Choose start card: $C(10, 1)$
Choose suit of first: $C(4, 1)$
Choose suit of second: $C(4, 1)$
Choose suit of next three: $C(4,1)^3$
Total: straight flushes
$$C(10,1) \cdot C(4,1)^5 - C(10,1) \times C(4,1) = 10,200$$

45. Step 1: Select Boondoggle as a Do-nothing member: $C(1, 1)$ choices
Step 2: Select the Chief Investigator from the Party Party (without Pork barrel): $C(9, 1)$ choices
Step 3: Select the Assistant Investigators from the Study Party (without Boondoggle): $C(9,2)$ choices
Step 4: Select the Rabble Rousers: $C(15, 2)$ choices
Step 5: Select the other two Do-nothings: $C(13, 2)$ choices
Thus, we get
$$C(1, 1) \cdot C(9, 1) \cdot C(9, 2) \cdot C(15, 2) \cdot C(13, 2)$$

47–51. Each step involves choosing slots for groups of either. The order of the letters selected is the order in which they appear on the given words (skipping repeats).

47. Mississippi: 1 m, 4 s's, 4 i's, 2 p's
Step 1: choose a place for the m: $C(11, 1)$ choices
Step 2: choose a place for the s's: $C(10, 4)$ choices
Step 3: choose a place for the i's: $C(6, 4)$ choices
Step 4: choose a place for the p's: $C(2, 2)$ choices
Total: $C(11, 1) \cdot C(10, 4) \cdot C(6, 4) \cdot C(2, 2)$

49. Megalomania:

m's	e	g	a's	l	o	n	i
$C(11, 2)$ $\cdot$	$C(9, 1)$ $\cdot$	$C(8, 1)$ $\cdot$	$C(7, 3)$ $\cdot$	$C(4, 1)$ $\cdot$	$C(3, 1)$ $\cdot$	$C(2, 1)$ $\cdot$	$C(1, 1)$

51. Casablanca:

c's	a's	s	b	l	n
$C(10, 2)$ $\cdot$	$C(8, 4)$ $\cdot$	$C(4, 1)$ $\cdot$	$C(3, 1)$ $\cdot$	$C(2, 1)$ $\cdot$	$C(1, 1)$

53. $6 + 6 \cdot 1 = 12$ different symmetries

55. a. $C(20, 2) = 190$

b. $C(n, 2)$

57. Answers may vary.

59. $C(50, 3)$ choices are possible

61. $(a+b)^4 = a^4 + 4a^3b + 6a^2b^2 + 4ab^3 + b^4$
$C(4, 4) = 1$
$C(4, 3) = 4$
$C(4, 2) = 6$
$C(4, 1) = 4$
$C(4, 0) = 1$

You're the Expert
Page 385

1. alternative 1: 3 in a row
   three choices
   alternative 2: 3 adjacent
   three choices
   total = 6 choices · 2 cubes per choice = 12 additional cubes

3. Notice that all color arrangements are the same: 1 choice.
   total = 1 choice · 2 blocks per choice = 2 blocks

Chapter 5 Review Exercises
Pages 385–387

1. The set I of all positive even integers no larger than 10.
   $I = \{2, 4, 6, 8, 10\}$

3. $A = \{n \mid n \in \mathbf{N}, n \text{ even and } 2 \leq n \leq 10 \}$
   $A = \{3, 5, 7, 9\}$

5. The set of all outcomes of tossing a coin five times.
   $S = \{HHHHH, HHHHT, HHHTH, HHHTT, HHTHH, HHTHT, HHTTH, HHTTT, HTHHH, HTHHT, HTHTH, HTHTT,$
   $HTTHH, HTTHT, HTTTH, HTTTT, THHHH, THHHT, THHTH, THHTT, THTHH, THTHT, THTTH, THTTT, TTHHH,$
   $TTHHT, TTHTH, TTHTT, TTTHH, TTTHT, TTTTH, TTTTT\}$

7. The set of all outcomes of tossing two dice such that the two numbers are different.
   $S = \{(1, 2), (1, 3), (1, 4), (1, 5), (1, 6), (2, 1), (2, 3), (2, 4), (2, 5), (2, 6), (3, 1), (3, 2), (3, 4), (3, 5), (3, 6), (4, 1), (4, 2),$
   $(4, 3), (4, 5), (4, 6), (5, 1), (5, 2), (5, 3), (5, 4), (5, 6), (6, 1), (6, 2), (6, 3), (6, 4), (6, 5)\}$

9. $A \cup C = \{1, 2, 3, 4, 5, 6, 7\}$

11. $A \cup (B \cup C) = \{1, 2, 3, 4, 5, 6, 7\}$

13. $A \cap B = \{3, 4, 5\}$

15. $A \cap \varnothing = \varnothing$

17. $(A \cap B) \cap C = \{5\}$ (refer to Exercise 13)

19–23 S is the set of outcomes when two distinguishable dice are thrown
   E = the subset of outcomes in which at most one die shows an even number
   $= \{(1, 1), (1, 2), (1, 3), (1, 4), (1, 5), (1, 6), (2, 1), (2, 3), (2, 5), (3, 1), (3, 2), (3, 3), (3, 4), (3, 5), (3, 6), (4, 1), (4, 3),$
   $(4, 5), (5, 1), (5, 2), (5, 3), (5, 4), (5, 5), (5, 6), (6, 1), (6, 3), (6, 5)\}$
   F = set of outcomes in which the sum of the numbers is 7
   $= \{(6, 1), (5, 2), (4, 3), (3, 4), (2, 5), (1, 6)\}$

19. $E' = \{(2, 2), (2, 4), (2, 6), (4, 2), (4, 4), (4, 6), (6, 2), (6, 4), (6, 6)\}$
   = the set of outcomes in which both dice show an even number

21. $(E \cup F)'$ = the set of outcomes in which both dice show an even number = E'

23. $E' \cup F'$ = the set of outcomes in which both dice show an even number or the sum of the numbers is not 7

25. $C(4, 2) \cdot 2 \cdot 1 = 12$

27. $26 \times 26 \times 26 = 26^3 = 17,576$ companies

29. logo $\times$ two letters $\times$ two digits $= 1 \times 26 \times 25 \times 9 \times 10 = 58,500$ different license plates

31. 500,000 to 700,000 customers:

$$500,000 \le 26^x 10^{5-x} \le 700,000$$

$$5 \le 2.6^x \le 7$$
$$\ln 5 \le x \ln 2.6 \le \ln 7$$
$$\frac{\ln 5}{\ln 2.6} \le x \le \frac{\ln 7}{\ln 2.6}$$
$$1.68 \le x \le 2.04$$
$$x = 2$$

Thus, the business should use 2 letters and 3 digits.

33. 800,000 words and each word contains the same number of letters:

$$26^x = 8 \times 10^5$$

$$x = \frac{\log 8 \times 10^5}{\log 26} \approx 4.17$$

Thus, $x = 5$ and each word would have to be 5 letters long.

35. a. two R's and two D's

$$C(4, 2) \cdot C(2, 2) = 6$$

b. 6

37–47. bag containing 12 marbles: 4 red, 2 green, 1 transparent, 3 yellows, 2 orange

37. possible groups of five marbles:

$$C(12, 5) = 792$$

39. possible groups of five marbles which include all red ones:

$$C(4, 4) \cdot C(8, 1) = 8$$

41. possible groups of five marbles which do *not* include all red ones:

Alternative 1, 0 red ones: $C(8, 5) \cdot C(4, 0) = 56$
Alternative 2, 1 red one: $C(8, 4) \cdot C(4, 1) = 280$
Alternative 3, 2 red ones: $C(8, 3) \cdot C(4, 2) = 336$
Alternative 4, 3 red ones: $C(8, 2) \cdot C(4, 3) = 112$
Total: 784
or
$C(12, 5) - C(4, 4) \cdot C(8, 1) = 784$

43. possible groups of six marbles which do not include one of each color:

$$C(12, 6) - (C(4, 2) \cdot C(2, 1) \cdot C(1, 1) \cdot C(3, 1) \cdot C(2, 1) + C(4, 1) \cdot C(2, 2) \cdot C(1, 1) \cdot C(3, 1) \cdot C(2, 1)$$
$$+ C(4, 1) \cdot C(2, 1) \cdot C(1, 1) \cdot C(3, 2) \cdot C(2, 1) + C(4, 1) \cdot C(2, 1) \cdot C(1, 1) \cdot C(3, 1) \cdot C(2, 2)) = 924 - 168 = 756$$

45. possible groups of five marbles which includes at least two yellow ones:

$$C(3, 2) \cdot C(9, 3) + C(3, 3) \cdot C(9, 2) = 288$$

47. possible groups of five marbles which include at most one of the red ones but no yellow ones:

$$C(4, 1) \cdot C(5, 4) + C(4, 0) \cdot C(5, 5) = 21$$

49. a full house with either two Kings and three Queens or two Queens and three Kings:

$$\begin{array}{cccc} 2K & 3Q & 2Q & 3K \\ \downarrow & \downarrow & \downarrow & \downarrow \end{array}$$
$$C(4, 2) \cdot C(4, 3) + C(4, 2) \cdot C(4, 3) = 2C(4, 2) \cdot C(4, 3)$$

51. Two of a kind with no Aces:

$$C(12, 1) \cdot C(4, 2) \cdot C(11, 3) \cdot C(4, 1)^3$$

53. $C(9.1)$ · $C(1, 1)$ · $C(8, 1)$ · $C(7, 3)$

   ↑   ↑   ↑   ↑

   Supreme   ·   Vladimir is Semi   ·   others are Semi   ·   3 Demis

55.   (Nikolai is supreme)   (Ivana is semi-supreme)
   $C(1, 1)$ · $C(1, 1) \cdot C(1, 1)$ · $C(7, 3)$

   (Nikolai is Supreme)   (Ivana is Demi or not on the committee)
   $C(1, 1)$ · $C(8, 2)$ · $(C(6, 2) + C(6, 3))$

   Total: $C(1, 1) \cdot C(1, 1) \cdot C(1, 1) \cdot C(7, 3) + C(1, 1) \cdot C(8, 2) \cdot (C(6, 2) + C(6, 3))$
   $= C(7, 3) + C(8, 2) \cdot C(7, 3)$

Chapter 6

1. The sample space is $S = \{HH, HT, TH, TT\}$.
   The event that the result is at most one tail is the subset $E = \{HH, HT, TH\}$.

3. The sample space is $S = \{HHH, HHT, HTH, HTT, THH, THT, TTH, TTT\}$.
   The event that the result is at most one head is the subset $E = \{HTT, THT, TTH, TTT\}$.

5. The sample space is
   $S = \{HHHH, HHHT, HHTH, HHTT, HTHH, HTTH, HTHT, HTTT, THHH, THHT, THTH, THTT, TTHH, TTHT,$
   $TTTH, TTTT\}$.
   The event that the result is that a head is never followed by a tail is the subset
   $E = \{HHHH, THHH, TTHH, TTTH, TTTT\}$.

7. The sample space is $S = \begin{Bmatrix} (1,1) & (1,2) & (1,3) & (1,4) & (1,5) & (1,6) \\ (2,1) & (2,2) & (2,3) & (2,4) & (2,5) & (2,6) \\ (3,1) & (3,2) & (3,3) & (3,4) & (3,5) & (3,6) \\ (4,1) & (4,2) & (4,3) & (4,4) & (4,5) & (4,6) \\ (5,1) & (5,2) & (5,3) & (5,4) & (5,5) & (5,6) \\ (6,1) & (6,2) & (6,3) & (6,4) & (6,5) & (6,6) \end{Bmatrix}$.

   The event that the numbers add to 5 is the subset $E = \{(1, 4), (2, 3), (3, 2), (4, 1)\}$.

9. The sample space is the same as in Exercise 7.
   The event that the numbers add to 9 is the subset $E = \varnothing$.

11. The sample space is the same as in Exercise 7.
    The event that the result is that both of the numbers are prime is the subset
    $E = \{(2, 2), (2, 3), (2, 5), (3, 2), (3, 3), (3, 5), (5, 2), (5, 3), (5, 5)\}$.
    (Note that the number 1 is not prime.)

13. The sample space is $S = \{M, o, z, a, r, t\}$.
    The event that the letter is a vowel is the subset $E = \{o, a\}$.

15. The sample space is $S = \{so, sr, se, os, or, oe, rs, ro, re, es, eo, er\}$.
    The event that the first letter is a vowel is the subset $E = \{os, or, oe, es, eo, er\}$.

17. The sample space is $S = \{so, sr, se, os, or, oe, rs, ro, re, es, eo, er\}$.
    The event that at most one letter is a vowel is the subset $E = \{so, sr, se, os, or, rs, ro, re, es, er\}$.

19. The sample space is $S = \{01, 02, 03, 04, 10, 12, 13, 14, 20, 21, 23, 24, 30, 31, 32, 34, 40, 41, 42, 43\}$.
    The event that the first digit is larger than the second is the subset $E = \{10, 20, 21, 30, 31, 32, 40, 41, 42, 43\}$.

21. $E =$ event that you buy a van or an antique truck

23. $E =$ event that the specialist is chosen from the set
    {Physical/Rehabilitation, Pulmonary, Hematology/Oncology, Cardiology, Nephrology}

25. $E \cup F$ is the event that the specialist generates revenues of at least \$1 million per year or that the specialist is neither a
    cardiologist nor a nephrologist. $E \cap F$ is the event that the specialist will generate revenues of at least \$1 million per year
    and that he/she is neither a cardiologist nor a nephrologist.
    $E \cup F = $ {Physical/Rehabilitation, Pulmonary, Hematology/Oncology, Cardiology, Nephrology, Endocrine,
    Infectious Diseases}
    $E \cap F = $ {Physical/Rehabilitation, Pulmonary, Hematology/Oncology}

27. a. E = event of hiring a specialist from {Physical/Rehabilitation, Hematology/Oncology, Cardiology}
F = event of hiring a specialist from {Pulmonary, Endocrine, Infectious Diseases, Nephrology}
E and F are mutually exclusive.

b. E = event of hiring a specialist from {Physical/Rehabilitation, Hematology/Oncology, Cardiology}
F = event of hiring a specialist from
{Pulmonary, Hematology/Oncology, Endocrine, Cardiology, Infectious Diseases, Nephrology}
E and F are not mutually exclusive, since the sets both contain Hematology/Oncology and Cardiology.

29. E = event of buying from {White Sox, Bulls, Blackhawks} (top three in gate receipts)
F = event of buying from {Blackhawks, Bulls, Bears} (lowest three in player costs)
$E \cap F$ = event of buying from {Bulls, Blackhawks}

31. E = event of buying from {White Sox, Bulls, Blackhawks, Cubs} $\cap$ {Bulls, White Sox, Bears, Blackhawks}
= event of buying from {White Sox, Bulls, Blackhawks} (teams from top four in both gate receipts and operating income)
F = event of buying from {Bulls, Blackhawks} (teams with player costs not exceeding $20 million)
$E \cap F$ = event of buying from {Bulls, Blackhawks}
(teams from top four in gate receipts and operating income and with player costs not exceeding $20 million)
F' = event of buying from {Bears, White Sox, Cubs} (teams with player costs exceeding $20 million)

33. a. E = event of buying from {Bears, Cubs, White Sox} (top three in media revenues)
F = event of buying team from {Bears, Bulls, Blackhawks} (operating expenses less than $50 million)
E and F are not mutually exclusive.

b. E = event of buying from {Bears, Cubs} (top two in media revenues)
F = event of buying from {Bulls, Blackhawks} (operating expenses less than $40 million)
E and F are mutually exclusive.

35. a. The sample space is the set of all hands of five cards chosen from a standard deck of 52 cards.

b. The event "a full house" is the set of all hands of five cards consisting of three cards of one denomination and two cards of another denomination. Example: {Q, Q, Q, 5, 5}

37. a. E = {Lindner Dividend, Vanguard Wellesley Income, Prudential Flex-A-Fund Con. Mgd. A}

b. F = {Franklin Income, Lindner Dividend}

c. $G = F'$ = {Berwyn Income, National Multi-Sector FIA, Seligman Income A, Putnam Diversified Inc. A, Vanguard Wellesley Income, Income Fund of America, Vanguard Preferred Stock, Pru. Flex-A-Fund Con. Mgd. A}

39. a. $E' \cap H$ = event that dog's "flight" drive is not strongest and its "fight" drive is weakest

b. $E \cup H$ = event that dog's "flight" drive is strongest or its "fight" drive is weakest

c. $E' \cap G'$ [or $(E \cup G)'$] = event that neither dog's "flight" drive nor "fight" drive is strongest

41. a. {9}

b. {6}

43. a. {1, 4, 7}: The dog's "fight" drive is weakest.

b. {1, 9}: The dog's "flight" drive and its "fight" drive are either both weakest or both strongest.

c. {3, 6, 7, 8, 9}: The dog's "flight" drive is strongest or its "fight" drive is strongest.

45. The number of possible outcomes is $C(6, 4) = \dfrac{6!}{4!2!} = \dfrac{6 \cdot 5}{2 \cdot 1} = 15$. If one of the marbles is red, the number of outcomes that includes the red marble is $C(1, 1)C(5, 3) = 1 \cdot 10 = 10$.

47. a. $n(S)$ = size of set of sample space S of all possible finishes of the race = $7 \cdot 6 \cdot 5 = 210$.
$n(S)$ can also be written as $P(7, 3)$, the number of permutations of three elements taken from a set of seven elements.

b. E = event that Quickest Blade is in second or third place

F = event that Boom Towner is the winner

$E \cap F$ = event that Boom Towner is the winner and Quickest Blade is in second or third place

Sample space = {(BT, QB, one of 5 others)} $\cup$ {(BT, one of 5 not including QB, QB)}

$n(E \cap F) = 1 \cdot 1 \cdot 5 + 1 \cdot 5 \cdot 1 = 10$

49. The set is $\{R_1, R_2, R_3, R_4, G_1, G_2, Y_1, Y_2\}$.

$$C(8, 3) = \frac{8!}{3!5!} = \frac{8 \cdot 7 \cdot 6}{3 \cdot 2 \cdot 1} = 56$$

51. $C(4, 1)C(2, 1)C(2, 1) = 4 \cdot 2 \cdot 2 = 16$

53. To determine a sample space for an experiment, you must have the total set of elements involved, the number of elements in the experiment, and any restrictions as to how the element are chosen.

55. An event is a subset of the sample space. If E and F are events, then $(E \cap F)'$ is the event that E and F do not both occur (at least one of the two events does not occur.)

Section 6.2 Introduction to Probability
Pages 417–421

1. $P(E) = \dfrac{fr(E)}{N} = \dfrac{40}{100} = 0.4$

3. $P(E) = \dfrac{fr(E)}{N} = \dfrac{687}{800} = 0.85875$

5. $N = 1621 + 856 + 1521 + 1002 = 5000$

Outcome	HH	HT	TH	TT
Frequency	1621	856	1521	1002
Experimental probability	$\dfrac{1621}{5000} = 0.3242$	$\dfrac{856}{5000} = 0.1712$	$\dfrac{1521}{5000} = 0.3042$	$\dfrac{1002}{5000} = 0.2004$

7. $P(E) = \dfrac{1621 + 1521}{5000} = \dfrac{3142}{5000} = 0.6284$

9. Let E = event second coin lands heads up.

Let F = event second coin lands tails up.

$$P(E) = \frac{1621 + 1521}{5000} = 0.6284$$

$$P(F) = \frac{856 + 1002}{5000} = 0.3716$$

The second coin appears to be biased toward landing heads up since $P(E)$ is a fair amount larger than $P(F)$.

11. $P(E) = \dfrac{n(E)}{n(S)} = \dfrac{5}{20} = \dfrac{1}{4}$

13. $P(E) = \dfrac{n(E)}{n(S)} = \dfrac{10}{10} = 1$

15. $P(E) = \dfrac{n(E)}{n(S)} = \dfrac{3}{4}$

17. $S = \{HH, HT, TH, TT\}$; $E = \{HH, HT, TH\}$

$$P(E) = \frac{n(E)}{n(S)} = \frac{3}{4}$$

19. $S = \{HHH, HHT, HTH, HTT, THH, THT, TTH, TTT\}$; $E = \{HTT, THT, TTH, TTT\}$

$$P(E) = \frac{n(E)}{n(S)} = \frac{4}{8} = \frac{1}{2}$$

21. $S = \{HHHH, HHHT, HHTH, HHTT, HTHH, HTHT, HTTH, HTTT, THHH, THHT, THTH, THTT, TTHH, TTHT, TTTH, TTTT\}$

$E = \{HHHH, THHH, TTHH, TTTH, TTTT\}$

$$P(E) = \frac{n(E)}{n(S)} = \frac{5}{16}$$

23. $S = $ set given in Example 4; $E = \{(1, 4), (4, 1), (2, 3), (3, 2)\}$

$$P(E) = \frac{n(E)}{n(S)} = \frac{4}{36} = \frac{1}{9}$$

25. $S = $ set given in Example 4; $E = \varnothing$

$$P(E) = \frac{n(E)}{n(S)} = \frac{0}{36} = 0$$

27. $S = $ set given in Example 4; $E = \{(2, 3), (3, 2), (2, 5), (5, 2), (3, 5), (5, 3), (2, 2), (3, 3), (5, 5)\}$

$$P(E) = \frac{n(E)}{n(S)} = \frac{9}{36} = \frac{1}{4}$$

29. $n(S) = 1{,}113{,}079 + 254{,}623 + 621{,}412 = 1{,}989{,}114$

$$P(E) = \frac{n(E)}{n(S)} = \frac{1{,}113{,}079}{1{,}989{,}114} \approx 0.5596$$

31. $n(S) = 74{,}945 + 36{,}918 + 187{,}739 = 299{,}602$

$$P(E) = 1 - \frac{187{,}739}{299{,}602} \approx 0.3734$$

33. $P(E) = \dfrac{187{,}739}{621{,}412} \approx 0.3021$

35. See Problem 6.1, 45

$$P(E) = \frac{n(E)}{n(S)} = \frac{10}{15} = \frac{2}{3}$$

37.

Outcome	a	b	c	d	e
Probability	0.1	0.05	0.6	0.04	0.2

(The sum of the probabilities = 1, so $P(\{e\}) = 1 - (0.1 + 0.05 + 0.6 + 0.04) = 0.2$.)

a. $P(\{a, c, e\}) = 0.1 + 0.6 + 0.2 = 0.9$

b. Let $E = \{a, c, e\}$ and $F = \{b, c, e\}$.
$P(E \cup F) = P\{(a, b, c, e)\} = 0.95$

c. Let $E = \{a, c, e\}$. $P(E') = 1 - (0.1 + 0.6 + 0.2) = 0.1$

d. Let E and F be as in part (b).
Then $P(E \cap F) = P(\{c, e\}) = 0.6 + 0.2 = 0.8$.

39. $S = \{$favor the law, oppose the law, undecided$\}$

Outcome	favor	oppose	undecided
Probability	0.85	0.13	0.02

P(definite opinion) $= 0.85 + 0.13 = 0.98$

41.

Outcome	Medicaid	Medicare	Private	Other
Probability	0.49	0.05	0.45	0.01

P(\$1 originated from either Medicare or Medicaid) $= 0.49 + 0.05 = 0.54$

43.

Outcome	Index ≤ 3220	$3220 <$ Index ≤ 3260	Index > 3260
Probability	0.5	0.2	0.3

45.

Outcome	Those with AIDS		Those without AIDS	
	Positive	Negative	Positive	Negative
Probability	0.975	0.025	0.05	0.95

P(false negative) $= 0.025$; P(false positive) $= 0.05$

47. Let x be the probability of rolling a 1.

$P(1) = P(6) = x$ and $P(2) = P(3) = P(4) = P(5) = 2x$

Since the sum of the probabilities is 1, $x + 2x + 2x + 2x + 2x + x = 1$ or $10x = 1$. Thus $x = \dfrac{1}{10}$.

Outcome	1	2	3	4	5	6
Probability	$\dfrac{1}{10}$	$\dfrac{1}{5}$	$\dfrac{1}{5}$	$\dfrac{1}{5}$	$\dfrac{1}{5}$	$\dfrac{1}{10}$

$P(\text{odd}) = P(1) + P(3) + P(5) = \dfrac{1}{2}$

49. The dice are weighted so that the probabilities are twice as likely for a pair of mismatching numbers to come up as a pair of matching numbers. There are six possible matching pairs and thirty mismatching pairs. Let a mismatch count as 2 and a match count as 1. Total count $= 60 + 6 = 66$.

Outcome	Mismatch	Match
Probability	$\dfrac{60}{66} = \dfrac{10}{11}$	$\dfrac{6}{66} = \dfrac{1}{11}$

$P(\text{odd sum}) = P(1, 2) + P(1, 4) + P(1, 6) + P(2, 1) + P(2, 3) + P(2, 5) + P(3, 2) + P(3, 4) + P(3, 6) + P(4, 1) + P(4, 3)$

$+ P(4, 5) + P(5, 2) + P(5, 4) + P(5, 6) + P(6, 1) + P(6, 3) + P(6, 5) = 18\left(\dfrac{1}{30} \cdot \dfrac{10}{11}\right) = \dfrac{6}{11}$

51. Deuce (2) is five times as likely to come up as 4 and three times as likely to come up as any of 1, 3, 5, 6. Assign 5 to 1, 3, 5, 6; 15 to 2; and 3 to 4. Total count = 20 + 15 + 3 = 38.

Outcome	1	2	3	4	5	6
Probability	$\frac{5}{38}$	$\frac{15}{38}$	$\frac{5}{38}$	$\frac{3}{38}$	$\frac{5}{38}$	$\frac{5}{38}$

$P(\text{odd}) = \frac{15}{38}$

53. Random digits will vary from one experiment to another. In one sample, the digits are 4, 0, 5, 8, 0, 9, 6, 4, 1, 7, 6, 9, 7, 3, 3, 8, 1, 2, 3, 1, 1, 2, 4, 8, 0, 4, 3, 9, 9, 1, 9, 8, 7, 4, 9, 3, 3, 3, 9, 3, 6, 2, 5, 2, 0, 0, 1, 5, 9, 9, 5, 4, 6, 6, 3, 4, 1, 0, 8, 2, 2, 0, 0, 3, 4, 0, 2, 6, 1, 7, 8, 9, 7, 9, 8, 0, 7, 0, 1, 0, 0, 8, 9, 6, 9, 5, 1, 8, 9, 8, 3, 3, 2, 6, 2, 1, 0, 8, 1, 6
$P(\text{random number in range } 0 - 4) = 0.55$

55. Answers will vary. For example, we generated the following data: 53, 72, 38, 55, 64, 55, 17, 57, 24, 29, 1, 50, 49, 34, 90, 35, 1, 89, 74, 38, 55, 59, 86, 87, 98, 48, 15, 40, 76, 28, 82, 76, 36, 82, 82, 37, 74, 9, 3, 35, 59, 80, 10, 43, 83, 36, 25, 69, 42, 44, 33, 47, 83, 87, 35, 92, 49, 98, 71, 38, 28, 91, 48, 18, 28, 77, 94, 11, 50, 79, 98, 49, 14, 69, 78, 1, 54, 10, 41, 4, 25, 90, 85, 8, 18, 19, 70, 91, 86, 41, 15, 33, 69, 39, 83, 11, 80, 68, 91, 37.
$P(1 \le x \le 49) = 0.51$
$P(50 \le x \le 54) = 0.04$
$P(55 \le x \le 99) = 0.45$
$P(100 \le x \le 100) = 0$

57. Answers will vary. For example, for the first 25 trials we get the following: 4, 3, 4, 6, 5, 3, 6, 3, 2, 2, 2, 4, 4, 3, 3, 2, 1, 4, 4, 6, 1, 1, 1, 2, 4, 2, 1, 3, 6, 3, 6, 4, 3, 6, 1, 6, 3, 1, 4, 6, 4, 3, 2, 5, 1, 5, 4, 4, 2, 6.
For the second 25 trials we get the following: 6, 1, 2, 2, 4, 2, 4, 4, 2, 6, 1, 1, 4, 4, 2, 4, 1, 4, 5, 3, 4, 3, 4, 4, 3, 6, 6, 5, 6, 1, 3, 2, 2, 5, 1, 5, 2, 5, 6, 5, 4, 4, 6, 4, 6, 2, 3, 2, 1, 2.
For the third 25 trials, we get the following: 4, 5, 4, 6, 1, 2, 6, 3, 5, 1, 5, 4, 2, 3, 1, 5, 1, 6, 6, 5, 6, 5, 4, 2, 3, 5, 3, 5, 5, 2, 3, 4, 1, 4, 1, 3, 4, 3, 5, 5, 5, 6, 1, 4, 2, 4, 4, 3, 3, 5.
For the last 25 trials we get the following: 4, 5, 1, 3, 4, 4, 6, 5, 1, 1, 4, 3, 6, 6, 6, 1, 4, 4, 2, 2, 2, 2, 6, 4, 4, 4, 6, 6, 3, 3, 3, 6, 1, 5, 2, 3, 5, 4, 3, 2, 4, 6, 1, 1, 3, 4, 4, 2, 3, 1.

   a. Using 25 trials gives four cases in which a pair of dice add to 10.

   b. Using 50 trials gives five cases adding to 10.

   c. Using 75 trials gives seven cases adding to 10.

   d. Using 100 trials gives nine cases adding to 10. With this last value, the experimental probability of getting 10 is $\frac{9}{100} = 0.09$. This value may be compared with the theoretical probability of $\frac{3}{36} = \frac{1}{12} \approx 0.083$

59. Answers will vary. For example, we generated the following data for 100 trials: 2, 4, 4, 3, 3, 3, 6, 4, 1, 4, 1, 4, 4, 4, 4, 4, 2, 5, 2, 3, 4, 4, 5, 4, 4, 1, 2, 1, 5, 4, 3, 6, 1, 6, 3, 4, 5, 4, 1, 5, 1, 3, 4, 2, 6, 6, 5, 1, 1, 4, 2, 1, 1, 5, 5, 2, 5, 1, 5, 3, 2, 5, 5, 4, 5, 1, 6, 3, 2, 5, 4, 5, 4, 5, 3, 6, 3, 3, 1, 1, 4, 6, 4, 3, 5, 2, 3, 3, 3, 3, 5, 6, 3, 4, 5, 4, 6, 2, 2, 1.
Probability distribution (from experiment) is

Outcome	1	2	3	4	5	6
Number of times	16	12	18	25	19	10
Experimental Probability	0.16	0.12	0.18	0.25	0.19	0.10

$P(\text{odd}) = 0.53$

The experiment was repeated. For example, we generated the following data for 100 trials: 2, 1, 5, 3, 4, 5, 4, 5, 2, 6, 4, 3, 5, 1, 3, 4, 4, 2, 5, 6, 4, 5, 4, 6, 3, 2, 3, 4, 3, 2, 2, 3, 1, 2, 5, 3, 1, 3, 4, 2, 4, 3, 6, 1, 2, 2, 3, 5, 6, 4, 2, 5, 3, 3, 4, 3, 6, 5, 5, 6, 3, 5, 4, 4, 5, 2, 2, 5, 5, 2, 5, 5, 5, 5, 3, 4, 5, 2, 6, 2, 3, 4, 4, 5, 5, 4, 4, 3, 3, 1, 6, 2, 6, 5, 3, 4, 3, 2, 4, 6.

Outcome	1	2	3	4	5	6
Number of times	6	18	21	21	23	11
Experimental Probability	0.06	0.18	0.21	0.21	0.23	0.11

$P(\text{odd}) = 0.50$

61. experimental; theoretical; the sample size increases

63. It suggests theoretical probability. The experimental probability is not known until after the fact.

65. Determine the GPA's for an entire group. Divide the set into subsets, such as
{below 50%, 51% – 60%, ..., 91% – 100%}.
Then find the experimental probability for your GPA by comparing it with the data for the group.

Section 6.3 Principles of Probability

Pages 427–429

1. $E \cap F = \varnothing$, $P(E) = 0.3$, $P(F) = 0.4$
$P(E \cup F) = P(E) + P(F) - P(E \cap F)$
$= 0.3 + 0.4 - 0 = 0.7$

3. $E \cap F = \varnothing$, $P(E) = 0.3$, $P(E \cup F) = 0.4$
$P(F) = P(E \cup F) - P(E) = 0.4 - 0.3 = 0.1$

5. $P(E) = 0.1$, $P(F) = 0.6$, $P(E \cap F) = 0.05$
$P(E \cup F) = P(E) + P(F) - P(E \cap F)$
$= 0.1 + 0.6 - 0.05 = 0.65$

7. $P(E \cup F) = 0.9$, $P(F) = 0.6$, $P(E \cap F) = 0.1$
$P(E) = P(E \cup F) - P(F) + P(E \cap F)$
$= 0.9 - 0.6 + 0.1 = 0.4$

9. $P(E) = 0.75$, $P(E') = 1 - 0.75 = 0.25$

11. $P(E) = 0.3$, $P(F) = 0.4$, $P(G) = 0.3$
E, F, and G are mutually disjoint.
$P(E \cup F \cup G) = P(E) + P(F) + P(G)$
$= 0.3 + 0.4 + 0.3 = 1.0$

13. $E \cap F = \varnothing$, $P(E) = 0.3$, $P(F) = 0.4$
$P((E \cup F)') = 1 - P(E \cup F)$
$= 1 - (0.3 + 0.4) = 1 - 0.7 = 0.3$

15. $E \cup F = S$, $E \cap F = \varnothing$
$P(E) + P(F) = P(S) = 1.0$

17. $P(E) = P(\text{one tail}) = \dfrac{1}{2}$, $P(F) = P(\text{one head}) = \dfrac{1}{2}$
$P(E \cup F) = P(E) + P(F) - P(E \cap F)$
$= \dfrac{1}{2} + \dfrac{1}{2} - \dfrac{1}{2} = \dfrac{1}{2} = 0.5$

19. $P(E) = P(\text{at least one tail}) = \dfrac{7}{8} = 0.875$,
$P(F) = P(\text{at least one head}) = 0.875$
$P(E \cup F) = P(E) + P(F) - P(E \cap F)$
$= 0.875 + 0.875 - 0.75 = 1.75 - 0.75 = 1.0$

21. $P(E) = P(\text{event that numbers add to 6 or 7}) = \dfrac{11}{36}$,
$P(F) = P(\text{event that numbers add to 7 or 2}) = \dfrac{7}{36}$
$P(E \cup F) = P(E) + P(F) - P(E \cap F)$
$= \dfrac{11}{36} + \dfrac{7}{36} - \dfrac{6}{36} = \dfrac{12}{36} = \dfrac{1}{3}$

23.

Outcome	Experimental Probability
favor law	0.85
oppose law	0.13
undecided	0.02

$P(\text{not opposed}) = 0.85 + 0.02 = 0.87$, or
$1 - 0.13 = 0.87$

25. Let E = event of failed to learn alphabet backward;
$P(E) = \dfrac{2}{3}$
Let F = event of failed to learn alphabet sideways;
$P(F) = \dfrac{3}{4}$
Let G = event of failed to learn alphabet backward or sideways;
$P(E \cap F) = P(G) = \dfrac{5}{12}$
$P(E \cup F) = P(E) + P(F) - P(E \cap F)$
$= \dfrac{8}{12} + \dfrac{9}{12} - \dfrac{5}{12} = 1$
Fraction disqualified $= 1$

27. Let E = event of transportation by rail; $P(E) = 0.38$
Let F = event of transportation by boat or truck;
$P(F) = 0.53$
$P(E \cup F) = 0.80 = P(E) + P(F) - P(E \cap F)$
$P(E \cap F) = 0.38 + 0.53 - 0.80 = 0.11$

29. Let E = event of acquittal by judge; $P(E) = 0.17$
Let F = event of acquittal by jury; $P(F) = 0.33$
$P(E \cap F) = 0.14$

a. $P(E') = 1 - P(E) = 1 - 0.17 = 0.83$

b. $P(F') = 1 - P(F) = 1 - 0.33 = 0.67$

c. $P((E \cup F)') = 1 - P(E \cup F)$
$= 1 - [P(E) + P(F) - P(E \cap F)]$
$= 1 - [0.17 + 0.33 - 0.14] = 1 - 0.36 = 0.64$

d. $P(E' \cap F)$; since $F = (E' \cap F) \cup (E \cap F)$, and
$(E' \cap F)$ and $(E \cap F)$ are mutually exclusive,
$P(F) = P(E' \cap F) + P(E \cap F)$;
$P(E' \cap F) = P(F) - P(E \cap F)$
$= 0.33 - 0.14 = 0.19$

31. Let E = event of being vaccinated against the Venusian flu; $P(E) = 0.80$.
Let F = event of getting the flu.
$P(E \cap F) = (0.02)(0.80) = 0.016$
$P(F) = 0.10$
$P(E \cup F) = P(E) + P(F) - P(E \cap F)$
$= 0.80 + 0.10 - 0.016 = 0.884$

33. The events are mutually exclusive.

35. For example, let $S = \{HH, HT, TH, TT\}$ be the possible outcomes of tossing a coin twice. Let $E = \{HH, TT\}$ and $F = \{HH\}$. Thus, E and F are not mutually exclusive, since $E \cap F = \{HH\}$. Now suppose the coin has been made so $P(H) = 0$. Then $P(E \cup F) = P(E) + P(F)$, since $P(HH) = 0$.

37. Since $P(E \cup E') = P(S) = 1$, we have $P(E \cup E') = P(E) + P(E') - P(E \cap E')$.
$1 = P(E) + P(E')$, or $P(E') = 1 - P(E)$.

Section 6.4 Counting Techniques and Probability

Pages 437–441

1. Number of marbles $= 4 + 3 + 2 + 1 = 10$
$n(S) = C(10, 5) = 252$
$E = $ the collection of 4 red and 1 non-red marbles
$n(E) = C(4, 4)C(6, 1) = 1 \cdot 6 = 6$
$$P(E) = \frac{n(E)}{n(S)} = \frac{6}{252} = \frac{1}{42}$$

3. $E = $ the collection of sets of 5 marbles, of which at least one is white
$n(E) = C(2, 1)C(8, 4) + C(2, 2)C(8, 3) = 2 \cdot 70 + 1 \cdot 56 = 196$
$$P(E) = \frac{n(E)}{n(S)} = \frac{196}{252} = \frac{7}{9}$$
Also, $P(E) = 1 - \dfrac{n(E')}{n(S)} = 1 - \dfrac{C(8,5)}{C(10,5)} = 1 - \dfrac{56}{252} = \dfrac{7}{9}$.

5. $E = $ event of two red marbles, one green, one white, and one purple marble
$n(E) = C(4, 2)C(3, 1)C(2, 1)C(1, 1) = 6 \cdot 3 \cdot 2 \cdot 1 = 36$
$$P(E) = \frac{n(E)}{n(S)} = \frac{36}{252} = \frac{1}{7}$$

7. $E = $ event of at most one green marble
$n(E) = C(7, 5) + C(3, 1)C(7, 4) = 21 + 3 \cdot 35 = 21 + 105 = 126$
$$P(E) = \frac{n(E)}{n(S)} = \frac{126}{252} = \frac{1}{2}$$

9. $E = $ event of not all the red marbles $= $ event of less than four red marbles
$n(E) = C(6, 5) + C(4, 1)C(6, 4) + C(4, 2)C(6, 3) + C(4, 3)C(6, 2) = 6 + 4 \cdot 15 + 6 \cdot 20 + 4 \cdot 15$
$= 6 + 60 + 120 + 60 = 246$
$$P(E) = \frac{n(E)}{n(S)} = \frac{246}{252} = \frac{41}{42}$$
Also, $P(E) = 1 - \dfrac{n(E')}{n(S)} = 1 - \dfrac{C(4, 4)C(6, 1)}{C(10, 5)} = \dfrac{41}{42}$.

11. $E = $ event of selecting three companies, all of which had quarterly earnings of at least 60¢ per share
$n(E) = C(5, 3) = 10; \; n(S) = C(6, 3) = 20$
$$P(E) = \frac{n(E)}{n(S)} = \frac{10}{20} = \frac{1}{2}$$

13. $E = $ event of selecting four stocks, which include Sears, Roebuck & Co., but not Abbott Labs
$n(E) = C(4, 3) = 4$
$$P(E) = \frac{n(E)}{n(S)} = \frac{4}{15}$$

15. E = event of owning 100 shares of Motorola, followed by purchasing 100 shares each of any two companies, one of which is Motorola

$n(E) = C(1, 1)C(5, 1) = 1 \cdot 5 = 5$

$n(S) = C(6, 2) = 15$

$P(E) = \dfrac{n(E)}{n(S)} = \dfrac{5}{15} = \dfrac{1}{3}$

17. We want two cards of one denomination, say two 4's, and three other cards of differing denominations. Choose 2 cards of the same denomination from 4 suits and 3 cards from 12 other denominations.

$n(E) = C(13, 1)C(4, 2)C(12, 3)C(4, 1)C(4, 1)C(4, 1)$
$= 1,098,240$

$P(E) = \dfrac{n(E)}{n(S)} = \dfrac{1,098,240}{C(52, 5)} = \dfrac{1,098,240}{2,598,960} \approx 0.4226$

19. $P(E) = \dfrac{n(E)}{n(S)} = \dfrac{C(13, 2)C(4, 2)C(4, 2)C(44, 1)}{C(52, 5)}$

$= \dfrac{123,552}{2,598,960} \approx 0.0475$

21. $P(E) = \dfrac{n(E)}{n(S)} = \dfrac{C(4, 1)[C(13, 5) - 10]}{2,598,960}$

$= \dfrac{5108}{2,598,960} \approx 0.0020$

23. Select five different numbers from $\{0, 1, \ldots, 49\}$ in any order.

E = event of big winner if your selection agrees with drawing

$P(E) = \dfrac{n(E)}{n(S)} = \dfrac{1}{C(50, 5)} = \dfrac{1}{2,118,760}$

≈ 0.000000472

E = event of being a small-fry winner

$= \dfrac{C(5, 4)C(45, 1)}{C(50, 5)}$

$P(E) = \dfrac{5(45)}{2,118,760} \approx 0.000106194$

Probability of being either a big winner or a small-fry winner = 0.000000472 + 0.000106194 ≈ 0.000106666 Note: Since the events are mutually exclusive, we may add their probabilities to get that of the union of the events.

25. Transfer 400 from {100 managers, 100 factory workers, 500 miscellaneous staff}.

a. E = event that all managers are offered transfer

$P(E) = \dfrac{n(E)}{n(S)} = \dfrac{C(100, 100)C(600, 300)}{C(700, 400)}$

$= \dfrac{(1)600!300!400!}{300!300!700!} = \dfrac{400 \cdot 399 \cdots 301}{700 \cdot 699 \cdots 601}$

b. E = event of personnel manager being offered transfer

$P(E) = \dfrac{n(E)}{n(S)} = \dfrac{C(1, 1)C(699, 399)}{C(700, 400)}$

$= \dfrac{699!400!300!}{399!300!700!} = \dfrac{4}{7}$

27. Three digits in range 0 - 9 are selected at random, not necessarily different.

E = event that all three are different

$P(E) = \dfrac{n(E)}{n(S)} = \dfrac{10 \cdot 9 \cdot 8}{10^3} = \dfrac{72}{100} = \dfrac{18}{25}$

or, $P(E) = \dfrac{n(E)}{n(S)} = \dfrac{P(10, 3)}{10^3} = \dfrac{72}{100} = \dfrac{18}{25}$.

29. E = event of perfect progression in a Big Eight Conference

Assume that "Won" scores were chosen at random in range 0 - 7.

Number of "Won" permutations = 8!

Number of possibilities (making the unreasonable assumption) $= 8^8$ since each team has 8 possible numbers for "won."

$P(E) = \dfrac{8!}{8^8}$

31. E = event of striking "to be or not to be that is the question"

There are 26 letter keys and a space bar. There are 39 total characters and spaces in the quotation.

$P(E) = \dfrac{1}{27^{39}}$

33. E = event of choosing 4 semifinalists from 7 contestants, without Tyler or Gabriella

$P(E) = \dfrac{n(E)}{n(S)} = \dfrac{C(5, 4)}{C(7, 4)} = \dfrac{5}{35} = \dfrac{1}{7}$

35. Let E be the event of going from the Start node to the Finish node in two moves. $n(E) = 2$ since you can go right then down or down then right. There are four possible moves from the Start node. After the first move, there are again four possible moves. If S is the sequence of two random moves, $n(S) = 4 \cdot 4 = 16$.

$$P(E) = \frac{n(E)}{n(S)} = \frac{2}{16} = \frac{1}{8}$$

37. Test has three parts: 8 true-false questions, 5 multiple-choice questions with 5 choices each, and 5 match question-answer problems.
E = event of answering all questions correctly with random answers

$$P(E) = \frac{n(E)}{n(S)} = \frac{1}{2^8 5^5 5!}$$

39. E = event that North Carolina will beat Central Connecticut but lose to Virginia
Probability of N.C. over C.C. = 0.5.
Probability of Virginia over Syracuse = 0.5
Probability of N.C. playing Virginia = 0.5 × 0.5 = 0.25
Probability of N.C. losing to Virginia = 0.5

Overall probability of $E = 0.125 = \dfrac{1}{8}$.

41. $P(E) = \dfrac{n(E)}{n(S)} = \dfrac{1 + 9C(4,1)}{10^4} = \dfrac{1 + 36}{10^4} = \dfrac{37}{10^4}$

Note that there are nine incorrect digits, one of which is to be entered into one of four positions.

43. Choose 5 bulbs randomly from 20. Reject lot if any bulb is defective.
E = event of 3 bulbs in crate being defective, and at least one being in the sample

Probability of not getting a bad bulb $= \dfrac{C(17, 5)}{C(20, 5)}$

Probability of rejecting crate $= P(E) = 1 - \dfrac{C(17, 5)}{C(20, 5)} = 1 - \dfrac{6188}{15,504} = \dfrac{9316}{15,504} \approx 0.6009$

45. E = event of committee consisting of a Chief Investigator (a Royal Party member), an Assistant Investigator (a Birthday Party member), two At-Large Investigators (either party), and five Ordinary Members (either party)
Committee is to be selected at random from 12 candidates, including Larry Sifford and Otis Taylor, half of whom are from each party.

a. Number of different committees = $C(6, 1)C(6, 1)C(10, 2)C(8, 5) = 90{,}720$

b. Number of committees in which Larry Sifford hopes to avoid serving unless he is Chief Investigator and Otis Taylor is the Assistant Investigator. They satisfy the party requirements.
$n(E) = C(1, 1)C(1, 1)C(10, 2)C(8, 5) + C(5, 1)C(6, 1)C(9, 2)C(7, 5)$
$= 1 \cdot 1 \cdot 45 \cdot 56 + 5 \cdot 6 \cdot 36 \cdot 21 = 2520 + 22{,}680 = 25{,}200$
Note that the second product covers the case that Larry does not serve on the committee but Otis does.

c. $P(E) = \dfrac{n(E)}{n(S)} = \dfrac{25{,}200}{90{,}720} = \dfrac{5}{18}$

Section 6.5 Conditional Probability and Independence

Pages 453–457

1. The sum is 5, given that the green die is not a 1. There are three ways of getting a sum of 5 without a green 1. They are $(1, 4)$; $(2, 3)$; and $(3, 2)$. The total number in the set is $5 \cdot 6 = 30$.

Then, $P(E|F) = \dfrac{n(E \cap F)}{n(F)} = \dfrac{3}{30} = \dfrac{1}{10}$.

3. The red one is 5, given that the sum is 6.
$n(E \cap F) = 1$: (5, 1)
$n(F) = n(\text{sum} = 6) = 5$
$$P(E|F) = \frac{n(E \cap F)}{n(F)} = \frac{1}{5}$$

5. The sum is 5, given that the dice have opposite parity. The possible outcomes are (1, 4), (4, 1), (2, 3), and (3, 2).
$n(E \cap F) = 4$
$n(F) = n(\text{dice have opposite parity}) = 18$
$$P(E|F) = \frac{n(E \cap F)}{n(F)} = \frac{4}{18} = \frac{2}{9}$$

7. The dice have the same parity, given that the green die is odd.
The possible outcomes are {red odd, green odd} = $C(3, 1)C(3, 1) = 9$
$$P(E|F) = \frac{n(E \cap F)}{n(F)} = \frac{9}{C(6, 1)C(3, 1)} = \frac{9}{18} = \frac{1}{2}$$

9. The sum is 7, given that the sum is 8.
$n(E \cap F) = 0$
$n(F) = 5$
$$P(E|F) = \frac{n(E \cap F)}{n(F)}$$
$$P(E|F) = \frac{0}{5} = 0$$

11. She gets all the red ones, given that she gets the fluorescent pink one.
$n(E \cap F) = 1$
$n(F) = C(1, 1)C(9, 3) = 84$
$$P(E|F) = \frac{n(E \cap F)}{n(F)} = \frac{1}{84}$$

13. She gets none of the red ones, given that she gets the fluorescent pink one.
$n(E \cap F) = C(1, 1)C(6, 3) = 1 \cdot 20 = 20$
$n(F) = C(1, 1)C(9, 3) = 1 \cdot 84 = 84$
$$P(E|F) = \frac{n(E \cap F)}{n(F)} = \frac{20}{84} = \frac{5}{21}$$

15. She gets one of each color other than fluorescent pink, given that she gets at least one red one.
$n(E \cap F) = C(3, 1)C(2, 1)C(2, 1)C(2, 1) = 24$
$n(F) = C(10, 4) - C(7, 4)$
All combinations – Those with no red ones = Set with at least one red one
$$P(E|F) = \frac{n(E \cap F)}{n(F)} = \frac{24}{210 - 35} = \frac{24}{175}$$
Another possibility for $n(F) = C(3, 1)C(7, 3) + C(3, 2)C(7, 2) + C(3, 3)C(7, 1)$, representing sets with one red marble, two red ones, and three red ones, thus fulfilling the condition of "at least one red marble."

17. She gets no more than one of any color, given that she gets two red ones. Since the outcome is incompatible with the condition, $P(E|F) = 0$.

19. She gets no more than one of any color and not the fluorescent pink one, given that she gets one yellow and one orange one.
$n(E \cap F) = C(3, 1)C(2, 1)C(2, 1)C(2, 1) = 24$
$n(F) = C(2, 1)C(2, 1)C(6, 2) = 2 \cdot 2 \cdot 15 = 60$
$$P(E|F) = \frac{n(E \cap F)}{n(F)} = \frac{24}{60} = \frac{2}{5}$$

21. Find the probability that one tax dollar spent on Superfund will be recovered, given that it is recoverable.

$$P(E|F) = \frac{P(E \cap F)}{P(F)} = \frac{0.119}{1 - 0.402} = 0.199$$

23. Percentage of Superfund tax dollars that were neither recovered nor sought in litigation that were ultimately written off

$$= \frac{0.038}{1 - (0.119 + 0.117)} \times 100 = 0.0497 \times 100 = 4.97\%$$

25. Probability that a student loan due for repayment in 1992 was given to a student in a 2-year college

$$= \frac{n(E \cap F)}{n(F)} = \frac{254,623}{1,113,079 + 254,623 + 621,412}$$

$$= \frac{254,623}{1,989,114} \approx 0.1280$$

27. Probability that a student loan at a 4-year college due for repayment in 1992 was defaulted in 1992-1993

$$= \frac{74,945}{1,113,079} \approx 0.0673$$

29. Probability that a defaulted loan was given to a student in a trade school $= \dfrac{187,739}{74,945 + 36,918 + 187,739} = \dfrac{187,739}{299,602} \approx 0.6266$

31. $P(E|F) = \dfrac{n(E \cap F)}{n(F)} = \dfrac{10,160}{15,441} = 0.6580$

33. $P(E|F) = \dfrac{n(E \cap F)}{n(F)} = \dfrac{15,777 - 10,160}{54,039 - 15,441} = \dfrac{5617}{38,598} \approx 0.1455$

35. $P(E|F) = \dfrac{n(E \cap F)}{n(F)} = \dfrac{459}{4220} \approx 0.1088$

37. $P(E|F) = \dfrac{n(E \cap F)}{n(F)} = \dfrac{5348}{19,722 - 2393} = \dfrac{5348}{17,329} \approx 0.3086$

39. For men:

$$P(E|F) = \frac{n(E \cap F)}{n(F)} = \frac{1022}{8307} \approx 0.1230$$

For women:

$$P(E|F) = \frac{n(E \cap F)}{n(F)} = \frac{2016}{5137} \approx 0.3924$$

The claim is correct.

41. The total number of five card combinations is $C(52, 5)$. The number of combinations that results in banishment is $C(13, 2)C(13, 1)C(26, 2)$. The number of combinations that results in marriage is $C(12, 2)C(12, 1)C(26, 1)$. The probability that a suitor will win her hand in marriage, assuming that he is not banished, is $\dfrac{C(12, 2)C(12, 1)C(26, 1)}{C(52, 5) - C(13, 2)C(13, 1)C(26, 2)}$.

43. In order for the suitor's hand to include a 5 of diamonds, 5 of hearts and jack of spades, he needs the Queen of Hearts and another heart.

$$P(E|F) = \frac{C(1, 1)C(11, 1)}{C(49, 2)}$$

45.

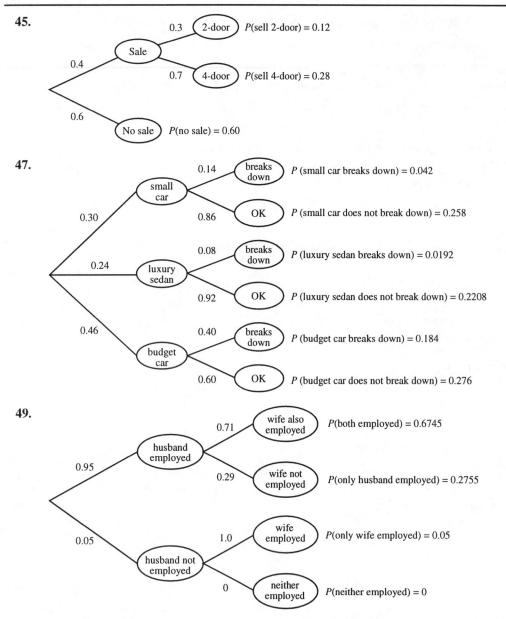

47.

49.

51. $E|F$ = event of being successful given one die even and the other odd

$$P(E|F) = \frac{n(E \cap F)}{n(F)} = \frac{8}{8} = 1$$

E = event of anything but double

$$P(E) = \frac{12}{16} = \frac{3}{4}$$

Since $P(E|F) \neq P(E)$, the events are not independent.

53. You fail; both dice are even.

$E|F$ = event of failing, given both dice are even

$$P(E|F) = \frac{n(E \cap F)}{n(F)} = \frac{2}{4} = \frac{1}{2}$$

E = event of failing

$$n(E) = \frac{4}{16} = \frac{1}{4}$$

Since $P(E|F) \neq P(E)$, the events are not independent.

55. You are successful; the red die is a 4.

$E|F$ = event of being successful, given red die is a 4

$$P(E|F) = \frac{C(3, 1)}{C(4, 1)} = \frac{3}{4}$$

E = event of being successful

$$P(E) = \frac{12}{16} = \frac{3}{4}$$

Since $P(E|F) = P(E)$, the events are independent.

57.

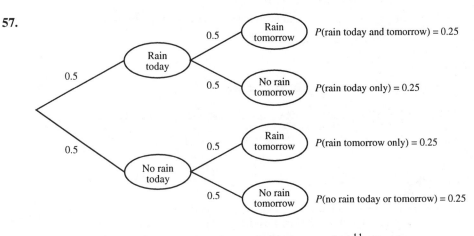

59. Probability of sequence $(H, T, T, H, H, H, T, H, H, T, T) = \left(\frac{1}{2}\right)^{11} = \frac{1}{2048}$

61. If 90% of athletes who test positive use steroids, and 10% of all athletes use steroids and test positive, what percentage of athletes test positive?

Let E = event of using steroids, F = event of testing positive.

$$P(E|F) = 0.90 = \frac{P(E \cap F)}{P(F)} = \frac{0.10}{P(F)}$$

$$P(F) = \frac{0.10}{0.90} = \frac{1}{9} = 11.11\%$$

63. False positive rate is 5%; false negative rate is 1%.
Suppose that 1% of the population is infected. What is the probability that a randomly selected person both has the disease and tests positive?
Let E = event of having AIDS, F = event of testing positive.

$P(F|E) = 1 - 0.01 = 0.99$

$P(E \cap F) = P(F|E)P(E) = 0.99(0.01) = 0.0099$

What is the probability that a randomly selected person does not have the disease but tests positive anyway?

$P(E' \cap F) = P(F|E')P(E') = 0.99(0.05) = 0.0495$

65. 40% of population uses Brand X laundry detergent.
5% of population gave up doing its laundry last year.
4% of population used Brand X and then gave up doing laundry.
Are events of using Brand X and giving up doing laundry independent?
Let E = event of giving up doing laundry, F = event of using Brand X detergent.
$E|F$ = event of giving up laundry having used Brand X detergent

$$P(E|F) = \frac{P(E \cap F)}{P(F)} = \frac{0.04}{0.40} = 0.1$$

Since $P(E|F) = 0.1 > P(E)0.05$, the events are not independent.

Is a user of Brand X detergent more or less likely to give up doing laundry than a user of another brand?
Let $E|F$ = event of not using Brand X and giving up, F = event of using another detergent.

$$P(E|F) = \frac{P(E \cap F)}{P(F)} = \frac{0.01}{0.60} = \frac{1}{60} < \frac{1}{10}$$

Therefore, a user of Brand X is more likely to give up than a user of another detergent.

67. Conditional probability is the probability that a certain event occurs in a reduced sample space containing that event.

Thus, $P(E|F) = \dfrac{P(E \cap F)}{P(F)}$ is the probability that event E exists in the sample space described by event F.

69. 10,000 people saw ad and purchased product.
2,000 people did not see ad but purchased product.
We need to know the number of people who saw the ad and did not purchase the product so we can calculate the conditional probabilities associated with seeing the ad and not seeing the ad. If the probability of not purchasing, given that the ad was seen is greater than the probability of purchasing, given that the ad was seen, the campaign was a failure.

71. Given that $A \subseteq B$ and $P(B) \neq 0$, then $P(A|B) = \dfrac{P(A)}{P(B)}$, because $P(A \cap B) = P(A)$, since $(A \cap B) = A$.

73. If A and B are mutually exclusive events, then $A \cap B = \varnothing$, and $A \cup B = A + B$. If A and B are independent events, then
$P(A|B) = \dfrac{P(A \cap B)}{P(B)} = P(A)$. If $P(B) \neq 0$, then $P(A \cap B) = P(A)P(B)$.
But $P(A \cap B) = 0$, since $A \cap B = \varnothing$. Therefore, $P(A) = 0$, or $A = \varnothing$. Your friend's hypothesis is correct.

Section 6.6 Bayes' Theorem and Applications
Pages 464–467

1. $P(A|B) = 0.8$, $P(B) = 0.2$, $P(A|B') = 0.3$

$$P(B|A) = \frac{P(A|B)P(B)}{P(A|B)P(B) + P(A|B')P(B')} = \frac{(0.8)(0.2)}{(0.8)(0.2) + (0.3)(0.8)} = \frac{0.16}{0.16 + 0.24} = \frac{0.16}{0.40} = 0.4000$$

3. $P(X|Y) = 0.8$, $P(Y') = 0.3$, $P(X|Y') = 0.5$

$$P(Y|X) = \frac{P(X|Y)P(Y)}{P(X|Y)P(Y) + P(X|Y')P(Y')} = \frac{(0.8)(0.7)}{(0.8)(0.7) + (0.5)(0.3)} = \frac{0.56}{0.56 + 0.15} = \frac{0.56}{0.71} \approx 0.7887$$

5. Y_1, Y_2, Y_3 form a partition of S. $P(X|Y_1) = 0.4$, $P(X|Y_2) = 0.5$, $P(X|Y_3) = 0.6$, $P(Y_1) = 0.8$ $P(Y_2) = 0.1$.

$$P(Y_1|X) = \frac{P(X|Y_1)P(Y_1)}{P(X|Y_1)P(Y_1) + P(X|Y_2)P(Y_2) + P(X|Y_3)P(Y_3)} = \frac{(0.4)(0.8)}{(0.4)(0.8) + (0.5)(0.1) + (0.6)(1 - 0.8 - 0.1)}$$

$$= \frac{0.32}{0.32 + 0.05 + 0.06} = \frac{0.32}{0.43} \approx 0.7442$$

7. Y_1, Y_2, Y_3 form a partition of S. $P(X|Y_1) = 0.4$, $P(X|Y_2) = 0.5$, $P(X|Y_3) = 0.6$, $P(Y_1) = 0.8$, $P(Y_2) = 0.1$

$$P(Y_2|X) = \frac{P(X|Y_2)P(Y_2)}{P(X|Y_1)P(Y_1) + P(X|Y_2)P(Y_2) + P(X|Y_3)P(Y_3)} = \frac{(0.5)(0.1)}{(0.4)(0.8) + (0.5)(0.1) + (0.6)(1 - 0.8 - 0.1)}$$

$$= \frac{0.05}{0.32 + 0.05 + 0.06} = \frac{0.05}{0.43} \approx 0.1163$$

9. The probability that Mona was justifiably dropped may be calculated from

$$1 - P(\text{Mona was unjustifiably dropped}) = 1 - P(\text{fit}|\text{failed test}) = 1 - \frac{P(\text{failed}|\text{fit})P(\text{fit})}{P(\text{failed}|\text{fit})P(\text{fit}) + P(\text{failed}|\text{not fit})P(\text{not fit})}$$

$$= 1 - \frac{(0.50)(0.45)}{(0.50)(0.45) + (1)(0.55)} = 1 - \frac{0.225}{0.225 + 0.55} = 1 - 0.29 = 0.71$$

11. a. Probability that a New Yorker who has a positive blood test is infected with HIV $= P(I|T) = \dfrac{P(T|I)P(I)}{P(T|I)P(I) + P(T|I')P(I')}$

where T = event of positive test result, I = event of being infected

$$= \frac{(0.95)(0.01)}{(0.95)(0.01) + (0.05)(0.99)} = \frac{0.0095}{0.0095 + 0.0495} \approx 0.16$$

b. If one in every five New Yorkers is infected, then $P(I|T) = \dfrac{(0.95)(0.20)}{(0.95)(0.20) + (0.05)(0.80)} \approx 0.826$

13. Probable number of representatives in contingent of ten who were former students of Professor A $= 10P(A|R)$
where A = event of having been a student of Professor A and R = event of being representative in contingent,

$$10P(A|R) = \frac{10P(R|A)P(A)}{P(R|A)P(A) + P(R|A')P(A')} = \frac{10(1.0)\left(\frac{3}{4}\right)}{(1.0)\left(\frac{3}{4}\right) + \left(\frac{1}{3}\right)\left(\frac{1}{4}\right)} = \frac{10(0.75)}{0.75 + \frac{1}{12}} = \frac{7.5}{\frac{5}{6}} = 9$$

Note that all of Professor A's former students have grades of C– or less, whereas $\left(\dfrac{1}{3} \cdot \dfrac{1}{4}\right) = \dfrac{1}{12}$ of total students have grades of C– or less, but did not have Professor A in previous semester.

15. a. Percentage of homes owned by married couples having pools $= 100\left[P(\text{Pool}|\text{MC})\right]$

$$= 100\left[\frac{P(\text{MC}|\text{Pool})P(\text{Pool})}{P(\text{MC}|\text{Pool})P(\text{Pool}) + P(\text{MC}|\text{Pool}')P(\text{Pool}')}\right] = 100\left[\frac{(0.86)(0.15)}{(0.86)(0.15) + (0.90)(0.85)}\right]$$

$$= 100\left[\frac{0.129}{0.129 + 0.765}\right] \approx 14.43\%$$

b. Percentage of homes owned by singles having pools $= 100\left[P(\text{Pool}|\text{S})\right] = 100\left[\dfrac{P(\text{S}|\text{Pool})P(\text{Pool})}{P(\text{S}|\text{Pool})P(\text{Pool}) + P(\text{S}|\text{Pool}')P(\text{Pool}')}\right]$

$$= 100\left(\frac{(0.14)(0.15)}{(0.14)(0.15) + (0.10)(0.85)}\right) = 19.81\%$$

It appears that the pool manufacturers should concentrate on the single homeowners, since they are more likely to install a pool.

17. a. Percentage of those for whom family values was the number-one issue favored Bush

$= 100 \times P(\text{favoring Bush}|\text{family values \#1})$

$= 100 \times \dfrac{P(\text{family values \#1}|\text{favoring Bush})P(\text{favoring Bush})}{\begin{bmatrix} P(\text{family values \#1}|\text{favoring Bush})P(\text{favoring Bush}) \\ +P(\text{family values \#1}|\text{favoring Clinton})P(\text{favoring Clinton}) \\ +P(\text{family values \#1}|\text{undecided voters})P(\text{undecided}) \end{bmatrix}}$

$= 100 \times \dfrac{(0.40)(0.32)}{(0.40)(0.32)+(0.20)(0.48)+(0.30)(0.20)} \approx 45.1\%$

b. The first thought is "no," since only 45% of the people who considered family values as the number-one issue favored Bush. Most of them (the other 55%) did not favor Bush. On the other hand, you can calculate that 34% of the "family values" voters favored Clinton, so that an increased emphasis on family values would wind up decreasing Clinton's support by a wider margin and swelling the undecided category.

19. Probability of having an accident if in "not at risk" group $= P(A|R') = \dfrac{P(R'|A)P(A)}{P(R'|A)P(A)+P(R'|A')P(A')}$

$= \dfrac{(0.40)(0.05)}{(0.40)(0.05)+(0.80)(0.95)} = \dfrac{0.02}{0.02+0.76} \approx 0.02564$

Percentage $= 100 \times P(A|R') \approx 2.564\%$

Note: A = event of having accident, R = event of being at risk

21. Probability of having a college degree and being employed in a private household $= P(H|D)$

$= \dfrac{P(D|H)P(H)}{P(D|H)P(H)+P(D|S)P(S)+P(D|O)P(O)} = \dfrac{(0.05)(0.069)}{(0.05)(0.069)+(0.25)(0.05)+(0.30)(0.881)}$

$= \dfrac{0.00345}{0.00345+0.0125+0.2643} = \dfrac{0.00345}{0.28025} \approx 0.0123$

where D = event of having a college degree, H = event of being a household worker,
S = event of being employed in a protective capacity, and O = event of being employed in another industry.

23. Probability that the husband of an employed woman was also employed $= P(H|W)$

$= \dfrac{P(W|H)P(H)}{P(W|H)P(H)+P(W|H')P(H')} = \dfrac{(0.71)(0.95)}{(0.71)(0.95)+(1.0)(0.05)} = \dfrac{0.6745}{0.6745+0.05} = \dfrac{0.6745}{0.7245} \approx 0.931$

where H = event of husband being employed, W = event of wife being employed.

25. Percentage of population in South in 1988 who moved there from Northeast $= 100 \times P(NE|S)$

$= 100 \times \dfrac{P(S|NE)P(NE)}{P(S|NE)P(NE)+P(S|MW)P(MW)+P(S|S)P(S)+P(S|W)P(W)}$

$= 100 \times \dfrac{(0.009)(49.3)}{(0.009)(49.3)+(0.0089)(59.1)+(0.9884)(82.2)+(0.0048)(48.2)}$

$= 100 \times \dfrac{0.4437}{0.4437+0.52599+81.2465+0.23136} = 100 \times \dfrac{0.4437}{82.44755} \approx 0.54\%$

27. Probability that a middle-aged man with diabetes is very active $= P(A|D) = \dfrac{P(D|A)P(A)}{P(D|A)P(A) + P(D|A')P(A')}$

$P(D|A') = \dfrac{202}{5,990}$ and $P(D|A) = \dfrac{1}{2}\left(\dfrac{202}{5,990}\right)$.

So $P(A|D) = \dfrac{\frac{1}{2}\left(\frac{202}{5990}\right)\left(\frac{1}{3}\right)}{\frac{1}{2}\left(\frac{202}{5990}\right)\left(\frac{1}{3}\right) + \left(\frac{202}{5990}\right)\left(\frac{2}{3}\right)} = \dfrac{1}{5} = 0.2$

where D = event of having diabetes, A = event of being very active.

29. Consider the case $P(A) = 0.7$, $P(B) = 0.4$, $P(A \cap B) = 0.1$.

Then, $P(A|B) = \dfrac{P(A \cap B)}{P(B)} = \dfrac{0.1}{0.4} = \dfrac{1}{4}$

$P(B|A) = \dfrac{P(B \cap A)}{P(A)} = \dfrac{0.1}{0.7} = \dfrac{1}{7}$

31. Suppose that 9% of the athletes are using steroids. Let R = event that the athlete uses steroids, and let T = event that the test comes up positive. Then $P(T|R') = 0.01$.

Say $P(T|R) = 0.001$.

$P(R|T) = \dfrac{0.001(0.09)}{0.001(0.09) + 0.01(0.91)} = 0.0098$

If an athlete tests positive, the chance that he/she has used steroids is 0.98%.

33. Given that $R = R_1 \cup R_2 \cup R_3$ and $R_1 \cap R_2 \cap R_3 = \varnothing$, we may write

$P(R_1|T) = \dfrac{P(R_1 \cap T)}{P(T)}$ (definition of conditional probability)

$= \dfrac{P(R_1)P(T|R_1)}{P(T)}$

$= \dfrac{P(R_1)P(T|R_1)}{P(R_1)P(T|R_1) + P(R_2)P(T|R_2) + P(R_3)P(T|R_3)}$

Section 6.7 Bernoulli Trials

Pages 472–473

In exercises 1–9, $n = 8$, $p = \dfrac{2}{3}$, $q = \dfrac{1}{3}$.

1. $P(2 \text{ successes, } 6 \text{ failures}) = C(8, 2)\left(\dfrac{2}{3}\right)^2\left(\dfrac{1}{3}\right)^6$

3. $P(\text{no successes}) = C(8, 0)\left(\dfrac{2}{3}\right)^0\left(\dfrac{1}{3}\right)^8$

5. $P(\text{all successes}) = C(8, 8)\left(\dfrac{2}{3}\right)^8\left(\dfrac{1}{3}\right)^0$

7. $P(\text{at least 5 failures}) = P(5 \text{ failures}) + P(6 \text{ failures}) + \cdots + P(8 \text{ failures})$

$$= C(8, 3)\left(\frac{2}{3}\right)^3\left(\frac{1}{3}\right)^5 + C(8, 2)\left(\frac{2}{3}\right)^2\left(\frac{1}{3}\right)^6 + C(8, 1)\left(\frac{2}{3}\right)^1\left(\frac{1}{3}\right)^7 + C(8, 0)\left(\frac{2}{3}\right)^0\left(\frac{1}{3}\right)^8$$

9. $P(\text{at most 3 successes}) = P(0 \text{ success}) + \cdots + P(3 \text{ successes})$

$$= C(8, 0)\left(\frac{2}{3}\right)^0\left(\frac{1}{3}\right)^8 + C(8, 1)\left(\frac{2}{3}\right)^1\left(\frac{1}{3}\right)^7 + C(8, 2)\left(\frac{2}{3}\right)^2\left(\frac{1}{3}\right)^6 + C(8, 3)P\left(\frac{2}{3}\right)^3\left(\frac{1}{3}\right)^5$$

In Exercises 11–15, a die is used that has a 6 on two of its faces and no 1. The die is rolled six times.

11. $P(6 \text{ comes up exactly twice}) = C(6, 2)\left(\frac{2}{6}\right)^2\left(\frac{4}{6}\right)^4 = (15)\left(\frac{1}{9}\right)\left(\frac{16}{81}\right) = \frac{80}{243}$

13. $P(6 \text{ comes up at least four times}) = P(6 \text{ four times}) + P(6 \text{ five times}) + P(6 \text{ six times})$

$$= C(6, 4)\left(\frac{2}{6}\right)^4\left(\frac{4}{6}\right)^2 + C(6, 5)\left(\frac{2}{6}\right)^5\left(\frac{4}{6}\right)^1 + C(6, 6)\left(\frac{2}{6}\right)^6\left(\frac{4}{6}\right)^0 \approx 0.1001$$

15. $P(6 \text{ comes up no more than five times}) = P(6 \text{ zero times}) + \cdots + P(6 \text{ five times})$

$$C(6,0)\left(\frac{2}{6}\right)^0\left(\frac{4}{6}\right)^6 + C(6,1)\left(\frac{2}{6}\right)^1\left(\frac{4}{6}\right)^5 + C(6,2)\left(\frac{2}{6}\right)^2\left(\frac{4}{6}\right)^4 + C(6,3)\left(\frac{2}{6}\right)^3\left(\frac{4}{6}\right)^3 + C(6,4)\left(\frac{2}{6}\right)^4\left(\frac{4}{6}\right)^2 + C(6,5)\left(\frac{2}{6}\right)^5\left(\frac{4}{6}\right)^1$$

≈ 0.9986

Note: The result can also be obtained from $1 - C(6, 6)\left(\frac{2}{6}\right)^6\left(\frac{4}{6}\right)^0 = 1 - \frac{1}{729} = \frac{728}{729} \approx 0.9986$

17. With $m = 5$, $p = \frac{1}{4}$, $q = \frac{3}{4}$, the probabilities are:

Success

0	$C(5, 0)\left(\frac{1}{4}\right)^0\left(\frac{3}{4}\right)^5 \approx 0.237305 = \dfrac{243}{1024}$	
1	$C(5, 1)\left(\frac{1}{4}\right)^1\left(\frac{3}{4}\right)^4 \approx 0.39551 = \dfrac{405}{1024}$	
2	$C(5, 2)\left(\frac{1}{4}\right)^2\left(\frac{3}{4}\right)^3 \approx 0.26367 = \dfrac{270}{1024}$	
3	$C(5, 3)\left(\frac{1}{4}\right)^3\left(\frac{3}{4}\right)^2 \approx 0.087891 = \dfrac{90}{1024}$	
4	$C(5, 4)\left(\frac{1}{4}\right)^4\left(\frac{3}{4}\right)^1 \approx 0.014648 = \dfrac{15}{1024}$	
5	$C(5, 5)\left(\frac{1}{4}\right)^5\left(\frac{3}{4}\right)^0 \approx 0.0009766 = \dfrac{1}{1024}$	

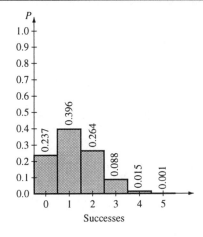

19. P(at least 1 male) $= 1 - P$(0 males)

$$= 1 - C(3, 0)\left(\frac{1}{2}\right)^0\left(\frac{1}{2}\right)^3 = 1 - \frac{1}{8} = \frac{7}{8} = 0.875$$

21. P(at least 2 false positive tests) $= 1 - P$(0 false positive) $- P$(1 false positive)

$$= 1 - C(100, 0)(0.05)^0(0.95)^{100} - C(100, 1)(0.05)^1(0.95^{99})$$

$$= 1 - 0.00592 - 0.03116 = 1 - 0.037 \approx 0.96$$

23. P(at least 3 people infected but not sick)

$$= 1 - P(0 \text{ infected but not sick}) - P(1 \text{ infected but not sick}) - P(2 \text{ infected but not sick})$$

$$= 1 - C(20, 0)(0.20)^0(0.80)^{20} - C(20, 1)(0.20)^1(0.80)^{19} - C(20, 2)(0.20)^2(0.80)^{18}$$

$$= 1 - 0.011529 - 0.057646 - 0.136909 \approx 1 - 0.2061 = 0.794$$

25. P(one woman developing breast cancer) $= C(10, 1)\left(\frac{1}{24}\right)^1\left(\frac{23}{24}\right)^9 = 0.284$

27. P(answering between 75 and 85 questions, inclusive)

$$= C(100, 75)(0.80)^{75}(0.20)^{25} + C(100, 76)(0.80)^{76}(0.20)^{24} + \cdots + C(100, 85)(0.80)^{85}(0.20)^{15}$$

$$\approx 0.0439 + 0.0577 + 0.0720 + 0.0849 + 0.0946 + 0.0993 + 0.0981 + 0.0909 + 0.0789 + 0.0638 + 0.0481 \approx 0.8321$$

29. P(finding at least one bad bulb from set of N bulbs) $= 1 - P(N \text{ good}) = 1 - C(N, N)(0.01)^0(0.99)^N = 1 - (0.99)^N \geq 0.50$

Using 0.50, we get $(0.99)^N = 0.50$.

$N \ln (0.99) = \ln (0.50)$

$$N = \frac{\ln(0.50)}{\ln(0.99)} = 68.97$$

Therefore, test at least 69 bulbs.

You're the Expert
Page 475

1. If Monty Hall opens door B or door C randomly and the prize is not there, the probability that the prize is behind door A is equal to the probability that it is behind the door that he did not open. Switching does not increase the probability of winning. It is one-half for both doors. In terms of Bayes' theorem, let A, B, C, and F be defined as in the original example in the text, i.e., $A =$ event that prize is behind door A, ..., and F is the event that Monty has opened door B and revealed that the prize is not there.

Then $P(C|F) = \dfrac{P(F|C)P(C)}{P(F|A)P(A) + P(F|B)P(B) + P(F|C)P(C)} = \dfrac{\left(\frac{1}{2}\right)\left(\frac{1}{3}\right)}{\left(\frac{1}{2}\right)\left(\frac{1}{3}\right) + (0)\left(\frac{1}{3}\right) + \left(\frac{1}{2}\right)\left(\frac{1}{3}\right)} = \dfrac{\frac{1}{6}}{\frac{2}{6}} = \dfrac{1}{2}$

The formula for $P(A|F)$ gives the same result. Note that $P(F|C)$ is now $\dfrac{1}{2}$, rather than 1.

Chapter 6 Review Exercises
Pages 475–477

1. Three coins are tossed, and the result is one or more tails.
   Sample space $S = \{HHH, HHT, HTH, HTT, THH, THT, TTH, TTT\}$
   $E = \{HHT, HTH, HTT, THH, THT, TTH, TTT\}$

   $P(E) = \dfrac{n(E)}{n(S)} = \dfrac{7}{8}$

3. Four coins are tossed, and the result is fewer heads than tails.
   $S = \{HHHH, HHHT, HHTH, HTHH, THHH, HHTT, HTHT, HTTH, THTH, THHT, TTHH, HTTT, THTT, TTHT, TTTH,$
   $\qquad TTTT\}$
   $E = \{HTTT, THTT, TTHT, TTTH, TTTT\}$

   $P(E) = \dfrac{n(E)}{n(S)} = \dfrac{5}{16}$

5. Two dice are thrown, and the numbers add to 7.
   $S = \{(1, 1), (1, 2), \ldots, (1, 6), (2, 1), \ldots, (2, 6), \ldots, (6, 6)\}$, the sample space of Example 1 in Section 6.1
   $E = \{(1, 6), (2, 5), (3, 4), (4, 3), (5, 2), (6, 1)\}$

   $P(E) = \dfrac{n(E)}{n(S)} = \dfrac{6}{36} = \dfrac{1}{6}$

7. Two dice are thrown, and both numbers have the same parity.
   $S =$ space given in Exercise 5
   $E = \{(1, 1), (1, 3), (1, 5), (2, 2), (2, 4), (2, 6), (3, 1), (3, 3), (3, 5), (4, 2), (4, 4), (4, 6), (5, 1), (5, 3), (5, 5),$
   $\qquad (6, 2), (6, 4), (6, 6)\}$

   $P(E) = \dfrac{n(E)}{n(S)} = \dfrac{18}{36} = \dfrac{1}{2}$

9. A die is weighted so that each of 2, 3, 4, and 5 is half as likely to come up as either 1 or 6.
   Let total probability $= 1$. $P(1) = P(6) = 2x$, $P(2), \ldots, P(5) = x$.
   $2x + 4x + 2x = 1$

   $x = \dfrac{1}{8}$

   $P(1) = P(6) = \dfrac{1}{4}$; $P(2) = P(3) = P(4) = P(5) = \dfrac{1}{8}$

11. A die is constructed in such a way that 6 is twice as likely to come up as 1 and 1.5 times as likely to come up as any of 2, 3, 4, and 5.
    Let $x = P(2), \ldots, P(5)$; $P(6) = 1.5x$, $P(1) = 0.75x$
    $0.75x + 4x + 1.5x = 1$
    $6.25x = 1$

    $x = \dfrac{1}{6.25} = \dfrac{4}{25}$

    $P(1) = \dfrac{3}{25}$, $P(2) = P(3) = P(4) = P(5) = \dfrac{4}{25}$, $P(6) = \dfrac{6}{25}$

    Total probability $= \dfrac{25}{25} = 1$, check.

13. Let E be the event of a tornado and let F be the event of a monsoon. Thus, $P(E) = 0.5$, $P(F) = 0.2$, and $P(E \cap F) = 0.1$. $P(E \cup F) = P(E) + P(F) - P(E \cap F) = 0.6$, so the probability that we will be lucky is $1 - P(E \cup F) = 0.4$.

15. Coin collection = {5 Maple Leafs, 2 Kruggerands, 1 Golden Eagle}
If one takes three at random, number of outcomes = $C(8, 3) = 56$.

Of these outcomes, $C(6, 1) = 6$ include the Kruggerands. $P(\text{get both Kruggerands}) = \dfrac{6}{56} = \dfrac{3}{28}$

17. There are a total of 31 pieces of furniture to choose from. Since furniture is randomly assigned, order does not matter in placing the furniture. Let S be the set of all arrangements, and let E be the set of arrangements with the bookcase in one of the bathrooms. Compute $n(S)$ by first choosing furniture for the bathrooms: choose 2 out of 31 pieces for one bathroom, then choose 2 out of 29 for the other bathroom, and then select from the the remaining 27 pieces for the other rooms. Let S' be the set of all arrangements of 27 pieces for the remaining rooms. Thus $n(S) = C(31, 2)C(29, 2)n(S')$.

Next compute $n(E)$: choose 1 of 2 bathrooms, then choose 1 out of 1 bookcase for the bathroom, then choose 1 out of 30 pieces for the bathroom, then choose 2 out of 29 for the other bathroom, and then select from the remaining 27 pieces for the other rooms. Thus $n(E) = C(2, 1)C(1, 1)C(30, 1)C(29, 2)n(S')$.

Therefore, $P(E) = \dfrac{n(E)}{n(S)} = \dfrac{C(2, 1)C(1, 1)C(30, 1)}{C(31, 2)} = \dfrac{4}{31}$.

19. Let S be the set of all arrangements, and let E be the set of arrangements with no beds in any of the bedrooms. Compute $n(S)$ by first choosing furniture for the bedrooms: choose 3 out of 31 pieces for one bedroom, then choose 3 out of 28 pieces for the other bedroom and then select from the remaing 25 pieces for the remaining rooms. Let S' be the set of all arrangements of 25 pieces for the remaining rooms. Thus $n(S) = C(31, 3)C(28, 3)n(S')$. Next compute $n(E)$: choose 3 out of 28 pieces (does not include the 3 beds) for one bedroom, then choose 3 out of 25 pieces (does not include the 3 beds) for the other bedroom, and then select from the remaining 25 pieces for the other rooms. Thus $n(E) = C(28, 3)C(25, 3)n(S')$.

Therefore, $P(E) = \dfrac{n(E)}{n(S)} = \dfrac{C(25, 3)}{C(31, 3)} = \dfrac{460}{899}$.

21. Let S be the set of all arrangements, and let E be the set of arrangements with the desk in one bathroom and a bed in the other. We interpret this to mean that there is exactly one bed in the other bathroom and no bed in the bathroom with the desk. Answers for other interpretations are given below. Compute $n(S)$ by first choosing furniture for the bathrooms: choose 2 out of 31 pieces for one bathroom, then choose 2 out of 29 pieces for the other, and select from the remaining 27 pieces for the remaining rooms. Let S' be the set of all arrangements of 27 pieces for the remaining rooms. Thus, $n(S) = C(31, 2)C(29, 2)n(S')$. Next compute $n(E)$:

Choose 1 out of 1 desk to be in one bathroom, then choose 1 out of 3 beds to be in the other, then choose 1 out of 27 pieces (does not include the remaining 2 beds) for the bathroom with the bed, then choose 1 out of 26 (does not include the remaining 2 beds) for the bathroom with the desk, and then select from the remaining 27 pieces for the rooms. Thus $n(E) = C(1, 1)C(3, 1)C(27, 1)C(26, 1)n(S')$. Therefore, $P(E) = \dfrac{n(E)}{n(S)} = \dfrac{C(1, 1)C(3, 1)C(27, 1)C(26, 1)}{C(31, 2)C(29, 2)}$

$= \dfrac{351}{31,465}$.

If we interpret the problem to mean that there is exactly one bed in the other bathroom and there may be a bed with the desk, then the probability is $\dfrac{C(1, 1)C(3, 1)C(28, 1)C(27, 1)}{C(31, 2)C(29, 2)} = \dfrac{54}{4495}$.

If we interpret the problem to mean that there is at least one bed in the other bathroom and there may be a bed with a desk, then the probability is $\dfrac{C(1, 1)C(3, 1)C(29, 1)C(28, 1)}{C(31, 2)C(29, 2)} = \dfrac{2}{155}$.

23. Let S be the set of all arrangements, and let E be the set of arrangements with a bed in each bathroom. We interpret this to mean that each bathroom has exactly one bed. Compute $n(S)$ by first choosing furniture for the bathrooms: choose 2 out of 31 pieces for one bathroom, choose 2 out of 29 pieces for the other bathroom, and then select from the remaining 27 pieces for the remaining rooms. Let S' be the set of all arrangements of 27 pieces for the remaining rooms. Thus $n(S) = C(31, 2)C(29, 2)n(S')$. Next compute $n(E)$: Choose 1 out of 3 beds for one bathroom, then choose 1 out of 2 beds for the other, then choose 1 out of 28 pieces (does not include the remaining bed) for one bathroom, then choose 1 out of 27 pieces (does not include the remaining bed) for the other, and then select from the remaining 27 pieces for the other rooms. Thus $n(E) = C(3, 1)C(2, 1)C(28, 1)C(27, 1)n(S')$.

Therefore, $P(E) = \dfrac{n(E)}{n(S)} = \dfrac{C(3, 1)C(2, 1)C(28, 1)C(27, 1)}{C(31, 2)C(29, 2)} = \dfrac{108}{4495}$.

If we interpret this to mean that a bathroom may have two beds, then the probability is

$\dfrac{C(3, 1)C(2, 1)C(29, 1)C(28, 1)}{C(31, 2)C(29, 2)} = \dfrac{4}{155}$.

25. $P(\text{5 cards being Kings or Queens}) = \dfrac{C(8, 5)}{C(52, 5)}$

27. $P(\text{All cards are 10 and up}) = \dfrac{C(20, 5)}{C(52, 5)}$

29. $P(\text{Prime Full House}) = P(\text{each card from } \{2, 3, 5, 7, J, K\} \text{ and full house}) = \dfrac{C(6, 1)C(4, 2)C(5, 1)C(4, 3)}{C(52, 5)}$

31. $P(\text{sum} = 5, \text{ given green die not a 1 and yellow die is a 1})$

$$P(A|B) = \frac{P(A \cap B)}{P(B)} = \frac{\left(\frac{2}{216}\right)}{\left(\frac{30}{216}\right)} = \frac{1}{15}$$

Note: The restricted sample space has $6 \cdot 5 \cdot 1 = 30$ elements.

33. $P(\text{yellow} = 4, \text{ given that green die is 4}) = \dfrac{P(A \cap B)}{P(B)} = \dfrac{\frac{1}{6} \cdot \frac{1}{6}}{\frac{1}{6}} = \dfrac{1}{6}$

35. $P(\text{dice all have same parity, given that at least two are odd})$
$n(\text{all odd combinations}) = 3 \cdot 3 \cdot 3 = 27$
$n(\text{two odd dice}) = C(3, 2) \cdot 3 \cdot 3 \cdot 3 = 81$

$$P(\text{three odd}|\text{at least two odd}) = \frac{27}{81 + 27} = \frac{27}{108} = \frac{1}{4}$$

37. Sum $= 5$, given that green die is not a 1 and yellow die is a 1.

Test to see if $P(A) \stackrel{?}{=} P(A|B) = \dfrac{1}{15}$. $P(A) = P(\text{sum} = 5)$

$n(A) = 6$ arrangements each of $(1, 2, 2)$, and $(1, 1, 3) = 6(2) = 12$.

$P(A) = \dfrac{12}{216} = \dfrac{1}{18} \neq \dfrac{1}{15}$

The events are dependent.

39. P(yellow die = 4, given that green die is 4)

$P(A|B) = \dfrac{1}{6}$ (from Exercise 33)

$P(A) = P$(yellow die = 4) $= \dfrac{1}{6}$

Since $P(A|B) = P(A)$, the events are independent.

41. P(dice have same parity, given two or more that are odd)

$P(A|B) = \dfrac{1}{4}$ (from Exercise 35)

$P(A) = P$(all same parity) $= \dfrac{1}{8} + \dfrac{1}{8} = \dfrac{1}{4}$

Since $P(A|B) = P(B)$, the events are independent.

43. **a.** P(infected at end of two years) $= 1 - (0.95)^2 = 1 -$ uninfected after two years $= 1 - 0.9025 = 0.0975$

 b. P(infected at end of three years) $= 1 - (0.95)^3 = 0.142625$

 c. P(infected at end of four years) $= 1 - (0.95)^4 \approx 0.1855$

45. **a.** P(moon in Seventh House when Jupiter and Mars align) $= P(A|B)$, where A = event of moon being in Seventh House and B = event of Jupiter aligning with Mars.

$$P(A|B) = \frac{P(B|A)P(A)}{P(B|A)P(A) + P(B|A')P(A')} = \frac{0.10\left(\frac{1}{30}\right)}{0.10\left(\frac{1}{30}\right) + (0.05)\left(\frac{29}{30}\right)} = \frac{\frac{1}{300}}{\frac{1}{300} + \frac{29}{600}} = \frac{2}{31}$$

 b. Probability of moon being in Seventh House and Jupiter aligning with Mars $= \dfrac{1}{300}$ (from part a)

47. P(6 comes up exactly once) $= C(4, 1)\left(\dfrac{1}{11}\right)^1\left(\dfrac{10}{11}\right)^3 = 4\left(\dfrac{1}{11}\right)^1\left(\dfrac{10}{11}\right)^3$

Note that the probability of getting a 6 $= \dfrac{1}{11}$ and that of any other number $(1-5) = \dfrac{2}{11}$.

49. P(6 comes up at most twice) $= C(4, 0)\left(\dfrac{1}{11}\right)^0\left(\dfrac{10}{11}\right)^4 + C(4, 1)\left(\dfrac{1}{11}\right)^1\left(\dfrac{10}{11}\right)^3 + C(4, 2)\left(\dfrac{1}{11}\right)^2\left(\dfrac{10}{11}\right)^2$

51. P(cancer between 60 and 70) $= \dfrac{1}{28}\left(1 - \dfrac{1}{24}\right) = \dfrac{23}{24 \cdot 28} \approx 0.0342$

Chapter 7

Problem Set 7.1 Random Variables and Distributions
Pages 490–495

1. **a.** $S = \{HH, HT, TH, TT\}$

   b. X is the rule that assigns to each outcome the number of tails.

   c. $X(HH) = 0$, $X(HT) = 1$, $X(TH) = 1$, $X(TT) = 2$

3. **a.** $S = \left\{ \begin{array}{l} (1, 1), (1, 2), (1, 3), (1, 4), (1, 5), (1, 6), \\ (2, 1), (2, 2), (2, 3), (2, 4), (2, 5), (2, 6), \\ (3, 1), (3, 2), (3, 3), (3, 4), (3, 5), (3, 6), \\ (4, 1), (4, 2), (4, 3), (4, 4), (4, 5), (4, 6), \\ (5, 1), (5, 2), (5, 3), (5, 4), (5, 5), (5, 6), \\ (6, 1), (6, 2), (6, 3), (6, 4), (6, 5), (6, 6) \end{array} \right\}$

   b. X is the rule that assigns to each outcome the sum of the two numbers.

   c.

Outcome	Value of X
(1, 1)	2
(1, 2), (2, 1)	3
(1, 3), (2, 2), (3, 1)	4
(1, 4), (2, 3), (3, 2), (4, 1)	5
(1, 5), (2, 4), (3, 3), (4, 2), (5, 1)	6
(1, 6), (2, 5), (3, 4), (4, 3), (5, 2), (6, 1)	7
(2, 6), (3, 5), (4, 4), (5, 3), (6, 2)	8
(3, 6), (4, 5), (5, 4), (6, 3)	9
(4, 6), (5, 5), (6, 4)	10
(5, 6), (6, 5)	11
(6, 6)	12

5. **a.** $S = \{(4, 0), (3, 1), (2, 2)\}$
   (listed in order (red, green))

   b. X is the rule that assigns to each outcome the number of red marbles.

   c. $X(4, 0) = 4$, $X(3, 1) = 3$, $X(2, 2) = 2$

7. **a.** $S = $ the set of students in your study group

   b. X is the rule that assigns to each student his or her final exam score.

   c. The values of X, in the order given, are 89%, 85%, 95%, 63%, 92%, 80%.

9. A red die and a green die are rolled, and X is the number on the red die minus the number on the green die.

$$S = \begin{Bmatrix} (1, 1), (1, 2), (1, 3), (1, 4), (1, 5), (1, 6), \\ (2, 1), (2, 2), (2, 3), (2, 4), (2, 5), (2, 6), \\ (3, 1), (3, 2), (3, 3), (3, 4), (3, 5), (3, 6), \\ (4, 1), (4, 2), (4, 3), (4, 4), (4, 5), (4, 6), \\ (5, 1), (5, 2), (5, 3), (5, 4), (5, 5), (5, 6), \\ (6, 1), (6, 2), (6, 3), (6, 4), (6, 5), (6, 6) \end{Bmatrix}$$

x	-5	-4	-3	-2	-1	0	1	2	3	4	5
$P(X = x)$	$\dfrac{1}{36}$	$\dfrac{2}{36}$	$\dfrac{3}{36}$	$\dfrac{4}{36}$	$\dfrac{5}{36}$	$\dfrac{6}{36}$	$\dfrac{5}{36}$	$\dfrac{4}{36}$	$\dfrac{3}{36}$	$\dfrac{2}{36}$	$\dfrac{1}{36}$

11. Two dice are rolled, and X is the larger of the two numbers obtained.

x	1	2	3	4	5	6
$P(X = x)$	$\dfrac{1}{36}$	$\dfrac{3}{36}$	$\dfrac{5}{36}$	$\dfrac{7}{36}$	$\dfrac{9}{36}$	$\dfrac{11}{36}$

Example: For $x = 3$, $S_3 = \{(1, 3), (3, 1), (2, 3), (3, 2), (3, 3)\}$, $n(S_3) = 5$.

13. Three coins are tossed, and X is the number of heads minus the number of tails.
$S = \{HHH, HHT, HTH, THH, HTT, THT, TTH, TTT\}$

x	-3	-1	1	3
$P(X = x)$	$\dfrac{1}{8}$	$\dfrac{3}{8}$	$\dfrac{3}{8}$	$\dfrac{1}{8}$

15.

$1.01 - 2.0$	$2.01 - 3.0$	$3.01 - 4.0$
4	7	9

Probability distribution using the midpoint values of X:

1.5	2.5	3.5
$\dfrac{4}{20}$	$\dfrac{7}{20}$	$\dfrac{9}{20}$

17. Five darts are thrown at a dartboard. The probability of hitting the bull's eye is $\frac{1}{10}$. Let X be the number of bull's eyes hit. Find $P(X \geq 3)$.

$P(X = 0) = C(5, 0)(0.1)^0(0.9)^5 \approx 0.59049$

$P(X = 1) = C(5, 1)(0.1)^1(0.9)^4 \approx 0.32805$

$P(X = 2) = C(5, 2)(0.1)^2(0.9)^3 \approx 0.07290$

$P(X = 3) = C(5, 3)(0.1)^3(0.9)^2 \approx 0.00810$

$P(X = 4) = C(5, 4)(0.1)^4(0.9)^1 \approx 0.00045$

$P(X = 5) = C(5, 5)(0.1)^5(0.9)^0 \approx 0.00001$

$P(X \geq 3) = P(X = 3) + P(X = 4) + P(X = 5)$
$\approx 0.00810 + 0.00045 + 0.00001 = 0.00856$

19. Select five cards without replacement from a standard deck of 52, and let X be the number of Queens you draw. Find $P(X \leq 2)$.

$P(X = 0) = \dfrac{C(4, 0)C(48, 5)}{C(52, 5)} = \dfrac{1,712,304}{2,598,960}$
≈ 0.658842

$P(X = 1) = \dfrac{C(4, 1)C(48, 4)}{C(52, 5)} = \dfrac{778,320}{2,598,960}$
≈ 0.299474

$P(X = 2) = \dfrac{C(4, 2)C(48, 3)}{C(52, 5)} = \dfrac{103,776}{2,598,960}$
≈ 0.039930

$P(X = 3) = \dfrac{C(4, 3)C(48, 2)}{C(52, 5)} = \dfrac{4512}{2,598,960}$
≈ 0.001736

$P(X = 4) = \dfrac{C(4, 4)C(48, 1)}{C(52, 5)} = \dfrac{48}{2,598,960}$
≈ 0.000018

$P(X \leq 2) = P(X = 0) + P(X = 1) + P(X = 2)$
$\approx 0.658842 + 0.299474 + 0.039930 = 0.998246$

21. A bag contains 4 red marbles, 2 green ones, and 1 yellow one. Suzy grabs four of them at random. Let X be the number of red marbles she holds. Find $P(X \geq 2)$.

$P(X = 1) = \dfrac{C(4, 1)C(3, 3)}{C(7, 4)} = \dfrac{4}{35} \approx 0.114286$

$P(X = 2) = \dfrac{C(4, 2)C(3, 2)}{C(7, 4)} = \dfrac{18}{35} \approx 0.514286$

$P(X = 3) = \dfrac{C(4, 3)C(3, 1)}{C(7, 4)} = \dfrac{12}{35} \approx 0.342857$

$P(X = 4) = \dfrac{C(4, 4)C(3, 0)}{C(7, 4)} = \dfrac{1}{35} \approx 0.028571$

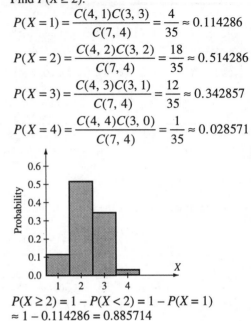

$P(X \geq 2) = 1 - P(X < 2) = 1 - P(X = 1)$
$\approx 1 - 0.114286 = 0.885714$

For Exercises 23–29, the following probability distribution is used.

X = Number of wet days	Number of cities	$P(X)$
0	2	$\frac{2}{28} \approx 0.07143$
1	1	$\frac{1}{28} \approx 0.03571$
2	2	$\frac{2}{28} \approx 0.07143$
3	0	$\frac{0}{28} = 0$
4	2	$\frac{2}{28} \approx 0.07143$
5	1	$\frac{1}{28} \approx 0.03571$
6	1	$\frac{1}{28} \approx 0.03571$
7	2	$\frac{2}{28} \approx 0.07143$
8	2	$\frac{2}{28} \approx 0.07143$
9	3	$\frac{3}{28} \approx 0.10714$
10	2	$\frac{2}{28} \approx 0.07143$
11	1	$\frac{1}{28} \approx 0.03571$
12	5	$\frac{5}{28} \approx 0.17857$
13	4	$\frac{4}{28} \approx 0.14286$
	28	

See page 10 for graph of distribution.

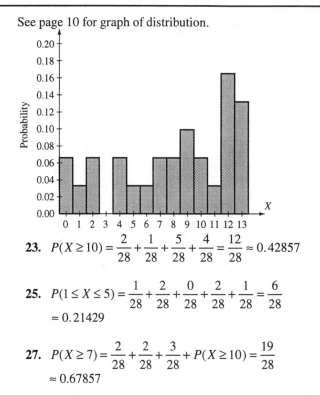

23. $P(X \geq 10) = \frac{2}{28} + \frac{1}{28} + \frac{5}{28} + \frac{4}{28} = \frac{12}{28} \approx 0.42857$

25. $P(1 \leq X \leq 5) = \frac{1}{28} + \frac{2}{28} + \frac{0}{28} + \frac{2}{28} + \frac{1}{28} = \frac{6}{28}$
≈ 0.21429

27. $P(X \geq 7) = \frac{2}{28} + \frac{2}{28} + \frac{3}{28} + P(X \geq 10) = \frac{19}{28}$
≈ 0.67857

29. Most likely number of wet days experienced is $x = 12$.

31.

X = Age of Car in Years	Number of Cars	P(Car is of age X)
0	140	$\frac{7}{100}$
1	350	$\frac{7}{40}$
2	450	$\frac{9}{40}$
3	650	$\frac{13}{40}$
4	200	$\frac{1}{10}$
5	120	$\frac{3}{50}$
6	50	$\frac{1}{40}$
7	10	$\frac{1}{200}$
8	5	$\frac{1}{400}$
9	15	$\frac{3}{400}$
10	10	$\frac{1}{200}$
	2000	

Find P(a car on campus is less than 6 years old) $= \dfrac{7}{100} + \dfrac{7}{40} + \dfrac{9}{40} + \dfrac{13}{40} + \dfrac{1}{10} + \dfrac{3}{50} = 0.955$

Claim: 95.5% of students have newer cars.

33.

x	0	1.5	7.5	36	90
$P(X = x)$	0.67	0.11	0.08	0.10	0.04

X = midpoint, in months, of time expected to be spent in nursing home
$P(X = x)$ = probability of U.S. male being in range of given midpoint

35.

X = midpoint of firm size	Percentage of workers
1	10
5.5	12
17	8
62	12
299.5	14
749.5	6
3,000	38
	100

Note that the scale on the *X*-axis is not uniform.

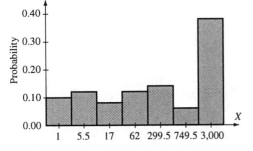

37. a.

X = midpoint of salary bracket	Percentage of college presidents in bracket
$150,000	6.897
$250,000	44.828
$350,000	41.379
$450,000	6.897
	100.001

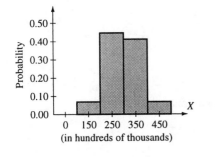

(in hundreds of thousands)

b. $P(X \geq \$200,000) = \dfrac{27}{29} \approx 0.931$

The shaded region is the last three columns in part (a).

39. $n = 4$, $p = \dfrac{1}{2}$, $q = \dfrac{1}{2}$

Binomial distribution

$P(X = 0) = C(4, 0)(0.5)^0(0.5)^4 = 0.0625$

$P(X = 1) = C(4, 1)(0.5)^1(0.5)^3 = 0.2500$

$P(X = 2) = C(4, 2)(0.5)^2(0.5)^2 = 0.3750$

$P(X = 3) = C(4, 3)(0.5)^3(0.5)^1 = 0.2500$

$P(X = 4) = C(4, 4)(0.5)^4(0.5)^0 = \underline{0.0625}$

$\qquad\qquad\qquad\qquad\qquad\qquad\quad 1.0000$

41. $n = 5$, $p = \dfrac{1}{3}$, $q = \dfrac{2}{3}$

Binomial distribution

$$P(X = x) = C(5, X)\left(\frac{1}{3}\right)^X\left(\frac{2}{3}\right)^{5-X} = \frac{1}{243}C(5, x)2^{5-X}$$

x	0	1	2	3	4	5
$P(X = x)$	0.131687	0.329218	0.329218	0.164609	0.041152	0.004115

43. Answers will vary.

45. Data for histogram

Midpoint x of range	1.2	1.7	2.2	2.7
Number of funds within range	1	4	10	5
$P(X = x)$	0.05	0.20	0.50	0.25

The distribution is approximately normal.

47. Examples will vary. For example, the color (red or black) of a card drawn from a standard deck of cards

49. Answers will vary.

Problem Set 7.2 Mean, Median, and Mode
Pages 509–516

1. mean $= \dfrac{2 + 5 + 6 + 8 + (-1) + (-3)}{6} = \dfrac{17}{6} = 2.8\overline{3}$

3. mean $= \dfrac{\frac{1}{2} + \frac{3}{2} + (-4) + \frac{5}{4}}{4} = \dfrac{-\frac{3}{4}}{4} = \dfrac{-3}{16} = -0.1875$

5. mean $= \dfrac{2.5 + (-5.4) + 4.1 + (-0.1) + 1.1}{5} = \dfrac{2.2}{5} = 0.44$

7.

x	10	20	30	40
$P(X = x)$	$\dfrac{15}{50}$	$\dfrac{20}{50}$	$\dfrac{10}{50}$	$\dfrac{5}{50}$
$x \cdot P(X = x)$	$\dfrac{150}{50}$	$\dfrac{400}{50}$	$\dfrac{300}{50}$	$\dfrac{200}{50}$
$P(X \le x)$	$\dfrac{15}{50}$	$\dfrac{35}{50}$	$\dfrac{45}{50}$	$\dfrac{50}{50}$

expected value $= E(X) = \dfrac{150}{50} + \dfrac{400}{50} + \dfrac{300}{50} + \dfrac{200}{50}$

$= \dfrac{1,050}{50} = 21$

median $= 20$, since $P(X \le 20) = \dfrac{35}{50} \ge \dfrac{1}{2}$ and $P(X \ge 20) = \dfrac{35}{50} \ge \dfrac{1}{2}$

mode $= 20$, since
$P(X = 20) > P(X = 10) > P(X = 30) > P(X = 40)$

9.

x	−5	−1	0	2	5	10
$P(X = x)$	0.2	0.3	0.2	0.1	0.2	0.0
$x \cdot P(X = x)$	−1	−0.3	0	0.2	1	0.0
$P(X \le x)$	0.2	0.5	0.7	0.8	1.0	1.0

expected value $= E(X) = (-1) + (-0.3) + 0 + 0.2 + 1 + 0.0 = -0.1$

median $= -1$, since $P(X \le -1) = 0.5 \ge \dfrac{1}{2}$ and $P(X \ge -1) = 0.8 \ge \dfrac{1}{2}$

mode $= -1$, since
$P(X = -1) > P(X = -5) = P(X = 0) = P(X = 5)$
$> P(X = 2) > P(X = 10)$

11.

Income Midpoint (x)	$10,000	$25,000	$35,000	$45,000	$55,000
$P(X = x)$	0.2	0.3	0.2	0.2	0.1
$x \cdot P(X = x)$	2000	7500	7000	9000	5500
$P(X \le x)$	0.2	0.5	0.7	0.9	1.0

expected value $= E(X) = 2000 + 7500 + 7000 + 9000 + 5500 = \$31,000$

median $= \$25,000$, since $P(X \le \$25,000) = 0.5 \ge \dfrac{1}{2}$ and $P(X \ge \$25,000) = 0.8 \ge \dfrac{1}{2}$

mode $= \$25,000$, since $P(X = \$25,000) > P(X = \$10,000) = P(X = \$35,000) = P(X = \$45,000) > P(X = \$55,000)$

13.

Population Midpoint (x)	2499.5	7499.5	12,499.5	17,499.5	22,499.5
$P(X = x)$	$\dfrac{4}{20} = 0.20$	$\dfrac{7}{20} = 0.35$	$\dfrac{2}{20} = 0.10$	$\dfrac{5}{20} = 0.25$	$\dfrac{2}{20} = 0.10$
$x \cdot P(X = x)$	499.9	2624.825	1249.95	4374.875	2249.95
$P(X \le x)$	0.20	0.55	0.65	0.90	1.00

expected value $= E(X) = 499.9 + 2624.825 + 1249.95 + 4374.875 + 2249.95 = 10,999.5$
median $= 7499.5$
mode $= 7499.5$

15. $S = \{HH, HT, TH, TT\}$, $X =$ number of tails

x	0	1	2
$P(X = x)$	$\dfrac{1}{4}$	$\dfrac{2}{4}$	$\dfrac{1}{4}$
$x \cdot P(X = x)$	0	0.5	0.5
$P(X \leq x)$	0.25	0.75	1.00

expected value $= E(X) = 0 + 0.5 + 0.5 = 1$ or $E(X) = np = 2 \cdot \dfrac{1}{2} = 1$

median $= 1$
mode $= 1$

17. See page 502 for the sample space. $X =$ higher number of the two dice

x	1	2	3	4	5	6
$P(X = x)$	$\dfrac{1}{36}$	$\dfrac{3}{36}$	$\dfrac{5}{36}$	$\dfrac{7}{36}$	$\dfrac{9}{36}$	$\dfrac{11}{36}$
$x \cdot P(X = x)$	$\dfrac{1}{36}$	$\dfrac{6}{36}$	$\dfrac{15}{36}$	$\dfrac{28}{36}$	$\dfrac{45}{36}$	$\dfrac{66}{36}$
$P(X \leq x)$	$\dfrac{1}{36}$	$\dfrac{4}{36}$	$\dfrac{9}{36}$	$\dfrac{16}{36}$	$\dfrac{25}{36}$	$\dfrac{36}{36}$

expected value $= E(X) = \dfrac{1}{36} + \dfrac{6}{36} + \dfrac{15}{36} + \dfrac{28}{36} + \dfrac{45}{36} + \dfrac{66}{36} = \dfrac{161}{36} = 4.47\overline{2}$

median $= 5$
mode $= 6$

19.

	$x = 4$	$x = 3$	$x = 2$
$P(X = x)$	$\dfrac{C(4, 4) \cdot C(2, 0)}{C(6, 4)} = \dfrac{1}{15}$	$\dfrac{C(4, 3) \cdot C(2, 1)}{C(6, 4)} = \dfrac{8}{15}$	$\dfrac{C(4, 2) \cdot C(2, 2)}{C(6, 4)} = \dfrac{6}{15}$
$x \cdot P(X = x)$	$\dfrac{4}{15}$	$\dfrac{24}{15}$	$\dfrac{12}{15}$
$P(X \geq x)$	$\dfrac{1}{15}$	$\dfrac{9}{15}$	$\dfrac{15}{15}$

expected value $= \dfrac{4}{15} + \dfrac{24}{15} + \dfrac{12}{15} = \dfrac{40}{15} = 2.\overline{6}$

median $= 3$
mode $= 3$

21. See page 502 for the sample space, where the first number represents the outcome of the red die and the second number represents the outcome of the green die.

X = the number on the red die minus the number on the green die

x	-5	-4	-3	-2	-1	0	1	2	3	4	5
$P(X = x)$	$\dfrac{1}{36}$	$\dfrac{2}{36}$	$\dfrac{3}{36}$	$\dfrac{4}{36}$	$\dfrac{5}{36}$	$\dfrac{6}{36}$	$\dfrac{5}{36}$	$\dfrac{4}{36}$	$\dfrac{3}{36}$	$\dfrac{2}{36}$	$\dfrac{1}{36}$
$x \cdot P(X = x)$	$\dfrac{-5}{36}$	$\dfrac{-8}{36}$	$\dfrac{-9}{36}$	$\dfrac{-8}{36}$	$\dfrac{-5}{36}$	0	$\dfrac{5}{36}$	$\dfrac{8}{36}$	$\dfrac{9}{36}$	$\dfrac{8}{36}$	$\dfrac{5}{36}$
$P(X \le x)$	$\dfrac{1}{36}$	$\dfrac{3}{36}$	$\dfrac{6}{36}$	$\dfrac{10}{36}$	$\dfrac{15}{36}$	$\dfrac{21}{36}$	$\dfrac{26}{36}$	$\dfrac{30}{36}$	$\dfrac{33}{36}$	$\dfrac{35}{36}$	$\dfrac{36}{36}$

$$\text{expected value} = \left(-\frac{5}{36}\right) + \left(-\frac{8}{36}\right) + \left(-\frac{9}{36}\right) + \left(-\frac{8}{36}\right) + \left(-\frac{5}{36}\right) + 0 + \frac{5}{36} + \frac{8}{36} + \frac{9}{36} + \frac{8}{36} + \frac{5}{36} = 0$$

median = 0

mode = 0

23. X = number of bull's eyes hit

$$P(X = x) = C(20, x) \cdot \left(\frac{1}{10}\right)^x \cdot \left(\frac{9}{10}\right)^{20-x}, \text{ binomial}$$

$$\text{expected value} = E(X) = np = 20 \cdot \left(\frac{1}{10}\right) = 2$$

median $\approx E(X) = 2$

mode: $(n+1)p = 21 \cdot \left(\frac{1}{10}\right) = 2.1$; therefore, the mode = 2

25. X = number of queens drawn

$$P(X = x) = \frac{C(4, x) \cdot C(48, 5 - x)}{C(52, 5)}$$

x	0	1	2	3	4
$P(X = x)$	0.658842	0.299474	0.039930	0.001736	0.000018
$x \cdot P(X = x)$	0	0.299474	0.079860	0.005208	0.000074
$P(X \le x)$	0.658842	0.958316	0.998245	0.999982	1

$$\text{expected value} = E(X) = \sum_{i=1}^{5} x_i \cdot P(X = x_i) \approx 0.384615$$

median = 0

mode = 0

27. X = annual yield in %

$$\text{expected annual yield} = E(X) = \sum_{i=1}^{3} x_i \cdot P(X = x_i) = 10 \cdot (0.28) + 15 \cdot (0.54) + 20 \cdot (0.18) = 14.5\%$$

29.

Age Midpoint (x)	7.5	20	30	40	50	60	70	85
$P(X = x)$	$\dfrac{459}{2210}$	$\dfrac{265}{2210}$	$\dfrac{247}{2210}$	$\dfrac{319}{2210}$	$\dfrac{291}{2210}$	$\dfrac{291}{2210}$	$\dfrac{212}{2210}$	$\dfrac{126}{2210}$
$x \cdot P(X = x)$	$\dfrac{3442.5}{2210}$	$\dfrac{5300}{2210}$	$\dfrac{7410}{2210}$	$\dfrac{12,760}{2210}$	$\dfrac{14,550}{2210}$	$\dfrac{17,460}{2210}$	$\dfrac{14,840}{2210}$	$\dfrac{10,710}{2210}$
$P(X \leq x)$	$\dfrac{459}{2210}$	$\dfrac{724}{2210}$	$\dfrac{971}{2210}$	$\dfrac{1290}{2210}$	$\dfrac{1581}{2210}$	$\dfrac{1872}{2210}$	$\dfrac{2084}{2210}$	$\dfrac{2210}{2210}$

average age $= E(X) = \left(\dfrac{1}{2210} \right) \cdot (3442.5 + 5300 + 7410 + 12,760 + 14,550 + 17,460 + 14,840 + 10,710)$

$= \dfrac{86,472.5}{2210} \approx 39.1$

median age $= 40$

mode $= 7.5$

31.

Age (x)	0	1	2	3	4	5	6	7	8	9	10
$P(X = x)$	0.07	0.175	0.225	0.325	0.1	0.06	0.025	0.005	0.0025	0.0075	0.005
$x \cdot P(X = x)$	0	0.175	0.45	0.975	0.4	0.3	0.15	0.035	0.02	0.0675	0.05
$P(X \leq x)$	0.07	0.245	0.47	0.795	0.895	0.955	0.98	0.985	0.9875	0.995	1

average age $= E(X) = 0 + 0.175 + 0.45 + 0.975 + 0.4 + 0.3 + 0.15 + 0.035 + 0.02 + 0.0675 + 0.05 = 2.6225$

median age $= 3$

mode $= 3$

Carmen's jalopy is $(6 - 2.6225) \approx 3.4$ years older than the average campus car, 3 years older than the median, and 3 years older than the mode.

33.

Firm Size Midpoint (x)	1	5.5	17	62	299.5	749.5	3,000
$P(X = x)$	0.10	0.12	0.08	0.12	0.14	0.06	0.38
$x \cdot P(X = x)$	0.10	0.66	1.36	7.44	41.93	44.97	1140
$P(X \leq x)$	0.10	0.22	0.30	0.42	0.56	0.62	1.00

average firm size $= E(X) = 0.10 + 0.66 + 1.36 + 7.44 + 41.93 + 44.97 + 1,140 = 1,236.46$

median firm size $= 299.5$

most common firm size $= 3,000$

35. a.

Income Midpoint (x)	\$150,000	\$250,000	\$350,000	\$450,000
$P(X = x)$	$\dfrac{2}{29}$	$\dfrac{13}{29}$	$\dfrac{12}{29}$	$\dfrac{2}{29}$
$x \cdot P(X = x)$	$\dfrac{300,000}{29}$	$\dfrac{3,250,000}{29}$	$\dfrac{4,200,000}{29}$	$\dfrac{900,000}{29}$

average salary $= E(X) = \left(\dfrac{1}{29} \right) \cdot (300,000 + 3,250,000 + 4,200,000 + 900,000) = \dfrac{8,650,000}{29} \approx \$298,276$

b. average salary $= E(X) = \left(\dfrac{1}{29} \right) \cdot (300,000 + 3,250,000 + 4,200,000 + \mathbf{800,000}) = \dfrac{8,550,000}{29} \approx \$294,828$

The average salary is lower when the top two salaries are known to be \$400,000.

37. Let X be the profit (or loss). If the driver does not wreck, Acme collects the annual insurance premium as profit ($X = \$5,000$). If the driver wrecks, Acme must pay and incur a loss ($X = \$5,000 - \$100,000 = -\$95,000$). Using a probability tree, we get the following.

$$P(\text{no wreck}) = 0.9^4 = 0.6561$$

$$P(\text{wreck}) = 0.1 + (0.9)(0.1) + (0.9)^2(0.1) + (0.9)^3(0.1) = 0.3439$$

expected profit (or loss) $= E(X) = 5,000 \cdot (0.6561) + (-95,000) \cdot (0.3439) = -\$29,390$

Acme can expect to lose \$29,390 per insured driver each year.

39. Let X = amount won on a \$1 bet. $X = 1$ when red, and $X = -1$ when not red.

$$P(\text{red}) = \frac{18}{38} \quad \text{and} \quad P(\text{not red}) = \frac{20}{38}$$

$$E(X) = 1 \cdot \left(\frac{18}{38}\right) + (-1) \cdot \left(\frac{20}{38}\right) = \frac{-2}{38} \approx -0.0526$$

You can expect to lose 5.3¢ for each \$1 bet on red.

41.

Passes Caught Midpoint (x)	54.5	64.5	74.5	84.5	94.5	104.5
$P(X = x)$	$\dfrac{2}{31}$	$\dfrac{4}{31}$	$\dfrac{13}{31}$	$\dfrac{7}{31}$	$\dfrac{3}{31}$	$\dfrac{2}{31}$
$x \cdot P(X = x)$	$\dfrac{109}{31}$	$\dfrac{258}{31}$	$\dfrac{968.5}{31}$	$\dfrac{591.5}{31}$	$\dfrac{283.5}{31}$	$\dfrac{209}{31}$

$$E(X) = \left(\frac{1}{31}\right) \cdot (109 + 258 + 968.5 + 591.5 + 283.5 + 209) = \frac{2,419.5}{31} \approx 78$$

Expect an NFL conference leader to catch 78 passes in a season.

43. a. Using a probability tree, we get the following.

Let X = number of trips until a fatal accident.

$P(X = 1) = 0.00000165$

$P(X \le 2) = P(X = 1) + P(X = 2) = 0.00000165 + (0.99999835) \cdot (0.00000165) \approx 0.00000330$

$P(X \le 3) = P(X = 1) + P(X = 2) + P(X = 3)$

$\quad = 0.00000165 + (0.99999835) \cdot (0.00000165) + (0.99999835)^2 \cdot (0.00000165)$

$\quad \approx 0.00000495$

$P(X \le n) = 0.00000165 + (0.99999835) \cdot (0.00000165) + (0.99999835)^2 \cdot (0.00000165) + \ldots$

$$+ (0.99999835)^{n-1}(0.00000165)$$

$$= \frac{0.00000165[1 - (0.99999835)^n]}{1 - 0.99999835} = 1 - (0.99999835)^n$$

b. $E(X) = \displaystyle\sum_{i=1}^{\infty} x_i P(X = x_i) = 1 \cdot P(X = 1) + 2 \cdot P(X = 2) + \cdots + n \cdot P(X = n) + \cdots$

$= 1 \cdot (0.00000165) + 2 \cdot (0.99999835) \cdot (0.00000165) + \cdots + n \cdot (0.99999835)^{n-1}(0.00000165) + \cdots$

$$= \frac{0.00000165}{(1 - 0.99999835)^2} \approx 606,061 \text{ trips}$$

45. Examples will vary.

47. Examples will vary.

Problem Set 7.3 Variance and Standard Deviation
Pages 528–534

1.

x_i	$(x_i - \overline{X})^2$
0	$(-5)^2 = 25$
2	$(-3)^2 = 9$
4	$(-1)^2 = 1$
6	$(1)^2 = 1$
8	$(3)^2 = 9$
10	$(5)^2 = 25$
30	70

$$\overline{X} = \frac{30}{6} = 5$$

$$\sigma^2 = \frac{70}{6} \approx 11.7$$

$$\sigma = \sqrt{\frac{70}{6}} \approx 3.42$$

3.

x_i	$(x_i - \overline{X})^2$
3.0	$(-0.6)^2 = 0.36$
3.0	$(-0.6)^2 = 0.36$
4.0	$(0.4)^2 = 0.16$
3.5	$(-0.1)^2 = 0.01$
4.5	$(0.9)^2 = 0.81$
18.0	1.70

$$\overline{X} = \frac{18.0}{5} = 3.6$$

$$\sigma^2 = \frac{1.70}{5} = 0.34$$

$$\sigma = \sqrt{0.34} \approx 0.583$$

5.

x_i	$(x_i - \overline{X})^2$
$\frac{1}{2}$	$\left(\frac{-1}{8}\right)^2 = \frac{1}{64}$
$\frac{3}{2}$	$\left(\frac{7}{8}\right)^2 = \frac{49}{64}$
$-\frac{5}{4}$	$\left(-\frac{15}{8}\right)^2 = \frac{225}{64}$
$\frac{1}{2}$	$\left(-\frac{1}{8}\right)^2 = \frac{1}{64}$
$\frac{5}{2}$	$\left(\frac{15}{8}\right)^2 = \frac{225}{64}$
0	$\left(-\frac{5}{8}\right)^2 = \frac{25}{64}$
$\frac{15}{4}$	$\frac{526}{64}$

$$\overline{X} = \frac{\frac{15}{4}}{6} = \frac{5}{8} = 0.625$$

$$\sigma^2 = \frac{\frac{526}{64}}{6} = \frac{263}{192} \approx 1.37$$

$$\sigma = \sqrt{\frac{263}{192}} \approx 1.170$$

7.

x_i	$(x_i - \overline{X})^2$
1	$0^2 = 0$
1	$0^2 = 0$
1	$0^2 = 0$
1	$0^2 = 0$
1	$0^2 = 0$
1	$0^2 = 0$
1	$0^2 = 0$
7	0

$$\overline{X} = \frac{7}{7} = 1$$

$$\sigma^2 = \frac{0}{7} = 0$$

$$\sigma = \sqrt{0} = 0$$

9. $E(X) = 21$ (from 7.2 Exercise 7)

x	10	20	30	40
$P(X = x)$	$\dfrac{15}{50}$	$\dfrac{20}{50}$	$\dfrac{10}{50}$	$\dfrac{5}{50}$
x^2	100	400	900	1600
$x^2 P(X = x)$	$\dfrac{1,500}{50}$	$\dfrac{8,000}{50}$	$\dfrac{9,000}{50}$	$\dfrac{8,000}{50}$

$E(X^2) = \dfrac{26,500}{50} = 530$

$\text{var}(X) = E(X^2) - [E(X)]^2 = 530 - (21)^2 = 89$

$\sigma(X) = \sqrt{89} \approx 9.43$

11. $E(X) = -0.1$ (from 7.2 Exercise 9)

x	-5	-1	0	2	5
$P(X = x)$	0.2	0.3	0.2	0.1	0.2
x^2	25	1	0	4	25
$x^2 P(X = x)$	5	0.3	0	0.4	5

$E(X^2) = 10.7$

$\text{var}(X) = 10.7 - (-0.1)^2 = 10.69$

$\sigma(X) = \sqrt{10.69} \approx 3.270$

13. $E(X) = \$31,000$ (from 7.2 Exercise 11)

Income Midpoint (x)	$10,000	$25,000	$35,000	$45,000	$55,000
$P(X = x)$	0.2	0.3	0.2	0.2	0.1
x^2	100,000,000	625,000,000	1,225,000,000	2,025,000,000	3,025,000,000
$x^2 P(X = x)$	20,000,000	187,500,000	245,000,000	405,000,000	302,500,000

$E(X^2) = 1,160,000,000$

$\text{var}(X) = 1,160,000,000 - (31,000)^2 = 199,000,000$

$\sigma(X) = \sqrt{199,000,000} \approx \$14,107$

15. $E(X) = 10,995.5$ (from 7.2 Exercise 13)

Population Midpoint (x)	2,499.5	7,499.5	12,499.5	17,499.5	22,499.5
$P(X = x)$	0.20	0.35	0.10	0.25	0.10
x^2	6,247,500.25	56,242,500.25	156,237,500.3	306,232,500.3	506,227,500.3
$x^2 P(X = x)$	1,249,500.05	19,684,875.09	15,623,750.03	76,558,125.06	50,622,750.03

$E(X^2) = 163,739,000.3$

$\text{var}(X) = 163,739,000.3 - (10,999.5)^2 = 42,750,000$

$\sigma(X) = \sqrt{42,750,000} \approx 6538$

17. Refer to 7.2 Exercise 15.

$$E(X) = np = 2 \cdot \frac{1}{2} = 1$$

$$\text{var}(X) = npq = 2 \cdot \frac{1}{2} \cdot \frac{1}{2} = \frac{1}{2} = 0.5$$

$$\sigma(X) = \sqrt{0.5} \approx 0.707$$

19. Refer to 7.2 Exercise 17.

x	1	2	3	4	5	6
$P(X = x)$	$\dfrac{1}{36}$	$\dfrac{3}{36}$	$\dfrac{5}{36}$	$\dfrac{7}{36}$	$\dfrac{9}{36}$	$\dfrac{11}{36}$
x^2	1	4	9	16	25	36
$x^2 P(X = x)$	$\dfrac{1}{36}$	$\dfrac{12}{36}$	$\dfrac{45}{36}$	$\dfrac{112}{36}$	$\dfrac{225}{36}$	$\dfrac{396}{36}$

$$E(X) = \frac{161}{36} = 4.47\overline{2}$$

$$E(X^2) = \frac{791}{36}$$

$$\text{var}(X) = \frac{791}{36} - \left(\frac{161}{36}\right)^2 = \frac{2555}{1296} \approx 1.97$$

$$\sigma(X) = \sqrt{\frac{2555}{1296}} \approx 1.40$$

21. Refer to 7.2 Exercise 19.

x	2	3	4
$P(X = x)$	$\dfrac{6}{15}$	$\dfrac{8}{15}$	$\dfrac{1}{15}$
x^2	4	9	16
$x^2 P(X = x)$	$\dfrac{24}{15}$	$\dfrac{72}{15}$	$\dfrac{16}{15}$

$$E(X) = \frac{40}{15} = 2.\overline{6}$$

$$E(X^2) = \frac{112}{15}$$

$$\text{var}(X) = \frac{112}{15} - \left(\frac{40}{15}\right)^2 = \frac{16}{45} = 0.3\overline{5}$$

$$\sigma(X) = \sqrt{\frac{16}{45}} \approx 0.596$$

23. Refer to 7.2 Exercise 21.

x	-5	-4	-3	-2	-1	0	1	2	3	4	5
$P(X=x)$	$\dfrac{1}{36}$	$\dfrac{2}{36}$	$\dfrac{3}{36}$	$\dfrac{4}{36}$	$\dfrac{5}{36}$	$\dfrac{6}{36}$	$\dfrac{5}{36}$	$\dfrac{4}{36}$	$\dfrac{3}{36}$	$\dfrac{2}{36}$	$\dfrac{1}{36}$
x^2	25	16	9	4	1	0	1	4	9	16	25
$x^2 P(X=x)$	$\dfrac{25}{36}$	$\dfrac{32}{36}$	$\dfrac{27}{36}$	$\dfrac{16}{36}$	$\dfrac{5}{36}$	0	$\dfrac{5}{36}$	$\dfrac{16}{36}$	$\dfrac{27}{36}$	$\dfrac{32}{36}$	$\dfrac{25}{36}$

$E(X) = 0$

$E(X^2) = \dfrac{210}{36} = \dfrac{35}{6}$

$\text{var}(X) = \dfrac{35}{6} - (0)^2 = \dfrac{35}{6} = 5.8\overline{3}$

$\sigma(X) = \sqrt{\dfrac{35}{6}} \approx 2.42$

25. Refer to 7.2 Exercise 23.

$E(X) = np = 20 \cdot \left(\dfrac{1}{10}\right) = 2$

$\text{var}(X) = npq = 20 \cdot \left(\dfrac{1}{10}\right) \cdot \left(\dfrac{9}{10}\right) = \dfrac{9}{5} = 1.8$

$\sigma(X) = \sqrt{\dfrac{9}{5}} \approx 1.34$

27. Refer to 7.2 Exercise 25.

x	0	1	2	3	4
$P(X=x)$	0.658842	0.299474	0.039930	0.001736	0.000018
x^2	0	1	4	9	16
$x^2 P(X=x)$	0	0.299474	0.159719	0.015625	0.000296

$E(X) \approx 0.384615$

$E(X^2) \approx 0.475113$

$\text{var}(X) \approx 0.475113 - (0.384615)^2 \approx 0.327$

$\sigma(X) = \sqrt{\text{var}(X)} \approx 0.572$

29. $P(E(X) - k\sigma(X) \le X \le E(X) + k\sigma(X)) \ge 1 - \dfrac{1}{k^2}$

$P(10.0 - k(0.3) \le X \le 10.0 + k(0.3)) \ge 1 - \dfrac{1}{k^2}$

$10.0 - k(0.3) = 9.4$ and $10.0 + k(0.3) = 10.6$

$k = 2$

$1 - \dfrac{1}{k^2} = 1 - \dfrac{1}{2^2} = \dfrac{3}{4}$

$P(9.4 \le X \le 10.6) \ge \dfrac{3}{4}$

31. $P(0 - k(3.1) \le X \le 0 + k(3.1)) \ge 1 - \dfrac{1}{k^2}$

$-k(3.1) = -12.4$ and $k(3.1) = 12.4$

$k = 4$

$1 - \dfrac{1}{k^2} = 1 - \dfrac{1}{4^2} = \dfrac{15}{16}$

$P(-12.4 \le X \le 12.4) \ge \dfrac{15}{16}$

For Exercises 33 - 35:

$1 - \dfrac{1}{k^2} = 0.9 \Rightarrow 0.1 = \dfrac{1}{k^2} \Rightarrow k^2 = 10 \Rightarrow k = \sqrt{10}$

33. $P(E(X) - k\sigma(X) \le X \le E(X) + k\sigma(X)) \ge 1 - \dfrac{1}{k^2}$

$P\left(10.0 - \sqrt{10}(0.3) \le X \le 10.0 + \sqrt{10}(0.3)\right) \ge 0.90$

$9.05 \le X \le 10.95$

35. $P\left(0 - \sqrt{10}(3.1) \le X \le 0 + \sqrt{10}(3.1)\right) \ge 0.90$

$-9.80 \le X \le 9.80$

37. Find $P(2.00 \le X \le 4.00)$

$3.00 - k(0.25) = 2.00$ and $3.00 + k(0.25) = 4.00$

$\Rightarrow k = 4$

$P(3.00 - 4(0.25) \le X \le 3.00 + 4(0.25)) \ge 1 - \dfrac{1}{4^2}$

$P(2.00 \le X \le 4.00) \ge \dfrac{15}{16} = 0.9375$

At least 93.75% of Julian's credits were 2.00 or better.

39. Given $P(\$80,000 \le X \le \$280,000) \ge 0.75$, find X_1 and X_2 such that $P(X_1 \le X \le X_2) \ge 0.90$.

$0.75 = 1 - \dfrac{1}{k^2} \Rightarrow k = \sqrt{4} = 2$

$\left.\begin{array}{l} E(X) - 2\sigma(X) = 80,000 \\ E(X) + 2\sigma(X) = 280,000 \end{array}\right\}$ solve to get $\left\{\begin{array}{l} E(X) = 180,000 \\ \sigma(X) = 50,000 \end{array}\right.$

$0.90 = 1 - \dfrac{1}{k^2} \Rightarrow k = \sqrt{10}$

$P\left(180,000 - \sqrt{10} \cdot 50,000 \le X \le 180,000 + \sqrt{10} \cdot 50,000\right) \ge 0.90$

$\$21,886 \le X \le \$338,114$

41.

x	x^2
2.59	6.7081
2.53	6.4009
2.53	6.4009
2.58	6.6564
2.78	7.7284
2.48	6.1504
2.53	6.4009
2.38	5.6644
2.48	6.1504
2.94	8.6436
2.48	6.1504
2.48	6.1504
2.28	5.1984
3.00	9.0000
2.53	6.4009
2.43	5.9049
2.79	7.7841
2.58	6.6564
46.39	120.499

$$E(X) = \frac{46.39}{18} \approx 2.58\%$$

$$\text{var}(X) = \frac{1}{n}\sum_{i=1}^{n} x_i^2 - \overline{X}^2 = \frac{1}{18}\cdot(120.1499) - \left(\frac{46.39}{18}\right)^2$$

$$\approx 0.0329$$

$$\sigma(X) = \sqrt{\text{var}(X)} \approx 0.18\%$$

43.

x	x^2
4.16	17.3056
4.93	24.3049
4.42	19.5364
4.37	19.0969
4.97	24.7009
4.92	24.2064
5.22	27.2484
4.52	20.4304
4.52	20.4304
5.56	30.9136
5.27	27.7729
4.27	18.2329
5.22	27.2484
4.50	20.2500
3.97	15.7609
4.55	20.7025
4.46	19.8916
5.92	35.0464
85.75	413.0795

$$E(X) = \frac{85.75}{18} \approx 4.76\%$$

$$\text{var}(X) = \frac{1}{18}\cdot(413.0795) - \left(\frac{85.75}{18}\right)^2 \approx 0.2542$$

$$\sigma(X) = \sqrt{\text{var}(X)} \approx 0.50\%$$

45. Refer to Exercise 41.
$$P(E(X) - 2\sigma(X) \le X \le E(X) + 2\sigma(X)) \ge 0.75$$
$$2.58 - 2(0.18) \le X \le 2.58 + 2(0.18)$$
$$2.22\% \le X \le 2.94\%$$

47. Refer to Exercise 41.
$$1 - \frac{1}{k^2} = 0.90 \Rightarrow k = \sqrt{10}$$
$$P\left(2.58 - \sqrt{10}(0.18) \le X \le 2.58 + \sqrt{10}(0.18)\right) \ge 0.90$$
$$2.01\% \le X \le 3.15\%$$

49.

x	$P(X = x)$	$x \cdot P(X = x)$	x^2	$x^2 \cdot P(X = x)$
1981	0.031414	62.230366	3924361	123278.36
1982	0.031414	62.261780	3928324	123402.85
1983	0.034031	67.484293	3932289	133821.35
1984	0.073298	145.424084	3936256	288521.38
1985	0.073298	145.497382	3940225	288812.30
1986	0.109948	218.356021	3944196	433655.06
1987	0.094241	187.256545	3948169	372078.75
1988	0.136126	270.617801	3952144	537988.19
1989	0.133508	265.547120	3956121	528173.22
1990	0.086387	171.910995	3960100	342102.88
1991	0.062827	125.089005	3964081	249052.21
1992	0.057592	114.722513	3968064	228527.25
1993	0.075916	151.301047	3972049	301542.99
Totals	**1**	**1987.698953**	**51326379**	**3950956.79**

a. $E(X) = \sum_{i=1}^{13} x_i \cdot P(X = x_i) \approx 1987.7$

$\text{var}(X) = E(X^2) - [E(X)]^2 = 3,950,956.79 - (1987.698953)^2 \approx 9.6622$

$\sigma(X) = \sqrt{\text{var}(X)} \approx 3.1$ years

b. $P(E(X) - 2\sigma(X) \le X \le E(X) + 2\sigma(X)) \ge 0.75$
$P(1987.7 - 2(3.1) \le X \le 1987.7 + 2(3.1)) \ge 0.75$
$1981.5 \le X \le 1993.9$
1981 to 1994

51. Refer to 7.2 Exercise 33.

$E(X^2) = \sum_{i=1}^{7} x_i^2 \cdot P(X = x_i) = 0.10 + 3.63 + 23.12 + 461.28 + 12{,}558.035 + 33{,}705.015 + 3{,}420{,}000 = 3{,}466{,}751.18$

$\text{var}(X) = 3,466,751.18 - (1236.46)^2 \approx 1,937,917.85$
$\sigma(X) \approx 1392.09$

$1 - \dfrac{1}{k^2} = 0.80 \Rightarrow k = \sqrt{5}$

$P(E(X) - k\sigma(X) \le X \le E(X) + k\sigma(X)) \ge 1 - \dfrac{1}{k^2}$

$P\left(1236.46 - \sqrt{5}(1392.09) \le X \le 1236.46 + \sqrt{5}(1392.09)\right) \ge 0.80$

$P(-1876 \le X \le 4349) \ge 0.80$
The range is 0 to 4350.

53. Refer to 7.2 Exercise 35.

$$E(X^2) = \frac{2,732,500,000,000}{29} = 94,224,137,930$$

$$\text{var}(X) \approx 94,224,137,930 - (298,276)^2 \approx 5,255,648,038$$

$$\sigma(X) \approx 72,495.85$$

$$1 - \frac{1}{k^2} = 0.50 \Rightarrow k = \sqrt{2}$$

$$P\left(298,276 - \sqrt{2}\,(72,495.85) \le X \le 298,276 + \sqrt{2}\,(72,495.85)\right) \ge 0.50$$

$$P(195,751 \le X \le 400,801) \ge 0.50$$

The range is \$196,000 to \$401,000.

55. Let X = number of women, out of 100, who will get breast cancer

Find $P(21 \le X \le 100) = 1 - P(0 \le X \le 20)$.

$$E(X) = np = 100 \cdot \left(\frac{1}{8}\right) = 12.5$$

$$\text{var}(X) = npq = 100 \cdot \left(\frac{1}{8}\right) \cdot \left(\frac{7}{8}\right) = 10.9375$$

$$\sigma(X) \approx 3.307$$

$$E(X) - k\sigma(X) = 12.5 - k(3.307) = 0$$

$$\Rightarrow k \approx 3.78 \Rightarrow 1 - \frac{1}{k^2} = 0.93$$

$$E(X) + k\sigma(X) = 12.5 + k(3.307) = 20 \Rightarrow k \approx 2.27$$

$$\Rightarrow 1 - \frac{1}{k^2} \approx 0.806$$

$$P(0 \le X \le 20) \ge 0.806$$

$$1 - P(0 \le X \le 20) \le 0.194$$

$$P(21 \le X \le 100) \le 0.194$$

57. Answers will vary.

59. A distribution of values of 10 exclusively.

Section 7.4 Normal Distributions
Pages 545–547

1. $P(0 \le X \le 0.5) = P(0 \le Z \le 0.5) = 0.1915$

3. $P(-0.71 \le X \le 0.71) = P(-0.71 \le Z \le 0) + P(0 \le Z \le 0.71) = 0.2611 + 0.2611 = 0.5222$

5. $P(-0.71 \le X \le 1.34) = P(-0.71 \le Z \le 0) + P(0 \le Z \le 1.34) = 0.2611 + 0.4099 = 0.6710$

7. $P(0.5 \le X \le 1.5) = P(0 \le Z \le 1.5) - P(0 \le Z \le 0.5)$
    $= 0.4332 - 0.1915 = 0.2417$

9. $P(35 \le X \le 65) = P\left(\dfrac{35 - 50}{10} \le Z \le \dfrac{65 - 50}{10}\right)$
    $= P(-1.5 \le Z \le 1.5) = P(-1.5 \le Z \le 0) + P(0 \le Z \le 1.5) = 0.4332 + 0.4332 = 0.8664$

11. $P(30 \le X \le 62) = P\left(\dfrac{30 - 50}{10} \le Z \le \dfrac{62 - 50}{10}\right)$
    $= P(-2 \le Z \le 1.2) = 0.4772 + 0.3849 = 0.8621$

13. $P(110 \le X \le 130) = P\left(\dfrac{110 - 100}{15} \le Z \le \dfrac{130 - 100}{15}\right)$
    $= P(0.67 \le Z \le 2) = 0.4772 - 0.2486 = 0.2286$

15. $P(\mu - 0.5\sigma \le X \le \mu + 0.5\sigma) = P\left(\dfrac{(\mu - 0.5\sigma) - \mu}{\sigma} \le Z \le \dfrac{(\mu + 0.5\sigma) - \mu}{\sigma}\right)$

$= P(-0.5 \le Z \le 0.5) = 2(0.1915) = 0.3830$

17. $P\left(X \le \mu - \dfrac{2}{3}\sigma\right) + P\left(X \ge \mu + \dfrac{2}{3}\sigma\right) = 1 - P\left(\mu - \dfrac{2}{3}\sigma \le X \le \mu + \dfrac{2}{3}\sigma\right)$

$= 1 - P\left(\dfrac{\left(\mu - \frac{2}{3}\sigma\right) - \mu}{\sigma} \le Z \le \dfrac{\left(\mu + \frac{2}{3}\sigma\right) - \mu}{\sigma}\right) = 1 - P\left(-\dfrac{2}{3} \le Z \le \dfrac{2}{3}\right)$

$= 1 - P(-0.67 \le Z \le 0.67) = 1 - 2(0.2486) = 0.5028$

19. X = number of 1's in 100 rolls. Find $P(10 \le X \le 15)$.

$\mu = E(X) = np = 100 \cdot \left(\dfrac{1}{6}\right) = \dfrac{50}{3}$

$\sigma = \sqrt{\text{var}(X)} = \sqrt{npq} = \sqrt{100 \cdot \left(\dfrac{1}{6}\right)\left(\dfrac{5}{6}\right)} = \sqrt{\dfrac{125}{9}}$

$P(10 \le X \le 15) \approx P(9.5 \le Y \le 15.5)$

$= P\left(\dfrac{9.5 - \frac{50}{3}}{\sqrt{\frac{125}{9}}} \le Z \le \dfrac{15.5 - \frac{50}{3}}{\sqrt{\frac{125}{9}}}\right)$

$= P(-1.92 \le Z \le -0.31) = 0.4726 - 0.1217 = 0.3509$

21. X = number of 1's in 200 rolls. Find $P(X < 25)$.

$\mu = 200 \cdot \left(\dfrac{1}{6}\right) = \dfrac{100}{3}$

$\sigma = \sqrt{200 \cdot \left(\dfrac{1}{6}\right)\left(\dfrac{5}{6}\right)} = \sqrt{\dfrac{250}{9}}$

$P(X < 25) \approx P(Y \le 24.5) = P\left(Z \le \dfrac{24.5 - \frac{100}{3}}{\sqrt{\frac{250}{9}}}\right)$

$= P(Z \le -1.68) = 0.5 - 0.4535 = 0.0465$

23. X = IQ score. Find $P(X > 120) \cdot 250,000,000$.

$P(X > 120) = P\left(Z \ge \dfrac{120 - 100}{16}\right) = P(Z \ge 1.25)$

$= 0.5 - 0.3944 = 0.1056$

$P(X > 120) \cdot 250,000,000 = 0.1056 \cdot 250,000,000$

$= 26,400,000$

25. X = time (in months) between failures. Find $P(X \le 1)$.

$P(X \le 1) = P\left(Z \le \dfrac{1 - 6}{1}\right) = P(Z \le -5)$

This is surprising, because the time between failures was more than 5 standard deviations away from the mean, which happens with an extremely small probability.

27. X = batting average. Find $250 \cdot P(X \ge 0.400)$.

$P(X \ge 0.400) = P\left(Z \ge \dfrac{0.400 - 0.250}{0.01}\right)$

$= P(Z \ge 15) \approx 0$

No batters are expected to average 0.400 or better.

29. X = spiciness rating. Find $100,000 \cdot P(X \ge 9)$.

$P(X \ge 9) = P\left(Z \ge \dfrac{9 - 7.5}{1}\right) = P(Z \ge 1.5) = 0.0668$

$100,000 \cdot P(X \ge 9) = 100,000 \cdot (0.0668) = 6680$

Expect to sell about 6680 jars with a spiciness of 9 or above.

31. X = number of crashes in 100,000,000 flights

Find $P(X < 180)$.

$E(X) = 100,000,000 \cdot (0.00000165) = 165$

$\sigma(X) = \sqrt{100,000,000 \cdot (0.00000165) \cdot (0.99999835)}$

≈ 12.845

$P(X < 180) \approx P(Y \le 179.5)$

$= P\left(Z \le \dfrac{179.5 - 165}{12.845}\right) = P(Z \le 1.13) = 0.5 + 0.3708$

$= 0.8708$

33. X = number of crashes in 100,000,000 flights

$E(X) = 100,000,000 \cdot (0.00000165) = 165$

$\sigma(X) = \sqrt{100,000,000 \cdot (0.00000165) \cdot (0.99999835)}$

≈ 12.845

The company collects \$20 per flight and pays out \$10,000,000 for each crash. Solve to find the break-even number of crashes:

$20 \cdot (100,000,000) - 10,000,000X = 0 \Rightarrow X = 200$

The company loses money when $X > 200$.

$P(X > 200) \approx P(Y \ge 200.5) = P\left(Z \ge \dfrac{200.5 - 165}{12.845}\right)$

$= P(Z \ge 2.76) = 0.5 - 0.4971 = 0.0029$

35. X = number of women, out of 100, who develop breast cancer

$$E(X) = 100 \cdot \left(\frac{1}{8}\right) = 12.5$$

$$\sigma(X) = \sqrt{100 \cdot \left(\frac{1}{8}\right) \cdot \left(\frac{7}{8}\right)} \approx 3.307$$

$$P(X \geq 16) \approx P(Y \geq 15.5) = P\left(Z \geq \frac{15.5 - 12.5}{3.307}\right) = P(Z \geq 0.91) = 0.5 - 0.3186 = 0.1814$$

You should expect 16 or more women to develop breast cancer in about 18% of all such neighborhoods, so this is not an unusual event.

37. X = number of times, out of 10,000 bets, the casino loses

$$E(X) = 10,000 \cdot \left(\frac{1}{38}\right) = \frac{5000}{19}$$

$$\sigma(X) = \sqrt{10,000 \cdot \left(\frac{1}{38}\right) \cdot \left(\frac{37}{38}\right)} \approx 16.007$$

Solve for the break-even point:

$$1 \cdot (10,000) - 36X = 0 \Rightarrow X = \frac{2500}{9}$$

$$P\left(X > \frac{2500}{9}\right) = P\left(Z \geq \frac{\frac{2500}{9} - \frac{5000}{19}}{16.007}\right) = P(Z \geq 0.91) = 0.5 - 0.3186 = 0.1814$$

39. X = number, out of 1000, that will choose Goode
Probability a person will choose Goode = $(0.9) \cdot (0.55) + (0.1) \cdot (0.45) = 0.54$
$E(X) = 1000(0.54) = 540$

$$\sigma(X) = \sqrt{1000(0.54)(0.46)} \approx 15.761$$

For 52%, $X = (0.52)1000 = 520$

$$P(X > 520) \approx P(Y \geq 520.5) = P\left(Z \geq \frac{520.5 - 540}{15.761}\right) = P(Z \geq -1.24) = 0.5 + 0.3925 = 0.8925$$

41. $\left.\begin{array}{l} P(X \geq 148) = 0.02 \\ P(Z \geq 2.055) \approx 0.02 \end{array}\right\} \dfrac{148 - 100}{\sigma} = 2.055 \Rightarrow \sigma \approx 23.4$

43. area $= 1 = (b - a)h \Rightarrow h = \dfrac{1}{b - a}$

You're the Expert
Page 549

Answers will vary.

Chapter 7 Review Exercises
Pages 550–551

1. $S = \{BB, BG, GB, GG\}$

x	0	1	2
$P(X = x)$	$\dfrac{1}{4}$	$\dfrac{2}{4}$	$\dfrac{1}{4}$

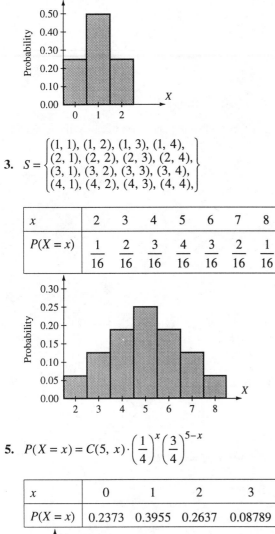

3. $S = \begin{Bmatrix} (1, 1), (1, 2), (1, 3), (1, 4), \\ (2, 1), (2, 2), (2, 3), (2, 4), \\ (3, 1), (3, 2), (3, 3), (3, 4), \\ (4, 1), (4, 2), (4, 3), (4, 4), \end{Bmatrix}$

x	2	3	4	5	6	7	8
$P(X = x)$	$\dfrac{1}{16}$	$\dfrac{2}{16}$	$\dfrac{3}{16}$	$\dfrac{4}{16}$	$\dfrac{3}{16}$	$\dfrac{2}{16}$	$\dfrac{1}{16}$

5. $P(X = x) = C(5, x) \cdot \left(\dfrac{1}{4}\right)^x \left(\dfrac{3}{4}\right)^{5-x}$

x	0	1	2	3	4	5
$P(X = x)$	0.2373	0.3955	0.2637	0.08789	0.01465	0.00098

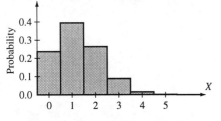

7. $P(X = x) = \dfrac{C(5, x) \cdot C(10, 5-x)}{C(15, 5)}$

x	$P(X = x)$
0	0.08392
1	0.34965
2	0.39960
3	0.14985
4	0.01665
5	0.00033

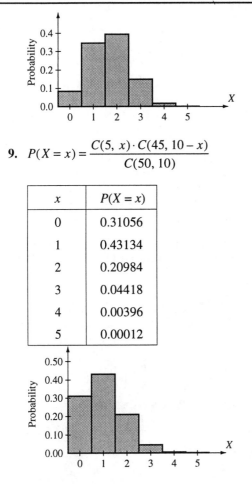

9. $P(X = x) = \dfrac{C(5,\ x) \cdot C(45,\ 10 - x)}{C(50,\ 10)}$

x	$P(X = x)$
0	0.31056
1	0.43134
2	0.20984
3	0.04418
4	0.00396
5	0.00012

11. Refer to Exercise 1.

$E(X) = 0 \cdot \dfrac{1}{4} + 1 \cdot \dfrac{2}{4} + 2 \cdot \dfrac{1}{4} = 1$

median = 1, since $P(X \le 1) = \dfrac{3}{4} \ge \dfrac{1}{2}$ and $P(X \ge 1) = \dfrac{3}{4} \ge \dfrac{1}{2}$

mode = 1

$\text{var}(X) = 0^2 \cdot \left(\dfrac{1}{4}\right) + 1^2 \cdot \left(\dfrac{2}{4}\right) + 2^2 \cdot \left(\dfrac{1}{4}\right) - 1^2 = \dfrac{2}{4} = 0.5$

$\sigma(X) \approx 0.7071$

13. Refer to Exercise 3.

$E(X) = 2 \cdot \left(\dfrac{1}{16}\right) + 3 \cdot \left(\dfrac{2}{16}\right) + \cdots + 8 \cdot \left(\dfrac{1}{16}\right) = \dfrac{80}{16} = 5$

median = 5, since $P(X \le 5) = \dfrac{10}{16} \ge \dfrac{1}{2}$ and $P(X \ge 5) = \dfrac{10}{16} \ge \dfrac{1}{2}$

mode = 5

$\text{var}(X) = 2^2 \cdot \left(\dfrac{1}{16}\right) + 3^2 \cdot \left(\dfrac{2}{16}\right) + \cdots + 8^2 \cdot \left(\dfrac{1}{16}\right) - 5^2 = \dfrac{440}{16} - 5^2 = 2.5$

$\sigma(X) \approx 1.581$

15. Refer to Exercise 5.
$E(X) = 0 \cdot (0.2373) + \cdots + 5(0.00098) = 1.25$
median = 1
mode = 1
$\text{var}(X) = 0^2 (0.2373) + \cdots + 5^2 (0.00098) - 1.25^2$
$= 2.5 - 1.5625 = 0.9375$
$\sigma(X) \approx 0.9682$

17. Refer to Exercise 7.
$E(X) = 0 \cdot (0.08392) + \cdots + 5 \cdot (0.00033) \approx 1.667$
median = 2
mode = 2
$\text{var}(X) = 0^2 (0.08392) + \cdots + 5^2 (0.00033) - 1.667^2$
$\approx 3.571 - 2.778 \approx 0.7937$
$\sigma(X) \approx 0.8909$

19. Refer to Exercise 9.
$E(X) = 0 \cdot (0.31056) + \cdots + 5(0.00012) = 1$
median = 1
mode = 1
$\text{var}(X) = 0^2 \cdot (0.31056) + \cdots + 5^2 (0.00012) - 1^2$
$\approx 1.735 - 1 \approx 0.7347$
$\sigma(X) \approx 0.8571$

For Exercises 21–25:
$1 - \dfrac{1}{k^2} = 0.90 \Rightarrow k = \sqrt{10}$

21. $P(E(X) - k\sigma(X) \le X \le E(X) + k\sigma(X)) \ge 1 - \dfrac{1}{k^2}$

$P\left(100 - \sqrt{10} \cdot 16 \le X \le 100 + \sqrt{10} \cdot 16\right) \ge 0.90$

$49.4 \le X \le 150.6$

23. $P\left(0 - \sqrt{10} \cdot 1 \le X \le 0 + \sqrt{10} \cdot 1\right) \ge 0.90$

$-3.16 \le X \le 3.16$

25. $P\left(-1 - \sqrt{10} \cdot (0.5) \le X \le -1 + \sqrt{10} \cdot (0.5)\right) \ge 0.90$

$-2.58 \le X \le 0.58$

27. $P(80 \le X \le 120) = P\left(\dfrac{80 - 100}{16} \le Z \le \dfrac{120 - 100}{16}\right)$

$= P(-1.25 \le Z \le 1.25) = 0.7888$

29. $P(-1 \le X \le 3) = P\left(\dfrac{-1 - 0}{2} \le Z \le \dfrac{3 - 0}{2}\right)$

$= P(-0.5 \le Z \le 1.5) = 0.6247$

31. $P(-1 \le X \le 0) = P\left(\dfrac{-1 - (-1)}{0.5} \le Z \le \dfrac{0 - (-1)}{0.5}\right)$

$= P(0 \le Z \le 2) = 0.4772$

For Exercises 33–37:
$P(\mu - k\sigma \le X \le \mu + k\sigma) = 0.90$

$P\left(\dfrac{(\mu - k\sigma) - \mu}{\sigma} \le Z \le \dfrac{(\mu + k\sigma) - \mu}{\sigma}\right) = 0.90$

$P(-k \le Z \le k) = 0.90$

$k = 1.645$

33. Refer to Exercise 27.
$P(\mu - k\sigma \le X \le \mu + k\sigma) = 0.90$
$P(100 - (1.645)(16) \le X \le 100 + (1.645)(16)) = 0.90$
$73.68 \le X \le 126.32$

35. Refer to Exercise 29.
$P(0 - (1.645)(2) \le X \le 0 + (1.645)(2)) = 0.90$
$-3.29 \le X \le 3.29$

37. Refer to Exercise 31.
$P(-1 - (1.645)(0.5) \le X \le -1 + (1.645)(0.5)) = 0.90$
$-1.8225 \le X \le -0.1775$

39. Let X be the number of heads in 100 tosses.
Find $P(45 \le X \le 50)$.

$E(X) = np = 100 \cdot \left(\dfrac{1}{2}\right) = 50$

$\sigma(X) = \sqrt{npq} = \sqrt{100 \cdot \left(\dfrac{1}{2}\right) \cdot \left(\dfrac{1}{2}\right)} = 5$

$P(45 \le X \le 50) \approx P(44.5 \le Y \le 50.5)$

$= P\left(\dfrac{44.5 - 50}{5} \le Z \le \dfrac{50.5 - 50}{5}\right) = P(-1.1 \le Z \le 0.1)$

$= 0.4041$

41. Grouped data

SAT scores	frequency
700 - 749	0
750 - 799	1
800 - 849	2
850 - 899	2
900 - 949	6
950 - 999	1
1000 - 1049	2
1050 - 1099	1
1100 - 1149	4
1150 - 1199	0
1200 - 1249	1

From frequency table:

$$E(X) = 725\left(\frac{0}{20}\right) + 775\left(\frac{1}{20}\right) + \cdots + 1225\left(\frac{1}{20}\right) = 977.5$$

median = 925, since $P(X \le 925) = \frac{11}{20} \ge \frac{1}{2}$ and $P(X \ge 925) = \frac{15}{20} \ge \frac{1}{2}$

mode = 925

$$\text{var}(X) = (725)^2\left(\frac{0}{20}\right) + (775)^2\left(\frac{1}{20}\right) + \cdots + (1225)^2\left(\frac{1}{20}\right) - (977.5)^2 = 969,875 - 955,506.25 = 14,368.75$$

$\sigma(X) \approx 119.87$

From original data:

$$E(X) = \frac{1}{20}(920 + 1000 + \cdots + 1140) = \frac{19,440}{20} = 972$$

median = 930

mode = 930

$$\text{var}(X) = \frac{1}{20}[(920 - 972)^2 + (1000 - 972)^2 + \cdots + (1140 - 972)^2] = \frac{1}{20}(292,920) = 14,646$$

$\sigma(X) \approx 121.02$

43. $E(X) = 69.5\left(\frac{2}{12}\right) + 89.5\left(\frac{1}{12}\right) + 109.5\left(\frac{3}{12}\right) + 129.5\left(\frac{2}{12}\right) + 149.5\left(\frac{3}{12}\right) + 169.5\left(\frac{1}{12}\right) = \frac{1434}{12} = 119.5$

$\text{var}(X) = (69.5)^2\left(\frac{2}{12}\right) + (89.5)^2\left(\frac{1}{12}\right) + \cdots + (169.5)^2\left(\frac{1}{12}\right) - (119.5)^2 = \frac{182,963}{12} - 14,280.25 \approx 966.67$

$\sigma(X) \approx 31.09$

45. $E(X) = 18\left(\dfrac{2678}{12,356}\right) + 22.5\left(\dfrac{4786}{12,356}\right) + 27.5\left(\dfrac{1928}{12,356}\right) + 32.5\left(\dfrac{1201}{12,356}\right) + 42.5\left(\dfrac{1763}{12,356}\right) \approx 26.13$

median = 22.5
mode = 22.5

$\mathrm{var}(X) = (18)^2\left(\dfrac{2678}{12,356}\right) + \cdots + (42.5)^2\left(\dfrac{1763}{12,356}\right) - (26.13)^2 \approx 744.71 - (26.13)^2 \approx 61.90$

$\sigma(X) \approx 7.87$

47. $P(Z \geq 3) = 0.5 - 0.4987 = 0.0013$
$E(X) = np = 250,000,000(0.0013) = 325,000$
Expect 325,000 people in the U.S. to qualify for the 3 Sigma Club.

49. X = number of women, out of 500, that develop breast cancer by age 50

$E(X) = 500 \cdot \left(\dfrac{1}{50}\right) = 10$

$\sigma(X) = \sqrt{500 \cdot \left(\dfrac{1}{50}\right) \cdot \left(\dfrac{49}{50}\right)} \approx 3.13$

$P(X \geq 20) \approx P(Y \geq 19.5) = P\left(Z \geq \dfrac{19.5 - 10}{3.13}\right) = P(Z \geq 3.03) = 0.0012$

It is very unlikely that a neighborhood of 500 women will have 20 or more women who develop breast cancer. There may be an unusual cause.

51. X = number of hits in 400 at bats
$E(X) = 400(0.300) = 120$
$\sigma(X) = \sqrt{400(0.3)(0.7)} \approx 9.165$

$P(X < (0.280)(400)) = P(X < 112) \approx P(Y \leq 111.5) = P\left(Z \leq \dfrac{111.5 - 120}{9.165}\right) = P(Z \leq -0.93) = 0.1762$

The probability the batter will average less than 0.280 is 0.1762.

Chapter 8

Section 8.1 Simple Interest
Pages 700–702

1. $P = 2000$, $r = 0.06$, $t = 1$
$I = 2000(0.06)(1) = \$120$

3. $P = 20{,}200$, $r = 0.05$, $t = 6$ (months) $= 0.5$ (years)
$I = 20{,}200(0.05)(0.5) = \$505$

5. $P = 10{,}000$, $r = 0.03$, $t = 10$ (months) $= \dfrac{5}{6}$ (years)

$I = 10{,}000(0.03)\left(\dfrac{5}{6}\right) = \$250$

7. $I = \$125 - \$100 = \$25$
$P = \$100$ (present value)
$r = ?$
$t = 5$ years
$I = Prt$
$25 = 100(r)(5)$
$r = \dfrac{25}{500} = 0.05 = 5\%$

9. $P = 1000$, $r = 0.08$, $I = 640$, $t = ?$
$I = Prt$
$640 = 1{,}000(0.08)t$
$t = \dfrac{640}{80} = 8$ years

11. $P = ?$, $r = 0.045$, $I = 1000 - P$, $t = 6$ years
$I = Prt$
$1000 - P = P(0.045)(6)$
$1000 - P = 0.27P$
$P = \dfrac{1{,}000}{1.27} \approx \$787.40$

13. $\dfrac{653 - 547}{547} \approx 0.1938$; 19.38% increase

15. $\dfrac{1}{3}\left(\dfrac{902 - 547}{547}\right) \approx 0.2163$; 21.63% increase

17. 1990–1991 :
$\dfrac{828 - 653}{653} \approx 0.2680$; 26.80% increase

19. No. The net income rose by 106 billion yen (653 – 547) from 1989 to 1990, but by 175 billion yen (828 – 653) from 1990 to 1991. Since the amount of growth changed, this was not simple-interest growth.

21. We use the formula:
$A = P(1 + rt)$
Jan. 1993: 2.5 billion dollars
Jan. 1994: 6 billion dollars
$P = 2.5$, $A = 6$, $r = ?$, $t = 12$ months
$6 = 2.5(1 + 12r)$
$r = \dfrac{1}{12}\left(\dfrac{6}{2.5} - 1\right) \approx 0.1167 \approx 12\%$

23. P (Jan. 1993) = 2.5 billion dollars, A (Jan. 1995) = ?,
$r \approx 0.12$, $t = 24$ months
$A = P(1 + rt)$
$= 2.5(1 + 0.12(24))$
$= 9.7$
Average value in January 1995: $9.5 billion

25.

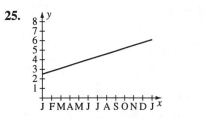

27. (A) represents a linear equation which is consistent with simple interest.
$A = P(1 + rt) = P + Prt$

29.

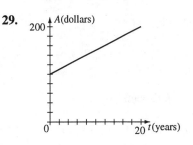

$A = 100 + 5t$

31. The interest is added to your account, which changes your present value.

Section 8.2 Compound Interest
Pages 713–718

1. $P = 10{,}000$
$r = 0.03$
$m = 1$
$t = 10$ (years)
$A = P\left(1 + \dfrac{r}{m}\right)^{mt}$
$A = 10{,}000(1 + 0.03)^{10}$
$= 10{,}000(1.03)^{10}$
$= \$13{,}439.16$

3. $P = 10,000$
$r = 0.025$;
compounded quarterly: $m = 4$
$t = 5$ (years)

$A = P\left(1 + \dfrac{r}{m}\right)^{mt}$

$A = 10,000\left(1 + \dfrac{0.025}{4}\right)^{4 \cdot 5}$

$= 10,000(1.00625)^{20}$

$= \$11,327.08$

5. $P = 10,000$
$r = 0.065$;
compounded daily: $m = 365$
$t = 10$ (years); $mt = 365 \cdot 10 = 3650$

$A = 10,000\left(1 + \dfrac{0.065}{365}\right)^{3,650} = \$19,154.30$

7. $P = 10,000$
$i = 0.002$;
compounded monthly: $m = 12$
$t = 10$ (years); $mt = 12 \cdot 10 = 120$
$A = 10,000(1 + 0.002)^{120} = \$12,709.44$

9. $A = 1000$, $P = ?$
$r = 0.05$; compounded annually: $m = 1$
$t = 10$ (years)
$1000 = P(1 + 0.05)^{10}$
$P = 1000(1.05)^{-10} = \$613.91$

11. $A = 1000$, $P = ?$
$r = 0.042$;
compounded weekly: $m = 52$
$t = 5$ (years); $mt = 52 \cdot 5 = 260$

$1000 = P\left(1 + \dfrac{0.042}{52}\right)^{260}$

$P = 1000\left(1 + \dfrac{0.042}{52}\right)^{-260} = \$810.65$

13. $A = 1000$, $P = ?$
$r = -0.05$; $m = 1$
$t = 4$ (years); $mt = 4$
$P = 1000(1 - 0.05)^{-4} = \$1227.74$

15. $r = 0.05$;
compounded quarterly: $m = 4$

$r_e = \left(1 + \dfrac{r}{m}\right)^{m} - 1$

$= \left(1 + \dfrac{0.05}{4}\right)^{4} - 1$

$= (1.0125)^{4} - 1$

$\approx 0.0509 = 5.09\%$

17. $r = 0.10$;
compounded monthly: $m = 12$

$r_e = \left(1 + \dfrac{0.10}{12}\right)^{12} - 1 \approx 0.1047 = 10.47\%$

19. $r = 0.10$;
compounded hourly: $m = 24 \cdot 365 = 8760$

$r_e = \left(1 + \dfrac{0.10}{8760}\right)^{8760} - 1 \approx 0.1052 = 10.52\%$

21. $P = 100$
$r = 0.06$;
compounded quarterly: $m = 4$
$t = 4$ (years); $mt = 16$

$A = 100\left(1 + \dfrac{0.06}{4}\right)^{16} \approx 126.90$

$A - P = 126.90 - 100 = 26.90$
Your investment will have grown $26.90 after four years.

23. $P = 3000$
$r = -0.068$; $m = 1$
$t = 5$ (years)
$A = 3000(1 - 0.068)^5 = 3000(0.932)^5 = \$2109.60$

25. $P_1 = 5000$ $P_2 = 5000$
$r_1 = 0.10$ $r_2 = 0.05$
$m = 1$ $m = 2$
$t = 10$ (years) $t = 10$ (years)
     $mt = 20$

$A_1 = 5000(1 + 0.10)^{10}$ $A_2 = 5000\left(1 + \dfrac{0.05}{2}\right)^{20}$

$= 5000(1.10)^{10}$ $= 5000(1.025)^{20}$

$= 12,968.71$ $= 8193.08$

$A_1 + A_2 = \$21,161.79$

27. $P = 200,000$
$r = -0.025$
$m = 2$
$t = 10$
$A = 200,000(1 - 0.025)^{20}$
$= 200,000(0.975)^{20} = \$120,537.54$

29. $A = 100,000$, $P = ?$
$r = 0.04$
$m = 1$, $t = 15$
$100,000 = P(1 + 0.04)^{15}$
$P = 100,000(1.04)^{-15}$
$P = 55,526.45$
I should request $55,526.45 per year

31. $A = 150$, $P = ?$
$r = -5\%$ every 4 months $= -0.05$, $m = 3$
$t = 6$; $mt = 18$
$P = 150(1 - 0.05)^{-18} = 150(0.95)^{-18} = \$377.63$

33. $A = 10,000$
$r_e = 0.05$ per year
$t = 5$
$P = 10,000(1 + 0.05)^{-5} = \$7835.26$

35. Value in account in two years:
$P = 1000$
$r = 0.05; t = 2$
$A = 1000(1.05)^2 = \$1102.50$
Present value based on inflation
$A = 1102.50$
$r_e = 0.03; \ t = 2$
$P = 1102.50(1.03)^{-2} \approx \$1039.21$

37. Investment 1:
$r = 0.12, m = 1$
$A = P(1.12) = 1.12P$

Investment 2:
$r = 0.119, m = 12$
$A = P\left(1 + \dfrac{0.119}{12}\right)^{12} \approx 1.13P$
Investment 2 (earning 11.9% compounded monthly)
is the better investment.

39. $P = 100$ cruzados
$r = 11.32$
$t = 5$
$A = 100(1 + 11.32)^5 = 28,382,689$ cruzados

41. $A = 1000$ pesos, $P = ?$
$r = 0.14$
$t = 10$
$P = 1000(1.14)^{-10} = 270$ pesos

43. $P = 1000$ pesos
$r = 0.28, m = 2; t = 10$
$A = 1000(1.14)^{20} \approx 13,743$
Inflation: $r = 0.26$
$P \approx 13,743\,(1.26)^{-10} \approx 1363$ pesos

45. Investment 1: Colombia
$P = 1$ peso
$r = 27.5\%, m = 1; t = 1$
$A = 1(1 + 0.275)$
$= 1.275$
Inflation: $r = 26\%$
$P = 1.275(1.26)^{-1}$
$P \approx 1.01190$ units of currency

Investment 2: Ecuador
$P = 1$ sucre
$r = 68.5\%, m = 2, t = 1$
$A = 1(1 + 0.3425)^2$
≈ 1.802
Inflation: $r = 66\%$
$P = 1.802(1.66)^{-1}$
$P \approx 1.086$ units of currency

The Ecuadorian investment is better.

47. $P = 500$
$r = 0.05, m = 1$
$0 \le t \le 40$
$A = 500(1.05)^t$

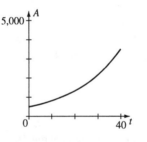

49. $P = 500$
$r = 0.05$ compounded daily; $m = 365$
$0 \le t \le 40$
$A = 500\left(1 + \dfrac{0.05}{365}\right)^{365t}$

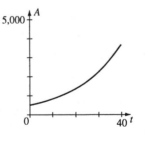

51. $A = 1000(1.05)^t$, $t = 2, 3, 4, 5, 6, 7$

Years (t)	1	2	3	4	5	6	7
Value (A)	$1050	$1103	$1158	$1216	$1276	$1340	$1407

$t = 2$, $A = 1102.50 \approx 1103$
$t = 3$, $A = 1157.63 \approx 1158$
$t = 4$, $A = 1215.51 \approx 1216$
$t = 5$, $A = 1276.28 \approx 1276$
$t = 6$, $A = 1340.10 \approx 1340$
$t = 7$, $A = 1407.10 \approx 1407$

53. My investment:
$P = 5000$
$r = 5.4\%$ every 6 months; $m = 2$
$A = 5000(1.054)^{2t}$
My friend's investment:
$P = 6000$
$r = 4.8\%$ every 6 months; $m = 2$
$A = 6000(1.048)^{2t}$
The investments will be the same when:
$5000(1.054)^{2t} = 6000(1.048)^{2t}$

$\left(\dfrac{1.054}{1.048}\right)^{2t} = \dfrac{6}{5}$

$2t \ln\left(\dfrac{1.054}{1.048}\right) = \ln 1.2$

$t = \dfrac{1}{2}\dfrac{(\ln 1.2)}{\ln\left(\frac{1.054}{1.048}\right)}$

≈ 16

$A \approx 5000(1.054)^{32} \approx 26,900$

The investments will be the same in about 16 years.
The value of the investment will be about $26,900.

55. Let $t = 0$ be 1985.
$A = 40,000 \cdot 2^{2t}$
$40,000 \cdot 4^t = 1,000,000$
$4^t = 25$
$t \approx 2.3$ years

57. $10,000(1.0516)^t = 15,000$
$1.0516^t = 1.5$
$t \approx 8$ years

59. $10,400\left(1 + \dfrac{0.0597}{12}\right)^{12t} = 20,000$
$t \approx 10.98$ years ≈ 132 months

61. $\left(1 + \dfrac{0.137}{2}\right)^{2t} = 2$
$(1.0685)^{2t} = 2$
$t \approx 5.23$ years ≈ 5 years

63. Canadian: $1000(1.0658)^t$

Mexican: $800(1.137)^t$

$1000(1.0658)^t = 800(1.137)^t$

$\left(\dfrac{1.0658}{1.137}\right)^t = 0.8$

$t \approx 3.45 \approx 3$ years

65. $\left(1 + \dfrac{0.0516}{365}\right)^{365t} = 2$

$t \approx 13.4$ years ≈ 14 years

67. $(1.159)^t = 3$

$t \approx 7.4$ years

69. The formula for the future value with respect to t is not a linear expression.

71. Since $r_e = \left(1 + \dfrac{r}{m}\right)^m - 1$, we have $r_e + 1 = \left(1 + \dfrac{r}{m}\right)^m$,

so $A = P\left(1 + \dfrac{r}{m}\right)^{mt} = P(r_e + 1)^t$. Thus, the graphs of the future value as a function of time are identical.

73. If a compound-interest investment is compounded more than once a year, the effective interest rate will always be greater than the nominal rate since

$\left(1 + \dfrac{r}{m}\right)^m - 1 > r$ for $m > 1$.

If interest is compounded less often than once per year, the effective interest rate will be less than the nominal rate.

75. Answers will vary.

77. There is a noticeable difference between compounding once a year and 10 times a year, but from 10 times a year to 10,000 times a year, there is no significant difference. The graphs seem to be approaching one shape.

Section 8.3 Annuities and Loans
Pages 732–734

1. $M = 100$

$r = 0.05, m = 12, t = 10$

$i = \dfrac{0.05}{12}, n = mt = 120$

$A = 100\dfrac{\left(1 + \frac{0.05}{12}\right)^{120} - 1}{\frac{0.05}{12}} \approx \$15,528.23$

3. $M = 1000, r = 0.07, m = 4, t = 20$

$i = \dfrac{0.07}{4} = 0.0175, n = mt = 80$

$A = 1000\dfrac{(1.0175)^{80} - 1}{0.0175} \approx \$171,793.82$

5. $A = 10,000, M = ?$

$r = 0.05, m = 12, t = 5$

$i = \dfrac{0.05}{12}, n = mt = 60$

$10,000 = M\dfrac{\left(1 + \frac{0.05}{12}\right)^{60} - 1}{\frac{0.05}{12}}$

$M \approx \$147.05$

7. $A = 75,000, M = ?$

$r = 0.06, m = 4, t = 20$

$i = \dfrac{0.06}{4} = 0.015, n = mt = 80$

$75,000 = M\dfrac{(1.015)^{80} - 1}{0.015}$

$M \approx \$491.12$

9. $W = 500, P = ?$

$r = 0.03, m = 12, t = 20$

$i = \dfrac{0.03}{12} = 0.0025, n = mt = 240$

$P = 500\dfrac{1 - (1.0025)^{-240}}{0.0025} \approx \$90,155.46$

11. $W = 1500, P = ?$

$r = 0.06, m = 4, t = 20$

$i = \dfrac{0.06}{4} = 0.015, n = mt = 80$

$P = 1500\dfrac{1 - (1.015)^{-80}}{0.015} \approx \$69,610.99$

13. $P = 100,000, W = ?$

$r = 0.03, m = 12, t = 20$

$i = \dfrac{0.03}{12} = 0.0025, n = mt = 240$

$100,000 = W\dfrac{1 - (1.0025)^{-240}}{0.0025}$

$W = 100,000\dfrac{0.0025}{1 - (1.0025)^{-240}} \approx \$554.60$

15. $P = 75,000, W = ?$

$r = 0.04, m = 4, t = 20$

$i = \dfrac{0.04}{4} = 0.01, n = mt = 80$

$W = 75,000\dfrac{0.01}{1 - (1.01)^{-80}} \approx \$1,366.41$

17. $P = 10,000, W = ?$

$r = 0.09, m = 12, t = 4$

$i = \dfrac{0.09}{12} = 0.0075, n = mt = 48$

$W = 10,000\dfrac{0.0075}{1 - (1.0075)^{-48}} \approx \$248.85$

19. $P = 100,000, W = ?$
$r = 0.05, m = 4, t = 20$
$i = \dfrac{0.05}{4} = 0.0125, n = mt = 80$
$W = 100,000\dfrac{0.0125}{1-(1.0125)^{-80}} \approx \$1,984.65$

21. We need to find out how large a decreasing annuity is needed to support the payment of \$5,000 per month for 20 years.
$P = ?, W = 5,000$
$r = 0.03, m = 12, t = 20$
$i = \dfrac{0.03}{12} = 0.0025, n = mt = 240$
$P = 5,000\dfrac{1-(1.0025)^{-240}}{0.0025} \approx \$901,554.57$
This total must be accumulated in the increasing annuity during the 40 years before retiring.
$A = 901,554.57, M = ?$
$i = 0.0025, m = 12, t = 40, n = mt = 480$
$M = 901,554.57\dfrac{0.0025}{(1.0025)^{480}-1}$
$\approx \$973.54$ per month.

23. Increasing annuity of \$1,200 per quarter during the 40 years before retiring:
$A = ?, M = 1,200$
$r = 0.04, m = 4, t = 40$
$i = \dfrac{0.04}{4} = 0.01, n = mt = 160$
$A = 1200\dfrac{(1.01)^{160}-1}{0.01} = \$469,659.16$
Decreasing annuity to be paid quarterly based on a 25-year payment:
$P = 469,659.16, W = ?$
$i = 0.01, m = 4, t = 25, n = mt = 100$
$W = 469,659.16\dfrac{0.01}{1-(1.01)^{-100}}$
$\approx \$7,451.49$ per quarter.

25. Solid S & L:
$P = 10,000, W = ?$
$r = 0.09, m = 12, t = 4$
$i = \dfrac{0.09}{12} = 0.0075, n = 48$
$W = 10,000\dfrac{0.0075}{1-(1.0075)^{-48}} \approx \$248.85$

Fifth Federal B & T:
$P = 10,000, W = ?$
$r = 0.07, m = 12, t = 3$
$i = \dfrac{0.07}{12}, n = 36$
$W = 10,000\dfrac{\frac{0.07}{12}}{1-\left(1+\frac{0.07}{12}\right)^{-36}} \approx \$308.77$
You should take the loan from Solid Savings & Loan; it will have payments of \$248.85 per month. The payments on the other loan would be \$308.77, more than the \$250 per month you can afford.

27. \$50,000 decreasing annuity, to be paid out over 15 years:
$P = 50,000, W = ?$
$r = 0.08, m = 12, t = 15$
$i = \dfrac{0.08}{12}, n = 180$
$W = 50,000\dfrac{\frac{0.08}{12}}{1-\left(1+\frac{0.08}{12}\right)^{-180}} \approx \$477.83$
Thus, the monthly payments will be \$477.83.
The total paid in one year will be
$477.83(12) = \$5733.96$.
To determine the amount of the loan at the end of one year, you can use the decreasing annuity formula to determine the present value of the annuity:
$P = 477.83\dfrac{1-\left(1+\frac{0.08}{12}\right)^{-168}}{\frac{0.08}{12}} \approx \$48,201.48$
This is the principal that remains unpaid. So, the amount paid on the principal the first year is $50,000 - 48,201.48 = \$1,798.52$. The rest of the year's payment was interest,
$5733.96 - 1798.52 = \$3935.44$.
The results are shown on the amortization table which follows: Calculations for the remaining 14 years are not shown here.

Year	Interest	Payments on Principal
1	\$3935.44	\$1798.52
2	3785.71	1948.25
3	3624.00	2109.96
4	3448.88	2285.08
5	3259.23	2474.73
6	3053.82	2680.14
7	2831.36	2902.60
8	2590.46	3143.50
9	2329.55	3404.41
10	2046.98	3686.98
11	1740.96	3993.00
12	1409.55	4324.41
13	1050.62	4683.24
14	661.91	5072.05
15	240.93	5493.03

29. Monthly payments for first five years:
$75,000 decreasing annuity, to be paid out over 30 years at 5% rate.
$P = 75,000, \ W = ?$
$r = 0.05, \ m = 12, \ t = 30$
$i = \dfrac{0.05}{12}, \ n = 360$

$W = 75,000 \dfrac{\frac{0.05}{12}}{1 - \left(1 + \frac{0.05}{12}\right)^{-360}} \approx \$402.62$ per month.

The amount still owed after 5 years:
$n = 25 \cdot 12 = 300$

$P = 402.62 \dfrac{1 - \left(1 + \frac{0.05}{12}\right)^{-300}}{\frac{0.05}{12}} \approx \$68,872.20$

Now, the monthly payment for the last 25 years can be determined:
$n = 300,$
$r = 0.095, \ i = \dfrac{0.095}{12}$

$W = 68,872.20 \dfrac{\frac{0.095}{12}}{1 - \left(1 + \frac{0.095}{12}\right)^{-300}} \approx \$601.73$

31. Original monthly payments:
$P = 96,000, \ W = ?$
$r = 0.0975, \ m = 12, \ t = 30$
$i = \dfrac{0.0975}{12}, \ n = 360$

$W = 96,000 \dfrac{\frac{0.0975}{12}}{1 - \left(1 + \frac{0.0975}{12}\right)^{-360}} \approx \$824.79$

Amount still owed after 4 years $n = 26(12) = 312$

$P = 824.79 \dfrac{1 - \left(1 + \frac{0.0975}{12}\right)^{-312}}{\frac{0.0975}{12}} \approx \$93,383.71$

New monthly payments:
$P = 93,383.71, W = ?$
$r = 0.06875, \ i = \dfrac{0.06875}{12}, \ n = 30(12) = 360$

$W = 93,383.71 \dfrac{\frac{0.06875}{12}}{1 - \left(1 + \frac{0.06875}{12}\right)^{-360}} \approx \$613.46$

Amount paid in 30 years at $613.46 per month:
220,845.60
Interest paid: 127,461.89
Amount paid in 26 years at $824.79 per month:
257,334.48
Interest paid: 163,950.77
You will save 163,950.77 − 127,461.89
= $36,488.88 in interest.

33. Increasing annuity:
$A = 100,000, \ M = 500$
$r = 0.04, \ m = 12, \ t = ?$
$i = \dfrac{0.04}{12}, \ n = 12t$

$100,000 = 500 \dfrac{\left(1 + \frac{0.04}{12}\right)^{12t} - 1}{\frac{0.04}{12}}$

$\left(1 + \frac{0.04}{12}\right)^{12t} = \frac{5}{3}$

$t = \dfrac{1}{12} \dfrac{\ln\left(1.\overline{6}\right)}{\ln\left(1.00\overline{3}\right)}$

$t \approx 12.79 \approx 13$ years

The graph of $y = 500 \dfrac{\left(1 + \frac{0.04}{12}\right)^{12x} - 1}{\frac{0.04}{12}}$ verifies this.

35. $P = 2000, \ W = 50$
$r = 0.15, \ m = 12, t = ?$
$i = \dfrac{0.15}{12} = 0.0125, \ n = 12t$

$2000 = 50 \dfrac{1 - (1.0125)^{-12t}}{0.0125}$

$(1.0125)^{-12t} = 0.5$

$t = -\dfrac{1}{12} \dfrac{\ln 0.5}{\ln 1.0125} \approx 4.6\overline{5} \approx 4.5$ years

The graph of $y = 50 \dfrac{1 - (1.0125)^{-12x}}{0.0125}$ verifies this.

37. Your account: Lucinda's account:
$A = ?, \ M = 100$ $A = ?, M = 75$
$r = 0.045, m = 12, t = ?$ $r = 0.065, m = 12, t = ?$
$i = \dfrac{0.045}{12} = 0.00375, \ n = 12t$

$\qquad\qquad\qquad i = \dfrac{0.065}{12}, \ n = 12t$

$100 \dfrac{(1.00375)^{12t} - 1}{\frac{0.045}{12}} \ \le \ 75 \dfrac{\left(1 + \frac{0.065}{12}\right)^{12t} - 1}{\frac{0.065}{12}}$

$\dfrac{1200}{0.045}\left[(1.00375)^{12t} - 1\right]$

$\le \ \dfrac{900}{0.065}\left[\left(1 + \frac{0.065}{12}\right)^{12t} - 1\right]$

Graph each side of the equation on a graphing utility or computer. Compare when the graph of Lucinda's account passes yours. From the graph, Lucinda's balance will pass yours after about 24 years.

39. Simon is wrong because he does not consider the amount made due to interest.

41. She is not correct because she does not account for the added interest from doubling the period. Consider the following example. Suppose that you take a $100,000 loan at 10% interest for 10 years. Then your monthly payment is

$$W = 100,000 \frac{\frac{0.10}{12}}{1 - \left(1 + \frac{0.10}{12}\right)^{-120}} \approx \$1,321.51$$

Suppose that you take a $100,000 loan at 10% interest for 20 years. Then your monthly payment is

$$W = 100,000 \frac{\frac{0.10}{12}}{1 - \left(1 + \frac{0.10}{12}\right)^{-240}} \approx \$965.02$$

You're the Expert
Pages 736–737

1. For the first car you will be paying off a loan of $3000 at 8% per year for 2 years. Use the decreasing annuity formula to calculate the monthly payment:

$$W = 3,000 \frac{\frac{0.08}{12}}{1 - \left(1 + \frac{0.08}{12}\right)^{-24}} = \$135.68$$

Therefore you will have 200 − 135.68 = $64.32 to deposit in your account for two years. Use the increasing annuity formula to determine the worth of the deposits after two years.

$$A = 64.32 \frac{\left(1 + \frac{0.04}{12}\right)^{24} - 1}{\frac{0.04}{12}} = \$1604.33.$$

This money will remain in the account for one more year. Calculate its value at the end of the third year using the compound interest formula.

$$A_1 = 1604.33\left(1 + \frac{0.04}{12}\right)^{12} = \$1669.69$$

During the third year, you will deposit $200 per month into the savings account. Calculate the amount produced:

$$A_2 = 200 \frac{\left(1 + \frac{0.04}{12}\right)^{12} - 1}{\frac{0.04}{12}} = \$2,444.49$$

The total amount that you will have in your account after three years is
$$A_T = A_1 + A_2 = 1669.69 + 2444.49 = \$4114.18$$
For the second car you will be paying off a loan of $2,800 at 8.5% per year for 3 years. Calculate the monthly payment:

$$W = 2,800 \frac{\frac{0.085}{12}}{1 - \left(1 + \frac{0.085}{12}\right)^{-36}} = \$88.39$$

Therefore you will have 200 − 88.39 = $111.61 to deposit in your account for three years. Determine the value of the deposits after three years:

$$B = 111.61 \frac{\left(1 + \frac{0.04}{12}\right)^{36} - 1}{\frac{0.04}{12}} = \$4261.44.$$

The total amount that you will have in your account after three years is $4261.44. Thus the second car will still cost less.

3. For the second car you will be paying off a loan of $3500 at 10% per year for 3 years. As in the original discussion, the value of the $700 after 3 years is: $B_1 = \$789.09$. Calculate the monthly payment:

$$W = 3500 \frac{\frac{0.10}{12}}{1 - \left(1 + \frac{0.10}{12}\right)^{-36}} = \$112.94$$

Therefore you will have 200 − 112.94 = $87.06 to deposit in your account for 3 years. Determine the worth of the deposits after 3 years.

$$B_2 = 87.06 \frac{\left(1 + \frac{0.04}{12}\right)^{36} - 1}{\frac{0.04}{12}} = \$3324.09$$

The total amount that you will have in your account after 3 years is
$$B_T = B_1 + B_2 = 789.09 + +3324.09 = \$4113.18$$
The second car will still cost less.

5. For the first car, you will pay 24(158.30) = $3799.20 for the loan, so you will pay 3799.20 − 3500 = $299.20 in interest. For the second car, you will pay 36(110.49) = $3977.64 for the loan, so you will pay 3977.64 − 3500 = $477.64 in interest. While you will pay more interest on the second car, the second car will end up costing less.

7. Compare the balance in your account after two years if you buy your car in cash and if you pay the $500 down payment and take out the loan. Suppose you pay in cash. Then you will deposit $200 into your account each month for 2 years, so calculate the amount produced:

$$A = 200 \frac{\left(1 + \frac{0.10}{12}\right)^{24} - 1}{\frac{0.10}{12}} = \$5289.38$$

The amount in your account will be $5289.38. Suppose you will be paying off a loan of $3500 at 8% per year for 2 years. The $500 down payment leaves $3500 in your account. Compute the value of the $3500 after 2 years:

$$B_1 = 3500\left(1 + \frac{0.10}{12}\right)^{24} = \$4271.37$$

As in the original discussion and in the previous problem, the monthly payment is $W = \$158.30$. Therefore you will have 200 − 158.30 = $41.70 to deposit in your account for 2 years. Determine the value of the deposits after 2 years:

$$B_2 = 41.70 \frac{\left(1 + \frac{0.10}{12}\right)^{24} - 1}{\frac{0.10}{12}} = \$1102.84$$

The total amount that you will have in your account after two years is
$$B_T = B_1 + B_2 = 4271.37 + 1102.84 = \$5374.21$$
It is better to take out a loan.

Chapter 8 Review Exercises
Pages 737–739

1. $P = 5000$, $r = 0.0475$, $t = 5$
$I = 5000(0.0475)(5) = \$1187.50$

3. $1000 = P(0.0475)(5)$
$P = \$4210.53$

5. $1000 = P[1 + 0.08(10)]$
$P = \dfrac{1000}{1.8} = \$555.56$

7. 1992 (start) $15
1994 (end) $30
doubled in 3 years
$30 = 15(1 + r \cdot 3)$
$r = \dfrac{1}{3} = 0.3333 = 33.33\%$

9. $A = 5000(1.0475)^5 = 6305.80$
Amount earned: $\$6305.80 - \$5000 = \$1305.80$

11. $5000 = P\left(1 + \dfrac{0.04}{12}\right)^{12(3)}$
$P = \dfrac{5000}{\left(1 + \frac{0.04}{12}\right)^{36}} = \$4435.49$

13. $r_e = (1 + 0.014)^{12} - 1$
$= (1.014)^{12} - 1 \approx 0.1816 = 18.16\%$

15. $r_e = \left(1 + \dfrac{0.055}{12}\right)^{12} - 1 \approx 0.0564 = 5.64\%$

17. $0.06 = \left(1 + \dfrac{r}{12}\right)^{12} - 1$
$\left(1 + \dfrac{r}{12}\right)^{12} = 1.06$
$r = 12\left[(1.06)^{1/12} - 1\right] \approx 0.0584 = 5.84\%$

19. CD's:
$A = 10,000\left(1 + \dfrac{0.06}{12}\right)^{12(5)}$
$= 10,000(1.005)^{60} = \$13,488.50$
Stock market:
$A = 13,488.50(1 - 0.01)^{12}$
$= 13,488.50(0.99)^{12} = \$11,956.00$

21. Equivalent car cost $= 10,000(1.04)^{24} = \$25,633.04$

23. $2 = (1.03)^{12t}$
$t = \dfrac{1}{12}\dfrac{\ln 2}{\ln 1.03} = 1.95$ years $= 23.4$ months

25. $A = ?$, $M = 150$
$r = 0.06$, $m = 12$, $t = 3$
$i = \dfrac{0.06}{12} = 0.005$, $n = 36$
$A = 150\dfrac{(1.005)^{36} - 1}{0.005} = \$5900.42$
Amount deposited: $150(36) = 5400$
Amount of interest earned:
$5900.42 - 5400 = \$500.42.$

27. $M = ?$, $A = 20,000$
$r = 0.05$, $m = 12$, $t = 7$
$i = \dfrac{0.05}{12}$, $n = 84$
$M = 20,000\dfrac{\frac{0.05}{12}}{1 + \frac{0.05}{12}^{84} - 1} = \$199.34$

29. Loan amount: $12,000 - 4000 = \$8000$
$W = ?$, $P = 8000$
$r = 0.07$, $m = 12$, $t = 4$
$i = \dfrac{0.07}{12}$, $n = 48$
$W = 8000\dfrac{\frac{0.07}{12}}{1 - \left(1 + \frac{0.07}{12}\right)^{-48}} = \$191.57$

31. Monthly payment: $W = ?$, $P = 120,000$
$r = 0.08$, $m = 12$, $t = 30$
$i = \dfrac{0.08}{12}$, $n = 360$
$W = 120,000\dfrac{\frac{0.08}{12}}{1 - \left(1 + \frac{0.08}{12}\right)^{-360}} = \$880.52$
Amount still owed after 1 year: $n = 29(12) = 348$
$P = 880.52\dfrac{1 - \left(1 + \frac{0.08}{12}\right)^{-348}}{\frac{0.08}{12}} = \$118,997.90$
Amount paid during first year:
$880.52(12) = \$10,566.24$
Amount paid toward principal first year:
$120,000 - 118,997.90 = \$1002.10$
Amount paid in interest during first year:
$10,566.24 - 1002.10 = \$9564.14$
Amount paid during 30 years:
$880.52(360) = \$316,987.20$
Amount paid in interest over life of mortgage:
$316,987.20 - 120,000 = \$196,987.20$

33. $W = ?$, $P = 4000$
$r = 0.16$, $m = 12$, $t = 2$
$i = \dfrac{0.16}{12}$, $n = 24$
$W = 4000\dfrac{\frac{0.16}{12}}{1 - \left(1 + \frac{0.16}{12}\right)^{-24}} = \$195.85$

35. $P = ?$, $W = 2000$
$r = 0.08$, $m = 12$, $t = 20$
$i = \dfrac{0.08}{12}$, $n = 240$

$$P = 2000 \frac{1 - \left(1 + \frac{0.08}{12}\right)^{-240}}{\frac{0.08}{12}} = \$239,108.58$$

37. Decreasing annuity to support payment of $4000 per month for 20 years.
$P = ?$, $W = 4000$
$r = 0.04$, $m = 12$, $t = 20$
$i = \dfrac{0.04}{12}$, $n = 240$

$$P = 4000 \frac{1 - \left(1 + \frac{0.04}{12}\right)^{-240}}{\frac{0.04}{12}} = 660,087.43$$

Increasing annuity during the 30 years before retiring:
$A = 660,087.43$, $M = ?$
$i = \dfrac{0.04}{12}$, $m = 12$, $t = 30$, $n = 360$

$$M = 660,087.43 \frac{\frac{0.04}{12}}{\left(1 + \frac{0.04}{12}\right)^{360} - 1} = \$951.07$$

CHAPTER 9
EXPONENTIAL AND LOGARITHMIC FUNCTIONS

9.1 EXERCISES

1. $f(x) = 2^x$

3. $f(x) = 3^{-x}$

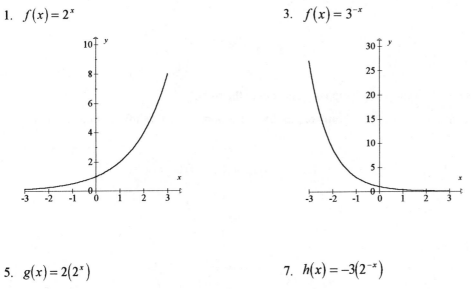

5. $g(x) = 2(2^x)$

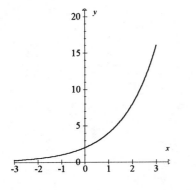

7. $h(x) = -3(2^{-x})$

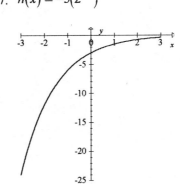

9. $r(x) = 1 - 2^x$ 11. $s(x) = 2^{x-1}$

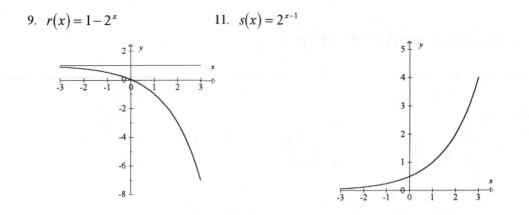

13. The graph of s is obtained by shifting the graph of f one unit to the right.

15. The graph of r is obtained by inverting the graph of f in the y-direction, and then shifting it up one unit.

17. $f_1(x) = 1.6^x$, $f_2(x) = 1.8^x$ 19. $f_1(x) = 300(1.1^x)$, $f_2(x) = 300(1.1^{2x})$

21. $f_1(x) = 1000(1.045^{-3x})$
    $f_2(x) = 1000(1.045^{3x})$

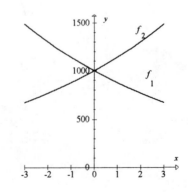

In Exercises 23 - 29 use $A = P\left(1 + \dfrac{r}{m}\right)^{mt}$ with $P = 10{,}000$.

23. (a) $A(t) = 10{,}000\left(1 + \dfrac{0.03}{1}\right)^{1t}$

$\qquad = 10{,}000(1.03)^{t}$

(b) $A(10) = 10{,}000(1.03)^{10}$

$\qquad = \$13{,}436.16$

25. (a) $A(t) = 10{,}000\left(1 + \dfrac{0.025}{4}\right)^{4t}$

$\qquad = 10{,}000(1.00625)^{4t}$

(b) $A(5) = 10{,}000(1.00625)^{4(5)}$

$\qquad = 10{,}000(1.00625)^{20}$

$\qquad = \$11{,}327.08$

27. (a) $A(t) = 10{,}000\left(1 + \dfrac{0.065}{}\right)^{365t}$

(b) $A(10) = 10{,}000\left(1 + \dfrac{0.065}{365}\right)^{365(10)}$

$\qquad = 10{,}000\left(1 + \dfrac{0.065}{365}\right)^{3650}$

$\qquad = \$19{,}154.30$

29. (a) $A(t) = 10{,}000(1 + 0.002)^{12t}$

$\qquad = 10{,}000(1.002)^{12t}$

(b) $A(10) = 10{,}000(1.002)^{12(10)}$

$\qquad = 10{,}000(1.002)^{120}$

$\qquad = \$12{,}709.44$

31. $\quad A(t) = 1000(1.056)^{t}$

(a) $A(11) = 1000(1.056)^{11}$

$\qquad = \$1{,}820.97$

(b) $A(21) = 1000(1.056)^{21}$

$\qquad = \$3{,}140.09$

33. $V(t) = 6000(1.1)^{2t}$

$\quad y_1 = 6000(1.1)^{2t}$

$\quad y_2 = 15{,}000$

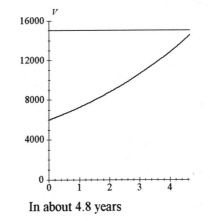

In about 4.8 years

233

35.
$$A(t) = P\left(1 + \frac{r}{m}\right)^{mt}$$

$$A(4) = 100\left(1 + \frac{0.06}{4}\right)^{4(4)}$$

$$= 100(1.015)^{16}$$

$$= 126.90$$

$$A(4) - P = 126.90 - 100$$

$$= \$26.90$$

In Exercises 37 - 39 use $A = P\left(1 - \dfrac{r}{m}\right)^{mt}$

37.
$$A(5) = 3000\left(1 - \frac{0.068}{1}\right)^{1(5)}$$

$$= 3000(0.932)^{5}$$

$$= \$2,109.60$$

39.
$$A(t) = 200,000\left(1 - \frac{0.05}{1}\right)^{2t}$$

$$A(t) = 200,000(0.95)^{2t}$$

$$A(10) = 200,000(0.95)^{2(10)}$$

$$= \$71,697.18$$

41. $A(t) = 550(1.015)^{t}$

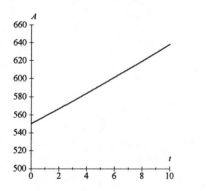

43. $A(t) = 550\left(1 + \frac{0.015}{365}\right)^{365t}$

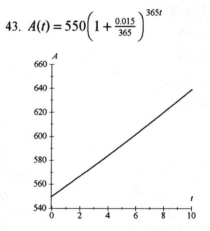

234

45. $A_1(t) = 5000\left(1 + \frac{0.054}{2}\right)^{2t}$

$A_2(t) = 6000\left(1 + \frac{0.048}{2}\right)^{2t}$

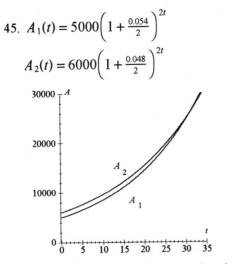

31 years, $26,300 to the nearest hundred

47. $A(t) = 40,000\left(1 + \frac{2}{2}\right)^{2t}$

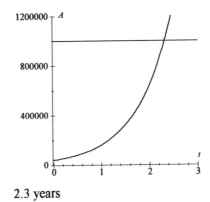

2.3 years

49. $A(r) = 1000(1+r)^{10}$

$A(0.1) = 1000(1+0.1)^{10}$

$\qquad = 2,593.74$

$A(1.15) = 1000(1+1.15)^{10}$

$\qquad = 2,110,496.32$

$A(r)$ is the amount that an item costing $1,000 now will cost in 10 years, given an annual rate of inflation r. $A(0.1) = 2,593.74$. Thus, at a rate of inflation of 10% per year, an item costing $1,000 now will cost $2,593.74 in 10 years. $A(1.15) = 2,110,496.32$. Thus, at a rate of inflation of 115% per year, an item costing $1,000 now will cost $2,110,496.32 in 10 years.

51.

x	0	1	2
$f(x)$	500	225	101.25

$f(x) = Ca^x$

$f(0) = 500 = Ca^0 \Rightarrow C = 500$

$f(1) = 225 = 500a^1 \Rightarrow a = 0.45$

$f(2) = 500(0.45)^2 = 101.25$

$f(x) = 500(0.45)^x$

53. (b) $r(x) = 2^x$

55. Exponential functions of the form $f(x) = a^x (a > 0)$ increase rapidly for large values of x. In real-life situations, such as population growth, this model is only reliable for relatively short periods of growth. Eventually, population growth tapers off due to such pressures as limited resources and overcrowding.

9.2 EXERCISES

1. $e^3 = 20.0855$

3. $e^{-1} = 0.3679$

5. $e^{30 \times 0.001} = e^{0.03} = 1.0305$

7. $100e^{0.0125} = 101.2578$

9. $10,200e^{-0.025 \times 20} = 10,200e^{-0.5}$
$= 6,186.6127$

11. $f(x) = e^x$

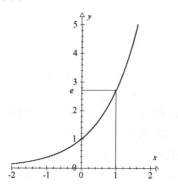

13. $g(x) = e^{-2x}$

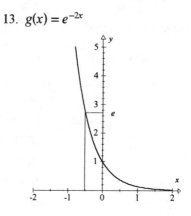

15. $h(x) = 2e^x$

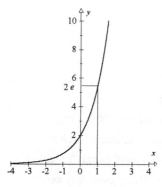

17. The graph of h is obtained from the graph of f by expanding it by a factor of 2 in the y-direction.

19. The graph of g is obtained from the graph of f by compressing it by a factor of 2 in the x-direction, and then reflecting it in the y-axis.

236

In Exercises 21 - 25 use $A = Pe^{rt}$ where $P = \$10,000$ and t is in years.

21. $r = 0.05$, $t = 10$

$A = 10{,}000e^{0.05(10)}$

$= \$16{,}487$

23. $r = 0.025$, $t = 50$

$A = 10{,}000e^{0.025(50)}$

$= \$34{,}903$

24.

25. $r = 0.025$, $t = 11.5$

$A = 10{,}000e^{0.025(11.5)}$

$= \$13{,}331$

In Exercises 27 - 59 use $A = Pe^{rt}$, and $A = P\left(1+\dfrac{r}{m}\right)^{mt}$ for growth and appreciation,

but use $A = Pe^{-kt}$ for decay and depreciation.

27. $r = 0.0235$, $P = 1000$

$A(t) = 1000e^{0.0235t}$

$A(5) = 1000e^{0.0235(5)}$

$= \$1{,}124.68$

29. $r = 0.04$, $t = 10$, $P = 1000$

$A = 1000e^{0.04(10)} = 1{,}491.82$

$\text{Interest} = A - P = 1{,}491.82 - 1000$

$= \$491.82$

In Exercises 31 - 35 use $r_e = \left(1+\dfrac{r}{m}\right)^{m} - 1$ and $r_e = e^r - 1$ for the effective yield.

31. $r = 0.04$

$r_e = e^{0.04} - 1 = 0.0408$

$= 4.08\%$

33. $r = 0.20$

$r_e = e^{0.20} - 1 = 0.2214$

$= 22.14\%$

35. Fifth Federal Bank; $r = 0.09$, $m = 12$

$r_e = \left(1+\dfrac{0.09}{12}\right)^{12} - 1 = 0.0938 = 9.38\%$

Ninth National Bank; $r = 0.089$, compounded continuously

$r_e = e^{0.089} - 1 = 0.0931 = 9.31\%$

Ninth National Bank has the lower effective rate.

37. (i) $A = 5000e^{0.055(10)} = 8,666.27$

(ii) $A = 1000e^{-0.30(10)} = 49.79$

(iii) $A = 10,000\left(1 + \dfrac{0.02}{4}\right)^{4(10)}$

$= 12,207.94$

Total value $= 8,666.27 + 49.79 + 12,207.94$

$= \$20,924.00$

39. $P = 100, \; k = 0.000121$

$A(t) = 100e^{-0.000121tr}$

$A(10,000) = 100e^{-0.000121(10,000)}$

$= 29.8g$

41. $\dfrac{2}{1} = 2 = 200\%$

$A = 1000\left(1 + \dfrac{2}{1}\right)^{12}$

$= 531,441,000$ bugs

43. Slope $= \dfrac{0.17 - 1.58}{5} = -0.282$

when $t = 0, \; r = 1.58$

(a) $r(t) = 1.58 - 0.282t$

(b) $\bar{r} = \dfrac{r(0) + r(t)}{2}$

$= \dfrac{1.58 + 1.58 - 0.282t}{2}$

$= 1.58 - 0.141t$

(c) $A(t) = Pe^{(1.58 - 0.141t)}$

(d) $A(4) = 100e^{[1.58 - 0.141(4)](4)}$

$= \$5,820$

45. Purchasing Power $= \left[1000e^{0.05(4)}\right]e^{-0.03(4)}$

$= \$1,083.29$

This is the same as an earning rate of 2% because

$e^{rt}e^{-kt} = e^{(r-k)t} = e^{(0.05-0.03)t}$

$= e^{0.02t}$

238

47. $f(x) = 100e^{0.3x}$; $-10 \le x \le 10$

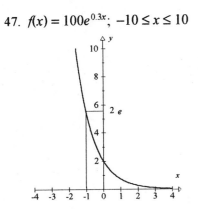

49. $f(x) = 1000e^{-0.005x}$; $-50 \le x \le 50$

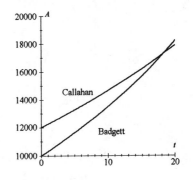

51. $k = 0.00121$, $P = 100$

$A(t) = 100e^{-0.000121t}$

$A = 1$ when $t = 38{,}059$ years old

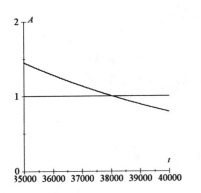

53.   Badgett: $A(t) = 10{,}000e^{0.03t}$

Callahan: $A(t) = 12{,}000e^{0.02t}$

The graphs intersect at $t = 18$ years

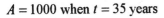

55. $k = 0.055$, $P = 7000$

$A(t) = 7000e^{-0.055t}$

$A = 1000$ when $t = 35$ years

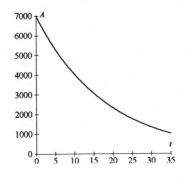

57. $r = 0.06$, $A = 1{,}000{,}000$, $t = 10$

$1{,}000{,}000 = Pe^{0.06(10)}$

$P = \$548{,}812$

239

59. $P = 28,131.7e^{-0.220800t}$

$P = 1000$ when $t = 15$ years

$1987 + 15 = 2002$

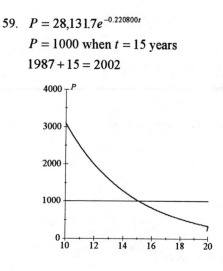

9.3 EXERCISES

1. $2^5 = 32$

$\log_2 32 = 5$

3. $3^{-2} = \dfrac{1}{9}$

$\log_3 \left(\dfrac{1}{9}\right) = -2$

5. $10^3 = 1000$

$\log_{10} 1000 = 3$

7. $y = e^x$

$\log_e y = x$

9. $y = x^{-3} \ (x \neq 1)$

$\log_x y = -3$

11. $1 = e^0$

$\log_e 1 = 0$

13. $\log_6 36 = 2$

$6^2 = 36$

15. $\log_2 \left(\dfrac{1}{4}\right) = -2$

$2^{-2} = \dfrac{1}{4}$

17. $\log 100,000,000 = 8$

$10^8 = 100,000,000$

19. $\ln \left(\dfrac{1}{e}\right) = -1$

$e^{-1} = \dfrac{1}{e}$

21. $\log_x y = 3$

$x^3 = y$

23. $y = \ln(-x), \ (x < 0)$

$e^y = -x$

Use $a = \ln 2$, $b = \ln 3$ and $c = \ln 5$ in Exercises 25-35.

25. $\ln 6 = \ln 2 \cdot 3 = \ln 2 + \ln 3 = a + b$

27. $\ln \left(\dfrac{2}{3}\right) = \ln 2 - \ln 3 = a - b$

29. $\ln 256 = \ln 2^8 = 8 \ln 2 = 8a$

31. $\ln \dfrac{3}{10} = \ln \dfrac{3}{2 \cdot 5} = \ln 3 - (\ln 2 + \ln 5) = b - (a + c)$

33. $\ln 0.02 = \ln \dfrac{2}{100} = \ln 2 - \ln 10^2 = \ln 2 - 2 \ln (2 \cdot 5)$

$\qquad = \ln 2 - 2(\ln 2 + \ln 5) = a - 2(a + c) = -(a + 2c)$

35. $\ln \dfrac{9}{e} = \ln 3^2 - \ln e = 2 \ln 3 - 1 = 2b - 1$

37. $\log_a 3 + \log_a 4 = \log_a (12)$

39. $2 \log_a x + \log_a y = \log_a \left(x^2 y\right)$

41. $2 \ln x + 4 \ln y - 6 \ln z = \ln \left(\dfrac{x^2 y^4}{z^6}\right)$

43. $\log_2 (x) + 2^x \log_2 y = \log_2 \left(xy^{2^x}\right)$

45. $\quad 4 = 2^x$

$\qquad 2^2 = 2^x$

$\qquad x = 2$

47. $2^t = 4^{2t}$

$\quad 2^t = \left(2^2\right)^{2t}$

$\quad 2^t = 2^{4t}$

$\quad t = 4t$

$\quad 3t = 0$

$\quad t = 0$

49. $\quad 2^t = 2^{-t^2}$

$\qquad t = -t^2$

$\quad t^2 + t = 0$

$\quad t(t + 1) = 0$

$\qquad t = 0 \quad t + 1 = 0$

$\qquad\qquad\qquad t = -1$

51.
$$100 = 50e^{3t}$$
$$2 = e^{3t}$$
$$\ln 2 = \ln e^{3t}$$
$$\ln 2 = 3t$$
$$t = \frac{\ln 2}{3}$$
$$= 0.23105$$

53.
$$10 = 1000e^{-2x}$$
$$0.01 = e^{-2t}$$
$$\ln 0.01 = \ln e^{-2t}$$
$$\ln 0.01 = -2t$$
$$t = -\frac{\ln 0.01}{2}$$
$$= 2.30259$$

55.
$$200 = 5(2^y)$$
$$40 = 2^y$$
$$\ln 40 = \ln 2^y$$
$$\ln 40 = y \ln 2$$
$$y = \frac{\ln 40}{\ln 2}$$
$$= 5.32193$$

57. $f(x) = \log_3 x;\ 0 < x \le 27$

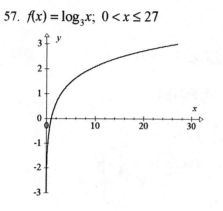

59. $g(x) = x - \log_2 x;\ 0 < x \le 8$

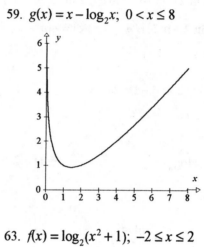

61. $h(x) = (\log_2 x)^2;\ 0 < x \le 8$

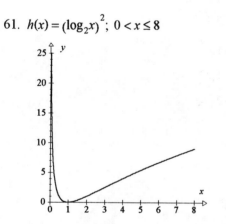

63. $f(x) = \log_2(x^2 + 1);\ -2 \le x \le 2$

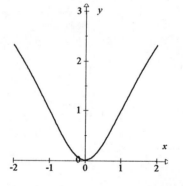

65. $f(x) = \ln |x|; \; -10 \le x \le 10, \; (x \ne 0)$

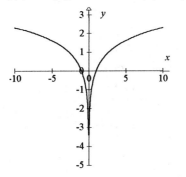

67.
$$R = \frac{2}{3}(\log E - 4.4)$$

(a) $R = 8.2$

$$8.2 = \frac{2}{3}\log E_1 - \frac{2}{3}(4.4)$$

$$\frac{2}{3}\log E_1 = 11.1333$$

$$\log E_1 = 16.7$$

$$E_1 = 10^{16.7}$$

$$= 5.012 \times 10^{16} \text{ joules}$$

(b) $R = 7.1$

$$7.1 = \frac{2}{3}\log E_2 - \frac{2}{3}(4.4)$$

$$\log E_2 = 15.05$$

$$E_2 = 10^{15.05}$$

$$= 1.122 \times 10^{15.05} \text{ joules}$$

$$\frac{E_2}{E_1} = \frac{1.122 \times 10^{15}}{5.012 \times 10^{16}} = 0.0223$$

$$= 2.23\%$$

(c) $R_1 = \frac{2}{3}(\log E_1 - 4.4)$

$$\frac{2}{3}R_1 = \log E_1 - 4.4$$

$$\log E_1 = \frac{2}{3}R_1 + 4.4$$

$$E_2 = 10^{\frac{3}{2}R_2 + 4.4}$$

$$\frac{E_2}{E_1} = \frac{10^{\frac{3}{2}R_2 + 4.4}}{10^{\frac{3}{2}R_1 + 4.4}}$$

$$= 10^{\frac{3}{2}R_2 + 4.4 - \left(\frac{3}{2}R_1 + 4.4\right)}$$

$$= 10^{1.5(R_2 - R_1)}$$

(d) $10^{1.5(1)} = 31.62$

69. $D = 10 \log\left(\dfrac{320 \times 10^7}{r^2}\right)$

(a) $r = 10$: $D = 10 \log\left(\dfrac{320 \times 10^7}{10^2}\right) = 75$ dB

$r = 20$: $D = 10 \log\left(\dfrac{320 \times 10^7}{20^2}\right) = 69$ dB

$r = 50$: $D = 10 \log\left(\dfrac{320 \times 10^7}{50^2}\right) = 61$ dB

(b) $D = 10 \log 320 \times 10^7 - 10 \log r^2$

$\quad = 95.05 - 20 \log r$

(c) $0 = 95.05 - 20 \log r$

$\log r = \dfrac{95.05}{20}$

$r = 10^{4.7525}$

$\quad = 56{,}559$ ft away

(c) $D_1 = 10 \log \dfrac{I_1}{I_0}$

$\dfrac{D_1}{10} = \log \dfrac{I_1}{I_0}$

$\dfrac{I_1}{I_0} = 10^{0.1 D_1}$

$I_1 = I_0 \, 10^{0.1 D_1}$

$I_2 = I_0 \, 10^{0.1 D_2}$

$\dfrac{I_2}{I_1} = \dfrac{I_0 \, 10^{0.1 D_2}}{I_0 \, 10^{0.1 D_1}}$

$\quad = 10^{0.1 D_2 - 0.1 D_1}$

(d) $10^{0.1(1)} = 1.259$

71.

(a) $m = \dfrac{\Delta q}{\Delta p} = \dfrac{3.290 - 4.826}{5.00 - 3.00} = -0.768$

$f(p) = mp + b$

$3.290 = f(5.00) = -0.768(5.00) + b$

$b = 7.13$

$f(p) = -0.768 p + 7.13$

(b) $f(p) = kp^r$

$3.290 = f(5) = k5^r$

$4.826 = f(3) = k3^r$

$\dfrac{3.290}{4.826} = \dfrac{k5^r}{k3^r}$

$\dfrac{3.290}{4.826} = \left(\dfrac{5}{3}\right)^r$

$\ln\left(\dfrac{3.290}{4.826}\right) = r \ln\left(\dfrac{5}{3}\right)$

$r = \dfrac{\ln\left(\dfrac{3.290}{4.826}\right)}{\ln\left(\dfrac{5}{3}\right)} = -0.750$

$3.290 = k5^{-0.750}$

$k = 3.290\left(5^{0.750}\right) = 11.0$

$f(p) = 11.0 p^{-0.750}$

(c) (i) $f(4) = -0.768(4) + 7.13 = 4.058$

(ii) $f(4) = 11.0(4)^{-0.750} = 3.889$

$f(p) = 11.0 p^{-0.750}$ predicts more accurately

(d) $f(3.50) = 11.0(3.50)^{-0.750}$

$\quad = 4.299$ servings per hour

9.4 EXERCISES

Use $A = Pe^{rt}$ and $A = P\left(1 + \dfrac{r}{m}\right)^{mt}$ for appreciation and growth, $A = Pe^{-kt}$ for depreciation and decay,

$t_D = \dfrac{\ln 2}{r}$ or $t_D = \dfrac{\ln 2}{k}$ for the time to double, $t_r = \dfrac{\ln 3}{r}$ for the time to triple, and $t_H = \dfrac{\ln 2}{k}$ for the half life.

1. $P = 10{,}000,\ A = 15{,}000,\ r = 0.0516,\ m = 1$

$15{,}000 = 10{,}000(1 + 0.0516)^t$

$1.5 = 1.0516^t$

$\ln 1.5 = t \ln 1.0516$

$t = \dfrac{\ln 1.5}{\ln 1.0516} = 8 \text{ years}$

3. $P = 10{,}400,\ A = 20{,}000,\ r = 0.0597,\ m = 12$

$20{,}000 = 10{,}400\left(1 + \dfrac{0.0597}{12}\right)^{12t}$

$1.923 = (1.004975)^{12t}$

$\ln 1.923 = 12t \ln 1.004975$

$t = \dfrac{\ln 1.923}{12 \ln 1.004975} = 10.98 \text{ years or}$

132 months

5. $A = 2P,\ r = 0.137,\ m = 2$

$2P = P\left(1 + \dfrac{0.137}{2}\right)^{2t}$

$2 = (1.0685)^{2t}$

$\ln 2 = 2t \ln 1.0685$

$t = \dfrac{\ln 2}{2 \ln 1.0685} = 5 \text{ years}$

7. $P_1 = 1000,\ r = 0.0658,\ m = 1$

$P_2 = 800,\ r_2 = 0.137,\ m = 1$

$A_1 = A_2$

$1000(1 + 0.0658)^t = 800(1 + 0.137)^t$

$\dfrac{1000}{800} = \left(\dfrac{1.137}{1.0658}\right)^t$

$\ln 1.25 = t \ln\left(\dfrac{1.137}{1.0658}\right)$

$t = \dfrac{\ln 1.25}{\ln\left(\dfrac{1.137}{1.0658}\right)} = 3 \text{ years}$

9. $r = 0.0516$

$t_D = \dfrac{\ln 2}{0.0516} = 13.43 \text{ years}$

11. $r = 0.159$

$t_T = \dfrac{\ln 3}{0.159} = 6.9 \text{ years}$

13. $t_D = 1$

$$k = \frac{\ln 2}{t_D} = \frac{\ln 2}{1} = \ln 2$$

$$500 = 1e^{(\ln 2)t}$$

$$500 = \left(e^{\ln 2}\right)^t$$

$$500 = 2^t$$

$$t = \frac{\ln 500}{\ln 2} = 9 \text{ days}$$

15. $A = 700,\ P = 500,\ r = 0.1$

$$700 = 500e^{0.1t}$$

$$1.4 = e^{0.1t}$$

$$\ln 1.4 = 0.1t$$

$$t = \frac{\ln 1.4}{0.1} = 3.36 \text{ years}$$

17. $r = 0.10$

$$t_T = \frac{\ln 3}{r} = \frac{\ln 3}{0.10} = 11 \text{ years}$$

19. $P = 1000,\ A = 666,\ k = 0.20$

$$666 = 1000e^{-0.20t}$$

$$0.666 = e^{-0.20t}$$

$$\ln 0.666 = -0.20t$$

$$t = \frac{\ln 0.666}{-0.20} = 2.03 \text{ years}$$

21. $t_H = 5730,\ A = 0.00125P$

$$t_H = \frac{\ln 2}{k}$$

$$k = \frac{\ln 2}{t_H} = \frac{\ln 2}{5730}$$

$$0.00125P = Pe^{-\left(\frac{\ln 2}{5730}\right)t}$$

$$0.00125 = -\frac{t}{5730}\ln 2$$

$$t = \frac{-5730\ln 0.00125}{\ln 2}$$

$$= 55{,}259 \text{ years old}$$

23. $t_T = 0.5$

$$t_T = \frac{\ln 3}{r}$$

$$r = \frac{\ln 3}{t_T} = \frac{\ln 3}{0.5}$$

$$t_D = \frac{\ln 2}{r} = \frac{\ln 2}{\ln 3 \,/\, 0.5}$$

$$= 0.315 \text{ years or } 3.8 \text{ months}$$

25. $P = 1000$, $t_D = 3$, $t = 48$

$$t_D = \frac{\ln 2}{r}$$

$$r = \frac{\ln 2}{t_D} = \frac{\ln 2}{3}$$

$$A = 1000e^{\frac{\ln 2}{3}(48)}$$

$$= 1000e^{16 \ln 2}$$

$$= 65{,}536{,}000 \text{ bacteria}$$

27. $P = 50{,}000$, $A = 75{,}000$, $t = 2$

$$75{,}000 = 50{,}000e^{r(2)}$$

$$1.5 = e^{2r}$$

$$\ln 1.5 = 2r$$

$$r = \frac{\ln 1.5}{2}$$

(ii) $P = 75{,}000$, $r = \dfrac{\ln 1.5}{2}$, $t = 1$

$$A = 75{,}000e^{\left(\frac{\ln 1.5}{2}\right)(1)}$$

$$= 91{,}856 \text{ tags}$$

29. (i) $P = 180{,}000{,}000$, $A = 250{,}000{,}000$, $t = 30$

$$250{,}000{,}000 = 180{,}000{,}000e^{r(30)}$$

$$\frac{25}{18} = e^{30r}$$

$$\ln \frac{25}{18} = 30r$$

$$r = \frac{\ln \frac{25}{18}}{30}$$

(ii) $P = 250{,}000{,}000$, $r = \dfrac{\ln \frac{25}{18}}{30}$, $t = 20$

$$A = 250{,}000{,}000e^{\left(\frac{\ln \frac{25}{18}}{30}\right)(20)}$$

$$= 311{,}000{,}000$$

31. $t_D = 1$, $P = 20$, $A = 250{,}000{,}000$

$$t_D = \frac{\ln 2}{r}$$

$$r = \frac{\ln 2}{t_D} = \frac{\ln 2}{1} = \ln 2$$

$$250{,}000{,}000 = 20e^{(\ln 2)t}$$

$$12{,}500{,}000 = 2^t$$

$$\ln 12{,}500{,}000 = t \ln 2$$

$$t = \frac{\ln 12{,}500{,}000}{\ln 2} = 24 \text{ semesters}$$

33. $t_H = 710 \times 10^6$, $P = 10$, $A = 1$

$$t_H = \frac{\ln 2}{k}$$

$$k = \frac{\ln 2}{t_H} = \frac{\ln 2}{710 \times 10^6}$$

$$1 = 10e^{-\left(\frac{\ln 2}{710 \times 10^6}\right)t}$$

$$0.1 = 2^{-t/710 \times 10^6}$$

$$\ln 0.1 = -\frac{t}{710 \times 10^6} \ln 2$$

$$t = -\frac{710 \times 10^6 \ln 0.1}{\ln 2}$$

$$= 2{,}360 \text{ million years}$$

35. $A = 0.7$, $P = 1$, $t = 2$

$$0.7 = 1e^{-k(2)}$$

$$\ln 0.7 = -2k$$

$$k = -\frac{\ln 0.7}{2}$$

$$t_H = \frac{\ln 2}{k} = \frac{\ln 2}{-\frac{\ln 0.7}{2}}$$

$$= 3.89 \text{ days}$$

37. $P = P_0 e^{-kt}$; Let $t = 1995 - 1987 = 8$,

$P_0 = 23{,}200$, $P = 3800$

$$3800 = 23{,}200e^{-k(8)}$$

$$0.1637931 = e^{-8k}$$

$$\ln 0.1637931 = -8k$$

$$k = 0.2261439$$

$$P = 23{,}200e^{-0.2261439t}$$

Let $t = 1997 - 1987 = 10$

$$P = 23{,}200e^{-0.2261439(10)}$$

$$\approx 2{,}400 \text{ people in 1997}$$

38. $P = P_0 e^{-kt}$; Let $t = 1994 - 1987 = 7$,

$P_0 = 23{,}200$, $P = 4800$

$$4800 = 23{,}200e^{-k(7)}$$

$$0.20689655 = e^{-7k}$$

$$\ln 0.20689655 = -7k$$

$$k = 0.2250766$$

$$P = 23{,}200e^{-0.2250766t}$$

Let $t = 1997 - 1987 = 10$

$$P = 23{,}200e^{-0.2250766(10)}$$

$$\approx 2{,}400 \text{ people in 1997}$$

39. $P = 28{,}131.7e^{-0.220800t}$

Let $t = 1997 - 1987 = 10$

$$P = 28{,}131.7e^{-0.220800(10)}$$

$$\approx 3{,}100 \text{ people in 1997}$$

The least squares model gives
a higher employment figure for 1997.

41. $P = 28{,}131.7e^{-0.220800t}$; $P = 1000$

$$1000 = 28{,}131.7e^{-0.220800t}$$

$$0.035547 = e^{-0.220800t}$$

$$\ln 0.035547 = -0.220800t$$

$$t = 15.1$$

In $1987 + 15 = 2002$ employment
will drop to 1000.

43. $t_H = 1, \ A = 40 - 35 = 5,$
$P = 70 - 35 = 35$

$t_H = \dfrac{\ln 2}{k}$

$k = \dfrac{\ln 2}{t_H} = \dfrac{\ln 2}{1} = \ln 2$

$5 = 35e^{-(\ln 2)t}$

$1/7 = -t \ln 2$

$t = -\dfrac{\ln 1/7}{\ln 2} = 2.8 \text{ hours}$

CHAPTER 9 REVIEW EXERCISES

1. $\quad 2^{10} = 1024$
   $\log_2 1024 = 10$

3. $\quad 10^{-4} = 0.0001$
   $\log 0.0001 = -4$

5. $\quad y = x^3, \ (x \neq 1)$
   $\log_x y = e$

7. $\log_3 81 = 4$
   $3^4 = 81$

9. $\log 1000 = 3$
   $10^3 = 1000$

11. $\log_x y = -1$
    $x^{-1} = y$

13. $\quad 4 = 3^x$
    $\ln 4 = \ln 3^x$
    $\ln 4 = x \ln 3$
    $x = \dfrac{\ln 4}{\ln 3}$
    $\quad = 1.262$

15. $\quad 9^x = 3^{-x^2}$
    $3^{2x} = 3^{-x^2}$
    $2x = -x^2$
    $x^2 + 2x = 0$
    $x(x + 2) = 0$
    $x = 0 \quad \text{or} \quad x + 2 = 0$
    $\qquad\qquad\qquad x = -2$

17. $1000 = 5e^{10x}$

$200 = e^{10x}$

$\ln 200 = \ln e^{10x}$

$\ln 200 = 10x$

$x = \dfrac{\ln 200}{10}$

$= 0.5298$

Use $A = Pe^{rt}$ and $A = P\left(1 + \dfrac{r}{m}\right)^{mt}$ in Exercises 19 - 47.

19. $P = 2000$, $r = 0.04$, $m = 12$, $t = 4$

$A = 2000\left(1 + \dfrac{0.04}{12}\right)^{12(4)}$

$= \$2,346.40$

21. $P = 2000$, $r = 0.0675$, $m = 365$, $t = 4$

$A = 2000\left(1 + \dfrac{0.0675}{365}\right)^{365(4)}$

$= \$2,619.86$

23. $P = 2000$, $r = 0.0375$, $t = 4$

$A = 2000e^{(0.0375)(4)}$

$= \$2,323.67$

25. $P = 2000$, $r = 0.04$, $m = 12$, $A = 3000$

$3000 = 2000\left(1 + \dfrac{0.04}{12}\right)^{12t}$

$1.5 = (1.00333)^{12t}$

$\ln 1.5 = 12t \ln 1.00333$

$t = \dfrac{\ln 1.5}{12 \ln 1.00333} = 10.2 \text{ years}$

27. $P = 2000$, $r = 0.0675$, $m = 365$, $A = 3000$

$3000 = 2000\left(1 + \dfrac{0.0675}{365}\right)^{365t}$

$1.5 = (1.000185)^{365t}$

$\ln 1.5 = 365t \ln 1.000185$

$t = \dfrac{\ln 1.5}{365 \ln 1.000185} = 6 \text{ years}$

29. $P = 2000$, $r = 0.0375$, $A = 3000$

$3000 = 2000e^{0.0375t}$

$1.5 = e^{0.0375t}$

$\ln 1.5 = 0.0375t$

$t = \dfrac{\ln 1.5}{0.0375} = 10.8 \text{ years}$

31. (i) $P = 10,000$, $r = 0.06$, $m = 12$, $t = 5$

$$A = 10,000\left(1 + \frac{0.06}{12}\right)^{12(5)}$$

$$= 13,488.50$$

(ii) $P = 13,488.50$, $r = -0.01$, $m = 1$, $t = 12$

$$A = 13,488.50(1 - 0.01)^{12}$$

$$= \$11,956.00 \text{ left}$$

33. $P = 10,000$, $r = 0.04$, $m = 1$, $t = 24$

$$A = 10,000(1 + 0.04)^{24}$$

$$= \$25,633.04$$

35. (i) $P = 10$, $A = 30$, $t = 4$

$$30 = 10e^{r(4)}$$

$$3 = e^{4r}$$

$$\ln 3 = 4r$$

$$r = \frac{\ln 3}{4}$$

(ii) $A = 1000$, $P = 30$, $r = \frac{\ln 3}{4}$

$$1000 = 30e^{\frac{\ln 3}{4}t}$$

$$33.333 = 3^{t/4}$$

$$\ln 33.333 = \frac{t}{4}\ln 3$$

$$t = \frac{4 \ln 33.333}{\ln 3} = 12.8 \text{ more months}$$

or 16.8 months total

37. $P = 10$, $A = 1$, $t = 1$

$$1 = 10e^{-k(1)}$$

$$0.1 = e^{-k}$$

$$\ln 0.1 = -k$$

$$k = -\ln 0.1$$

$$t_H = \frac{\ln 2}{-\ln 0.1} = 0.301 \text{ seconds}$$

39. (i) $r_e = \left(1 + \frac{r}{m}\right)^m - 1$

$$= \left(1 + \frac{0.25}{1}\right)^1 - 1$$

$$= 0.25 = 25\%$$

(ii) $r_e = e^r - 1$

$$= e^{0.24} - 1$$

$$= 0.2712 = 27.12\%$$

Erewhon is the better investment.

41. $t_H = 1.28 \times 10^9, \quad A = 0.05P$

$$t_H = \frac{\ln 2}{k}$$

$$k = \frac{\ln 2}{t_H} = \frac{\ln 2}{1.28 \times 10^9}$$

$$0.05P = Pe^{-\left(\frac{\ln 2}{1.28 \times 10^9}\right)t}$$

$$0.05 = 2^{-t/1.28 \times 10^9}$$

$$\ln 0.05 = -\frac{t}{1.28 \times 10^9} \ln 2$$

$$t = \frac{-1.28 \times 10^9 \ln 0.05}{\ln 2}$$

$$= 5.5 \times 10^9 \text{ years or}$$

$$5.5 \text{ billion years}$$

43. $t_D = 1$

$$t_D = \frac{\ln 2}{r}$$

$$r = \frac{\ln 2}{t_D} = \frac{\ln 2}{1}$$

(i) $A = 640, \quad r = \ln 2, \quad t = 6$

$$640 = Pe^{(\ln 2)(6)}$$

$$P = \frac{640}{2^6} = 10$$

(ii) $P = 10, \quad r = \ln 2, \quad t = 3$

$$A = 10e^{(\ln 2)(3)}$$

$$= 10(2)^3$$

$$= \$80$$

45. $t_D = 2, \quad P = 1000, \quad A = 500{,}000$

$$t_D = \frac{\ln 2}{r}$$

$$r = \frac{\ln 2}{t_D} = \frac{\ln 2}{2}$$

$$500{,}000 = 1000e^{\left(\frac{\ln 2}{2}\right)t}$$

$$500 = 2^{t/2}$$

$$\ln 500 = \frac{t}{2} \ln 2$$

$$t = \frac{2 \ln 500}{\ln 2} = 18 \text{ minutes}$$

47. $r = 0.10, \quad A = 1, \quad t = 2, \quad m = 1$

$$1 = P(1 + 0.1)^2$$

$$P = \frac{1}{1.1^2} = \$0.83$$

CHAPTER 10
Introduction to the Derivative

10.1 EXERCISES

In Exercises 1-7 use $\dfrac{\Delta f}{\Delta x} = \dfrac{f(x+h) - f(x)}{h}$.

1. $f(x) = 2x^2$; $x = 0$

$$\frac{\Delta f}{\Delta x} = \frac{2(x+h)^2 - 2x^2}{h}$$

$$= \frac{2x^2 + 4xh + 2h^2 - 2x^2}{h}$$

$$= 4x + 2h$$

At $x = 0$, $\dfrac{\Delta f}{\Delta x} = 4(0) + 2h = 2h$

h	$\dfrac{\Delta f}{\Delta x}$
1	2
0.1	0.2
0.01	0.02
0.001	0.002
0.0001	0.0002
−0.0001	− 0.0002
−0.001	− 0.002
−0.01	− 0.02
−0.1	− 0.2
−1	− 2

3. $f(x) = \dfrac{1}{x}$; $x = 2$

$$\frac{\Delta f}{\Delta x} = \frac{\dfrac{1}{x+h} - \dfrac{1}{x}}{h}$$

$$= \frac{x - (x+h)}{x(x+h)(h)}$$

$$= -\frac{1}{x(x+h)}$$

At $x = 2$, $\dfrac{\Delta f}{\Delta x} = -\dfrac{1}{2(2+h)}$

h	$\dfrac{\Delta f}{\Delta x}$
1	− 0.1667
0.1	− 0.2381
0.01	− 0.2488
0.001	− 0.2499
0.0001	− 0.24999
−0.0001	− 0.25001
−0.001	− 0.2501
−0.01	− 0.2513
−0.1	− 0.2632
−1	− 0.5

5. $f(x) = x^3;\ x = 1$

$$\frac{\Delta f}{\Delta x} = \frac{(x+h)^3 - x^3}{h}$$

$$= \frac{x^3 + 3x^2 h + 3xh^2 + h^3 - x^3}{h}$$

$$= 3x^2 + 2xh + h^2$$

At $x = 1$, $\dfrac{\Delta f}{\Delta x} = 3 + 3h + h^2$

h	$\dfrac{\Delta f}{\Delta x}$
1	7
0.1	3.31
0.01	3.0301
0.001	3.0030
0.0001	3.0003
−0.0001	2.9997
−0.001	2.9970
−0.01	2.9701
−0.1	2.71
−1	1

7. $f(x) = x^2 + 2x;\ x = 3$

$$\frac{\Delta f}{\Delta x} = \frac{\left[(x+h)^2 + 2(x+h)\right] - (x^2 + 2x)}{h}$$

$$= \frac{x^2 + 2xh + h^2 + 2x + 2h - x^2 - 2x}{h}$$

$$= 2x + 2 + h$$

At $x = 3$, $\dfrac{\Delta f}{\Delta x} = 2(3) + 2 + h = 8 + h$

h	$\dfrac{\Delta f}{\Delta x}$
1	9
0.1	8.1
0.01	8.01
0.001	8.001
0.0001	8.0001
−0.0001	7.9999
−0.001	7.999
−0.01	7.99
−0.1	7.9
−1	7

In Exercises 9-19 use $f'(x) \approx \dfrac{f(x+0.0001)-f(x)}{0.0001}$.

9. $f(x)=1-2x; \quad x=2$

$$f'(x) \approx \frac{1-2(x+0.0001)-(1-2x)}{0.0001}$$

$$\approx \frac{-0.0002}{0.0001}$$

$$\approx -2$$

$$f'(2) \approx -2$$

11. $f(x)=x^3; \quad x=-1$

$$f'(x) \approx \frac{(x+0.0001)^3 - x^3}{0.0001}$$

$$f'(-1) \approx \frac{(-1+0.0001)^3 - (-1)^3}{0.0001}$$

$$\approx \frac{(-0.9999)^3 + 1}{0.0001}$$

$$\approx 3$$

13. $h(x)=x^{1/4}; \quad x=16$

$$h'(x) \approx \frac{(x+0.0001)^{1/4} - x^{1/4}}{0.0001}$$

$$h'(16) \approx \frac{(16+0.0001)^{1/4} - 16^{1/4}}{0.0001}$$

$$\approx \frac{(16.0001)^{1/4} - 2}{0.0001}$$

$$\approx 0.03125$$

15. $f(x)=x^{-1/2}; \quad x=2$

$$f'(x) \approx \frac{(x+0.0001)^{-1/2} - x^{-1/2}}{0.0001}$$

$$f'(2) \approx \frac{(2+0.0001)^{-1/2} - 2^{-1/2}}{0.0001}$$

$$\approx \frac{(2.0001)^{-1/2} - 2^{-1/2}}{0.0001}$$

$$\approx -0.17677$$

17. $g(t)=\dfrac{1}{t^5}; \quad t=1$

$$g'(t) \approx \frac{\dfrac{1}{(t+0.0001)^5} - \dfrac{1}{t^5}}{0.0001}$$

$$g'(1) \approx \frac{\dfrac{1}{(1+0.0001)^5} - \dfrac{1}{1^5}}{0.0001}$$

$$\approx \frac{\dfrac{1}{(1.0001)^5} - 1}{0.0001}$$

$$\approx -5$$

19. $f(x)=\dfrac{x^2}{4} - \dfrac{x^3}{3}; \quad x=-1$

$$f'(x) \approx \frac{\dfrac{(x+0.0001)^2}{4} - \dfrac{(x+0.0001)^3}{3} - \left(\dfrac{x^2}{4} - \dfrac{x^3}{3}\right)}{0.0001}$$

$$f'(-1) \approx \frac{\dfrac{(-1+0.0001)^2}{4} - \dfrac{(-1+0.0001)^3}{3} - \left[\dfrac{(-1)^2}{4} - \dfrac{(-1)^3}{3}\right]}{0.0001}$$

$$\approx \frac{\dfrac{(-0.9999)^2}{4} - \dfrac{(-0.9999)^3}{3} - \dfrac{1}{4} - \dfrac{1}{3}}{0.001}$$

$$\approx -1.5$$

In Exercises 21-27 use $v_{ave} = \dfrac{\Delta s}{\Delta t} = \dfrac{s(t+h) - s(t)}{h}$ and $v_{inst} = \dfrac{ds}{dt} \approx \dfrac{s(t + 0.0001) - s(t)}{0.0001} = v_{ave}$ when $h = 0.0001$.

21. $s(t) = 50t - t^2$; $\quad t = 5$

$$v_{ave} = \frac{50(t+h) - (t+h)^2 - (50t - t^2)}{h}$$

$$= \frac{50t + 50h - t^2 - 2th - h^2 - 50t + t^2}{h}$$

$$= 50 - 2t - h$$

At $t = 5$ hours, $v_{ave} = 50 - 2(5) - h = 40 - h$

h(hour)	1	0.1	0.01
v_{ave} (mph)	39	39.9	39.99

$v_{inst} \approx 40 - 0.0001 \approx 40$ mph

23. $s(t) = 100 + 20t^3$; $\quad t = 1$

$$v_{ave} = \frac{100 + 20(t+h)^3 - (100 + 20t^3)}{h}$$

$$= \frac{100 + 20t^3 + 60t^2h + 60th^2 + 20h^3 - 100 - 20t^3}{h}$$

$$= 60t^2 + 60th + 20h^2$$

At $t = 1$ hour, $v_{ave} = 60 + 60h + 20h^2$

h(hour)	1	0.1	0.01
v_{ave} (mph)	140	66.2	60.602

$v_{inst} \approx 60 + 60(0.0001) + 20(0.0001)^2 \approx 60$ mph

25. $s(t) = 60\left(t + \dfrac{1}{t}\right)$; $\quad t = 10$

$$v_{ave} = \frac{60\left(t + h + \dfrac{1}{t+h}\right) - 60\left(t + \dfrac{1}{t}\right)}{h}$$

$$= \frac{60t + 60h + \dfrac{60}{t+h} - 60t - \dfrac{60}{t}}{h}$$

$$= 60 + \frac{60t - 60(t+h)}{ht(t+h)}$$

$$= 60 + \frac{-60}{t(t+h)}$$

$$= 60\left[1 - \frac{1}{t(t+h)}\right]$$

At $t = 10$ hours, $v_{ave} = 60\left[1 - \dfrac{1}{10(10+h)}\right] = 60 - \dfrac{6}{10+h}$

h(hour)	1	0.1	0.01
v_{ave} (mph)	59.45	59.41	59.40

$v_{inst} \approx 60 - \dfrac{6}{10 + 0.0001} \approx 59.4$ mph

27. $s(t) = 40\sqrt{t}; \quad t = 4$

$$v_{ave} = \frac{40\sqrt{t+h} - 40\sqrt{t}}{h}$$

$$= 40\left(\frac{\sqrt{t+h} - \sqrt{t}}{h} \cdot \frac{\sqrt{t+h} + \sqrt{t}}{\sqrt{t+h} + \sqrt{t}}\right)$$

$$= 40\left[\frac{t+h-t}{h(\sqrt{t+h} + \sqrt{t})}\right]$$

$$= \frac{40}{\sqrt{t+h} + \sqrt{t}}$$

At $t = 4$ hours, $v_{ave} = \dfrac{40}{\sqrt{4+h} + \sqrt{4}} = \dfrac{40}{\sqrt{4+h} + 2}$

h(hour)	1	0.1	0.01
v_{ave} (mph)	9.443	9.938	9.994

$$v_{inst} \approx \frac{40}{\sqrt{4+0.0001} + 2} \approx 10 \text{ mph}$$

In Exercises 29-33 use $C_{ave} = \dfrac{\Delta C}{\Delta x} = \dfrac{C(x+h) - C(x)}{h}$ and $C'(x) = \dfrac{dC}{dx} \approx \dfrac{C(x + 0.0001) - C(x)}{0.0001} = C_{ave}$ when $h = 0.0001$.

29. $C(x) = 10,000 + 5x - \dfrac{x^2}{10,000}; \quad x = 1000$

$$C_{ave} = \frac{10,000 + 5(x+h) - \frac{(x+h)^2}{10,000} - \left(10,000 + 5x - \frac{x^2}{10,000}\right)}{h}$$

$$= \frac{10,000 + 5x + 5h - \frac{x^2 + 2xh + h^2}{10,000} - 10,000 - 5x - \frac{x^2}{10,000}}{h}$$

$$= 5 - \frac{x}{5000} - \frac{h}{10,000}$$

when $x = 1000$, $\quad C_{ave} = 5 - \dfrac{1000}{5000} - \dfrac{h}{10,000} = \dfrac{48,000 - h}{10,000}$

h	50	10	1
C_{ave}	4.795	4.799	4.7999

$$C'(1000) \approx \frac{48,000 - 0.0001}{10,000} \approx 4.8$$

31. $C(x) = 1000 + 10x - \sqrt{x}; \quad x = 100$

$$C_{ave} = \frac{1000 + 10(x+h) - \sqrt{x+h} - \left(1000 + 10x - \sqrt{x}\right)}{h}$$

$$= \frac{1000 + 10x + 10h - \sqrt{x+h} - 1000 - 10x + \sqrt{x}}{h}$$

$$= 10 + \frac{\sqrt{x} - \sqrt{x+h}}{h}$$

$$= 10 + \frac{\sqrt{x} - \sqrt{x+h}}{h} \cdot \frac{\sqrt{x} + \sqrt{x+h}}{\sqrt{x} + \sqrt{x+h}}$$

$$= 10 + \frac{x - (x+h)}{h\left(\sqrt{x} + \sqrt{x+h}\right)}$$

$$= 10 - \frac{1}{\sqrt{x} + \sqrt{x+h}}$$

When $x = 100$, $C_{ave} = 10 - \dfrac{1}{\sqrt{100} + \sqrt{100+h}}$

$$= 10 - \frac{1}{10 + \sqrt{100+h}}$$

h	50	10	1
C_{ave}	9.955	9.951	9.950

$C'(100) \approx 10 - \dfrac{1}{10 + \sqrt{100 + 0.0001}} \approx 9.95$

33. $C(x) = 15{,}000 + 100x + \dfrac{1000}{x}; \quad x = 100$

$$C_{ave} = \frac{15{,}000 + 100(x+h) + \dfrac{1000}{x+h} - \left(15{,}000 + 100x + \dfrac{1000}{x}\right)}{h}$$

$$= \frac{15{,}000 + 100x + 100h + \dfrac{1000}{x+h} - 15{,}000 - 100x - \dfrac{1000}{x}}{h}$$

$$= 100 + \frac{1000x - 1000(x+h)}{x(x+h)h}$$

$$= 100 - \frac{1000}{x(x+h)}$$

when $x = 100$, $C_{ave} = 100 - \dfrac{1000}{100(100+h)} = 100 - \dfrac{10}{100+h}$

h	50	10	1
C_{ave}	99.93	99.91	99.90

$C'(100) \approx 100 - \dfrac{10}{100 + 0.0001} \approx 99.9$

35. $q = \dfrac{5{,}000{,}000}{p}$

$q(100) = \dfrac{5{,}000{,}000}{100} = 50{,}000$

$q'(100) \approx \dfrac{q(100 + 0.0001) - q(100)}{0.0001} = \dfrac{\dfrac{5{,}000{,}000}{100.0001} - 50{,}000}{0.0001} \approx -500$

A total of 50,000 sneakers can be sold at a price of \$100, but the demand decreases by 500 sneakers for each increase of \$1 in the price.

37. $P = 5n + \sqrt{n}, \ n = 50$

$P(50) = 5(50) + \sqrt{50} = 257.07$

$\left.\dfrac{dP}{dn}\right|_{n=50} \approx \dfrac{P(50 + 0.0001) - P(50)}{0.0001} = \dfrac{5(50,000) + \sqrt{50.0001} - 257.07}{0.0001} \approx 5.07$

Your current profit is \$257.07 per month, and this would increase \$5.07 for each increase by one magazine in sales.

39. $E = 20\sqrt[4]{g}$

$E(2.5) = 20\sqrt[4]{2.5} = 25.1$

$\left.\dfrac{dE}{dg}\right|_{g=2.5} \approx \dfrac{E(2.5 + 0.0001) - E(2.5)}{0.0001} = \dfrac{20\sqrt[4]{2.5001} - 25.1}{0.0001} \approx 2.51$

The professor's average class size is 25.1 students, and this would rise by 2.51 students per 1 point increase in grades awarded.

41. $A(t) = 1000 + 1500t - 800t^2 + 100t^3; \ t = 0.5$

$A(0.5001) = 1000 + 1500(0.5001) - 800(0.5001)^2 + 100(0.5001)^3 \approx 1562.5775$

$A(0.5) = 1000 + 1500(0.5) - 800(0.5)^2 + 100(0.5)^3 = 1562.5000$

$A'(0.5) \approx \dfrac{A(0.5 + 0.0001) - A(0.5)}{0.0001} \approx \dfrac{1562.5775 - 1562.5000}{0.0001} = 775 \text{ points per day}$

43.   (a) $p(10) = 60$ and $p'(10) = 18.2$

      60% of children can speak at the age of 10 months. Further, at the age of 10 months, this percentage is increasing by 18.2% per month.

  (b) As t increases, $p(t)$ approaches 100 (assuming all children eventually learn to speak) and $p'(t)$ approaches zero since the percentage stops increasing.

45. $R(p) = \dfrac{r}{1+kp} = \dfrac{45}{1+0.125p}$

$R'(4) \approx \dfrac{R(4+0.0001)-R(4)}{0.0001} = \dfrac{\dfrac{45}{1+0.125(4.0001)} - \dfrac{45}{1+0.125(4)}}{0.0001} \approx -2.5$

(a) $R'(4) = -2.5$ thousand organisms per hour, per 1,000 new organisms. This means that the reproduction rate of organisms in a culture containing 4,000 organisms is declining by 2,500 organisms per hour for each 1,000 new organisms.

47. $l(p) = l_o\sqrt{1-p^2} = 100\sqrt{1-p^2}$

$l(0.95) = 100\sqrt{1-(0.95)^2} \approx 31.22$

$l'(0.95) \approx \dfrac{l(0.95+0.0001)-l(0.95)}{0.0001} = \dfrac{100\sqrt{1-(0.9501)^2} - 100\sqrt{1-(0.95)^2}}{0.0001} \approx -304.41$

$l(0.95) = 31.22$ meters and $l'(0.95) = -304.41$ meters / warp. Thus, at a speed of warp 0.95, the rocket has an observed length of 31.22 meters and its length is decreasing at a rate of 304.41 meters per unit warp speed or 3.0441 meters per one percent increase in the speed (measured in warp).

In Exercises E49 - E52 use $f'(x) \approx \dfrac{f(x+0.0001)-f(x)}{0.0001}$.

E49. $f(x) = e^x$; $x = 0$

$\begin{aligned} f'(0) &\approx \dfrac{e^{(0+0.0001)} - e^0}{0.0001} \\ &\approx \dfrac{e^{0.0001}-1}{0.0001} \\ &\approx 1 \end{aligned}$

E51. $f(x) = \ln x$; $x = 1$

$f'(1) \approx \dfrac{\ln(1.0001) - \ln 1}{0.0001} \approx 1.000$

E53. $S(t) = 200 - 150e^{-t/10}$; $t = 5$

$S(5) = 200 - 150e^{-5/10} \approx 109$

$S'(5) \approx \dfrac{2000 - 150e^{-5.0001/10} - \left(200 - 150e^{-5/10}\right)}{0.0001} \approx 9.10$

$S(5) = 37$, $S'(5) = -74.4$. After 5 weeks, sales are 109 sneakers per week, and sales are increasing by 9.10 sneakers per week each week.

E55. $q(t) = \dfrac{N}{1+ke^{-rt}} = \dfrac{10,000}{1+0.5e^{-0.4t}}; \quad t = 2$

$q(2) = \dfrac{10,000}{1+0.5e^{-0.4(2)}} \approx 8,165$

$q'(2) \approx \dfrac{\dfrac{10,000}{1+0.5e^{-0.4(2.0001)}} - \dfrac{10,000}{1+0.5e^{-0.4(2)}}}{0.0001} \approx -599.2$

$q(2) = 8,165$, $q'(2) = 599.2$. Thus, two months after
the introduction, 8,165 video game units have sold,
and the demand is growing by 599.2 units per month.

10.2 EXERCISES

1. (a) R (b) P 3. (a) P (b) R 5. (a) Q (b) P

7. (a) P (b) Q 9. (a) Q (b) R (c) P 11. (a) R (b) Q (c) P

13. (a) R (b) Q (c) P

15. (a) $m_{\tan} = 0$ at $(1, 0)$
    (b) $m_{\tan} = 1$ at no point
    (c) $m_{\tan} = -1$ at $(-2, 1)$

17. (a) $m_{\tan} = 0$ at $(-2, 2)$, $(0, 1)$ and $(2, 0)$
    (b) $m_{\tan} = 1$ at no point
    (c) $m_{\tan} = -1$ at no point

19. The tangent to the graph of the function f at
the general point where $x = a$ is the line
passing through $f(a)$ with slope

$$f'(a) = \lim_{h \to 0} \frac{f(a+h) - f(a)}{h}.$$

21. If $m_{\tan} = f'(x)$ and $(x_1, y_1) = (a, f(a))$, then
the point - slope formula becomes

$$y - f(a) = f'(a)(x - a)$$
$$y = f(a) + (x - a)f'(a)$$

In Exercises 23 - 33 use (a) $m_{sec} = \dfrac{f(x+h) - f(x)}{h}$, (b) $f'(x) = m_{tan} = \lim\limits_{h \to 0} m_{sec}$ and

(c) $f'(a) = m_{tan}$ at $x = a$.

23. $f(x) = x^2 + 1;\ (2, 5)$

(a) $f(x+h) = (x+h)^2 + 1 = x^2 + 2xh + h^2 + 1$

$m_{sec} = \dfrac{x^2 + 2xh + h^2 + 1 - (x^2 + 1)}{h}$

$= 2x + h$

(b) $f'(x) = \lim\limits_{h \to 0}(2x + h) = 2x$

(c) $f'(2) = 2(2) = 4$

25. $f(x) = 2 - x^2;\ (-1, 1)$

(a) $f(x+h) = 2 - (x+h)^2 = 2 - x^2 - 2xh - h^2$

$m_{sec} = \dfrac{2 - x^2 - 2xh - h^2 - (2 - x^2)}{h}$

$= -2x - h$

(b) $f'(x) = \lim\limits_{h \to 0}(-2x - h) = -2x$

(c) $f'(-1) = -2(-1) = 2$

27. $h(x) = 3x;\ (2, 6)$

(a) $h(x+h) = 3(x+h) = 3x + 3h$

$m_{sec} = \dfrac{3x + 3h - 3x}{h} = 3$

(b) $f'(x) = \lim\limits_{h \to 0}(3) = 3$

(c) $f'(2) = 3$

29. $f(x) = 1 - 2x;\ (2, -3)$

(a) $f(x+h) = 1 - 2(x+h) = 1 - 2x - 2h$

$m_{sec} = \dfrac{1 - 2x - 2h - (1 - 2x)}{h} = -2$

(b) $f'(x) = \lim\limits_{h \to 0}(-2) = -2$

(c) $f'(2) = -2$

31. $f(x) = 3x^2 + 1;\ (-1, 2)$

(a) $f(x+h) = 3(x+h)^2 + 1 = 3x^2 + 6xh + 3h^2 + 1$

$m_{sec} = \dfrac{3x^2 + 6xh + 3h^2 + 1 - (3x^2 + 1)}{h}$

$= 6x + 3h$

(b) $f'(x) = \lim\limits_{h \to 0}(6x + 3h) = 6x$

(c) $f'(-1) = 6(-1) = -6$

33. $f(x) = x - x^2;\ (2, 0)$

(a) $f(x+h) = (x+h) - (x+h)^2$

$= x + h - x^2 - 2xh - h^2$

$m_{sec} = \dfrac{x + h - x^2 - 2xh - h^3 - (x - x^2)}{h}$

$= 1 - 2x - h$

(b) $f'(x) = \lim\limits_{h \to 0}(1 - 2x - h) = 1 - 2x$

(c) $f'(2) = 1 - 2(2) = -3$

In Exercises 35 - 39 use $m_{tan} = f'(a) = \lim\limits_{h \to 0} \dfrac{f(a+h) - f(a)}{h}$ and, from Exercise 21,
use the equation of the tangent line $y = f(a) + (x-a)f'(a)$.

35. $f(x) = x^3$; $x = -1$

(a) $f'(-1) = \lim\limits_{h \to 0} \dfrac{(-1+h)^3 - (-1)^3}{h}$

$= \lim\limits_{h \to 0} \dfrac{-1 + 3h - 3h^2 + h^3 + 1}{h}$

$= \lim 3 - 3h + h^2$

$= 3$

(b) $y = (-1)^3 + \left[x - (-1)\right](3)$

$= -1 + 3x + 3$

$= 3x + 2$

37. $f(x) = x + \dfrac{1}{x}$; $x = 2$

(a) $f'(2) = \lim\limits_{h \to 0} \dfrac{2 + h + \dfrac{1}{2+h} - \left(2 + \dfrac{1}{2}\right)}{h}$

$= \lim\limits_{h \to 0} \dfrac{\dfrac{h + 2 - (2+h)}{2(2+h)}}{h}$

$= \lim\limits_{h \to 0} 1 + \dfrac{-1}{2(2+h)}$

$= 1 - \dfrac{1}{4} = \dfrac{3}{4}$

(b) $y = 2 + \dfrac{1}{2} + (x-2)\left(\dfrac{3}{4}\right)$

$= \dfrac{3}{4}x + 1$

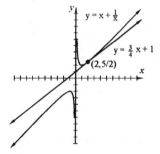

263

39. $f(x) = \sqrt{x}; \quad x = 4$

   (a) $f'(4) = \lim\limits_{h \to 0} \dfrac{\sqrt{4+h} - \sqrt{4}}{h}$

   $= \lim\limits_{h \to 0} \dfrac{\sqrt{4+h} - 2}{h} \cdot \dfrac{\sqrt{4+h} + 2}{\sqrt{4+h} + 2}$

   $= \lim\limits_{h \to 0} \dfrac{4+h-4}{h\left(\sqrt{4+h} + 2\right)}$

   $= \lim\limits_{h \to 0} \dfrac{1}{\sqrt{4+h} + 2}$

   $= \dfrac{1}{4}$

   (b) $y = \sqrt{4} + (x-4)\dfrac{1}{4}$

   $= 2 + \dfrac{1}{4}x - 1$

   $= \dfrac{1}{4}x + 1$

41. (b)   43. (c)   45. (a)

In Exercises 47 - 49 use $f(x) = \dfrac{1}{x^{1.1} - 4}$.

47. $f'(1) \approx m_{\text{segment}} = \dfrac{f(1.001) - f(0.999)}{1.001 - .999} \approx \dfrac{-0.3334556 - (-0.3332112)}{0.002} \approx -0.12$

49. $f'(1.2) \approx m_{\text{segment}} = \dfrac{f(1.201) - f(1.199)}{1.201 - 1.199} \approx \dfrac{-0.3601267 - (-0.3598364)}{0.002} \approx -0.15$

In Exercises 51 - 53 use $f(x) = \sqrt{1 - x^2}$.

51. $f'(0.5) \approx m_{\text{segment}} = \dfrac{f(0.501) - f(0.499)}{0.501 - 0.499} \approx \dfrac{0.8654473 - 0.8666020}{0.002} \approx -0.58$

53. $f'(-0.25) \approx m_{\text{segment}} = \dfrac{f(-0.249) - f(-0.251)}{-0.249 - (-0.251)} \approx \dfrac{0.9685035 - 0.9679871}{0.002} \approx 0.26$

55. $f(x) = x^{3.4} - x^{1.2}$; $\ 0 \le x \le 5$

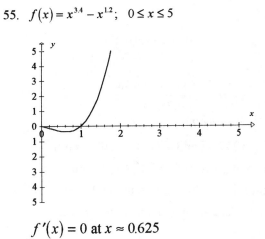

$f'(x) = 0$ at $x \approx 0.625$

57. $f(x) = (x-1)(x-2)(x-3)$; $\ -1 \le x \le 5$

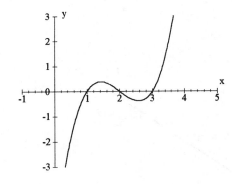

$f'(x) = 0$ at $x \approx 1.425,\ 2.575$

59. $q = \dfrac{5,000,000}{p}$

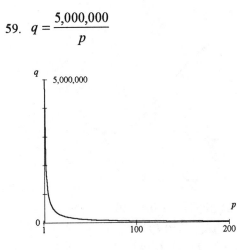

$$q'(100) \approx \frac{q(101) - q(99)}{101 - 99}$$
$$\approx \frac{49,505 - 50,505}{2}$$
$$\approx -500$$

$q(100) = 50,000$, $q'(100) = -500$. A total of 50,000 sneakers can be sold at a price of \$100, but the demand decreases by 500 sneakers for each increase of \$1 in the price.

61. $P = 5n + \sqrt{n}$

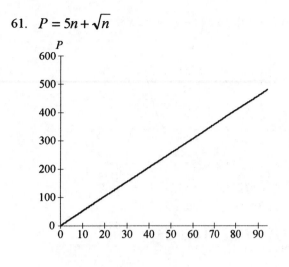

$$\left.\frac{dP}{dn}\right|_{n=50} \approx \frac{P(50.1) - P(49.9)}{50.1 - 49.9}$$

$$\approx \frac{257.578 - 256.564}{0.2}$$

$$\approx 5.07$$

$P = \$257.07$ and $dP/dn = 5.07$. Your current profit is $257.07 per month, and this would increase by $5.07 for each increase by one magazine in sales.

63. $E = 20\sqrt[4]{g}$

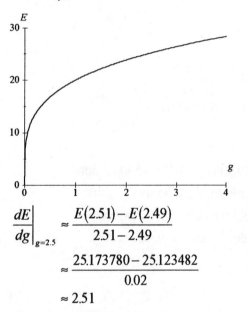

$$\left.\frac{dE}{dg}\right|_{g=2.5} \approx \frac{E(2.51) - E(2.49)}{2.51 - 2.49}$$

$$\approx \frac{25.173780 - 25.123482}{0.02}$$

$$\approx 2.51$$

$E = 25.1$, $dE/dg = 2.51$. The professor's average class size is 25.1 students, and this would rise by 2.51 students per 1 point increase in grades awarded.

65. $A(t) = 1000 + 1500t - 800t^2 + 100t^3$

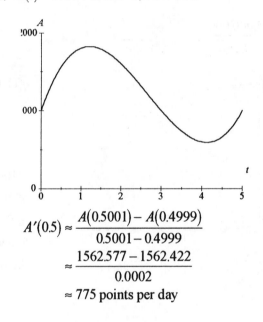

$$A'(0.5) \approx \frac{A(0.5001) - A(0.4999)}{0.5001 - 0.4999}$$

$$\approx \frac{1562.577 - 1562.422}{0.0002}$$

$$\approx 775 \text{ points per day}$$

67. $R(p) = \dfrac{r}{1+kp} = \dfrac{45}{1+0.125p}$

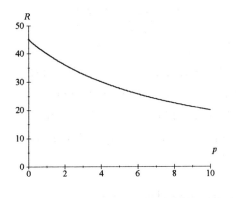

$$R'(4) \approx \frac{R(4.01) - R(3.99)}{4.01 - 3.99}$$
$$\approx \frac{29.9750 - 30.0250}{0.02}$$
$$\approx -2.5$$

$R'(4) = -2.5$ thousand organisms per hour, per 1,000 new organisms. This means that the reproduction rate of organisms in a culture containing 4,000 organisms is declining by 2,500 organisms per hour for each 1,000 new organisms.

69. $l(p) = l_0\sqrt{1-p^2} = 100\sqrt{1-p^2}$

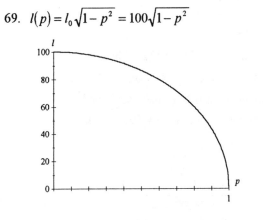

$$l'(0.95) \approx \frac{l(0.951) - l(0.949)}{0.951 - 0.949}$$
$$\approx \frac{30.9191 - 31.5276}{0.002}$$
$$\approx -304.25$$

$l(0.95) = 31.22$ meters and $l'(0.95) = -304.24$ meters/warp. Thus, at a speed of warp 0.95, the rocket has an observed length of 31.22 meters and its length is decreasing at a rate of 304.24 meters per unit warp for each 3.0424 meters per 1% increase in the speed (measured in warp).

71. $S(t) = 200 - 150e^{-t/10}$

73. $q(t) = \dfrac{N}{1 + ke^{-rt}} = \dfrac{10{,}000}{1 + 0.5e^{-0.4t}}$

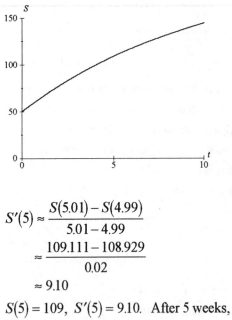

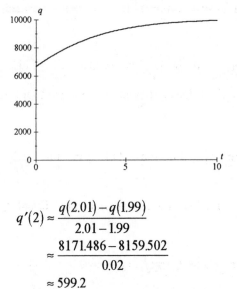

$$S'(5) \approx \frac{S(5.01) - S(4.99)}{5.01 - 4.99}$$
$$\approx \frac{109.111 - 108.929}{0.02}$$
$$\approx 9.10$$

$S(5) = 109$, $S'(5) = 9.10$. After 5 weeks, sales are 109 sneakers per week, and sales are increasing at a rate of 9.10 sneakers per week each week.

$$q'(2) \approx \frac{q(2.01) - q(1.99)}{2.01 - 1.99}$$
$$\approx \frac{8171.486 - 8159.502}{0.02}$$
$$\approx 599.2$$

$q(2) = 8165$, $q'(2) = 599.2$. Thus, two months after the introduction, 8,165 video game units have been sold, and the demand is growing by 599.2 units per month.

10.3 EXERCISES

1.

x	−0.1	−0.01	−0.001	0	0.001	0.01	0.1
$f(x) = \dfrac{x^2}{x+1}$	0.011	0.0001	0.000001		0.000000999	0.000099	0.00909

$$\lim_{x \to 0} \frac{x^2}{x+1} = 0$$

3.

x	1.9	1.99	1.999	1.9999	2	2.0001	2.001	2.01	2.1
$f(x) = \dfrac{x^2 - 4}{x - 2}$	3.9	3.99	3.999	3.9999		4.0001	4.001	4.01	4.1

$$\lim_{x \to 2} \frac{x^2 - 4}{x - 2} = 4$$

5. (a) $\lim\limits_{x\to 1^+} f(x)=2,\ \lim\limits_{x\to 1^-} f(x)=2 \Rightarrow \lim\limits_{x\to 1} f(x)=2$ (b) $\lim\limits_{x\to -1^+} f(x)=-1,\ \lim\limits_{x\to -1^-} f(x)=-1 \Rightarrow \lim\limits_{x\to -1} f(x)=-1$

7. (a) $\lim\limits_{x\to 0^+} f(x)=1,\ \lim\limits_{x\to 0^-} f(x)=1 \Rightarrow \lim\limits_{x\to 0} f(x)=1$ (b) $\lim\limits_{x\to 1^+} f(x)=1,\ \lim\limits_{x\to 1^-} f(x)=1 \Rightarrow \lim\limits_{x\to 1} f(x)=1$

9. (a) $\lim\limits_{x\to 2^+} f(x)=0,\ \lim\limits_{x\to 2^-} f(x)=0 \Rightarrow \lim\limits_{x\to 2} f(x)=0$ (b) $\lim\limits_{x\to 0^+} f(x)=2,$ (c) $\lim\limits_{x\to 0^-} f(x)=-1$
   (d) $\lim\limits_{x\to 0} f(x)$ does not exist, since the left and right limits disagree.   (e) $f(0)=2$

11. (a) $\lim\limits_{x\to -2^+} f(x)=1,\ \lim\limits_{x\to -2^-} f(x)=1 \Rightarrow \lim\limits_{x\to -2} f(x)=1$ (b) $\lim\limits_{x\to -1^+} f(x)=1,$ (c) $\lim\limits_{x\to -1^-} f(x)=2$

    (d) $\lim\limits_{x\to -1} f(x)$ does not exist, since the left and right limits disagree $x\to -1$.   (e) $f(-1)=1$

13. (a) $\lim\limits_{x\to -1^+} f(x)=1,\ \lim\limits_{x\to -1^-} f(x)=1 \Rightarrow \lim\limits_{x\to -1} f(x)=1$ (b) $\lim\limits_{x\to 0^+} f(x)$ does not exist (c) $\lim\limits_{x\to 0^-} f(x)$ does not exist
    (d) $\lim\limits_{x\to 0} f(x)$ does not exist (e) $f(0)$ is undefined

15. (a) $\lim\limits_{x\to -2^+} f(x)=1,\ \lim\limits_{x\to -2^-} f(x)=1 \Rightarrow \lim\limits_{x\to -2} f(x)=1$ (b) $\lim\limits_{x\to 0^+} f(x)$ does not exist (c) $\lim\limits_{x\to 0^-} f(x)=-1$
    (d) $\lim\limits_{x\to 0} f(x)$ does not exist (e) $f(0)=2$ (f) $f(-2)=0$

17. (a) $\lim\limits_{x\to -0^+} f(x)$ does not exist $\Rightarrow \lim\limits_{x\to 0} f(x)$ does not exist
    (b) $\lim\limits_{x\to 1^+} f(x)=0,\ \lim\limits_{x\to 1^-} f(x)=0 \Rightarrow \lim\limits_{x\to 1} f(x)=0$ (c) $f(0)=1.5$

19. (a) $\lim\limits_{x\to 1^+} f(x)=1,\ \lim\limits_{x\to 1^-} f(x)=1 \Rightarrow \lim\limits_{x\to 1} f(x)=1$ (b) $\lim\limits_{x\to 2^+} f(x)=0,\ \lim\limits_{x\to 2^-} f(x)=0 \Rightarrow \lim\limits_{x\to 2} f(x)=0$

21. continuous on its domain

23. continuous on its domain

25. not continuous at its domain; discontinuous at $x=0$

27. not continuous at its domain; discontinuous at $x=-1$

29. continuous on its domain

31. not continuous on its domain; discontinuous at $x=0$ and $x=-2$

33. not continuous on its domain; discontinuous at $x = 0$

35. continuous on its domain

37. $\lim\limits_{x \to 0} (x + 1) = 1$

39. $\lim\limits_{x \to 2} \dfrac{2 + x}{x} = 2$

41. $\lim\limits_{x \to -1} \dfrac{x + 1}{x} = 0$

43. $\lim\limits_{x \to 8} \left(x - \sqrt[3]{x} \right) = 6$

45. $\lim\limits_{h \to 1} \left(h^2 + 2h + 1 \right) = (1)^2 + 2(1) + 1 = 4$

47. $\lim\limits_{h \to 3} 2 = 2$

49. $\lim\limits_{h \to 0} \dfrac{h^2}{h + h^2} = \lim\limits_{h \to 0} \dfrac{h}{1 + h} = \dfrac{0}{1 + 0} = 0$

51. $\lim\limits_{x \to 1} \dfrac{x^2 - 2x + 1}{x^2 - x} = \lim\limits_{x \to 1} \dfrac{(x - 1)(x - 1)}{x(x - 1)} = \lim\limits_{x \to 1} \dfrac{x - 1}{x} = \dfrac{(1) - 1}{1} = 0$

53. $\lim\limits_{x \to 2} \dfrac{x^3 - 8}{x - 2} = \lim\limits_{x \to 2} \dfrac{(x - 2)(x^2 + 2x + 4)}{x - 2} = \lim\limits_{x \to 2} \left(x^2 + 2x + 4 \right) = 2^2 + 2(2) + 4 = 12$

In Exercises 55 - 77 use $f'(x) = \lim\limits_{h \to 0} \dfrac{f(x + h) - f(x)}{h}$.

55. $f(x) = -14$

$f'(x) = \lim\limits_{h \to 0} \dfrac{-14 - (-14)}{h} = \lim\limits_{h \to 0} \dfrac{0}{h} = 0$

57. $f(x) = 2x - 3$

$f'(x) = \lim\limits_{h \to 0} \dfrac{2(x + h) - 3 - (2x - 3)}{h} = \lim\limits_{h \to 0} \dfrac{2h}{h} = \lim\limits_{h \to 0} 2 = 2$

59. $g(x) = -4x - 1$

$g'(x) = \lim\limits_{h \to 0} \dfrac{-4(x + h) - 1 - (-4x - 1)}{h} = \lim\limits_{h \to 0} \dfrac{-4h}{h} = \lim\limits_{h \to 0} (-4) = -4$

270

61. $g(x) = x^2 - 2x$

$$g'(x) = \lim_{h \to 0} \frac{(x+h)^2 - 2(x+h) - (x^2 - 2x)}{h}$$

$$= \lim_{h \to 0} \frac{x^2 + 2xh + h^2 - 2x - 2h - x^2 + 2x}{h}$$

$$= \lim_{h \to 0} \frac{2xh + h^2 - 2h}{h}$$

$$= \lim_{h \to 0} (2x - 2 + h)$$

$$= 2x - 2$$

63. $h(x) = -5x^2 + 2x - 1$

$$h'(x) = \lim_{h \to 0} \frac{-5(x+h)^2 + 2(x+h) - 1 - (-5x^2 + 2x - 1)}{h}$$

$$= \lim_{h \to 0} \frac{-5x^2 - 10xh - 5h^2 + 2x + 2h - 1 + 5x^2 - 2x + 1}{h}$$

$$= \lim_{h \to 0} \frac{-10xh + 2h - 5h^2}{h}$$

$$= \lim_{h \to 0} -10x + 2 - 5h$$

$$= -10x + 2$$

65. $f(t) = t^3 + t$

$$f'(t) = \lim_{h \to 0} \frac{(t+h)^3 + (t+h) - (t^3 + t)}{h}$$

$$= \lim_{h \to 0} \frac{t^3 + 3t^2 h + 3th^2 + h^3 + t + h - t^3 - t}{h}$$

$$= \lim_{h \to 0} \frac{3t^2 h + 3th^2 + h^3 + h}{h}$$

$$= \lim_{h \to 0} (3t^2 + 1 + 3th + h^2)$$

$$= 3t^2 + 1$$

67. $g(t) = t^4 - t$

$$g'(t) = \lim_{h \to 0} \frac{(t+h)^4 - (t+h) - (t^4 - t)}{h}$$

$$= \lim_{h \to 0} \frac{t^4 + 4t^3 h + 6t^2 h^2 + 4th^3 + h^4 - t - h - t^4 + t}{h}$$

$$= \lim_{h \to 0} \frac{4t^3 h - h + 6t^2 h^2 + 4th^3 + h^4}{h}$$

$$= \lim_{h \to 0} (4t^3 - 1 + 6t^2 h + 4th^2 + h^3)$$

$$= 4t^3 - 1$$

69. $h(t) = \dfrac{6}{t}$

$$h'(t) = \lim_{h \to 0} \frac{\dfrac{6}{t+h} - \dfrac{6}{t}}{h} = \lim_{h \to 0} \frac{6t - 6(t+h)}{ht(t+h)} = \lim_{h \to 0} \frac{-6h}{ht(t+h)} = \lim_{h \to 0} \frac{-6}{t^2 + th} = \frac{-6}{t^2}$$

71. $f(x) = x + \dfrac{1}{x}$

$$f'(x) = \lim_{h\to 0} \frac{x+h+\dfrac{1}{x+h}-\left(x+\dfrac{1}{x}\right)}{h}$$

$$= \lim_{h\to 0} \frac{h+\dfrac{x-(x+h)}{x(x+h)}}{h}$$

$$= \lim_{h\to 0} \frac{h}{h} + \frac{-h}{hx(x+h)}$$

$$= \lim_{h\to 0} 1 + \frac{-1}{x^2+xh}$$

$$= 1 - \frac{1}{x^2}$$

73. $f(x) = \dfrac{1}{x-2}$

$$f'(x) = \lim_{h\to 0} \frac{\dfrac{1}{x+h-2}-\dfrac{1}{x-2}}{h}$$

$$= \lim_{h\to 0} \frac{(x-2)-(x+h-2)}{h(x-2)(x+h-2)}$$

$$= \lim_{h\to 0} \frac{-h}{h(x-2)(x+h-2)}$$

$$= \lim_{h\to 0} \frac{-1}{(x-2)(x+h-2)}$$

$$= -\frac{1}{(x-2)^2}$$

75. $f(x) = \sqrt{x+1}$

$$f'(x) = \lim_{h\to 0} \frac{\sqrt{x+h+1}-\sqrt{x+h}}{h}$$

$$= \lim_{h\to 0} \frac{\sqrt{x+h+1}-\sqrt{x+1}}{h} \cdot \frac{\sqrt{x+h+1}+\sqrt{x+1}}{\sqrt{x+h+1}+\sqrt{x+1}}$$

$$= \lim_{h\to 0} \frac{x+h+1-(x+1)}{h\left(\sqrt{x+h+1}+\sqrt{x+1}\right)}$$

$$= \lim_{h\to 0} \frac{h}{h\left(\sqrt{x+h+1}+\sqrt{x+1}\right)}$$

$$= \lim_{h\to 0} \frac{1}{\sqrt{x+h+1}+\sqrt{x+1}}$$

$$= \frac{1}{2\sqrt{x+1}}$$

77. $g(t) = \dfrac{1}{\sqrt{t}}$

$$g'(t) = \lim_{h\to 0} \frac{\dfrac{1}{\sqrt{t+h}}-\dfrac{1}{\sqrt{t}}}{h}$$

$$= \lim_{h\to 0} \frac{\sqrt{t}-\sqrt{t+h}}{h\sqrt{t}\sqrt{t+h}}$$

$$= \lim_{h\to 0} \frac{\sqrt{t}-\sqrt{t+h}}{h\sqrt{t}\sqrt{t+h}} \cdot \frac{\sqrt{t}+\sqrt{t+h}}{\sqrt{t}+\sqrt{t+h}}$$

$$= \lim_{h\to 0} \frac{t-(t+h)}{h\sqrt{t}\sqrt{t+h}\left(\sqrt{t}+\sqrt{t+h}\right)}$$

$$= \lim_{h\to 0} \frac{-h}{h\sqrt{t}\sqrt{t+h}\left(\sqrt{t}+\sqrt{t+h}\right)}$$

$$= \lim_{h\to 0} \frac{-1}{\sqrt{t}\sqrt{t+h}\left(\sqrt{t}+\sqrt{t+h}\right)}$$

$$= -\frac{1}{2t\sqrt{t}}$$

10.4 EXERCISES

In Exercises 1-13 use the rule for constant multiples and the power rule $f'(x) = \dfrac{d}{dx}(x^n) = nx^{n-1}$.

1. $f(x) = x^3$

   $f'(x) = 3x^2$

3. $f(x) = 2x^{-2}$

   $f'(x) = 2(-2x^{-3}) = -4x^{-3}$

5. $f(x) = -x^{1/4}$

   $f'(x) = -\dfrac{1}{4}x^{-3/4}$

7. $f(x) = 2x^4 + 3x^3 - 1$

   $f'(x) = 2(4x^3) + 3(3x^2) - 0 = 8x^3 + 9x^2$

9. $f(x) = -x + \dfrac{1}{x} + 1 = -x^1 + x^{-1} + 1$

   $f'(x) = -1(x^0) + (-1)x^{-2} + 0 = -1 - \dfrac{1}{x^2}$

11. $f(x) = 2\sqrt{x} = 2x^{1/2}$

   $f'(x) = 2\left(\dfrac{1}{2}x^{-1.2}\right) = \dfrac{1}{\sqrt{x}}$

13. $f(x) = 3\sqrt[3]{x} = 3x^{1/3}$

   $f'(x) = 3\left(\dfrac{1}{3}x^{-2/3}\right) = x^{-2/3}$

15. $y = 10 = 10x^0$

   $\dfrac{dy}{dx} = 10\dfrac{d}{dx}(x^0)$ (constant multiples)

   $= 10 \cdot 0 = 0$ (power rule)

17. $y = x^2 + x$

   $\dfrac{dy}{dx} = \dfrac{d}{dx}(x^2) + \dfrac{d}{dx}(x^1)$ (sum)

   $= 2x + 1$ (power rule)

19. $y = 4x^3 + 2x - 1$

   $\dfrac{dy}{dx} = \dfrac{d}{dx}(4x^3) + \dfrac{d}{dx}(2x^1) - \dfrac{d}{dx}(1x^0)$ (sum and difference)

   $= 4\dfrac{d}{dx}(x^3) + 2\dfrac{d}{dx}(x^1) - \dfrac{d}{dx}(x^0)$ (constant multiples)

   $= 12x^2 + 2$ (power rule)

21. $y = x^{104} - 99x^2 + x$

   $\dfrac{dy}{dx} = \dfrac{d}{dx}(x^{104}) - \dfrac{d}{dx}(99x^2) + \dfrac{d}{dx}(x^1)$ (sum and difference)

   $= \dfrac{d}{dx}(x^{104}) - 99\dfrac{d}{dx}(x^2) + \dfrac{d}{dx}(x^1)$ (constant multiples)

   $= 104x^{103} - 198x + 1$ (power rule)

273

23. $s = \sqrt{t}\left(t - t^3\right) + \dfrac{1}{t^3} = t^{1/2}\left(t - t^3\right) + t^{-3} = t^{3/2} - t^{7/2} + t^{-3}$

$\dfrac{ds}{dt} = \dfrac{d}{dt}\left(t^{3/2}\right) - \dfrac{d}{dt}\left(t^{7/2}\right) + \dfrac{d}{dt}\left(t^{-3}\right)$ (sum and difference)

$\qquad = \dfrac{3}{2}t^{1/2} - \dfrac{7}{2}t^{5/2} - 3t^{-4}$ (power rule)

25. $V = \dfrac{4\pi r^3}{3}$

$\dfrac{dV}{dr} = \dfrac{4\pi}{3}\dfrac{d}{dr}\left(r^3\right)$ (constant multiples)

$\qquad = 2\pi r^2$ (power rule)

27. $\dfrac{d}{dt}\left(t^2 + 4at^5\right) = \dfrac{d}{dt}\left(t^2\right) + \dfrac{d}{dt}\left(4at^5\right)$ (sum)

$\qquad = \dfrac{d}{dt}\left(t^2\right) + 4a\dfrac{d}{dt}\left(t^5\right)$ (constant multiples)

$\qquad = 2t + 20at^4$ (power rule)

29. $\dfrac{d}{dx}\left[\sqrt{x}\left(1 + x\right)\right] = \dfrac{d}{dx}\left(x^{1/2} + x^{3/2}\right) = \dfrac{d}{dx}\left(x^{1/2}\right) + \dfrac{d}{dx}\left(x^{3/2}\right)$ (sum)

$\qquad = \dfrac{1}{2}x^{-1/2} + \dfrac{3}{2}x^{1/2}$ (power rule)

In Exercises 31 - 41 use slope of tangent to $f(x)$ at $\left(x_0, y_0\right)$ is $f'\left(x_0\right)$.

31. $f(x) = x^3;\ (-1,\,-1)$ 33. $f(x) = 1 - 2x;\ (2,\,-3)$

$\quad f'(x) = 3x^2$ $\quad f'(x) = -2$

$\quad f'(-1) = 3(-1)^2 = 3$ $\quad f'(2) = -2$

35. $h(x) = x^{1/4}$; $(16, 2)$

$h'(x) = \frac{1}{4}x^{-3/4}$

$h'(16) = \frac{1}{4}(16)^{-3/4} = \frac{1}{32}$

37. $f(x) = x^{-1/2}$; $\left(2, 2^{-1/2}\right)$

$f'(x) = -\frac{1}{2}x^{-3/2}$

$f'(2) = -\frac{1}{2}(2)^{-3/2} = -\frac{1}{4\sqrt{2}}$

39. $g(t) = \frac{1}{t^5} = t^{-5}$; $(1, 1)$

$g'(t) = -5t^{-6}$

$g'(1) = -5(1)^{-6} = -5$

41. $f(x) = \frac{x^2}{4} - \frac{x^3}{3}$; $\left(-1, \frac{7}{12}\right)$

$f'(x) = \frac{1}{2}x - x^2$

$f'(-1) = \frac{1}{2}(-1) - (-1)^2 = -\frac{3}{2}$

In Exercises 43 - 47 use $y - y_1 = m(x - x_1)$ to find the equation of the tangent line.

43. $f(x) = x^3$, $x = -1$

$f'(x) = 3x^2$

$m = f'(-1) = 3(-1)^2 = 3$

$y = f(-1) = (-1)^3 = -1$

$y - (-1) = 3[x - (-1)]$

$y = 3x + 2$

45. $f(x) = x + \frac{1}{x} = x + x^{-1}$; $x = 2$

$f'(x) = 1 - x^{-2}$

$m = f'(2) = 1 - (2)^{-2} = \frac{3}{4}$

$y = f(2) = 2 + \frac{1}{2} = \frac{5}{2}$

$y - \frac{5}{2} = \frac{3}{4}(x - 2)$

$4y - 10 = 3x - 6$

$3x - 4y = -4$

47. $f(x) = \sqrt{x} = x^{1/2}$, $x = 4$

$f'(x) = \frac{1}{2}x^{-1/2}$

$m = f'(4) = \frac{1}{2}(4)^{-1/2} = \frac{1}{4}$

$y = f(4) = \sqrt{4} = 2$

$y - 2 = \frac{1}{4}(x - 4)$

$4y - 8 = x - 4$

$4y - 8 = x - 4$

$x - 4y = -4$

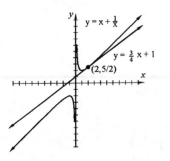

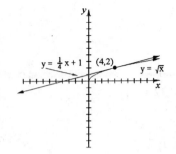

49. $f(x) = x^2 - 3x + 5$

$f'(x) = 2x - 3$

51. $f(x) = x + \sqrt{x} = x + x^{1/2}$

$f'(x) = 1 + \dfrac{1}{2}x^{-1/2}$

$= 1 + \dfrac{1}{2\sqrt{x}}$

53. $g(x) = \dfrac{1}{x^2} + \dfrac{2}{x^3} = x^{-2} + 2x^{-3}$

$g'(x) = -2x^{-3} - 6x^{-4}$

$= -\dfrac{2}{x^3} - \dfrac{6}{x^4}$

55. $h(x) = \dfrac{1}{x} + \dfrac{1}{x^2} + \dfrac{1}{x^3} = x^{-1} + x^{-2} + x^{-3}$

$h'(x) = -x^{-2} - 2x^{-3} - 3x^{-4}$

$= -\dfrac{1}{x^2} - \dfrac{2}{x^3} - \dfrac{3}{x^4}$

57. $r(x) = \sqrt{x} + \dfrac{1}{\sqrt{x}} = x^{1/2} + x^{-1/2}$

$r'(x) = \dfrac{1}{2}x^{-1/2} - \dfrac{1}{2}x^{-3/2}$

$= \dfrac{1}{2\sqrt{x}} - \dfrac{1}{2x\sqrt{x}}$

59. $f(x) = x(x^2 - 1/x) = x^3 - 1$

$f'(x) = 3x^2$

61. $g(x) = \dfrac{x^2 - 2x^3}{x} = x - 2x^2$

$g'(x) = 1 - 4x$

63. $\dfrac{d}{dx}\left(x + \dfrac{1}{x^2}\right) = \dfrac{d}{dx}\left(x + x^{-2}\right)$

$= 1 - 2x^{-3}$

$= 1 - \dfrac{2}{x^3}$

65. $\dfrac{d}{dx}\left(2x^{1.3} - x^{-1.2}\right) = 2.6x^{0.3} + 1.2x^{-2.2}$

67. $\dfrac{d}{dt}\left(at^3 - 4at\right) = 3at^2 - 4a$

69. $\dfrac{d}{dx}\left[\sqrt{x}(1+x)\right] = \dfrac{d}{dx}\left(x^{1/2} + x^{3/2}\right)$

$\qquad\qquad\qquad = \dfrac{1}{2}x^{-1/2} + \dfrac{3}{2}x^{1/2}$

$\qquad\qquad\qquad = \dfrac{1}{2\sqrt{x}} + \dfrac{3\sqrt{x}}{2}$

71. $y = \dfrac{x^{10.3}}{2} + 99x^{-1}$

$\qquad \dfrac{dy}{dx} = 5.15x^{9.3} - 99x^{-2}$

73. $s = 2.35 + \dfrac{2.1}{t^{1.1}} + t^{-1.2}$

$\qquad = 2.35 + 2.1t^{-1.1} + t^{-1.2}$

$\qquad \dfrac{ds}{dt} = -2.31t^{-2.1} - 1.2t^{-2.2}$

75. $V = \dfrac{4}{3}\pi r^3$

$\qquad \dfrac{dV}{dr} = 4\pi r^2$

In Exercises 77-83 use $m = \dfrac{dy}{dx} = 0$ for the slope of the tangent line to the graph when the tangent line is horizontal.

77. $y = 2x^2 + 3x - 1$

$\qquad m = \dfrac{dy}{dx} = 4x + 3$

$\qquad 0 = 4x + 3$

$\qquad x = -\dfrac{3}{4}$

78. $y = -3x^2 - x$

$\qquad m = \dfrac{dy}{dx} = -6x - 1$

$\qquad 0 = -6x - 1$

$\qquad x = -\dfrac{1}{6}$

79. $y = 2x + 8$

$\qquad m = \dfrac{dy}{dx} = 2$

$\qquad 0 \ne 2$

There is no such value.

81. $y = x + \dfrac{1}{x} = x + x^{-1}$

$m = \dfrac{dy}{dx} = 1 - x^{-2}$

$0 = 1 - \dfrac{1}{x^2}$

$0 = x^2 - 1$

$x = \pm 1$

83. $y = \sqrt{x} - x = x^{1/2} - x$

$m = \dfrac{dy}{dx} = \dfrac{1}{2}x^{-1/2} - 1$

$0 = \dfrac{1}{2\sqrt{x}} - 1$

$2\sqrt{x} = 1$

$\sqrt{x} = \dfrac{1}{2}$

$x = \dfrac{1}{4}$

85. $\dfrac{d}{dx}\left[x^4\right] = \lim\limits_{h \to 0} \dfrac{(x+h)^4 - x^4}{h} = \lim\limits_{h \to 0} \dfrac{x^4 + 4x^3h + 6x^2h^2 + 4xh^3 + h^4 - x^4}{h}$

$= \lim\limits_{h \to 0}\left(4x^3 + 6x^2h + 4xh^2 + h^3\right)$

$= 4x^3$

87. $\dfrac{d}{dx}\left[3x^2\right] = \lim\limits_{h \to 0} \dfrac{3(x+h)^2 - 3x^2}{h} = \lim\limits_{h \to 0} 3\left[\dfrac{(x+h)^2 - x^2}{h}\right] = 3\lim\limits_{h \to 0}\left[\dfrac{(x+h)^2 - x^2}{h}\right] = 3\dfrac{d}{dx}\left[x^2\right]$

89. $\dfrac{d}{dx}\left[x^2 + x^3\right] = \lim\limits_{h \to 0} \dfrac{(x+h)^2 + (x+h)^3 - (x^2 + x^3)}{h}$

$= \lim\limits_{h \to 0} \dfrac{(x+h)^2 - x^2 + (x+h)^3 - x^3}{h}$

$\lim\limits_{h \to 0}\left[\dfrac{(x+h)^2 - x^2}{h} + \dfrac{(x+h)^3 - x^3}{h}\right]$

$= \lim\limits_{h \to 0}\left[\dfrac{(x+h)^2 - x^2}{h}\right] + \lim\limits_{h \to 0}\left[\dfrac{(x+h)^3 - x^3}{h}\right]$

$= \dfrac{d}{dx}\left[x^2\right] + \dfrac{d}{dx}\left[x^3\right]$

91. $C_1(x) = 10,000 + 5x - x^2 / 10$

$C_1'(x) = 5 - x / 5x$

$C_2(x) = 20,000 + 10x - x^2 / 5$

$C_2'(x) = 10 - 2x / 5$

The rate of change of C_2 is twice the rate of change of C_1.

93. Profit = Revenue – Cost

$P(x) = R(x) - C(x)$

$P'(x) = R'(x) - C'(x)$

When the rate of change of profit is zero,

$0 = R'(x) - C'(x)$

$R'(x) = C'(x)$

The rates of change of cost and revenue must be equal. This means that the next case costs as much to make as it will bring in revenue.

95. $A(t) = 1000 + 1500t - 800t^2 + 100t^3$

$A'(t) = 1500 - 1600t + 300t^2$

$A'(0.5) = 1500 - 1600(0.5) + 300(0.5)^2$

$\qquad = 775$ points per day

97. $C(t) = -0.00271t^3 + 0.137t^2 - 0.892t + 0.149 \quad (8 \le t \le 30)$

$C'(t) = -0.00813t^2 + 0.274t - 0.892$

$C'(15) = -0.00813(15)^2 + 0.274(15) - 0.892 \approx 1.389$

$C'(30) = -0.00813(30)^2 + 0.274(30) - 0.892 \approx 0.011$

The hourly oxygen consumption is increasing at a slower rate when a chick hatches. (Notice that the oxygen consumption is still increasing, however).

99. $s = 100 - 16t^2$

(a) velocity $= v(t) = \dfrac{ds}{dt} = -32t$

$v(0) = -32(0) = 0$ ft / s

$v(1) = -32(1) = -32$ ft / s

$v(2) = -32(2) = -64$ ft / s

$v(3) = -32(3) = -96$ ft / s

$v(4) = -32(4) = -128$ ft / s

(b) At ground level, $s = 0$

$0 = 100 - 16t^2$

$t^2 = \dfrac{100}{16}$

$t = \dfrac{5}{2} = 2.5s$

$v\left(\dfrac{5}{2}\right) = -32\left(\dfrac{5}{2}\right) = -80$ ft / s

101. $V(r) = \dfrac{4}{3}\pi r^3$

$\quad V'(r) = 4\pi r^2$

$\quad V'(10) = 4(3.141592)(10)^2$

$\qquad \approx 1256.64$ cubic centimeters

per centimeter increase in r. When $r = 10$,
the volume increases by 1,256.63 cubic
centimeters for every 1 centimeter increase
in the radius.

10.5 EXERCISES

In Exercises 1-5 use $\dfrac{C(x+h)-C(x)}{h}$ to find the average cost $C_{ave}(x)$ and $C'(x)$ to find the marginal cost.

1. $\quad C(x) = 10,000 + 5x - \dfrac{x^2}{10,000};\ x = 1000$

h	50	10	1
C_{ave}	4.795	4.799	4.7999

$C_{ave}(x) = \dfrac{10,000 + 5(x+h) - \dfrac{(x+h)^2}{10,000} - \left[10,000 + 5x - \dfrac{x^2}{10,000}\right]}{h}$

$\qquad = \dfrac{5x + 5h - \dfrac{x^2 + 2xh + h^2}{10,000} - 5x + \dfrac{x^2}{10,000}}{h}$

$\qquad = \dfrac{5h - \dfrac{2xh}{10,000} - \dfrac{h^2}{10,000}}{h}$

$\qquad = 5 - \dfrac{x}{5000} - \dfrac{h}{10,000}$

$C_{ave}(1000) = 5 - \dfrac{1}{5} - \dfrac{h}{10,000}$

$C'(x) = 5 - \dfrac{x}{5000}$

$C'(1000) = 5 - \dfrac{1000}{5000} = 4.8$

3. $C(x) = 1000 + 10x - \sqrt{x};\ x = 100$

$$C_{ave}(x) = \frac{1000 + 10(x+h) - \sqrt{x+h} - \left(1000 + 10x - \sqrt{x}\right)}{h}$$

$$= \frac{10h - \sqrt{x+h} + \sqrt{x}}{h}$$

$$C_{ave}(100) = \frac{10h - \sqrt{100+h} + 10}{h}$$

h	50	10	1
C_{ave}	9.955	9.951	9.950

$$C'(x) = 10 - \frac{1}{2}x^{-1/2} = 10 - \frac{1}{2\sqrt{x}}$$

$$C'(100) = 10 - \frac{1}{2\sqrt{100}} = 9.950$$

5. $C(x) = 15{,}000 + 100x + \dfrac{1000}{x};\ x = 100$

$$C_{ave}(x) = \frac{15{,}000 + 100(x+h) + \frac{1000}{x+h} - \left(15{,}000 + 100x + \frac{1000}{x}\right)}{h}$$

$$= \frac{100h + \frac{1000}{x+h} - \frac{1000}{x}}{h}$$

$$C_{ave}(100) = \frac{100h + \frac{1000}{100+h} - 10}{h}$$

h	50	10	1
C_{ave}	99.93	99.91	99.90

$$C'(x) = 100 - \frac{1000}{x^2}$$

$$C'(100) = 100 - \frac{1000}{(100)^2} = 99.9$$

7. $C(x) = 4x;\ R(x) = 8x - \dfrac{x^2}{1000}$

$C'(x) = 4;\ R'(x) = 8 - \dfrac{x}{500}$

$$P'(x) = R'(x) - C'(x) = 8 - \frac{x}{500} - 4$$

$$= 4 - \frac{x}{500}$$

When $P'(x) = 0$

$$0 = 4 - \frac{x}{500}$$

$$x = 2000$$

At a production level of 2,000, the profit is stationary (neither increasing nor decreasing) with respect to the production level. This may indicate a maximum profit at a production level of 2,000.

9. $C(x) = 100 + 40x - 0.001x^2$

$C'(x) = 40 - 0.002x$

$C'(100) = 40 - 0.002(100) = 39.80$

$$C(101) = 100 + 40(101) - 0.001(101)^2$$
$$= 4{,}129.799$$

$$C(100) = 100 + 40(100) - 0.001(100)^2$$
$$= 4{,}090.000$$

The cost is going up at a rate of \$39.80 per teddy bear. The cost of producing the teddy is $C(101) - C(100) = \$39.799$.

11. $P(1000) = 3000$ and $P'(1000) = -3$

The profit on the sale of 1,000 videocassettes is $3,000 and is decreasing by $3.00 per additional videocassette sold.

13.
$$P = 5n + \sqrt{n} = 5n + n^{1/2}$$
$$P' = 5 + \frac{1}{2}n^{-1/2} = 5 + \frac{1}{2\sqrt{n}}$$
$$P(50) = 5(50) + \sqrt{50} = 257.07$$
$$P'(50) = 5 + \frac{1}{2\sqrt{50}} = 5.07$$

Your current profit is $257.07 per month, and this would increase by $5.07 for each increase by one magazine in sales.

15.
$$P = 400n - 0.5n^2$$
$$P' = 400 - n$$
$$P'(50) = 400 - 50 = 350$$

This means that, at an employment level of 50 workers, the firm's daily profit will increase by $350 per additional worker it hires.

17.
$$p(q) = \frac{50{,}000}{q^{1.5}}$$

(a) $p(500) = \dfrac{50{,}000}{(500)^{1.5}} = \$4.47$ per pound

(b) $R(q) = pq = \dfrac{50{,}000}{q^{1.5}}q = \dfrac{50{,}000}{q^{0.5}}$

(c) $R(500) = \dfrac{50{,}000}{(500)^{0.5}} = \$2{,}236.07$

This is the monthly revenue that results from setting the price at $4.47 per pound.

(d) $R(q) = 50{,}000q^{-0.5}$
$$R'(q) = -25{,}000q^{-1.5} = -\frac{25{,}000}{q^{1.5}}$$
$$R'(500) = -\frac{25{,}000}{(500)^{1.5}} = -2.24$$

At a demand level of 500 pounds of tuna, the revenue is decreasing by $2.24 per additional pound.

(e) The fishery should raise the price in order to decrease demand, since the revenue drops when the price is lowered, even though the sales increase. It would be more economical to store the excess tuna.

19. $q = 63.15 - 0.45p + 0.12b$

If $b = 45$, then
$$q = 63.15 - 0.45p + 0.12(45)$$
$$q = 68.55 - 0.45p$$
$$0.45p = 68.55 - q$$
$$p = 152.33 - 2.22q$$

(a) $R(q) = pq = (152.33 - 2.22q)q$
$$= 152.33q - 2.22q^2$$
$$R'(q) = 152.33 - 4.44q$$

(b) $R'(50) = 152.33 - 4.44(50)$
$$= -69.67 \text{ cents per pound}$$

If the price is raised, causing a decrease in demand of one pound of poultry, the revenue will increase by 69.67 cents.

(c) $P(q) = R(q) - C(q)$
$$= 152.33q - 2.22q^2 - 10q$$
$$= 142.33q - 2.22q^2$$
$$P'(q) = 142.33 - 4.44q$$
$$P'(50) = 142.33 - 4.44(50)$$
$$= -79.67 \text{ cents per pound}$$

This means that if the farmer lowers the price to increase the per capita demand by one pound the annual profit will decrease by 79.67 cents.

21. $q = -\dfrac{4p}{3} + 80$

(a) $R(p) = qp = \left(-\dfrac{4p}{3} + 80\right)p$
$$= -\dfrac{4p^2}{3} + 80p$$

(b) $R'(p) = -\dfrac{8p}{3} + 80$

(i) $R'(20) = -\dfrac{8(20)}{3} + 80$
$$= \$26.67 \text{ per \$1 increase in price}$$

(ii) $R'(30) = -\dfrac{8(30)}{3} + 80$
$$= 0 \text{ per \$1 increase in price}$$

(iii) $R'(40) = -\dfrac{8(40)}{3} + 80$
$$= -\$26.67 \text{ per \$1 increase in price}$$

(c) If it charges \$20 per ruby it will increase its revenue by approximately \$26.67 by raising the price \$1. If it charges \$30 per ruby, the revenue is not moving with the price. Thus, if it raises the price by \$1, the revenue will not change significantly, but if it charges \$40 per ruby, the revenue would drop by approximately \$26.67 if it were to raise the price another \$1.

23. $C(q) = 4000 + 100q^2$

   (a) $C'(q) = 200q$

   $C'(10) = 200(10)$

   $= \$2000$ per one pound
   reduction in emissions

   (b) $S(q) = 500q$

   $S'(q) = 500$

   If $S'(q) = C'(q)$

   $200q = 500$

   $q = \dfrac{5}{2}$ pounds per day reduction

   (c) $N(q) = C(q) - S(q)$

   $= 4000 + 100q^2 - 500q$

This is a parabola with lowest point (vertex) given by $q = -b/2a = 500/200 = 5/2$. The net cost at this production level is $N(5/2) = \$3,375$ per day. The value of q is the same as that for part (b). The net cost to the firm is minimized at the reduction level for which the cost of controlling emissions begins to increase faster than the subsidy. This is why we get the answer by setting these two marginal increases equal to each other.

25. $M'(x) = \dfrac{3600x^{-2} - 1}{\left(3600x^{-1} + x\right)^2}$

   $M'(10) = \dfrac{3600(10)^{-2} - 1}{\left[3600(10)^{-1} + 10\right]^2}$

   $= 0.0002557$ mpg / mph

   $M'(60) = \dfrac{3600(60)^{-2} - 1}{\left[3600(60)^{-1} + 60\right]^2}$

   $= 0$ mpg / mph

   $M'(70) = \dfrac{3600(70)^{-2} - 1}{\left[3600(70)^{-1} + 70\right]^2}$

   $= -0.00001799$ mpg / mph

At a speed of 10 mph, the fuel economy is increased by a rate of 0.0002557 miles per gallon for each 1 - mph increase in speed. At a speed of 60 mph, the fuel economy is neither increasing nor decreasing with increasing speed. At 70 mph, the fuel economy is decreasing by 0.00001799 miles per gallon for each 1 - mph increase in speed. Thus 60 mph is the most fuel - efficient speed for the car.

27.
$$C = 0.3 + \frac{10}{T} - \frac{200}{T^2}$$

(a) $C(1000) = 0.3 + \frac{10}{1000} - \frac{200}{(1000)^2}$

$$= 0.0398$$

$100[C(1000)] = 100(0.0398)$

$$\approx \$3.98 \text{ per } 1000 \text{ nautical miles}$$

(b) $C'(T) = -\frac{10}{T^2} + \frac{400}{T^3}$

$C'(1000) = \frac{-10}{(1000)^2} + \frac{400}{(1000)^3}$

$$= -0.000096$$

$100C'(1000) = -\$0.00096 \approx -0.1 \text{ cent}$

29. (c)   31. (d)

10.6 EXERCISES

1. (a) $\lim_{x \to +\infty} f(x)$ diverges to $+\infty$

(b) $\lim_{x \to -\infty} f(x) = 0$

3. (a) $\lim_{x \to +\infty} f(x)$ diverges to $-\infty$

(b) $\lim_{x \to -\infty} f(x) = 0$

5. (a) $\lim_{x \to 0^-} f(x)$ diverges to $+\infty$

(b) $\lim_{x \to 0^+} f(x)$ diverges to $+\infty$

(c) $\lim_{x \to 0} f(x)$ diverges to $+\infty$

7. (a) $\lim_{x \to 0^-} f(x) = -1$

(b) $\lim_{x \to 0^+} f(x)$ diverges to $-\infty$

(c) $\lim_{x \to 0} f(x)$ does not exist since the left and right limits do not agree.

9. (a) $\lim_{x \to 0} f(x)$ does not exist since the left and right limits do not agree.

(b) $\lim_{x \to -\infty} f(x)$ diverges to $+\infty$

(c) $\lim_{x \to +\infty} f(x) = -1$

11. (a) $\lim_{x \to -\infty} f(x)$ does not exist since the graph oscillates between -1 and 1

(b) $\lim_{x \to +\infty} f(x)$ does not exist since the graph oscillates between -1 and 1

13. $\displaystyle\lim_{x\to 0^+}\frac{1}{x^2}$ diverges to $+\infty$

15.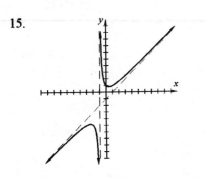

$\displaystyle\lim_{x\to -1}\frac{x^2+1}{x+1}$ does not exist since the

left and right limits do not agree.

17. $\displaystyle\lim_{x\to\infty}\frac{3x^2+10x-1}{2x^2-5x}=\lim_{x\to\infty}\frac{3x^2}{2x^2}=\frac{3}{2}$

19. $\displaystyle\lim_{x\to\infty}\frac{x^5-1000x^4}{2x^5+10,000}=\lim_{x\to\infty}\frac{x^5}{2x^5}=\frac{1}{2}$

21. $\displaystyle\lim_{x\to\infty}\frac{10x^2+300x+1}{5x+2}=\lim_{x\to\infty}\frac{10x^2}{5x}$

$=\displaystyle\lim_{x\to\infty}2x$ does not exist

23. $\displaystyle\lim_{x\to\infty}\frac{10x^2+300x+1}{5x^3+2}=\lim_{x\to\infty}\frac{10x^2}{5x^3}=\lim_{x\to\infty}\frac{2}{x}=0$

25. $\displaystyle\lim_{x\to -\infty}\frac{3x^2+10x-1}{2x^2-5x}=\lim_{x\to -\infty}\frac{3x^2}{2x^2}=\lim_{x\to -\infty}\frac{3}{2}=\frac{3}{2}$

27. $\displaystyle\lim_{x\to -\infty}\frac{x^5-1000x^4}{2x^5+10,000}=\lim_{x\to -\infty}\frac{x^5}{2x^5}=\lim_{x\to -\infty}\frac{1}{2}=\frac{1}{2}$

29. $\displaystyle\lim_{x\to -\infty}\frac{10x^2+300x+1}{5x+2}=\lim_{x\to -\infty}\frac{10x^2}{5x}=\lim_{x\to -\infty}2x$

Does not exist

31. $\displaystyle\lim_{x\to -\infty}\frac{10x^2+300x+1}{5x^3+2}=\lim_{x\to -\infty}\frac{10x^2}{5x^3}=\lim_{x\to -\infty}\frac{2}{x}=0$

33. $\lim\limits_{x \to 0} |x| = 0$

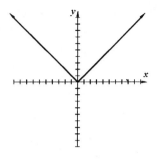

35.

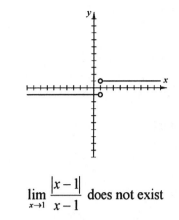

$\lim\limits_{x \to 1} \dfrac{|x-1|}{x-1}$ does not exist

37. $\lim\limits_{x \to 2} e^{x-2} = e^{2-2} = 1$

39. $\lim\limits_{x \to +\infty} xe^{-x} = 0$

x	10	100
xe^{-x}	0.000454	0

41. $\lim\limits_{x \to 2^+} \dfrac{e^{-1/(x-2)}}{x-2} = 0$

x	2.001	2.01	2.1	2.2
$\dfrac{e^{-1/(x-2)}}{x-2}$	0	0	.00045	0.034

43. $f(x) = \begin{cases} x+2 & \text{if } x < 0 \\ 2x-1 & \text{if } x \geq 0 \end{cases}$

$\lim\limits_{x \to 0^-} (x+2) = 2$

$\lim\limits_{x \to 0^+} (2x-1) = -1$

Jump discontinuity at $x = 0$

45. $g(x) = \begin{cases} x+2 & \text{if } x < 0 \\ 2x+2 & \text{if } x \geq 0 \end{cases}$

$\lim\limits_{x \to 0^-} (x+2) = 2$

$\lim\limits_{x \to 0^+} (2x+2) = 2$

$g(0) = 2$

Continuous everywhere

47. $h(x) = \begin{cases} x+2 & \text{if } x < 0 \\ 0 & \text{if } x = 0 \\ 2x+2 & \text{if } x > 0 \end{cases}$

$\lim_{x \to 0^-} (x+2) = 2$

$\lim_{x \to 0^+} (2x+2) = 2$

$h(0) = 0$

Removeable discontinuity at $x = 0$

49. $g(x) = \begin{cases} 1/x & \text{if } x < 0 \\ x & \text{if } x \geq 0 \end{cases}$

$\lim_{x \to 0^-} (1/x) = -\infty$

$\lim_{x \to 0^+} (x) = 0$

Discontinuity at $x = 0$

51. $f(x) = \begin{cases} x^2 & \text{if } x < 0 \\ x & \text{if } x \geq 0 \end{cases}$

$\lim_{x \to 0^-} \frac{d}{dx}(x^2) = \lim_{x \to 0^-} 2x = 0$

$\lim_{x \to 0^+} \frac{d}{dx}(x) = \lim_{x \to 0^+}(1) = 1$

Not differentiable at $x = 0$

53. $g(x) = x^{4/3}$

$g'(x) = \frac{4}{3} x^{1/3}$

Differentiable everywhere

55. $h(x) = |x-1|$

$\lim_{x \to 1^-} h'(x) = \lim_{x \to 1^-} \frac{d}{dx}(-x+1) = \lim_{x \to 1^-}(-1) = -1$

$\lim_{x \to 1^+} h'(x) = \lim_{x \to 1^+} \frac{d}{dx}(x-1) = \lim_{x \to 1^+}(1) = 1$

Not differentiable at $x = 1$

57. $n(t) = \dfrac{18{,}000}{(t+1)^{0.4}}$

$\lim_{t \to +\infty} n(t) = 0$

If the trend were to continue indefinitely, the annual number of DWI arrests in New Jersey will decrease to zero in the long term.

59. $p(t) = 100 \left(1 - \dfrac{12{,}196}{t^{4.478}}\right)$ $(t \geq 8.5)$

$p'(t) = -12{,}196(-4.478)t^{-5.478}$

$\qquad = 54{,}613.688 / t^{5.478}$

$\lim_{t \to +\infty} p(t) = 100$

$\lim_{t \to +\infty} p'(t) = 0$

This tells us that the percentage of children who learn to speak approaches 100% as their age increases, with the number of additional children learning to speak approaches zero.

61. $q(t) = \dfrac{N}{1 + ke^{-rt}}$

If $N = 10{,}000$, $k = 0.5$, and $r = 0.4$,

then $q(t) = \dfrac{10{,}000}{1 + 0.5e^{-0.4t}}$

$\lim_{t \to +\infty} q(t) = 10{,}000$

A total of 10,000 units are sold over the life of the game.

CHAPTER 10   REVIEW EXERCISES

1. (a) $\dfrac{dy}{dx} = 0$ at P (b) $\dfrac{dy}{dx} = 2$ at Q (c) $\dfrac{dy}{dx} = -1$ at R

3. (a) $\dfrac{dy}{dx} = 0$ at Q and R (b) $\dfrac{dy}{dx} = -1$ at P

5. (a) $\dfrac{dy}{dx} = 0$ at $(0, -1)$ (b) $\dfrac{dy}{dx} = 2$ at $(2, 1)$ (c) $\dfrac{dy}{dx} = -2$ at $(-2, 1)$

7. (a) $\dfrac{dy}{dx} = 0$ at $(0, 1)$ (b) $\dfrac{dy}{dx} = 2$ at $(-2, 0)$ (c) $\dfrac{dy}{dx} \neq -2$ at any point

9. $f(x) = 3x^2 - 4x^{-1}$

   $f'(x) = 6x + 4x^{-2}$

11. $f(x) = 1 + \dfrac{x^2}{2}$

   $f'(x) = x$

13. $g(x) = x^{1.2} + x^{2.3}$

   $g'(x) = 1.2x^{0.2} + 2.3x^{1.3}$

15. $g(s) = 3s^{0.2} - 44s$

   $g'(s) = 0.6s^{-0.8} - 44$

17. $h(t) = at^3 - bt^2 + c$

   $h'(t) = 3at^2 - 2bt$

19. $r(x) = 2\sqrt{x} + \sqrt[3]{x} = 2x^{1/2} + x^{1/3}$

   $r'(x) = x^{-1/2} + \dfrac{1}{3}x^{-2/3} = \dfrac{1}{\sqrt{x}} - \dfrac{1}{3\sqrt[3]{x^2}}$

21. $f(x) = 3x + \dfrac{1}{x^3} = 3x + x^{-3}$

   $f'(x) = 3 - 3x^{-4} = 3 - \dfrac{3}{x^4}$

23. $h(x) = \dfrac{x}{5} - \dfrac{5}{x} = \dfrac{x}{5} - 5x^{-1}$

   $h'(x) = \dfrac{1}{5} + 5x^{-2} = \dfrac{1}{5} + \dfrac{5}{x^2}$

25. $r(x) = \sqrt{9x} = \sqrt{9}\sqrt{x} = 3x^{1/2}$

   $r'(x) = \dfrac{3x^{-1/2}}{2} = \dfrac{3}{2\sqrt{x}}$

27. $g(x) = x^2(2x + 1) = 2x^3 + x^2$

   $g'(x) = 6x^2 + 2x$

29. $g(r) = \dfrac{1}{\sqrt{r}} + \dfrac{2}{r} = r^{-1/2} + 2r^{-1}$

$g'(r) = \dfrac{-r^{-3/2}}{2} - 2r^{-2} = -\dfrac{1}{2r\sqrt{r}} - \dfrac{2}{r^2}$

In Exercises 30 - 39 use $y - y_1 = m(x - x_1)$ for the equation of the tangent line.

31. $f(x) = x^2 + 2x - 1;\ x = -2$

$f'(x) = 2x + 2$

$y = f(-2) = (-2)^2 + 2(-2) - 1 = -1$

$m_{tan} = f'(-2) = 2(-2) + 2 = -2$

$y - (-1) = -2[x - (-2)]$

$y = -2x - 5$

33. $g(t) = \dfrac{1}{5t^4} = \dfrac{t^{-4}}{5};\ t = 1$

$g'(t) = \dfrac{-4t^{-5}}{5} = -\dfrac{4}{5t^5}$

$y = g(1) = \dfrac{1}{5(1)^4} = \dfrac{1}{5}$

$m_{tan} = g'(1) = -\dfrac{4}{5(1)^5} = -\dfrac{4}{5}$

$y - \dfrac{1}{5} = -\dfrac{4}{5}(t - 1)$

$y = -\dfrac{4}{5}t + 1$

35. $h(s) = \dfrac{1}{s} + s = s^{-1} + s;\ s = 2$

$h'(s) = -s^{-2} + 1 = -\dfrac{1}{s^2} + 1$

$y = h(2) = \dfrac{1}{2} + 2 = \dfrac{5}{2}$

$m_{tan} = h'(2) = -\dfrac{1}{2^2} + 1 = \dfrac{3}{4}$

$y - \dfrac{5}{2} = \dfrac{3}{4}(s - 2)$

$y = \dfrac{3}{4}s + 1$

37. $r(t) = \dfrac{t^2}{3} - \dfrac{2t^3}{6};\ t = -1$

$r'(t) = \dfrac{2t}{3} - t^2$

$y = r(-1) = \dfrac{(-1)^2}{3} - \dfrac{2(-1)^3}{6} = \dfrac{2}{3}$

$m_{tan} = r'(-1) = \dfrac{2(-1)}{3} - (-1)^2 = -\dfrac{5}{3}$

$y - \dfrac{2}{3} = -\dfrac{5}{3}[t - (-1)]$

$y = -\dfrac{5}{3}t - 1$

39. $h(t) = \dfrac{t^2 - 1}{t} = t - t^{-1}; \ t = 2$

$\quad h'(t) = 1 + t^{-2} = 1 + \dfrac{1}{t^2}$

$\qquad y = h(2) = \dfrac{(2)^2 - 1}{2} = \dfrac{3}{2}$

$\qquad m_{\tan} = h'(2) = 1 + \dfrac{1}{(2)^2} = \dfrac{5}{4}$

$\quad y - \dfrac{3}{2} = \dfrac{5}{4}(t - 2)$

$\qquad y = \dfrac{5}{4}t - 1$

In Exercises 41-47 when the tangent line to the graph is horizontal $\dfrac{dy}{dx} = m_{\tan} = 0$.

41. $y = -x^2 - 3x - 1$

$\quad \dfrac{dy}{dx} = -2x - 3$

$\qquad 0 = -2x - 3$

$\qquad x = -\dfrac{3}{2}$

43. $y = x^2 + \dfrac{1}{x^2} = x^2 + x^{-2}$

$\quad \dfrac{dy}{dx} = 2x - 2x^{-3} = 2x - \dfrac{2}{x^3}$

$\qquad 0 = 2x - \dfrac{2}{x^3}$

$\qquad 0 = 2x^4 - 2$

$\qquad x^4 = 1$

$\qquad x = \pm 1$

45. $y = \sqrt{x} - 1 = x^{1/2} - 1$

$\quad \dfrac{dy}{dx} = \dfrac{x^{-1/2}}{2} = \dfrac{1}{2\sqrt{x}}$

$\qquad 0 = \dfrac{1}{2\sqrt{x}}$

None exists

47. $y = x - \dfrac{1}{x^2} + 4 = x - x^{-2} + 4$

$\quad \dfrac{dy}{dx} = 1 + 2x^{-3} = 1 + \dfrac{2}{x^3}$

$\qquad 0 = 1 + \dfrac{2}{x^3}$

$\qquad 0 = x^3 + 2$

$\qquad x^3 = -2$

$\qquad x = \sqrt[3]{-2}$

$\qquad = -\sqrt[3]{2}$

49. $\lim\limits_{x \to 0} 3x - 2 = -2$

51. $\lim\limits_{x \to -1} \dfrac{1+x}{x} = 0$

53. $\lim\limits_{x \to 9} \left(x - \sqrt{x}\right) = 6$

55. $\lim\limits_{x \to -\infty} \left(x^2 - 3x + 1\right)$ diverges to $+\infty$

57.

x	-1.1	-1.01	-1.001	-1	-0.999	-0.99	-0.9
$\dfrac{x^3+1}{x+1}$	3.31	3.03	3.003		2.997	2.97	2.71

$\lim\limits_{x \to -1} \dfrac{x^3+1}{x+1} = 3$

59.

x	-0.1	-0.01	-0.001	-0.0001
$\dfrac{1}{x^2-2x}$	4.76	49.8	500	5000

$\lim\limits_{x \to 0^-} \dfrac{1}{x^2-2x}$ diverges to $+\infty$

61.

x	-0.999	-0.99	-0.9	-0.9999
$\dfrac{x^2+1}{x+1}$	1998	198	18.1	19,998

$\lim\limits_{x \to -1^+} \dfrac{x^2+1}{x+1}$ diverges to $+\infty$

63.

x	1	10	100
$x^2 e^{-x}$	0.368	0.0045	0

$\lim\limits_{x \to +\infty} x^2 e^{-x} = 0$

65. (a) $\lim\limits_{x \to 1} f(x) = -1$

    (b) $\lim\limits_{x \to -1^+} f(x) = 1$

    (c) $\lim\limits_{x \to -1} f(x)$ does not exist because left and right limits do not agree

    (d) $\lim\limits_{x \to 3} f(x)$ diverges to $+\infty$

    (e) $f(1) = 2$

    (f) $f(-1) = 1$

67. (a) $\lim\limits_{x \to 0} f(x) = -1$

    (b) $\lim\limits_{x \to +\infty} f(x)$ diverges to $-\infty$

    (c) $\lim\limits_{x \to -\infty} f(x) = 1$

69. (a) $\lim\limits_{x \to -1} f(x)$ diverges to $+\infty$

    (b) $\lim\limits_{x \to -3^+} f(x)$ diverges to $-\infty$

    (c) $\lim\limits_{x \to +\infty} f(x) = -1$

    (d) $f(-1) = 1$

71. (a) $\lim\limits_{x \to 1^-} f(x) = -1$

    (b) $\lim\limits_{x \to -\infty} f(x)$ diverges to $-\infty$

    (c) $\lim\limits_{x \to +\infty} f(x)$ diverges to $+\infty$

73. $f(x)$ is discontinuous at -1 because left and right limits disagree

    $f(x)$ is discontinuous at 1 because $\lim\limits_{x \to 1} f(x) \neq f(1)$

75. $f(x)$ is continuous

77. $f(x)$ is discontinuous at -1 because $\lim\limits_{x \to -1} f(x) \neq f(-1)$

79. $f(x)$ is discontinuous at every integer because the left and right limits disagree

81. $\lim\limits_{h \to -1^+} 4 = 4$

83. $\lim\limits_{h \to 0} \dfrac{h^2 - 2h}{h + h^3} = \lim\limits_{h \to 0} \dfrac{h(h-2)}{h(1+h^2)} = \lim\limits_{h \to 0} \dfrac{h-2}{1+h^2} = -2$

85. $\lim\limits_{x \to -3} \dfrac{2x^2 + 5x - 3}{x+3} = \lim\limits_{x \to -3} \dfrac{(2x-1)(x+3)}{x+3} = \lim\limits_{x \to -3} 2x - 1 = -7$

87. $\lim\limits_{x \to 0} |-x| = 0$

89. $\lim\limits_{x \to +\infty} \dfrac{1}{e^x - e^{-x}} = \lim\limits_{x \to +\infty} \dfrac{1}{e^{-x}(e^{2x} - 1)} = \lim\limits_{x \to +\infty} \dfrac{e^x}{e^{2x} - 1} = \lim\limits_{x \to +\infty} \dfrac{1}{e^x} = 0$

91. $s = 100t - 16t^2$

$$v_{average} = \frac{s(t+h) - s(t)}{h} = \frac{100(t+h) - 16(t+h)^2 - \left(100t - 16t^2\right)}{h}$$

$$= \frac{100h - 32th - 16h^2}{h}$$

$$= 100 - 32t - 16h$$

(a)

Interval	[1,2]	[1,1.1]	[1,1.01]	[1,1.001]
t,h	1,1	1,0.1	1,0.01	1,0.001
v_{ave} (ft / s)	52	66.4	67.84	67.984

(b) $v = \lim\limits_{h \to 0} 100 - 32t - 16h = 100 - 32t$

$v(1) = 68$ ft / s

(c) $100t - 16t^2 = 0$

$t(100 - 16t) = 0$

$t = 0$ or $100 - 16t = 0$

$t = 6.25s$

$v(6.25) = 100 - 32(6.25) = -100$ ft / s

93. $C(x) = 100 + 60x - 0.001x^2$

$C'(x) = 60 - 0.002x$

Marginal Cost:

$C'(50) = 60 - 0.002(50) = \$59.90$

Exact Cost:

$C(51) - C(50)$

$= 100 + 60(51) - 0.001(51)^2 - \left[100 + 60(50) - 0.001(50)^2\right]$

$= \$59.899$

95. $V = a^3$

If $a = rt$ where $r =$ the rate of growth of each edge

$V = rt^3$

$\dfrac{dV}{dt} = 3rt^2$

When $V = 1000$ and $r = 1$, $t = 10$

$\dfrac{dV}{dt} = 3(1)(10)^2 = 300$ cm^3 / s

97. $$C(s) = \frac{\sqrt{s}}{10}$$

(a) $R(s) = s - 1000 - \dfrac{sC(s)}{10 \cdot 100}$

$\quad = s - 1000 - \dfrac{s^{3/2}}{1000}$

(b) $R'(s) = 1 - \dfrac{3s^{1/2}}{2000}$

$R'(100,000) = 1 - \dfrac{3(100,000)^{1/2}}{2000}$

$\quad = 0.526$

(c) $R'(s) > 0$

$1 - \dfrac{3s^{1/2}}{2000} > 0$

$s^{1/2} < \dfrac{2000}{3}$

$s < \$44,444.44$

$R'(s) < 0$

$1 - \dfrac{3s^{1/2}}{2000} > 0$

$s > \$444,444.44$

(d) $R(444,444.44) = 444,444.44 - 1000 - \dfrac{(444,444.44)^{3/2}}{1000}$

$\quad = \$147,148.15$

99. $P = 10,000e^{0.5t}$

(a) $P'(t) = $ slope of the tangent line

$P'(5) = 60,912$ frogs per year

(b) $P'(t) = 1,000,000$ for $t \approx 11$ years
which is the year 2006.

(c) There will be an abundance of frogs
in 2006.

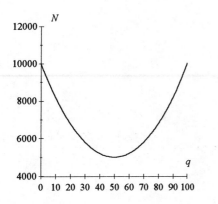

Lowest net cost for 50 employees which is the level in part (b). Therefore, the lowest net cost occurs where the marginal cost equals the marginal subsidy.

CHAPTER 11
TECHNIQUES OF DIFFERENTIATION

11.1 EXERCISES

Use the product and quotient rules, $\dfrac{d}{dx}[f(x)g(x)] = f'(x)g(x) + f(x)g'(x)$ and $\dfrac{d}{dx}\left(\dfrac{f(x)}{g(x)}\right) = \dfrac{f'(x)g(x) - f(x)g'(x)}{[g(x)]^2}$.

1. $f(x) = 3x$

    (a) $\dfrac{d}{dx}(3x) = 3$

    (b) $\dfrac{d}{dx}(3x) = 0 \cdot x + 3 \cdot 1 = 3$

3. $g(x) = x \cdot x^2$

    (a) $\dfrac{d}{dx}(x \cdot x^2) = 3x^2$

    (b) $\dfrac{d}{dx}(x \cdot x^2) = 1 \cdot x^2 + x \cdot 2x = 3x^2$

5. $h(x) = x(x+3)$

    (a) $\dfrac{d}{dx}[x(x+3)] = 2x + 3$

    (b) $\dfrac{d}{dx}[x(x+3)] = 1 \cdot (x+3) + x \cdot 1 = 2x + 3$

7. $r(x) = \dfrac{x^2}{3}$

    (a) $\dfrac{d}{dx}\left(\dfrac{x^2}{3}\right) = \dfrac{2x}{3}$

    (b) $\dfrac{d}{dx}\left(\dfrac{x^2}{3}\right) = \dfrac{2x \cdot 3 - x^2 \cdot 0}{9} = \dfrac{2x}{3}$

9. $s(x) = \dfrac{2}{x}$

    (a) $\dfrac{d}{dx}\left(\dfrac{2}{x}\right) = -\dfrac{2}{x^2}$

    (b) $\dfrac{d}{dx}\left(\dfrac{2}{x}\right) = \dfrac{0 \cdot x - 2 \cdot 1}{x^2} = -\dfrac{2}{x^2}$

11. $y = (x+1)(x^2 - 1)$

$$\begin{aligned}
\dfrac{dy}{dx} &= \dfrac{d}{dx}\left[(x+1)(x^2 - 1)\right] \\
&= 1(x^2 - 1) + (x+1)(2x) \\
&= x^2 - 1 + 2x^2 + 2x \\
&= 3x^2 + 2x - 1 \\
&= (x+1)(3x-1)
\end{aligned}$$

13. $y = (2x^{1/2} + 4x - 5)(x - x^{-1})$

$$\begin{aligned}
\dfrac{dy}{dx} &= \dfrac{d}{dx}\left[(2x^{1/2} + 4x - 5)(x - x^{-1})\right] \\
&= (x^{-1/2} + 4)(x - x^{-1}) + (2x^{1/2} + 4x - 5)(1 + x^{-2})
\end{aligned}$$

15. $y = \left(2x^2 - 4x + 1\right)^2 = \left(2x^2 - 4x + 1\right)\left(2x^2 - 4x + 1\right)$

$\dfrac{dy}{dx} = \dfrac{d}{dx}\left[\left(2x^2 - 4x + 1\right)\left(2x^2 - 4x + 1\right)\right] = 2(4x - 4)\left(2x^2 - 4x + 1\right) = 8(x - 1)\left(2x^2 - 4x + 1\right)$

17. $y = \left(x^2 - \sqrt{x}\right)\left(\sqrt{x} + \dfrac{1}{\sqrt{x}}\right) = \left(x^2 - x^{1/2}\right)\left(x^{1/2} + x^{-1/2}\right)$

$\dfrac{dy}{dx} = \dfrac{d}{dx}\left[\left(x^2 - x^{1/2}\right)\left(x^{1/2} + x^{-1/2}\right)\right] = \left(2x - \dfrac{1}{2}x^{-1/2}\right)\left(x^{1/2} + x^{-1/2}\right) + \left(x^2 - x^{1/2}\right)\left(\dfrac{1}{2}x^{-1/2} - \dfrac{1}{2}x^{-3/2}\right)$

$= \left(2x - \dfrac{1}{2\sqrt{x}}\right)\left(\sqrt{x} + \dfrac{1}{\sqrt{x}}\right) + \left(x^2 - \sqrt{x}\right)\left(\dfrac{1}{2\sqrt{x}} - \dfrac{1}{2x\sqrt{x}}\right)$

19. $y = \left(\sqrt{x} + 1\right)\left(\sqrt{x} + \dfrac{1}{x^2}\right) = \left(x^{1/2} + 1\right)\left(x^{1/2} + x^{-2}\right)$

$\dfrac{dy}{dx} = \dfrac{d}{dx}\left[\left(x^{1/2} + 1\right)\left(x^{1/2} + x^{-2}\right)\right] = \left(\dfrac{1}{2}x^{-1/2}\right)\left(x^{1/2} + x^{-2}\right) + \left(x^{1/2} + 1\right)\left(\dfrac{1}{2}x^{-1/2} - 2x^{-3}\right)$

$= \dfrac{1}{2\sqrt{x}}\left(\sqrt{x} + \dfrac{1}{x^2}\right) + \left(\sqrt{x} + 1\right)\left(\dfrac{1}{2\sqrt{x}} - \dfrac{2}{x^3}\right)$

21. $y = \dfrac{2x + 4}{3x - 1}$

$\dfrac{dy}{dx} = \dfrac{d}{dx}\left(\dfrac{2x + 4}{3x - 1}\right) = \dfrac{2(3x - 1) - (2x + 4)(3)}{(3x - 1)^2} = \dfrac{6x - 2 - 6x - 12}{(3x - 1)^2} = -\dfrac{14}{(3x - 1)^2}$

23. $y = \dfrac{2x^2 + 4x + 1}{3x - 1}$

$\dfrac{dy}{dx} = \dfrac{d}{dx}\left(\dfrac{2x^2 + 4x + 1}{3x - 1}\right) = \dfrac{(4x + 4)(3x - 1) - (2x^2 + 4x + 1)(3)}{(3x - 1)^2} = \dfrac{12x^2 + 8x - 4 - 6x^2 - 12x - 3}{(3x - 1)^2} = \dfrac{6x^2 - 4x - 7}{(3x - 1)^2}$

25. $y = \dfrac{x^2 - 4x + 1}{x^2 + x + 1}$

$$\dfrac{dy}{dx} = \dfrac{d}{dx}\left(\dfrac{x^2 - 4x + 1}{x^2 + x + 1}\right) = \dfrac{(2x-4)(x^2+x+1) - (x^2-4x+1)(2x+1)}{(x^2+x+1)^2}$$

$$= \dfrac{2x^3 - 2x^2 - 2x - 4 - 2x^3 + 7x^2 + 2x - 1}{(x^2+x+1)^2} = \dfrac{5x^2 - 5}{(x^2+x+1)^2}$$

27. $y = \dfrac{\sqrt{x}+1}{\sqrt{x}-1} = \dfrac{x^{1/2}+1}{x^{1/2}-1}$

$$\dfrac{dy}{dx} = \dfrac{d}{dx}\left(\dfrac{x^{1/2}+1}{x^{1/2}-1}\right) = \dfrac{\frac{1}{2}x^{-1/2}(x^{1/2}-1) - (x^{1/2}+1)\frac{1}{2}x^{-1/2}}{(x^{1/2}-1)^2} = \dfrac{\frac{1}{2} - \frac{1}{2}x^{-1/2} - \frac{1}{2} - \frac{1}{2}x^{-1/2}}{(x^{1/2}-1)^2}$$

$$= -\dfrac{x^{-1/2}}{(x^{1/2}-1)^2} = \dfrac{-1}{\sqrt{x}\left(\sqrt{x}-1\right)^2}$$

29. $y = \dfrac{\left(\dfrac{1}{x} + \dfrac{1}{x^2}\right)}{x + x^2} = \dfrac{x^{-1} + x^{-2}}{x + x^2}$

$$\dfrac{dy}{dx} = \dfrac{d}{dx}\left(\dfrac{x^{-1}+x^{-2}}{x+x^2}\right) = \dfrac{(-x^{-2} - 2x^{-3})(x+x^2) - (x^{-1}+x^{-2})(1+2x)}{(x+x^2)^2}$$

$$= \dfrac{-x^{-1} - 2x^{-2} - 1 - 2x^{-1} - x^{-1} - x^{-2} - 2 - 2x^{-1}}{(x+x^2)^2} = \dfrac{-6x^{-1} - 3x^{-2} - 3}{(x+x^2)^2}$$

$$y = \dfrac{\left(\dfrac{1}{x} + \dfrac{1}{x^2}\right)}{x + x^2} = \dfrac{\dfrac{x+1}{x^2}}{x(x+1)} = \dfrac{1}{x^3} = x^{-3}$$

$$\dfrac{dy}{dx} = \dfrac{d}{dx}\left(x^{-3}\right) = -3x^{-4} = -\dfrac{3}{x^4}$$

31. $y = \dfrac{(x+3)(x+1)}{3x-1}$

$\dfrac{dy}{dx} = \dfrac{d}{dx}\left(\dfrac{(x+3)(x+1)}{3x-1}\right) = \dfrac{[1(x+1)+(x+3)(1)](3x-1)-(x+3)(x+1)(3)}{(3x-1)^2} = \dfrac{(2x+4)(3x-1)-3(x+3)(x+1)}{(3x-1)^2}$

$= \dfrac{6x^2+10x-4-3x^2-12x-9}{(3x-1)^2} = \dfrac{3x^2-2x-13}{(3x-1)^2}$

33. $y = \dfrac{(x+3)(x+1)(x+2)}{3x-1}$

$\dfrac{dy}{dx} = \dfrac{d}{dx}\left(\dfrac{(x+3)(x+1)(x+2)}{3x-1}\right) = \dfrac{\{1(x+1)(x+2)+(x+3)[1(x+2)+(x+1)(1)]\}(3x-1)-(x+3)(x+1)(x+2)(3)}{(3x-1)^2}$

$= \dfrac{[(x+1)(x+2)+(x+3)(x+2)+(x+3)(x+1)](3x-1)-3(x+3)(x+1)(x+2)}{(3x-1)^2}$

35. $\dfrac{d}{dx}\left[(x^2+x)(x^2-x)\right] = (2x+1)(x^2-x)+(x^2+x)(2x-1) = 2x^3-x^2-x+2x^3+x^2-x = 4x^3-2x$

37. $\dfrac{d}{dx}\left[(x^3+2x)(x^2-x)\right]\Big|_{x=2} = \left[(3x^2+2)(x^2-x)+(x^3+2x)(2x-1)\right]\Big|_{x=2}$

$= \left[3(2)^2+2\right]\left(2^2-2\right)+\left[(2)^3+2(2)\right][2(2)-1] = 64$

39. $\dfrac{d}{dt}\left(\left(t^2-\sqrt{t}\right)\left(\sqrt{t}+\dfrac{1}{\sqrt{t}}\right)\right)\Big|_{t=1} = \dfrac{d}{dt}\left(\left(t^2-t^{1/2}\right)\left(t^{1/2}+t^{-1/2}\right)\right)\Big|_{t=1}$

$= \left[\left(2t-\dfrac{1}{2}t^{-1/2}\right)\left(t^{1/2}+t^{-1/2}\right)+\left(t^2-t^{1/2}\right)\left(\dfrac{1}{2}t^{-1/2}-\dfrac{1}{2}t^{3/2}\right)\right]\Big|_{t=1}$

$= \left[2(1)-\dfrac{1}{2}(1)^{-1/2}\right]\left(1^{1/2}+1^{-1/2}\right)+\left(1^2-1^{1/2}\right)\left[\dfrac{1}{2}(1)^{-1/2}-\dfrac{1}{2}(1)^{-3/2}\right] = 3$

41. $\dfrac{d}{dt}\left(\dfrac{t^2-\sqrt{t}}{\sqrt{t}+\dfrac{i}{\sqrt{t}}}\right)=\dfrac{d}{dt}\left(\dfrac{t^2-t^{1/2}}{t^{1/2}+t^{-1/2}}\right)=\dfrac{\left(2t-\dfrac{1}{2}t^{-1/2}\right)\left(t^{1/2}+t^{-1/2}\right)-\left(t^2-t^{1/2}\right)\left(\dfrac{1}{2}t^{-1/2}-\dfrac{1}{2}t^{-3/2}\right)}{\left(t^{1/2}+t^{-1/2}\right)^2}$

$\qquad\qquad =\dfrac{\left(2t-\dfrac{1}{2\sqrt{t}}\right)\left(\sqrt{t}+\dfrac{1}{\sqrt{t}}\right)-\left(t^2-\sqrt{t}\right)\left(\dfrac{1}{2\sqrt{t}}-\dfrac{1}{2t\sqrt{t}}\right)}{\left(\sqrt{t}+\dfrac{1}{\sqrt{t}}\right)^2}$

43. $f(x)=\left(x^2+1\right)\left(x^3+x\right),\ x_1=1$

$f'(x)=2x\left(x^3+x\right)+\left(x^2+1\right)\left(3x^2+1\right)$

$\qquad y_1=f(1)=\left(1^2+1\right)\left(1^3+1\right)=4$

$\qquad m=f'(1)=2(1)\left(1^3+1\right)+\left(1^2+1\right)\left(3\cdot 1^2+1\right)=12$

$y-y_1=m(x-x_1)$

$\quad y-4=12(x-1)$

$\qquad y=12x-8$

45. $f(x)=\dfrac{x+1}{x+2},\ x_1=0$

$\cdot f'(x)=\dfrac{1(x+2)-(x+1)1}{(x+2)^2}=\dfrac{1}{(x+2)^2}$

$\qquad y_1=f(0)=\dfrac{0+1}{0+2}=\dfrac{1}{2}$

$\qquad m=f'(0)=\dfrac{1}{(0+2)^2}=\dfrac{1}{4}$

$y-y_1=m(x-x_1)$

$\quad y-\dfrac{1}{2}=\dfrac{1}{4}(x-0)$

$\qquad y=\dfrac{x}{4}+\dfrac{1}{2}$

47. $f(x)=\dfrac{x^2+1}{x},\ x_1=-1$

$\qquad f'(x)=\dfrac{2x(x)-\left(x^2+1\right)(1)}{x^2}=\dfrac{x^2-1}{x^2}$

$\qquad y_1=f(-1)=\dfrac{(-1)^2+1}{-1}=-2$

$\qquad m=f'(-1)=\dfrac{(-1)^2-1}{(-1)^2}=0$

$y-y_1=m(x-x_1)$

$y-(-2)=0\left[x-(-1)\right]$

$\qquad y=-2$

49. $S(x) = 20x - x^2$

$p(x) = 1000 - x^2$

$R(x) = S(x)p(x) = (20x - x^2)(1000 - x^2)$

$S'(x) = 20 - 2x$

$p'(x) = -2x$

$R'(x) = (20 - 2x)(1000 - x^2) + (20x - x^2)(-2x)$

$S'(5) = 20 - 2(5) = 10$

Sales are increasing by 1000 units per month.

$p'(5) = -2(5) = -10$

The price is dropping by \$10 per month.

$R'(5) = [20 - 2(5)](1000 - 5^2) + [20(5) - 5^2](-2)(5) = 9000$ (times 100)

The revenue is increasing by \$900,000 per month.

51. Daily sales $= S(x) = 20 - 3x$

Price $= p(x) = 7 + x$

Daily revenue $R(x) = S(x)p(x) = (20 - 3x)(7 + x)$

$\qquad\qquad\qquad R'(x) = -3(7 + x) + (20 - 3x)(1) = -1 - 6x$

Current day is $x = 0$, $R'(0) = -1$

Daily revenue is currently decreasing by \$1 per day.

53. $\dfrac{\text{Cost}}{\text{Passenger}} = \dfrac{C(t)}{P(t)} = \dfrac{10{,}000 + t^2}{1000 + t^2}$

$\dfrac{d}{dt}\left(\dfrac{10{,}000 + t^2}{1000 + t^2}\right) = \dfrac{2t(1000 + t^2) - (10{,}000 + t^2)(2t)}{(1000 + t^2)^2}$

At $t = 6$, the rate is

$\dfrac{2(6)(1000 + 6^2) - (10{,}000 + 6^2)(2)(6)}{(1000 + 6^2)^2} = -0.10$

The cost per passenger is decreasing by \$0.10 per month.

55. $M(x) = \dfrac{1}{\left(x + \dfrac{3600}{x}\right)} = \dfrac{1}{x + 3600x^{-1}}$

$M'(x) = \dfrac{0\left(x + 3600x^{-1}\right) - 1\left(1 - 3600x^{-2}\right)}{\left(x + 3600x^{-1}\right)^2}$

$= \dfrac{3600x^{-2} - 1}{\left(3600x^{-1} + x\right)^2}$

$M'(10) = \dfrac{3600(10)^{-2} - 1}{\left(3600(10)^{-1} + 10\right)^2} \approx 0.0002557$

$M'(60) = \dfrac{3600(60)^{-2} - 1}{\left(3600(60)^{-1} + 60\right)^2} = 0$

$M'(70) = \dfrac{3600(70)^{-2} - 1}{\left(3600(70)^{-1} + 70\right)^2} \approx -0.00001799$

This means that, at a speed of 10 mph., the fuel economy is increasing at a rate of 0.0002557 miles per gallon for each 1 mph. increase in speed. At 60 mph., the fuel economy is neither increasing nor decreasing with increasing speed, and at 70 mph., the fuel economy is decreasing by 0.00001799 miles per gallon for each 1 mph. increase in speed. Thus 60 mph. is the most fuel‑efficient speed for the car.

57. Let x = number of years since 1984

Let $N(x)$ = number of stores

$N(0) = 171, \quad N(7) = 1032$

$m = \dfrac{1032 - 171}{10 - 0} = 86.1$

$N(x) = 86.1x + 171$

Let $r(x)$ = revenue at each store

$r(x) = 50{,}000x + 600{,}000$

$R(x) = r(x)N(x)$ = worldwide revenue

$R(x) = (50{,}000x + 600{,}000)(86.1x + 171)$

$R'(x) = 50{,}000(86.1x + 171) + (50{,}000x + 600{,}000)(86.1)$

$R'(6) = 50{,}000[86.1(6) + 171] + [50{,}000(6) + 600{,}000](86.1)$

$= \$111{,}870{,}000$

Worldwide revenue was increasing at $111,870,000 per year.

59. $R(p) = \dfrac{r}{1+kp} = \dfrac{45}{1+0.125p}$

$R'(p) = \dfrac{0(1+0.125p) - 45(0.125)}{(1+0.125)^2} = \dfrac{-5.625}{(1+0.125p)^2}$

$R'(4) = \dfrac{-5.625}{[1+0.125(4)]^2}$

$= -2.5$ thousand organisms per hour,
per 1000 new organisms. This means
that the reproduction rate of organisms
in a culture containing 4000 organisms
is declining by 2500 organisms per hour
for every 1000 new organisms.

61. $C(t) = -0.0163t^4 + 1.096t^3 - 10.704t^2 + 3.576t \quad (t \le 30)$

$C'(t) = -0.0652t^3 + 3.288t^2 - 21.408t + 3.576$

$C(25) = -0.0163(25)^4 + 1.096(25)^3 - 10.704(25)^2 + 3.576(25) = 4{,}157.2125$

$C'(25) = -0.0652(25)^3 + 3.288(25)^2 - 21.408(25) + 3.576 = 504.626$

Let $N(t) =$ number of eggs, $N(0) = 30$, $N(30) = 0$

$m = \dfrac{0-30}{30-0} = -1$

$N(t) = -t + 30$

$N'(t) = -1$

$N(25) = -25 + 30 = 5$

$N'(25) = -1$

Let $0(t) =$ total oxygen consumption $= C(t)N(t)$

$0'(t) = C'(t)N(t) + C(t)N'(t)$

$0'(25) = C'(25)N(25) + C(25)N(25)$

$= 504.626(5) + 4{,}157.2125(-1) = -1634$ milliliters per day

Oxygen consumption is decreasing by 1634 milliliters per day. This is due to the fact that the number of eggs is decreasing, since $C'(25)$ is positive.

11.2 EXERCISES

1. $f(x) = (2x+1)^2$

$$\frac{d}{dx}(2x+1)^2 = 4(2x+1)$$

3. $f(x) = (x-1)^{-1}$

$$\frac{d}{dx}(x-1)^{-1} = -(x-1)^{-2}$$

5. $f(x) = (2-x)^{-2}$

$$\frac{d}{dx}(2-x)^{-2} = 2(2-x)^{-3}$$

7. $f(x) = \sqrt{2x+1} = (2x+1)^{1/2}$

$$\frac{d}{dx}(2x+1)^{1/2} = \frac{1}{\sqrt{2x+1}}$$

9. $f(x) = \dfrac{1}{3x-1} = (3x-1)^{-1}$

$$\frac{d}{dx}(3x-1)^{-1} = -\frac{3}{(3x-1)^2}$$

11. $f(x) = (x^2+2x)^4$

$$\frac{d}{dx}(x^2+2x)^4 = 4(x^2+2x)^3(2x+2)$$

13. $f(x) = (2x^2-2)^{-1}$

$$\frac{d}{dx}(2x^2-2)^{-1} = -1(2x^2-2)^{-2}(4x)$$
$$= -4x^2(2x^2-2)^{-2}$$

15. $g(x) = (x^2-3x-1)^{-5}$

$$\frac{d}{dx}(x^2-3x-1)^{-5} = -5(x^2-3x-1)^{-6}(2x-3)$$

17. $h(x) = \dfrac{1}{(x^2+1)^3} = (x^2+1)^{-3}$

$$\frac{d}{dx}(x^2+1)^{-3} = -3(x^2+1)^{-4}(2x) = -\frac{6x}{(x^2+1)^4}$$

19. $s(t) = (t^2-\sqrt{t})^4 = (t^2-t^{1/2})^4$

$$\frac{d}{dt}(t^2-t^{1/2})^4 = 4(t^2-t^{1/2})^3\left(2t-\frac{1}{2}t^{-1/2}\right)$$
$$= 4(t^2-\sqrt{t})^3\left(2t-\frac{1}{2\sqrt{t}}\right)$$

21. $f(x) = \sqrt{1-x^2} = (1-x^2)^{1/2}$

$$\frac{d}{dx}(1-x^2)^{1/2} = \frac{1}{2}(1-x^2)^{-1/2}(-2x) = -\frac{x}{\sqrt{1-x^2}}$$

23.
$$f(x) = \sqrt{3\sqrt{x} - \frac{1}{\sqrt{x}}} = \left(3x^{1/2} - x^{-1/2}\right)^{1/2}$$

$$\frac{d}{dx}\left(3x^{1/2} - x^{-1/2}\right)^{1/2} = \frac{1}{2}\left(3x^{1/2} - x^{-1/2}\right)^{-1/2}\left(\frac{3}{2}x^{-1/2} + \frac{1}{2}x^{-3/2}\right) = \frac{\left(\dfrac{3}{2\sqrt{x}} + \dfrac{1}{2x\sqrt{x}}\right)}{2\sqrt{3\sqrt{x} - \dfrac{1}{x}}}$$

25.
$$r(x) = \left(\sqrt{2x+1} - x^2\right)^{-1} = \left[(2x+1)^{1/2} - x^2\right]^{-1}$$

$$\frac{d}{dx}\left[(2x+1)^{1/2} - x^2\right]^{-1} = -1\left[(2x+1)^{1/2} - x^2\right]^{-2}\left[\frac{1}{2}(2x+1)^{-1/2}(2) - 2x\right]$$

$$= -\frac{\left(\dfrac{1}{\sqrt{2x+1}} - 2x\right)}{\left(\sqrt{2x+1} - x^2\right)^2}$$

27.
$$s(x) = \left(\frac{2x+4}{3x-1}\right)^2$$

$$\frac{d}{dx}\left(\frac{2x+4}{3x-1}\right)^2 = 2\left(\frac{2x+4}{3x-1}\right)\left[\frac{2(3x-1) - (2x+4)(3)}{(3x-1)^2}\right]$$

$$= \frac{2(2)(x+2)(-14)}{(3x-1)(3x-1)^2} = -\frac{56(x+2)}{(3x-1)^3}$$

29.
$$h(r) = \left[(r+1)(r^2-1)\right]^{-1/2}$$

$$\frac{d}{dr}\left[(r+1)(r^2-1)\right]^{-1/2} = -\frac{1}{2}\left[(r+1)(r^2-1)\right]^{-3/2}\left[1(r^2-1) + (r+1)(2r)\right]$$

$$= -\frac{1}{2}\left[(r+1)(r^2-1)\right]^{-3/2}\left(3r^2 + 2r - 1\right)$$

31.
$$f(x) = (x^2 - 3x)\sqrt{1-x^2} = (x^2-3x)^{-2}(1-x^2)^{1/2}$$

$$\frac{d}{dx}(x^2-3x)^{-2}(1-x^2)^{1/2} = -2(x^2-3x)^{-3}(2x-3)(1-x^2)^{1/2} + (x^2-3x)^{-2}\left(\frac{1}{2}\right)(1-x^2)^{-1/2}(-2x)$$

$$= -2(x^2-3x)^{-3}(2x-3)\sqrt{1-x^2} - \frac{x(x^2-3x)^{-2}}{\sqrt{1-x^2}}$$

306

33. $y = x^{100} + 99x^{-1}$

$$\frac{dy}{dt} = \frac{d}{dx}\left(x^{100} + 99x^{-1}\right)\frac{dx}{dt}$$

$$= \left(100x^{99} - 99x^{-2}\right)\frac{dx}{dt}$$

35. $s = \dfrac{1}{r^3} + \sqrt{r} = r^{-3} + r^{1/2}$

$$\frac{ds}{dt} = \frac{d}{dr}\left(r^{-3} + r^{1/2}\right)\frac{dr}{dt}$$

$$= \left(-3r^{-4} + \frac{1}{2}r^{-1/2}\right)\frac{dr}{dt}$$

$$= \left(-\frac{3}{r^4} + \frac{1}{2\sqrt{r}}\right)\frac{dr}{dt}$$

37. $V = \dfrac{4}{3}\pi r^3$

$$\frac{dV}{dt} = \frac{d}{dt}\left(\frac{4}{3}\pi r^3\right)\frac{dr}{dt} = 4\pi r^2\,\frac{dr}{dt}$$

39. $y = x^3 + \dfrac{1}{x} = x^3 + x^{-1}$

$$\frac{dy}{dt} = \frac{d}{dx}\left(x^3 + x^{-1}\right)\frac{dx}{dt} = \left(3x^2 - x^{-2}\right)\frac{dx}{dt}$$

$$x = 2 \text{ when } t = 1, \quad \left.\frac{dx}{dt}\right|_{t=1} = -1$$

$$\left.\frac{dy}{dt}\right|_{t=1} = \left[3(2)^2 - (2)^{-2}\right](1) = -\frac{47}{4}$$

41.

$$q = -\frac{4p}{3} + 80 \Rightarrow p = \frac{-3q}{4} + 60$$

Thus, at a demand level of 60 rubies per week, the weekly revenue is decreasing by \$30 for each additional ruby, demanded.

(a) $R = qp = \left(-\frac{4p}{3} + 80\right)p = -\frac{4p^2}{3} + 80p$

$$\frac{dR}{dp} = \frac{d}{dp}\left(-\frac{4p^2}{3} + 80p\right) = -\frac{8p}{3} + 80$$

when $q = 60,\ p = \dfrac{-3(60)}{4} + 60 = 15$

$$\frac{dR}{dp}\bigg|_{q=60} = \frac{-8(15)}{3} + 80 = \$40$$

per \$1 increase in price

(b) $\dfrac{dp}{dq} = \dfrac{d}{dq}\left(\dfrac{-3q}{4} + 60\right) = -\dfrac{3}{4}$

$$\frac{dp}{dq}\bigg|_{q=60} = -\frac{3}{4} = -\$0.75 \text{ per ruby}$$

(c) $\dfrac{dR}{dq}\bigg|_{q=60} = \dfrac{dR}{dp}\bigg|_{q=60}\dfrac{dp}{dq}\bigg|_{q=60}$

$$= 40\left(-\frac{3}{4}\right) = -\$30 \text{ per ruby}$$

43. $P = -100{,}000 + 5000q - 0.25q^2,\ q = 30n + 0.01n^2$

$$\frac{dP}{dq} = \frac{d}{dq} = \left(-100{,}000 + 5000q - 0.25q^2\right) = 5000 - 0.5q$$

$$\frac{dq}{dn} = \frac{d}{dn}\left(30n + 0.01n^2\right) = 30 + 0.02n$$

when $n = 10,\ q = 30(10) + 0.01(10)^2 = 301,\ \dfrac{dq}{dn}\bigg|_{n=10} = 30 + 0.02(10) = 30.2,$

$$\frac{dP}{dq}\bigg|_{n=10} = 5000 - 0.5(301) = 4{,}849.5$$

$$\frac{dP}{dn}\bigg|_{n=10} = \frac{dP}{dq}\bigg|_{n=10}\frac{dq}{dn}\bigg|_{n=10} = (4{,}849.5)(30.2) = \$146{,}454.90 \text{ per additional engineer}$$

At an employment level of 10 engineers, it will increase its profit by \$146,454.90 per additional engineer hired.

45. $M(t) \approx 2.27t + 11.5 \,(0 \le t \le 16)$, $B(t) \approx 19{,}500t + 436{,}000 \; (0 \le t \le 16) \Rightarrow t \approx \dfrac{B - 436{,}000}{19{,}500}$

$$\frac{dM}{dt} = \frac{d}{dt}(2.27t + 11.5) = 2.27, \quad \frac{dt}{dB} = \frac{d}{dt}\left(\frac{B - 436{,}000}{19{,}500}\right) = \frac{1}{19{,}500}$$

$$\frac{dM}{dB} = \frac{dM}{dt}\frac{dt}{dB} \approx (2.27)\left(\frac{1}{19{,}500}\right) \approx 0.000116$$

This means that 1.16 manatees are killed each year for every 10,000 registered boats.

47. $A = \pi r^2$, $\dfrac{dr}{dt} = 2$

$$\frac{dA}{dt} = \frac{dA}{dr}\frac{dr}{dt} = \frac{d}{dr}\left(\pi r^2\right)\frac{dr}{dt} = 2\pi r\frac{dr}{dt}$$

$$\left.\frac{dA}{dt}\right|_{r=3} = 2\pi(3)(2) = 12\pi \ \text{mi}^2\,/\,\text{h}$$

49. $V = \dfrac{4}{3}\pi r^3$, $\dfrac{dC}{dV} = 1000$, $\dfrac{dr}{dt} = 0.5$ $\dfrac{dV}{dr} = \dfrac{d}{dr}\left(\dfrac{4}{3}\pi r^3\right) = 4\pi r^2$

$$\left.\frac{dC}{dt}\right|_{r=10} = \left.\frac{dC}{dV}\right|_{r=10} \left.\frac{dV}{dr}\right|_{r=10} \left.\frac{dr}{dt}\right|_{r=10} = 1000(4\pi)(10)^2(0.5) = \$628{,}318.53\,/\,\text{week}$$

51. $q(t) = \dfrac{10{,}000}{1 + 0.5e^{-0.4t}}$

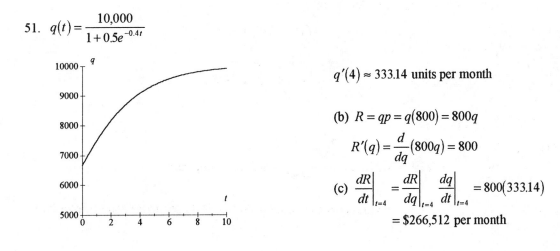

$q'(4) \approx 333.14$ units per month

(b) $R = qp = q(800) = 800q$

$$R'(q) = \frac{d}{dq}(800q) = 800$$

(c) $\left.\dfrac{dR}{dt}\right|_{t=4} = \left.\dfrac{dR}{dq}\right|_{t=4} \left.\dfrac{dq}{dt}\right|_{t=4} = 800(333.14)$

$\qquad\qquad = \$266{,}512$ per month

Use $M_d = 2y^{0.6}r^{-0.3}p$ and percentage rate of growth of f at x is $f'(x)/f(x)$ in Exercises 53-55.

53. If $\dfrac{dy/dt}{y} = 5$, $\dfrac{dr/dt}{r} = \dfrac{dp/dt}{t} = 0$

$$\frac{M'}{M} = \frac{2\dfrac{d}{dt}\left(y^{0.6}\right)r^{-0.3}p}{2y^{0.6}r^{-0.3}p}$$

$$= \frac{0.6y^{-0.4}dy/dt}{y^{0.6}}$$

$$= 0.6\frac{dy/dt}{y} = 0.6(5) = 3\% \text{ per year}$$

55. If $\dfrac{dy/dt}{y} = \dfrac{dp/dt}{p} = 5$, $\dfrac{dr/dt}{r} = 0$

$$\frac{M'}{M} = \frac{2\dfrac{d}{dt}\left(y^{0.6}\right)r^{-0.3}p + 2y^{0.6}r^{-0.3}\dfrac{d}{dt}(p)}{2y^{0.6}r^{-0.3}p}$$

$$= \frac{0.6y^{-0.4}dy/dt}{y^{0.6}} + \frac{dp/dt}{p}$$

$$= 0.6\frac{dy/dt}{y} + \frac{dp/dt}{p}$$

$$= 0.6(5) + 5 = 8\% \text{ per year}$$

11.3 EXERCISES

1. $f(x) = \ln(x-1)$

$\dfrac{d}{dx}\ln(x-1) = \dfrac{1}{x-1}$

3. $f(x) = \log_2 x$

$\dfrac{d}{dx}\log_2 x = \dfrac{1}{x\ln 2}$

5. $g(x) = \ln\left|x^2 + 3\right|$

$\dfrac{d}{dx}\ln\left|x^2 + 3\right| = \dfrac{2x}{x^2 + 3}$

7. $h(x) = e^{x+3}$

$\dfrac{d}{dx}e^{x+3} = e^{x+3}$

9. $f(x) = e^{-x}$

$\dfrac{d}{dx}e^{-x} = -e^{-x}$

11. $g(x) = 4^x$

$\dfrac{d}{dx}4^x = 4^x \ln 4$

13. $h(x) = 2^{x^2-1}$

$\dfrac{d}{dx} 2^{x^2-1} = \left(2^{x^2-1}\right) 2x \ln 2$

15. $\dfrac{d}{dx} x \ln x = 1 \ln x + x\left(\dfrac{1}{x}\right) = 1 + \ln x$

17. $f(x) = (x^2 + 1) \ln x$

$\dfrac{d}{dx}(x^2 + 1) \ln x = 2x \ln x + (x^2 + 1)\left(\dfrac{1}{x}\right)$

$\qquad\qquad = 2x \ln x + \dfrac{x^2 + 1}{x}$

19. $f(x) = (x^2 + 1)^5 \ln x$

$\dfrac{d}{dx}(x^2 + 1)^5 \ln x = 5(x^2 + 1)^4 (2x) \ln x + (x^2 + 1)^5\left(\dfrac{1}{x}\right) = 10x(x^2 + 1)^4 \ln x + \dfrac{(x^2 + 1)^5}{x}$

21. $g(x) = \ln |3x - 1|$

$\dfrac{d}{dx} \ln |3x - 1| = \dfrac{1}{3x - 1}(3) = \dfrac{3}{3x - 1}$

23. $g(x) = \ln |2x^2 + 1|$

$\dfrac{d}{dx} \ln |2x^2 + 1| = \dfrac{1}{2x^2 + 1}(4x) = \dfrac{4x}{2x^2 + 1}$

25. $g(x) = \ln \left(x^2 - \sqrt{x}\right) = \ln \left(x^2 - x^{1/2}\right)$

$\dfrac{d}{dx} \ln \left(x^2 - x^{1/2}\right) = \dfrac{1}{x^2 - x^{1/2}}\left(2x - \dfrac{1}{2}x^{-1/2}\right)$

$\qquad\qquad = \dfrac{2x - \dfrac{1}{2\sqrt{x}}}{x^2 - \sqrt{x}}$

27. $h(x) = \log_2 (x + 1)$

$\dfrac{d}{dx} \log_2 (x + 1) = \dfrac{1}{(x + 1) \ln 2}$

29. $r(t) = (t^2 + 1) \log_3 (t + 1/t) = (t^2 + 1) \log_3 (t + t^{-1})$

$\dfrac{d}{dt}(t^2 + 1) \log_3 (t + t^{-1}) = 2t \log_3 (t + t^{-1}) + (t^2 + 1)\dfrac{1}{(t + t^{-1}) \ln 3}(1 - t^{-2})$

$\qquad\qquad = 2t \log_3 (t + 1/t) + \dfrac{(t^2 + 1)\left(1 - \dfrac{1}{t^2}\right)}{\left(t + \dfrac{1}{t}\right) \ln 3}$

31. $f(x) = (\ln |x|)^2$

$$\frac{d}{dx}(\ln |x|)^2 = (2 \ln |x|)\left(\frac{1}{x}\right) = \frac{2 \ln |x|}{x}$$

33. $r(x) = \ln (x^2) - (\ln (x-1))^2$

$$\frac{d}{dx}\left[\ln (x^2) - (\ln (x-1))^2\right] = \frac{1}{x^2}(2x) - 2 \ln (x-1)\frac{1}{x-1} = \frac{2}{x} - \frac{2 \ln (x-1)}{x-1}$$

35. $\quad f(x) = xe^x$

$$\frac{d}{dx}xe^x = (1)e^x + xe^x$$
$$= e^x(1+x)$$

37. $r(x) = \ln (x+1) + 3x^3 e^x$

$$\frac{d}{dx}\left[\ln (x+1) + 3x^3 e^x\right] = \frac{1}{x+1} + 3(3x^2 e^x + x^3 e^x)$$
$$= \frac{1}{x+1} + 3e^x(x^3 + 3x^2)$$

39. $\quad f(x) = e^x \ln |x|$

$$\frac{d}{dx}e^x \ln |x| = e^x \ln |x| + e^x\left(\frac{1}{x}\right)$$
$$= e^x\left(\ln |x| + \frac{1}{x}\right)$$

41. $\quad f(x) = e^{2x+1}$

$$\frac{d}{dx}e^{2x+1} = e^{2x+1}(2) = 2e^{2x+1}$$

43. $h(x) = e^{x^2-x+1}$

$$\frac{d}{dx}e^{x^2-x+1} = e^{x^2-x+1}(2x-1)$$
$$= (2x-1)e^{x^2-x+1}$$

45. $s(x) = x^2 e^{2x-1}$

$$\frac{d}{dx}x^2 e^{2x-1} = 2xe^{2x-1} + x^2 e^{2x-1}(2)$$
$$= 2xe^{2x-1} + 2x^2 e^{2x-1}$$

47. $r(x) = \left(e^{2x-1}\right)^2$

$$\frac{d}{dx}\left(e^{2x-1}\right)^2 = 2\left(e^{2x-1}\right)e^{2x-1}(2)$$
$$= 4\left(e^{2x-1}\right)^2$$

49. $g(x) = \dfrac{e^x + e^{-x}}{e^x - e^{-x}}$

$\dfrac{d}{dx}\left(\dfrac{e^x + e^{-x}}{e^x - e^{-x}}\right) = \dfrac{\left[e^x + e^{-x}(-1)\right]\left(e^x - e^{-x}\right) - \left(e^x + e^{-x}\right)\left[e^x - e^{-x}(-1)\right]}{\left(e^x - e^{-x}\right)^2}$

$= \dfrac{e^{2x} - 2e^x e^{-x} + e^{-2x} - e^{2x} - 2e^x e^{-x} - e^{-2x}}{\left(e^x - e^{-x}\right)^2} = -\dfrac{4}{\left(e^x - e^{-x}\right)^2}$

51. $f(x) = \dfrac{1}{x(\ln x)^{1/2}}$

$\dfrac{d}{dx}\left[\dfrac{1}{x(\ln x)^{1/2}}\right] = \dfrac{(0)x(\ln x)^{1/2} - (1)\left[(1)(\ln x)^{1/2} + (x)\left(\dfrac{1}{2}\right)(\ln x)^{-1/2}\left(\dfrac{1}{x}\right)\right]}{\left[x(\ln x)^{1/2}\right]^2} = -\dfrac{2(\ln x)^{1/2} + (\ln x)^{-1/2}}{2x^2 \ln x}$

53. $r(x) = \dfrac{\sqrt{\ln x}}{x} = \dfrac{(\ln x)^{1/2}}{x}$

$\dfrac{d}{dx}\left[\dfrac{(\ln x)^{1/2}}{x}\right] = \dfrac{\dfrac{1}{2}(\ln x)^{-1/2}\left(\dfrac{1}{x}\right)(x) - (\ln x)^{1/2}(1)}{x^2}$

$= \dfrac{1 - 2\ln x}{2x^2(\ln x)^{1/2}}$

55. $f(x) = \ln|\ln x|$

$\dfrac{d}{dx}\ln|\ln x| = \dfrac{1}{\ln x}\left(\dfrac{1}{x}\right) = \dfrac{1}{x\ln x}$

57. $s(x) = \ln\sqrt{\ln x} = \ln(\ln x)^{1/2} = \dfrac{1}{2}\ln(\ln x)$

$\dfrac{d}{dx}\left[\dfrac{1}{2}\ln(\ln x)\right] = \left(\dfrac{1}{2}\right)\left(\dfrac{1}{\ln x}\right)\left(\dfrac{1}{x}\right) = \dfrac{1}{2x\ln x}$

59. $y = e^x \log_2 x, \ (1, 0)$

$$m = \left. \frac{dy}{dx} \right|_{x=1} = \left[e^x \log_2 x + e^x \left(\frac{1}{x \ln 2} \right) \right]_{x=1}$$

$$= e^1 \log_2(1) + e^1 \left(\frac{1}{1 \ln 2} \right) = \frac{e}{\ln 2}$$

$$y - y_1 = m(x - x_1)$$

$$y - 0 = \frac{e}{\ln 2}(x - 1)$$

$$y = \frac{e}{\ln 2}(x - 1) = 3.92(x - 1)$$

61. $y = \ln \sqrt{2x + 1}, \ x_1 = 0$

$$y_1 = \ln \sqrt{2(0) + 1} = \ln 1 = 0$$

$$m = \left. \frac{dy}{dx} \right|_{x=0} = \left[\frac{2}{\sqrt{2x+1}} \frac{1}{2}(2x+1)^{-1/2}(2) \right]_{x=0}$$

$$= \left(\frac{1}{2x+1} \right)_{x=0} = \frac{1}{2(0)+1} = 1$$

$$y - y_1 = m(x - x_1)$$

$$y - 0 = 1(x - 0)$$

$$y = x$$

63. $y = e^{x^2}, \ x_1 = 1$

$$y_1 = e^{(1)^2} = e$$

$$m = \left. \frac{dy}{dx} \right|_{x=1} \left[e^{x^2}(2x) \right]_{x=1} = e^{(1)^2}(2)(1) = 2e$$

$$m_\perp = -\frac{1}{m} = -\frac{1}{2e}$$

$$y - y_1 = m_\perp(x - x_1)$$

$$y - e = -\frac{1}{2e}(x - 1)$$

$$y = -\frac{1}{2e}(x - 1) + e$$

65. $P = 10,000, \ r = 0.04$

$$A = 10,000e^{0.04t}$$

$$\left. \frac{dA}{dt} \right|_{t=3} = \left[10,000 \left(e^{0.04t} \right)(0.04) \right]_{t=3}$$

$$= 400e^{0.04(3)}$$

$$= \$451.00 \text{ per year}$$

67. $P = 4,000,000, \ r = \dfrac{\ln 2}{t_D} = \dfrac{\ln 2}{10}$

$$A = 4,000,000e^{\frac{\ln 2}{10}t} = (4,000,000)2^{t/10}$$

$$\left. \frac{dA}{dt} \right|_{t=0} = \left[(4,000,000)2^{t/10} \ln 2 \left(\frac{1}{10} \right) \right]_{t=0}$$

$$= 400,000 \ln 2 \left(2^{0/10} \right)$$

$$\approx 277,000 \text{ people per year}$$

69. $P = 10, \ k = \dfrac{\ln 2}{t_H} = \dfrac{\ln 2}{24,400}$

$$A = 10e^{\frac{\ln 2}{24,400}t} = 10 \left(2^{t/24,400} \right)$$

$$\left. \frac{dA}{dt} \right|_{t=100} = \left[10 \left(2^{t/24,400} \right) \ln 2 \left(\frac{1}{24,400} \right) \right]_{t=100}$$

$$= \frac{\ln 2}{2440} 2^{100/24,400}$$

$$= 0.000285 \text{ grams per year}$$

71. $P = 10{,}000 \quad r = 0.04, \quad m = 2$

$$A = 10{,}000\left(1 + \frac{0.04}{2}\right)^{2t} = 10{,}000(1.02)^{2t}$$

$$\left.\frac{dA}{dt}\right|_{t=3} = \left[10{,}000(1.02)^{2t} \ln 1.02(2)\right]_{t=3}$$

$$= 20{,}000 \ln 1.02 \,(1.02)^{2(3)}$$

$$= \$446.02 \quad \text{per year}$$

73. $P(t) = 92e^{-0.0277t}$

$$P'(22) = \left[92e^{-0.0277t}(-0.0277)\right]_{t=22}$$

$$= -2.5484e^{-0.0277(22)}$$

$$\approx -1.3855$$

This indicates that, in ancient Rome, the percentage of people surviving was decreasing by 1.3855% per year at age 22.

75. $\displaystyle A = \frac{150{,}000{,}000}{1 + 14{,}999e^{-0.3466t}}$

$$A'(t) = \frac{0 - 150{,}000{,}000(14{,}999)e^{-0.3466t}(-0.3466)}{\left(1 + 14{,}999e^{-0.3466t}\right)^2}$$

$$A'(20) = \frac{779{,}798{,}010{,}000e^{-0.3466(20)}}{\left[1 + 14{,}999e^{-0.3466(20)}\right]^2}$$

$$\approx 3{,}110{,}000 \text{ cases per month}$$

$$A'(30) = \frac{779{,}798{,}010{,}000e^{-0.3466(30)}}{\left[1 + 14{,}999e^{-0.3466(30)}\right]^2}$$

$$\approx 11{,}200{,}000 \text{ cases per month}$$

$$A'(40) = \frac{779{,}798{,}010{,}000e^{-0.3466(40)}}{\left[1 + 14{,}999e^{-0.3466(40)}\right]^2}$$

$$\approx 722{,}000 \text{ cases per month}$$

77. $\displaystyle q(t) = \frac{2e^{0.69t}}{3 + 1.5e^{-0.4t}}$

$$q'(t) = \frac{2e^{0.69t}(0.69)(3 + 1.5e^{-0.4t}) - 2e^{0.69t}(1.5)e^{-0.4t}(-0.4)}{\left(3 + 1.5e^{-0.4t}\right)^2}$$

$$q'(6) = \frac{1.38e^{0.69(6)}\left[3 + 1.5e^{-0.4(6)}\right] + 1.2e^{0.69(6)}e^{-0.4(6)}}{\left[3 + 1.5e^{-0.4(6)}\right]^2} \approx 28.3309$$

Growing by 28.3309 billion packets per month each year.

79. $p(t) = \dfrac{0.80}{1 + e^{4.46 - 0.477t}}$

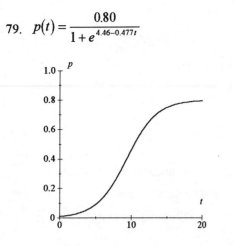

$p'(10) \approx 0.0931$, so that the percentage of firms using numeric control is increasing by 9.31% per year after 10 years.

(b) $\lim\limits_{t \to +\infty} p(t) = 0.80$ Thus, in the long term, 80% of all firms will be using numeric control.

(c) $p'(t) = \dfrac{0 - 0.80 e^{4.46 - 0.477t}(-0.477)}{\left(1 + e^{4.46 - 0.477t}\right)^2}$

$\quad = \dfrac{0.3816 e^{4.46 - 0.477t}}{\left(1 + e^{4.46 - 0.477t}\right)^2}$

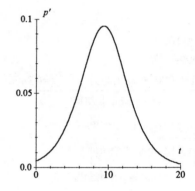

same as (a)

(d) $\lim\limits_{t \to +\infty} p'(t) = 0$ Thus, in the long run, the percentage of firms using numeric control will remain constant at 80%.

81. $n(t) = 5 + \dfrac{40 e^{0.002x}}{1 + 25 e^{3 - 0.5x}}$

$n'(t) = \dfrac{40 e^{0.002x}(0.002)\left[1 + 25 e^{3 - 0.5x}\right] - 40 e^{0.002x}(25) e^{3 - 0.5x}(-0.5)}{\left(1 + 25 e^{3 - 0.5x}\right)^2}$

$n'(t)$ is a maximum when $t \approx 12.47$. $\left(n'(12.47) \approx 5.1673\right)$.
This means that the rate of growth of enrollment in HMO's reached a maximum of 5.167 million new enrollments per year at about June 1987.

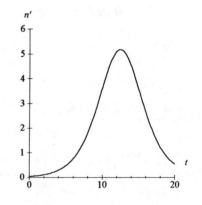

$$= \frac{0.08e^{0.002x}\left[1+25e^{3-0.5x}\right]+500e^{3-0.498x}}{\left(1+25e^{3-0.5x}\right)^2}$$

11.4 EXERCISES

1. $2x + 3y = 7$

(a) $\dfrac{d}{dx}(2x + 3y) = \dfrac{d}{dx}(7)$

$2 + 3\dfrac{dy}{dx} = 0$

$\dfrac{dy}{dx} = -\dfrac{2}{3}$

(b) $3y = -2x + 7$

$y = -\dfrac{2}{x}x + \dfrac{7}{3}$

$\dfrac{dy}{dx} = -\dfrac{2}{3}$

3. $x^2 - 2y = 6$

(a) $\dfrac{d}{dx}(x^2 - 2y) = \dfrac{d}{dx}(6)$

$2x - 2\dfrac{dy}{dx} = 0$

$\dfrac{dy}{dx} = x$

(b) $-2y = -x^2 + 6$

$y = \dfrac{x^2}{2} - 3$

$\dfrac{dy}{dx} = x$

5. $2x + 3y = xy$

(a) $\dfrac{d}{dx}(2x + 3y) = \dfrac{d}{dx}(xy)$

$3\dfrac{dy}{dx} - x\dfrac{dy}{dx} = y - 2$

$(3 - x)\dfrac{dy}{dx} = y - 2$

$\dfrac{dy}{dx} = \dfrac{y - 2}{3 - x}$

(b) $3y - xy = -2x$

$(3 - x)y = -2x$

$y = \dfrac{-2x}{3 - x}$

$\dfrac{dy}{dx} = \dfrac{-2(3 - x) + 2x(-1)}{(3 - x)^2}$

$= \dfrac{-2}{3 - x} + \dfrac{-2x(3 - x)}{3 - x} = \dfrac{y - 2}{3 - x}$

7. $e^x y = 1$

(a) $\dfrac{d}{dx}(e^x y) = \dfrac{d}{dx}(1)$

$e^x y + e^x \dfrac{dy}{dx} = 0$

$\dfrac{dy}{dx} = -y$

(b) $-y - xy = -x$

$-(1 + x)y = -x$

$y = \dfrac{x}{1 + x}$

$\dfrac{dy}{dx} = \dfrac{1 + x - x(1)}{(1 + x)^2}$

$= \dfrac{1}{1 + x} - \dfrac{x(1 + x)}{(1 + x)}$

$= \dfrac{1 - y}{1 + x}$

9. $y \ln x + y = 2$

(a) $\dfrac{d}{dx}(y \ln x + y) = \dfrac{d}{dx}(2)$

$\dfrac{dy}{dx} \ln x + y\left(\dfrac{1}{x}\right) + \dfrac{dy}{dx} = 0$

$(\ln x + 1)\dfrac{dy}{dx} = -\dfrac{y}{x}$

$\dfrac{dy}{dx} = -\dfrac{y}{x(1 + \ln x)}$

(b) $(1 + \ln x)y = 2$

$y = \dfrac{2}{1 + \ln x}$

$\dfrac{dy}{dx} = \dfrac{-2(1/x)}{(1 + \ln x)^2} = \dfrac{-[2/(1 + \ln x)]}{x(1 + \ln x)}$

$= -\dfrac{y}{x(1 + \ln x)}$

11. $x^2 + y^2 = 5$

$\dfrac{d}{dx}\left(x^2 + y^2\right) = \dfrac{d}{dx}(5)$

$2x + 2y\dfrac{dy}{dx} = 0$

$\dfrac{dy}{dx} = -\dfrac{x}{y}$

13. $x^2 y - y^2 = 4$

$\dfrac{d}{dx}\left(x^2 y - y^2\right) = \dfrac{d}{dx}(4)$

$2xy + x^2\dfrac{dy}{dx} - 2y\dfrac{dy}{dx} = 0$

$\left(x^2 - 2y\right)\dfrac{dy}{dx} = -2xy$

$\dfrac{dy}{dx} = -\dfrac{2xy}{x^2 - 2y}$

15. $3xy - \dfrac{y}{3} = \dfrac{2}{x}$

$\dfrac{d}{dx}\left(3xy - \dfrac{y}{3}\right) = \dfrac{d}{dx}\left(\dfrac{2}{x}\right)$

$3y + 3x\dfrac{dy}{dx} - \dfrac{1}{3}\dfrac{dy}{dx} = -\dfrac{2}{x^2}$

$9y + (x - 1)\dfrac{dy}{dx} = -\dfrac{6}{x^2}$

$(9x - 1)\dfrac{dy}{dx} = -\dfrac{6}{x^2} - 9y$

$\dfrac{dy}{dx} = \dfrac{-(6 + 9x^2 y)/x^2}{9x - 1}$

$= -\dfrac{6 + 9x^2 y}{9x^3 - x^2}$

17. $x^2 - 3y^2 = 8$

$\dfrac{d}{dx}\left(x^2 - 3y^2\right) = \dfrac{d}{dx}(8)$

$2x\dfrac{dy}{dx} - 6y = 0$

$\dfrac{dy}{dx} = \dfrac{3y}{x}$

19.
$$p^2 - pq = 5p^2q^2$$
$$\frac{d}{dq}(p^2 - pq) = \frac{d}{dq}(5p^2q^2)$$
$$2p\frac{dp}{dq} - \left[\frac{dp}{dq}(q) + p\right] = 10p\frac{dp}{dq}q^2 + 10p^2q$$
$$2p\frac{dp}{dq} - q\frac{dp}{dq} = -10pq^2\frac{dp}{dq} = p + 10p^2q$$
$$(2p - q - 10pq^2)\frac{dp}{dq} = p + 10p^2q$$
$$\frac{dp}{dq} = \frac{p + 10p^2q}{2p - q - 10pq^2}$$

21.
$$xe^y - ye^x = 1$$
$$\frac{d}{dx}(xe^y - ye^x) = \frac{d}{dx}(1)$$
$$e^y + xe^y\frac{dy}{dx} - \left(\frac{dy}{dx}e^x + ye^x\right) = 0$$
$$(xe^y - e^x)\frac{dy}{dx} = ye^x - e^y$$
$$\frac{dy}{dx} = \frac{ye^x - e^y}{xe^y - e^x}$$

23.
$$e^{st} = s^2$$
$$\frac{d}{dt}(e^{st}) = \frac{d}{dt}(s^2)$$
$$e^{st}\left(\frac{ds}{dt}t + s\right) = 2s\frac{ds}{dt}$$
$$te^{st}\frac{ds}{dt} - 2s\frac{ds}{dt} = -se^{st}$$
$$-(2s - te^{st})\frac{ds}{dt} = -se^{st}$$
$$\frac{ds}{dt} = \frac{se^{st}}{2s - te^{st}}$$

25.
$$\frac{e^x}{y^2} = 1 + e^y$$
$$\frac{d}{dx}\left(\frac{e^x}{y^2}\right) = \frac{d}{dx}(1 + e^y)$$
$$\frac{e^xy^2 - e^x2y\frac{dy}{dx}}{y^4} = e^y\frac{dy}{dx}$$
$$e^xy - 2e^x\frac{dy}{dx} = y^3e^y\frac{dy}{dx}$$
$$-2e^x\frac{dy}{dx} - y^3e^y\frac{dy}{dx} = -ye^x$$
$$-(2e^x + y^3e^y)\frac{dy}{dx} = -ye^x$$
$$\frac{dy}{dx} = \frac{ye^x}{2e^x + y^3e^y}$$

27.

$$\ln\left(y^2 - y\right) + x = y$$

$$\frac{d}{dx}\left[\ln\left(y^2 - y\right) + x\right] = \frac{d}{dx}(y)$$

$$\frac{1}{y^2 - y}\left(2y\frac{dy}{dx} - \frac{dy}{dx}\right) + 1 = \frac{dy}{dx}$$

$$2y\frac{dy}{dx} - \frac{dy}{dx} + y^2 - y = \left(y^2 - y\right)\frac{dy}{dx}$$

$$2y\frac{dy}{dx} - \frac{dy}{dx} - y^2\frac{dy}{dx} + y\frac{dy}{dx} = y - y^2$$

$$\left(-1 + 3y - y^2\right)\frac{dy}{dx} = y - y^2$$

$$\frac{dy}{dx} = \frac{y - y^2}{-1 + 3y - y^2}$$

29.

$$\ln\left(xy + y^2\right) = e^y$$

$$\frac{d}{dx}\left[\ln\left(xy + y^2\right)\right] = \frac{d}{dx}\left(e^y\right)$$

$$\frac{1}{xy + y^2}\left(y + x\frac{dy}{dx} + 2y\frac{dy}{dx}\right) = e^y\frac{dy}{dx}$$

$$y + x\frac{dy}{dx} + 2y\frac{dy}{dx} = \left(xy + y^2\right)e^y\frac{dy}{dx}$$

$$-xye^y\frac{dy}{dx} - y^2e^y\frac{dy}{dx} + x\frac{dy}{dx} + 2y\frac{dy}{dx} = -y$$

$$\left(x + 2y - xye^y - y^2e^y\right)\frac{dy}{dx} = -y$$

$$\frac{dy}{dx} = -\frac{y}{x + 2y - xye^y - y^2e^y}$$

31.

$$y = \left(x^3 + x\right)\sqrt{x^3 + 2}$$

$$\ln y = \ln\left[\left(x^3 + x\right)\left(x^3 + 2\right)^{1/2}\right]$$

$$\ln y = \ln\left(x^3 + x\right) + \ln\left(x^3 + 2\right)^{1/2}$$

$$\ln y = \ln\left(x^3 + x\right) + \frac{1}{2}\ln\left(x^3 + 2\right)$$

$$\frac{d}{dx}(\ln y) = \frac{d}{dx}\left[\ln\left(x^3 + x\right) + \frac{1}{2}\ln\left(x^3 + 2\right)\right]$$

$$\frac{1}{y}\frac{dy}{dx} = \frac{1}{x^3 + x}\left(3x^2 + 1\right) + \frac{1}{2\left(x^3 + 2\right)}\left(3x^2\right)$$

$$\frac{dy}{dx} = y\left[\frac{3x^2 + 1}{x^3 + x} + \frac{3x^2}{2\left(x^3 + 2\right)}\right]$$

$$= \left(x^3 + x\right)\sqrt{x^3 + 2}\left[\frac{3x^2 + 1}{x^3 + x} + \frac{3x^2}{2\left(x^3 + 2\right)}\right]$$

33.

$$y = x^x$$

$$\ln y = x \ln x$$

$$\frac{d}{dx}(\ln y) = \frac{d}{dx}(x \ln x)$$

$$\frac{1}{y}\frac{dy}{dx} = \ln x + x\left(\frac{1}{x}\right)$$

$$\frac{1}{y}\frac{dy}{dx} = \ln x + 1$$

$$\frac{dy}{dx} = y(1 + \ln x)$$

$$= x^x(1 + \ln x)$$

35.

$$4x^2 + 2y^2 = 12, \ (1, -2)$$

$$\frac{d}{dx}\left(4x^2 + 2y^2\right) = \frac{d}{dx}(12)$$

$$8x + 4y\frac{dy}{dx} = 0$$

$$\frac{dy}{dx} = -\frac{2x}{y} = -\frac{2(1)}{-2} = 1$$

321

37. $2x^2 - y^2 = xy, \ (-1, 2)$

$$\frac{d}{dx}\left(2x^2 - y^2\right) = \frac{d}{dx}(xy)$$

$$4x - 2y\frac{dy}{dx} = y + x\frac{dy}{dx}$$

$$-(2y + x)\frac{dy}{dx} = -(4x - y)$$

$$\frac{dy}{dx} = \frac{4x - y}{2y + x} = \frac{4(-1) - 2}{2(2) + (-1)} = -2$$

39. $3x^{0.3}y^{0.7} = 10, \ x = 20$

$3(20)^{0.3}y^{0.7} = 10$

$y^{0.7} = 1.3570$

$y = (1.3570)^{1/0.7}$

$= 1.5466$

$$\frac{d}{dx}\left(3x^{0.3}y^{0.7}\right) = \frac{d}{dx}(10)$$

$$3(0.3)x^{-0.7}y^{0.7} + 3x^{0.3}(0.7)y^{-0.3}\frac{dy}{dx} = 0$$

$$\frac{dy}{dx} = \frac{-3(0.3)x^{-0.7}y^{0.7}}{3x^{0.3}(0.7)y^{-0.3}}$$

$$= -\frac{0.3y}{0.7x}$$

$$= -\frac{0.3(1.5466)}{0.7(20)}$$

$$= -0.03314$$

41. $x^{0.4}y^{0.6} - 0.2x^2 = 100, \ x = 20$

$(20)^{0.4}y^{0.6} - 0.2(20)^2 = 100$

$y^{0.6} = \dfrac{180}{3.31445}$

$= 54.3077$

$= (54.3077)^{1/0.6}$

$= 778.82$

$$\frac{d}{dx}\left(x^{0.4}y^{0.6} - 0.2x^2\right) = \frac{d}{dx}(100)$$

$$0.4x^{-0.6}y^{0.6} + x^{0.4}(0.6)y^{-0.4}\frac{dy}{dx} - 0.4x = 0$$

$$0.6x^{0.4}y^{-0.4}\frac{dy}{dx} = 0.4x - 0.4x^{-0.6}y^{0.6}$$

$$\frac{dy}{dx} = \frac{0.4x - 0.4x^{-0.6}y^{0.6}}{0.6x^{0.4}y^{-0.4}}$$

$$= \frac{0.4x^{0.6}y^{0.4}}{0.6} - \frac{0.4y}{0.6x}$$

$$= \frac{0.4(20)^{0.6}(778.82)^{0.4}}{0.6} - \frac{0.4(778.82)}{0.6(20)}$$

$$= 31.7295$$

43. $e^{xy} - x = 4x, \ x = 3$

$\qquad e^{xy} = 5x$

$\qquad e^{3y} = 5(3)$

$\qquad \ln e^{3y} = \ln 15$

$\qquad\qquad y = \dfrac{\ln 15}{3} = 0.90268$

$\dfrac{d}{dx}\left(e^{xy}\right) = \dfrac{d}{dx}(5x)$

$e^{xy}\left(y + x\dfrac{dy}{dx}\right) = 5$

$xe^{xy}\dfrac{dy}{dx} = 5 - ye^{xy}$

$\dfrac{dy}{dx} = \dfrac{5 - ye^{xy}}{xe^{xy}}$

$\qquad = \dfrac{5 - (0.90268)e^{3(0.90268)}}{3e^{3(0.90268)}}$

$\qquad = -0.18798$

45. $\ln(x + y) - x = 3x^2, \ x = 0$

$\qquad \ln(0 + y) - 0 = 3(0)^2$

$\qquad\qquad \ln y = 0$

$\qquad\qquad\quad y = 1$

$\dfrac{d}{dx}\left[\ln(x + y) - x\right] = \dfrac{d}{dx}\left(3x^2\right)$

$\dfrac{1}{x+y}\left(1 + \dfrac{dy}{dx}\right) - 1 = 6x$

$1 + \dfrac{dy}{dx} - (x + y) = 6x(x + y)$

$\dfrac{dy}{dx} = 6x(x + y) + (x + y) - 1$

$\qquad = (6x + 1)(x + y) - 1$

$\qquad = \left[6(0) + 1\right][0 + 1] - 1$

$\qquad = 0$

47. $\quad pq - 2000 = q$

(a) $\ 5q - 2000 = q$

$\qquad\quad 4q = 2000$

$\qquad\qquad q = 500 \text{ T-shirts}$

(b) $\dfrac{d}{dp}(pq - 2000) = \dfrac{d}{dp}(q)$

$q + p\dfrac{dq}{dp} = \dfrac{dq}{dp}$

$(p - 1)\dfrac{dq}{dp} = -q$

$\dfrac{dq}{dp} = \dfrac{-q}{p - 1}$

$\left.\dfrac{dq}{dp}\right|_{p=5} = \dfrac{-500}{5 - 1}$

$\left.\dfrac{dq}{dp}\right|_{p=5} = -125$ T-shirts per dollar. Thus, when the price is set at \$5, the demand is dropping by 125 T-shirts per \$1 increase in price.

323

11.5 EXERCISES

Use $L(x) = f(a) + (x-a)f'(a)$ for the linear approximation of $f(x)$ near $x = a$ in Exercises 1-19.

1. $f(x) = 3x + 5$, $a = 2$
$f(2) = 3(2) + 5 = 11$
$f'(x) = 3$
$f'(2) = 3$
$L(x) = 11 + (x-2)3$
$\qquad = 3x + 5$

3. $f(x) = 3x^2 - 4x + 5$, $a = -1$
$f(-1) = 3(-1)^2 - 4(-1) + 5 = 12$
$f'(x) = 6x - 4$
$f'(-1) = 6(-1) - 4 = -10$
$L(x) = 12 + [x - (-1)](-10)$
$\qquad = -10x + 2$

5. $f(x) = \dfrac{x}{2x+1}$, $a = 0$
$f(0) = \dfrac{0}{2(0)+1} = 0$
$f'(x) = \dfrac{1(2x+1) - x(2)}{(2x+1)^2} = \dfrac{1}{(2x+1)^2}$
$f'(0) = \dfrac{1}{[2(0)+1]^2} = 1$
$L(x) = 0 + (x-0)(1)$
$\qquad = x$

7. $f(x) = e^x$, $a = 0$
$f(0) = e^0 = 1$
$f'(x) = e^x$
$f'(0) = e^0 = 1$
$L(x) = 1 + (x-0)(1)$
$\qquad = x + 1$

9. $f(x) = \ln(1+x)$, $a = 0$
$f(0) = \ln(1+0) = \ln 1 = 0$
$f'(x) = \dfrac{1}{1+x}(1) = \dfrac{1}{1+x}$
$f'(0) = \dfrac{1}{1+0} = 1$
$L(x) = 0 + (x-0)(1)$
$\qquad = x$

11. $f(x) = \sqrt{1+x}$, $a = 0$
$f(0) = \sqrt{1+0} = 1$
$f'(x) = \dfrac{1}{2}(1+x)^{-1/2} = \dfrac{1}{2\sqrt{1+x}}$
$f'(0) = \dfrac{1}{2\sqrt{1+0}} = \dfrac{1}{2}$
$L(x) = 1 + (x-0)\left(\dfrac{1}{2}\right)$
$\qquad = \dfrac{x}{2} + 1$

324

13. $f(x) = x^{1.3}, \ a = 1$

$f(1) = (1)^{1.3} = 1$

$f'(x) = 1.3x^{0.3}$

$f'(1) = 1.3(1)^{0.3} = 1.3$

$L(x) = 1 + (x-1)(1.3)$

$\quad = 1.3x + 0.3$

15. $f(x) = \dfrac{e^x + e^{-x}}{2}, \ a = 2$

$f(2) = \dfrac{e^2 + e^{-2}}{2}$

$f'(x) = \dfrac{e^x - e^{-x}}{2}$

$f'(2) = \dfrac{e^2 - e^{-2}}{2}$

$L(x) = \dfrac{e^2 + e^{-2}}{2} + (x-2)\left(\dfrac{e^2 - e^{-2}}{2}\right)$

$\quad = \dfrac{(e^2 - e^{-2})x + 3e^{-2} - e^2}{2}$

17. $f(x) = \dfrac{1}{1 + e^{0.2x}}, \ a = 0$

$f(0) = \dfrac{1}{1 + e^{0.2(0)}} = \dfrac{1}{2} = 0.5$

$f'(x) = 0\dfrac{(1 + e^{0.2x}) - e^{0.2x}(0.2)}{(1 + e^{0.2x})^2} = -\dfrac{0.2e^{0.2x}}{(1 + e^{0.2x})^2}$

$f'(0) = \dfrac{0.2e^{0.2(0)}}{\left[1 + e^{0.2(0)}\right]^2} = -0.05$

$L(x) = 0.5 + (x-0)(-0.05)$

$\quad = -0.05x + 0.5$

19. $S(r) = 4\pi r^2, \ a = 10$

$S(10) = 4\pi(10)^2 = 400\pi$

$S'(r) = 8\pi r$

$S'(10) = 8\pi(10) = 80\pi$

$L(r) = 400\pi + (r-10)80\pi$

$\quad = 80\pi(r-5)$

In Exercises 21 - 23 use the linear approximation to $f(x) = \sqrt{x}, \ L(x) = \sqrt{a} + (x-a)\dfrac{1}{2\sqrt{a}}$.

21. $\sqrt{16.3}, \ a = 16$

$\sqrt{x} = \sqrt{16.3} \approx \sqrt{16} + (16.3 - 16)\dfrac{1}{2\sqrt{16}} = 4.0375$

23. $\sqrt{48.82}, \ a = 49$

$\sqrt{x} = \sqrt{48.82} \approx \sqrt{49} + (48.82 - 49)\dfrac{1}{2\sqrt{49}}$

≈ 6.98714

25. $(8.1)^{2/3}$, $a = 8$

$$f(x) = x^{2/3}$$

$$f'(x) = \frac{2}{3x^{1/3}}$$

$$x^{2/3} = (8.1)^{2/3} \approx 8^{2/3} + (8.1-8)\frac{2}{3(8)^{1/3}}$$

$$\approx 4.0333$$

26. $(3.9)^{3/2}$, $a = 4$

$$f(x) = x^{3/2}$$

$$f'(x) = \frac{3x^{1/2}}{2}$$

$$x^{3/2} = (3.9)^{3/2} \approx 4^{3/2} + (3.9-4)\frac{3(4)^{1/2}}{2}$$

$$\approx 7.7$$

27. $e^{0.3}$, $a = 0$

$$f(x) = e^x$$

$$f'(x) = e^x$$

$$e^x = e^{0.3} \approx e^0 + (0.3-0)e^0$$

$$\approx 1.3$$

28. $e^{-0.2}$, $a = 0$

$$f(x) = e^x$$

$$f'(x) = e^x$$

$$e^x = e^{-0.2} \approx e^0 + (-0.2-0)e^0$$

$$\approx 0.8$$

29. $\ln(0.95)$, $a = 1$

$$f(x) = \ln x$$

$$f'(x) = \frac{1}{x}$$

$$\ln x = \ln(0.95) \approx \ln 1 + (0.95-1)\frac{1}{1}$$

$$\approx -0.05$$

30. $\ln(1.05)$, $a = 1$

$$f(x) = \ln x$$

$$f'(x) = \frac{1}{x}$$

$$\ln x = \ln(1.05) \approx \ln 1 + (1.05-1)\frac{1}{1}$$

$$\approx 0.05$$

In Exercises 31 - 36 use the relative change in $f \approx (b-a)\dfrac{f'(a)}{f(a)}$.

31. $f(x) = 2x^2 - x$; $a = 3$, $b = 3.5$

$$f'(x) = 4x - 1$$

$$\text{Percentage change} \approx (3.5-3)\left[\frac{4(3)-1}{2(3)^2-3}\right]$$

$$\approx 0.3667$$

$$\approx 36.67\%$$

32. $f(x) = 5x^2 + x$; $a = 1$, $b = 1.2$

$$f'(x) = 10x + 1$$

$$\text{Percentage change} \approx (1.2-1)\left[\frac{10(1)+1}{5(1)^2+1}\right]$$

$$\approx 0.3667$$

$$\approx 36.67\%$$

33. $f(x) = e^x$; $a = 0$, $b = -0.3$

$f'(x) = e^x$

Percentage change $\approx (-0.3 - 0) \dfrac{e^0}{e^0}$

≈ -0.3

$\approx -30\%$

35. $f(x) = \dfrac{1}{x}$; $a = 5$, $b = 6$

$f'(x) = -\dfrac{1}{x^2}$

Percentage change $\approx (6 - 5) \left[\dfrac{-1/(5)^2}{1/5} \right]$

≈ -0.2

$\approx -20\%$

37. $C(x) = 1000 + 150x - 0.01x^2$, $a = 100$

$C(100) = 1000 + 150(100) - 0.01(100)^2$

$= 15{,}900$

$C'(x) = 150 - 0.02x$

$C'(100) = 150 - 0.02(100) = 148$

(a) $L(x) = 15{,}900 + (x - 100)148$

$= 148x + 1100$

$C(105) \approx 148(105) + 1100 = \$16{,}640$

(b) $a = 100$, $b = 105$

Percentage increase $\approx (105 - 100) \dfrac{148}{15{,}900}$

$\approx 4.65\%$

39. $\overline{C}(Q) = 350 + \dfrac{9000}{Q}$; $a = 1000$, $b = 1050$

$\overline{C}(1000) = 350 + \dfrac{9000}{1000} = 359$

$\overline{C}'(Q) = -\dfrac{9000}{Q^2} = -\dfrac{9000}{Q^2}$

$\overline{C}'(1000) = -\dfrac{9000}{(1000)^2} = -0.009$

Percentage change $\approx (1050 - 1000)\left(\dfrac{-0.009}{359} \right)$

$\approx -0.13\%$ (decrease)

41. $C(q) = 5000 + 120q^2$, $S(q) = 600q$

(a) $N(q) = C(q) - S(q)$

$= 120q^2 - 600q + 5000$

(b) $a = 15$, $b = 1.10(15) = 16.5$

$N(15) = 120(15)^2 - 600(15) + 5000 = 23{,}000$

$N'(q) = 240q - 600$

$N'(15) = 240(15) - 600 = 3000$

Percentage change $\approx (16.5 - 15)\left(\dfrac{3000}{23{,}000} \right)$

Daily net cost increase $\approx 19.57\%$

43. $q = Kp^{-1.040}$; $q = 2000$, $p = 10$

    (a) $2000 = K(10)^{-1.040}$

        $K \approx 21,930$

        $q = 21,930p^{-1.040}$

    (b) $a = 10$, $b = 12$

        $q(10) = 2000$

        $q'(p) = -22,807p^{-2.040}$

        $q'(10) = -22,807(10)^{-2.040} = -208$

    Percentage change $\approx (12 - 10)\left(\dfrac{-208}{2000}\right)$

    Decrease in sales $\approx -20.8\%$

(c) $P(q) = R(q) - C(q)$

      $= qp - 7.50q$

      $= 21,930p^{-0.040} - 164,475p^{-1.040}$

  $P(10) = 21,930(10)^{-0.040} - 164,475(10)^{-1.040}$

      $= 5000$

  $P'(q) = -877.2p^{-1.040} + 171,054p^{-2.040}$

  $P'(10) = -877.2(10)^{-1.040} + 171,054(10)^{-2.040}$

      $= 1480$

$a = 10$, $b = 12$

Percentage change $\approx (12 - 10)\left(\dfrac{1480}{5000}\right)$

Increase in profits $\approx 59.2\%$

45. $A = \pi r^2 t$, $t = 0.1$, $a = 5$

    $A(r) = 0.1\pi r^2$

    $A(5) = 0.1\pi(5)^2 = 2.5\pi$

    $A'(r) = 0.2\pi r$

    $A'(5) = 0.2\pi(5) = \pi$

    $b = 5 \pm 0.01(5) = 5 \pm 0.05$

    Percentage change $\approx (5 \pm 0.05 - 5)\left(\dfrac{\pi}{2.5\pi}\right)$

                  $\approx \pm 2\%$

CHAPTER 12
APPLICATIONS OF THE DERIVATIVE

12.1 EXERCISES

1. Abs. min. of -1 at $x = 3$ and $x = -3$, abs. max. of 2 at $x = 1$. Increasing on $[-3, 1]$ decreasing on $[1, 3]$.

3. Abs min of 0 at $x = -3$ and $x = 1$, abs max of 2 at $x = -1$ and $x = 3$. Increasing on $[-3, -1]$ and $[1, 3]$, and decreasing on $[-1, 1]$.

5. Local min of 1 at $x = -1$. Increasing on $[-1, 0)$ and $(0, +\infty)$, decreasing on $(-\infty, -1]$.

7. Local max of 0 at $x = -3$, abs min of -1 at $x = -2$, stationary point at $(1, 1)$. Increasing on $[-2, +\infty)$, and decreasing on $[-3, -2]$.

9. $f(x) = (x-1)(x-2)$; with domain $[0, +\infty)$
   Local max at $(0.0, 2.0)$, abs min at $(1.5, -0.3)$

11. $f(x) = x^2$ with domain $[0.1, +\infty)$
    Local max at $(0.1, 0.8)$, abs min at $(0.4, 0.7)$

13. $f(x) = x(x-1)^{2/3}$ with domain all real numbers
    Local max at $(0.6, 0.3)$, local min at $(1.0, 0.0)$

15. $f(x) = \dfrac{e^{-x}}{1 + e^{-x}}$ with domain all real numbers
    No maxima or minima

17. $f(x) = x^2 - 4x + 1$ with domain $[0, 3]$

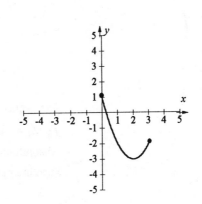

17. continued

$f'(x) = 2x - 4$; no singular points

Stationary points; $\quad 0 = 2x - 4$

$$x = 2$$

$f(2) = -3$

End points: $x = 0, \; x = 3$

$f(0) = 1, \; f(3) = -2$

Absolute maximum of 1 at 0; absolute minimum of -3 at 2; local maximum of -2 at 3.

19. $g(x) = x^3 - 12x$ with domain $[-4, 4]$

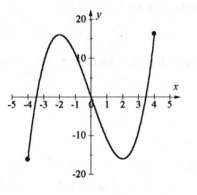

$g'(x) = 3x^2 - 12$; no singular points

Stationary points: $\quad 0 = 3x^2 - 12$

$$x^2 = 4$$

$$x = \pm 2$$

$$g(-2) = 16, \; g(2) = -16$$

End points: $x = -4, \; x = 4$

$g(-4) = -16, \; g(4) = 16$

Absolute minimum of -16 at -4; absolute maximum of 16 at -2; absolute minimum of -16 at 2; absolute maximum of 16 at 4.

21. $f(t) = t^3 + t$; with domain $[-2, 2]$

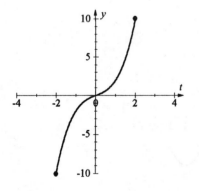

$f'(t) = 3t^2 + 1$; no singular points

Stationary points: $\quad 0 = 3t^2 + 1$

$$t^2 = -\frac{1}{3}$$

No real values

End points: $t = -2, \; t = 2$

$f(-2) = -10, f(2) = 10$

Absolute minimum of -10 at -2; absolute maximum of 10 at 2.

330

23. $h(t) = 2t^3 + 3t^2$; with domain $[-2, +\infty)$

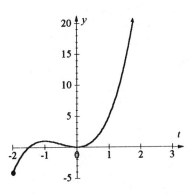

$h'(t) = 6t^2 + 6t$; no singular points

Stationary points: $0 = 6t^2 + 6t$

$$6t(t + 1) = 0$$
$$t = 0, \ t = -1$$
$$h(0) = 0, \ h(-1) = 1$$

End points: $t = -2$
$$h(-2) = -4$$

Absolute minimum of -4 at -2; local maximum of 1 at -1; local minimum of 0 at 0.

25. $f(x) = x^4 - 4x^3$; with domain $[-1, +\infty)$

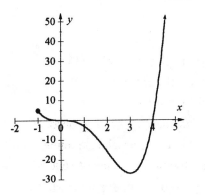

$f'(x) = 4x^3 - 12x^2$; no singular points

Stationary points: $0 = 4x^3 - 12x^2$

$$4x^2(x - 3) = 0$$
$$x = 0, \ x = 3$$
$$f(0) = 0, \ f(3) = -27$$

End points: $x = -1$
$$f(-1) = 5$$

Local max of 5 at -1; absolute min of -27 at 3.

27. $g(t) = \frac{1}{4}t^4 - \frac{2}{3}t^3 + \frac{1}{2}t^2$; with domain $(-\infty, +\infty)$

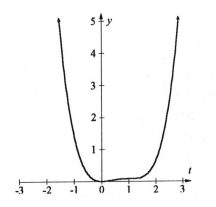

$g'(t) = t^3 - 2t^2 + t$; no singular points

Stationary points: $0 = t^3 - 2t^2 + t$

$$t(t^2 - 2t + 1) = 0$$
$$t(t - 1)^2 = 0$$
$$t = 0, \ t = 1$$
$$g(0) = 0, \ g(1) = \frac{1}{12}$$

No end points

Absolute min of 0 at 0.

29. $f(t) = \dfrac{t^2+1}{t^2-1}; \ -2 \le t \le 2, \ t \ne \pm 1$

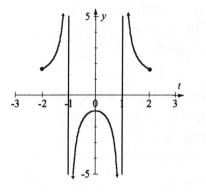

$$f'(t) = \frac{2t(t^2-1)-(t^2+1)(2t)}{(t^2-1)^2} = \frac{-4t}{(t^2-1)^2}$$

No singular points

Stationary points: $\ 0 = -\dfrac{4t}{(t^2-1)^2}$

$$4t = 0$$
$$t = 0$$
$$f(0) = -1$$

End points: $\ t = -2, \ t = 2$

$f(-2) = \dfrac{5}{3}, \ f(2) = \dfrac{5}{3}$

Local min of $5/3$ at -2; local max of -1 at 0; local min of $5/3$ at 2.

31. $f(x) = \sqrt{x}(x-1) = x^{1/2}(x-1), \ x \ge 0$

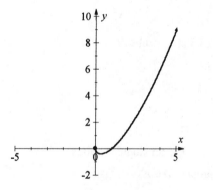

$$f'(x) = \frac{1}{2}x^{-1/2}(x-1) + x^{1/2}$$

$$= \frac{x-1}{2\sqrt{x}} + \sqrt{x}$$

$$= \frac{3x-1}{2\sqrt{x}}$$

Singular points: $\ x = 0, \ f'(0)$ doesn't exist

Stationary points: $\ 0 = \dfrac{3x-1}{2\sqrt{x}}$

$$3x - 1 = 0$$
$$x = 1/3$$
$$f(1/3) = -\frac{2}{3\sqrt{3}}$$

End points: $\ x = 0$

$$f(0) = 0$$

Local max of 0 at $x = 0$; absolute min of $-2/(3\sqrt{3})$ at $x = 1/3$.

33. $g(x) = x^2 - 4\sqrt{x} = x^2 - 4x^{1/2}$

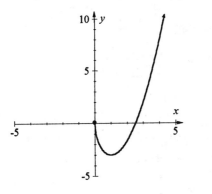

$g'(x) = 2x - 2x^{-1/2} = \dfrac{2x\sqrt{x} - 2}{\sqrt{x}}$

Singular points: $x = 0$, $f'(0)$ doesn't exist

Stationary points: $0 = \dfrac{2x\sqrt{x} - 2}{\sqrt{x}}$

$$2x\sqrt{x} - 2 = 0$$
$$x\sqrt{x} = 1$$
$$g(1) = -3$$

End points: $x = 0$

$$g(0) = 0$$

Local max of 0 at $x = 0$; absolute min of -3 at $x = 1$.

35. $g(x) = \dfrac{x^3}{x^2 + 3}$

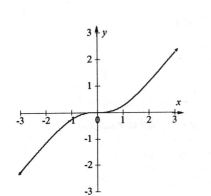

$g'(x) = \dfrac{3x^2(x^2 + 3) - x^3(2x)}{(x^2 + 3)^2} = \dfrac{x^4 + 9x^2}{(x^2 + 3)^2}$

No singular points

Stationary points: $0 = \dfrac{x^4 + 9x^2}{(x^2 + 3)^2}$

$$x^2(x^2 + 9) = 0$$
$$x^2 = 0, \quad x^2 + 9 = 0$$
$$x = 0 \qquad x^2 = -9$$
$$g(0) = 0 \quad \text{Not a real value}$$

No end points

No local extrema

37. $f(x) = x - \ln x$; with domain $(0, +\infty)$

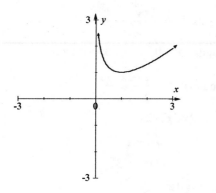

$$f'(x) = 1 - \frac{1}{x} = \frac{x-1}{x}$$

No singular points

Stationary points: $0 = \dfrac{x-1}{x}$

$$x - 1 = 0$$
$$x = 1$$
$$f(1) = 1$$

No end points

Absolute min of 1 at 1

39. $g(t) = e^t - t$; with domain $[-1, 1]$

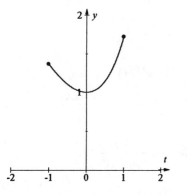

$$g'(t) = e^t - 1$$

No singular points

Stationary points: $0 = e^t - 1$

$$e^t = 1$$
$$t = 0$$
$$g(0) = 1$$

End points: $t = -1$, $t = 1$

$$g(-1) = 1 + \frac{1}{e}, \ \ g(1) = e - 1$$

Local max of $1 + 1/e$ at -1; absolute min of 1 at 0; absolute max of $e - 1$ at 1

41. $f(x) = \dfrac{2x^2 - 24}{x + 4}$

$$f'(x) = \frac{4x(x+4) - (2x^2 - 24)}{(x+4)^2} = \frac{2x^2 + 16x + 24}{(x+4)^2}$$

Singular points: $0 = 2x^2 + 16x + 24$

$$x^2 + 8x + 12 = 0$$
$$(x+2)(x+6) = 0$$
$$x = -2, \ x = -6$$

$f(-2) = -8, \ f(-6) = -24$

Local max at $(-6, -24)$, local min at $(-2, -8)$

43. $f(x) = xe^{1-x^2}$

$f'(x) = e^{1-x^2} + xe^{1-x^2}(-2x)$

$\quad = e^{1-x^2}(1-2x^2)$

Stationary points: $0 = e^{1-x^2}(1-2x^2)$

$\quad\quad\quad\quad\quad\quad 1-2x^2 = 0$

$\quad\quad\quad\quad\quad\quad\quad x^2 = \dfrac{1}{2}$

$\quad\quad\quad\quad\quad\quad\quad x = \pm\dfrac{1}{\sqrt{2}}$

$f\left(-\dfrac{1}{\sqrt{2}}\right) = -\sqrt{\dfrac{e}{2}},\ f\left(\dfrac{1}{\sqrt{2}}\right) = \sqrt{\dfrac{e}{2}}$

Abs max at $\left(\dfrac{1}{\sqrt{2}}, \sqrt{\dfrac{e}{2}}\right)$, abs min at $\left(-\dfrac{1}{\sqrt{2}}, -\sqrt{\dfrac{e}{2}}\right)$

45. Stationary points: $x = -1$

$x = -1$, minimum because the slope changes from negative to positive.

47. Stationary points: $x = -3, -1, 1$

$x = -3$, minimum because the slope changes from negative to positive.

$x = -1$, maximum because the slope changes from positive to negative.

$x = 1$, minimum because the slope changes from negative to positive.

49. Stationary points: $x = 1$

$x = 1$, non-extremum because the slope doesn't change sign.

Singular points: $x = 0$

$x = 0$, minimum because the slope changes from negative to positive.

51. Stationary points: $x = -3, -1, 1, 3$

$x = -3$, maximum because the slope changes from positive to negative.

$x = -1$, minimum because the slope changes from negative to positive.

$x = 1$, maximum because the slope changes from positive to negative.

$x = 3$, minimum because the slope changes from negative to positive

53. $y = x^2 + \dfrac{1}{x-2}$; with domain $(-3, 2) \cup (2, 6)$

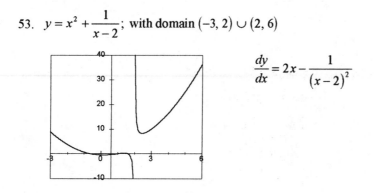

$$\frac{dy}{dx} = 2x - \frac{1}{(x-2)^2}$$

55. $f(x) = (x-5)^2(x+4)(x-2)$; $D[-5,6]$

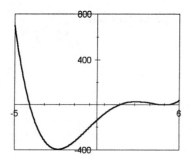

$$f'(x) = 2(x-5)(x+4)(x-2) + (x-5)^2(x-2)$$
$$+ (x-5)^2(x+4)$$
$$= 2(x-5)(2x^2 - 2x - 13)$$

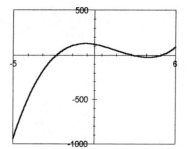

12.2 EXERCISES

1. $P = xy$

   with $x + y = 10$ or $y = 10 - x$.

   $P = x(10 - x) = 10x - x^2$

   Maximize P

   $\dfrac{dP}{dx} = 10 - 2x$

   $0 = 10 - 2x$

   $x = 5$ (critical point)

   $y = 10 - x = 10 - 5 = 5$

   Maximum $P = 10(5) - 5^2 = 25$

x	2	5	7
P	16	25	21

   $x = y = 5$; $P = 25$

3. $S = x + y$ with $xy = 9$ or $y = \dfrac{9}{x}$

   and $x > 0$, $y > 0$.

   $S = x + \dfrac{9}{x}$; $D(0, +\infty)$

   Minimize S

   $\dfrac{dS}{dx} = 1 - \dfrac{9}{x^2}$

   $0 = 1 - \dfrac{9}{x^2}$

   $x^2 = 9$

   $x = 3$ or $x = -3$

   (critical point)   (outside the domain)

   $y = \dfrac{9}{x} = \dfrac{9}{3} = 3$

   Minimum $S = 3 + \dfrac{9}{3} = 6$

x	1	3	9
S	10	6	10

   $x = y = 3$; $S = 6$

5. $F = x^2 + y^2$

   with $x + 2y = 10$ or $x = 10 - 2y$.

   $F = (10 - 2y)^2 + y^2 = 5y^2 - 40y + 100$

   Minimize F

   $\dfrac{dF}{dy} = 10y - 40$

   $0 = 10y - 40$

   $y = 4$ (critical point)

   $x = 10 - 2y = 10 - 2(4) = 2$

   Minimum $F = 5(4)^2 - 40(4) + 100 = 20$

y	2	4	6
F	40	20	40

   $x = 2$, $y = 4$; $F = 20$

7. $P = xyz$ with $x + y = 30$ or $x = 30 - y$,
$y + z = 30$ or $z = 30 - y$ and
$x \geq 0, \ y \geq 0, \ z \geq 0$.
$P = (30 - y)(30 - y) = y(30 - y)^2$; $D[0, 30]$
Maximize P
$\dfrac{dP}{dy} = (30 - y)^2 - 2y(30 - y) = (30 - y)(30 - 3y)$
$0 = (30 - y)(30 - 3y)$
$y = 30$ or $y = 10$ (critical points)

y	0	10	20	30
P	0	4000	2000	0

Maximum $P = 1.0(30 - 10)^2 = 4000$
$x = 30 - y = 30 - 10 = 20$
$z = 30 - y = 30 - 10 = 20$
$x = 20, \ y = 10, \ z = 20; \ P = 4000$

9. Let $x = $ width, $y = $ length, $A = $ area
$A = xy$ with $2x + 2y = 20$
or $x = 10 - y$ and $x \geq 0, \ y \geq 0$. If $x \geq 0$
then $10 - y \geq 0$ or $y \leq 10$
$A = (10 - y)y = 10y - y^2$; $D[0, 10]$
Maximize A
$\dfrac{dA}{dy} = 10 - 2y$
$0 = 10 - 2y$
$y = 5$ (critical points)
$x = 10 - y = 10 - 5 = 5$
Maximum $A = 10(5) - 5^2 = 25$

y	0	5	10
A	0	25	0

Dimensions: 5×5

11. Let $x = $ E& W sides, $y = $ N& S sides, $A = $ area
$A = xy$ with $2x(4) + 2y(2) = 80$
or $y = 20 - 2x$ and $x \geq 0, \ y \geq 0$.
If $y \geq 0$, then $20 - 2x \geq 0$ or $x \leq 10$.
$A = x(20 - 2x) = 20x - 2x^2$; Domain$[0, 10]$
Maximize A
$\dfrac{dA}{dx} = 20 - 4x$
$0 = 20 - 4x$
$x = 5$
$y = 20 - 2(5) = 10$
Maximum $A = 20(5) - 2(5)^2 = 50$

x	0	5	10
A	0	50	0

Area $= 5 \times 10$

13. $R = pq$ with $q = 200,000 - 10,000p$
and $p \geq 0$, $q \geq 0$.
If $q \geq 0$, then $200,000 - 10,000p \geq 0$
or $p \leq 20$.
$R = p(200,000 - 10,000p)$
$ = 200,000p - 10,000p^2$; $D[0, 20]$
Maximize R
$\dfrac{dR}{dp} = 200,000 - 20,000p$
$0 = 200,000 - 20,000p$
$p = 10$ (critical point)
Maximum $R = 200,000(10) - 10,000(10)^2$
$ = 1,000,000$

p	0	10	20
R	0	1,000,000	0

$p = \$10$

15. $R = pq$ with $q = -\dfrac{4}{3}p + 80$
and $p \geq 0$, $q \geq 0$. If $q \geq 0$,
then $-\dfrac{4}{3}p + 80 \geq 0$ or $p \leq 60$.
$R = p\left(-\dfrac{4}{3}p + 80\right) = -\dfrac{4}{3}p^2 + 80p$; $D[0, 60]$
Maximize R
$\dfrac{dR}{dp} = -\dfrac{8}{3}p + 80$
$0 = -\dfrac{8}{3}p + 80$
$p = 30$ (critical point)
Maximum $R = -\dfrac{4}{3}(30)^2 + 80(30) = 1200$

p	0	30	60
R	0	1200	0

$p = \$30$

17. (a) $R = pq$ with $p = \dfrac{500,000}{q^{1.5}}$
and $p \geq 0$, $q \geq 5000$.
$R = \dfrac{500,000}{q^{1.5}}(q) = 500,000q^{-1/2}$; $D[5000, +\infty)$
Maximize R
$\dfrac{dR}{dq} = -250,000q^{-3/2}$
$0 = \dfrac{-250,000}{q^{1.5}}$
There are no critical values.

q	5000	10,000	40,000
R	7,071.07	5000	2500

If $q = 5000$, $p = \dfrac{500,000}{(5000)^{1.5}} = 1.41$

$p_{max} = \$1.41$ per pound
(b) $q_{max} = 5000$ pounds
(c) $R_{max} = 7,071.07$

19. $p_1 = 0.25$ when $q_1 = 22$

   $p_2 = 0.14$ when $q_2 = 27.5$

   $m = \dfrac{27.5 - 22}{0.14 - 0.25} = -50$

   $q - q_1 = m(p - p_1)$

   $q - 22 = -50(p - 0.25)$

   $q = -50p + 34.5$

   where q is in pounds per year
   and p is in dollars.

   $R = pq$ with $q = -50p + 34.5$

   and $p \geq 0$, $q \geq 0$. If $q \geq 0$

   $-50p + 34.5 \geq 0$ or $p \leq 0.69$

   $R = p(-50p + 34.5) = -50p^2 + 34.5p$; $D[0, 0.69]$

   Maximize R

   $\dfrac{dR}{dp} = -100p + 34.5$

   $0 = -100p + 34.5$

   $p = 0.345$ (critical point)

   Maximum $R = -50(0.345)^2 + 34.5(0.345) = 5.95$

p	0	0.345	0.69
R	0	5.95	0

34.5 cents per pound, for an annual (per capita) revenue of $5.95.

21. $P = R - C = pq - 25q$

   with $q = -\dfrac{4}{3}p + 80$ and

   $p \geq 0$, $q \geq 0$. If $q \geq 0$, then

   $-\dfrac{4}{3}p + 80 \geq 0$ or $p \leq 60$.

   $P = p\left(-\dfrac{4}{3}p + 80\right) - 25\left(-\dfrac{4}{3}p + 80\right)$

   $= -\dfrac{4}{3}p^2 + \dfrac{340}{3}p - 2000$; $D[0, 60]$

   Maximize P

   $\dfrac{dP}{dp} = -\dfrac{8}{3}p + \dfrac{340}{3}$

   $0 = -\dfrac{8}{3}p + \dfrac{340}{3}$

   $p = 42.5$ (critical point)

   Maximum $P = -\dfrac{4}{3}(42.5)^2 + \dfrac{340}{3}(42.5) - 2000 = 408.33$

p	0	42.5	60
P	-2000	408.33	0

$42.50 per ruby, for a weekly profit of $408.33.

23. (a) $P = R - C = pq - 100q$ with

$p = \dfrac{1000}{q^{0.3}}$ and $p \geq 0$, $q \geq 0$.

$P = \dfrac{1000}{q^{0.3}}(q) - 100q = 1000q^{0.7} - 100q$; $D[0, +\infty)$

Maximize P

$\dfrac{dP}{dq} = 0.7(1000)q^{-0.3} - 100$

$= 700q^{-0.3} - 100$

$0 = \dfrac{700}{q^{0.3}} - 100$

$q^{0.3} = 7$

$q = 7^{1/0.3} = 656$ (critical point)

Maximum $P = 1000(656)^{0.7} - 100(656) = 28{,}120$

q	0	656	1000
P	0	28,120	25,892

656 headsets, for a profit of \$28,120.09

(b) $p = \dfrac{1000}{q^{0.3}} = \dfrac{1000}{(656)^{0.3}} = \$143$ per headset

25.

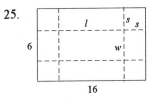

$V = lws$ with $w = 6 - 2s$,

$l = 16 - 2s$, and $l \geq 0$, $w \geq 0$, $s \geq 0$.

If $w \geq 0$, then $6 - 2s \geq 0$ or $s \leq 3$.

$V = (16 - 2s)(6 - 2s)s$; $D[0, 3]$

Maximize V

$\dfrac{dV}{ds} = -2s(6 - 2s) - 2s(16 - 2s) + (16 - 2s)(6 - 2s)$

$= -4s(11 - 2s) + (16 - 2s)(6 - 2s)$

$0 = 12s^2 - 88s + 96$

$0 = 3s^2 - 22s + 24$

$0 = (3s - 4)(s - 6)$

$s = 4/3$ or $s = 6$

(critical point) (not in domain)

s	0	4 / 3	3
V	0	1600 / 27	0

Maximum $V = [16 - 2(4/3)][6 - 2(4/3)](4/3)$

$= 1600/27$ cubic inches

27. $p = ve^{-0.05t}$ with $v = 3{,}000{,}000 + 1000t^2$ and $t \geq 0$.

   $p = (300{,}000 + 1000t^2)e^{-0.05t}$; $D[0, +\infty)$

   Maximize p

   $\dfrac{dp}{dt} = 2000te^{-0.05t} + (300{,}000 + 1000t^2)e^{-0.05t}(-0.05)$

   $\quad = e^{-0.05t}(-15{,}000 + 2000t - 50t^2)$

   $0 = -15{,}000 + 2000t - 50t^2$

   $0 = t^2 - 40t + 300$

   $0 = (t - 30)(t - 10)$

   $t = 30$ or $t = 10$ (critical points)

t	0	2.78	10	30	50
p	300,000	267,756	242,612	267,756	229,838

Local Maximum $p = 267{,}756$

Local Minimum $p = 242{,}612$

Sell within the next 2.78 years' or in 30 years time.

29. $R = pq$ with $p = 100 + 2t$,

   $q = 400{,}000 - 2500t$ and $t \geq 0$.

   $R = (100 + 2t)(400{,}000 - 2500t)$; $D[0, +\infty)$

   Maximize R

   $\dfrac{dR}{dt} = 2(400{,}000 - 2500t) + (-2500)(100 + 2t)$

   $\quad = 550{,}000 - 10{,}000t$

   $0 = 550{,}000 - 10{,}000t$

   $t = 55$ (critical point)

   Maximum $R = [100 + 2(55)][400{,}000 - 2500(55)]$

   $\quad = 55{,}125{,}000$

t	0	55	100
R	40,000,000	55,125,000	45,000,0000

Release in 55 days

31. Let $F = $ total yield, $y = $ yield per tree,

   $n = $ number of trees, and $t = $ additional trees planted.

   $F = yn$ with $y = 100 - t$, $n = 50 + t$ and $t \geq 0$.

   $F = (100 - t)(50 + t) = 5000 + 50t - t^2$; $D[0, +\infty)$

   Maximize F

   $\dfrac{dF}{dt} = 50 - 2t$

   $0 = 50 - 2t$

   $t = 25$ (critical point)

   Maximum $F = 5000 + 50(25) - (25)^2 = 5625$

t	0	25	50
F	5000	5625	5000

Plant 25 additional trees

33. $C(x) = 150,000 + 20x + \dfrac{x^2}{100}$

$150,000 is the fixed cost, $20x$ is the linear

variation in cost and $\dfrac{x^2}{100}$ is the quadratic

variation in cost.

$\overline{C}(x) = \dfrac{C(x)}{x} = \dfrac{150,000}{x} + 20 + \dfrac{x}{10,000}$

with $x > 0$.

Minimize $\overline{C}(x)$

$\overline{C}'(x) = -\dfrac{150,000}{x^2} + \dfrac{1}{10,000}$

$0 = -\dfrac{150,000}{x^2} + \dfrac{1}{10,000}$

$x^2 = 1,500,000,000$

$x = 38,730$ (critical point) or

$x = -38,730$ (not in the domain)

Minimum $\overline{C}(x) = \dfrac{150,000}{38,730} + 20 + \dfrac{38,730}{10,000} \approx 28$

x	10,000	38,730	100,000
$\overline{C}(x)$	36	28	32

Manufacture 38,730 CD players,

giving an average cost of $28 per player

35. Let N = net cost

$N = C - 500q$ with $C(q) = 4000 + 100q^2$

and $q \geq 0$.

$N(q) = 4000 + 100q^2 - 500q; \quad D[0, +\infty)$

Minimize N

$N'(q) = 200q - 500$

$0 = 200q - 500$

$q = 2.5$ (critical point)

q	0	2.5	5
$N(q)$	4000	3375	4000

Remove 2.5 pounds of pollutant per day

37. $V = lwh$ with $l + w = 45$ or $l = 45 - w$,

$h + w = 45$ or $h = 45 - w$, and $h \geq 0$,

$l \geq 0$, $w \geq 0$. If $h \geq 0$, then $45 - w \geq 0$

or $w \leq 45$.

$V = (45 - w)^2 w; \quad D[0, 45]$

Maximize V

$\dfrac{dV}{dw} = (45 - w)^2 + 2w(45 - w)(-1)$

$= (45 - w)(45 - 3w)$

$0 = (45 - w)(45 - 3w)$

$w = 45$ or $w = 15$ (critical points)

w	0	15	45
V	0	13,500	0

$l = 45 - w = 45 - 15 = 30$ inch

$h = 45 - w = 45 - 15 = 30$ inch

$w = 15$ inch

39. $V = ls^2$ with $l + 4s = 108$ or $l = 108 - 4s$, and $l \geq 0$, $s \geq 0$. If $l \geq 0$, then $108 - s \geq 0$ or $s \leq 27$.

$V = (108 - 4s)s^2$; D$[0, 27]$

Maximize V

$\dfrac{dV}{ds} = -4s^2(108 - 4s)(2s)$

$\qquad = 216s - 12s^2$

$\quad 0 = 12s(18 - s)$

$\quad s = 0$ or $s = 18$ (critical points)

s	0	18	27
V	0	11,664	0

Maximum $V = [108 - 4(18)](18)^2 = 11{,}664$ in^3

41. $\overline{P}(x) = \dfrac{P(x)}{x} = \dfrac{R - C}{x} = \dfrac{R}{x} - \dfrac{C}{x}$

with $\dfrac{R}{x} = \dfrac{500\sqrt{x}}{x} = \dfrac{500}{\sqrt{x}}$,

$\dfrac{C}{x} = \dfrac{10{,}000 + 2x}{x} = \dfrac{10{,}000}{x} + 2$ and $x > 0$.

$\overline{P}(x) = \dfrac{500}{\sqrt{x}} - \dfrac{10{,}000}{x} - 2$; D$[0, +\infty)$

Maximize $\overline{P}(x)$

$\overline{P}'(x) = -\dfrac{250}{x^{3/2}} + \dfrac{10{,}000}{x^2}$

$\quad 0 = -\dfrac{250}{x^{3/2}} + \dfrac{10{,}000}{x^2}$

$250x^{1/2} = 10{,}000$

$\qquad \sqrt{x} = 40$

$\qquad\quad x = 1600$ (critical point)

x	400	1600	2500
$\overline{P}(x)$	-2.00	4.25	4.00

Maximum $\overline{P}(x) = \overline{P}(1600) = \dfrac{500}{4} - \dfrac{10{,}000}{1600} - 2 = 4.25$

Marginal Profit $= P'(1600) = R'(1600) - C'(1600)$

$= \left[\dfrac{250}{\sqrt{x}} - 2 \right]_{x=1600} = 4.25$

Buy 1600 copies. At this value of x, average profit equals marginal profit; beyond this the marginal profit is smaller than the average.

344

43. $N(t) = 0.028234t^3 - 1.0922t^2 + 13.029t + 146.88 \quad (0 \le t \le 39)$

Find the rate of change of $N(t)$

$N'(t) = 0.084702t^2 - 2.1844t + 13.029 \quad (0 \le t \le 39)$

Minimize $N'(t)$

$\frac{d}{dt}[N'(t)] = 0.169404t - 2.1844$

$0 = 0.169404t - 2.1844$

$t = 12.9 \quad \text{(critical point)}$

t	0	12.9	39
$N'(t)$	13	−1	57

$t = 13$ represents the year 1963

$t = 39$ represents the year 1989

Decreasing the most rapidly in 1963;

increasing the most rapidly in 1989

45. $c(t) = -0.00271t^3 + 0.137t^2 - 0.892t + 0.149 \quad (8 \le t \le 30)$

$c'(t) = -0.00813t^2 + 0.274t - 0.892$

Maximize $c'(t)$

$\frac{dc'(t)}{dt} = -0.01626t + 0.274$

$0 = -0.01626t + 0.274$

$t = 16.85 \quad \text{(critical point)}$

t	8	16.85	30
$c'(t)$	0.8	1.4	0.01

Maximum when $t = 17$ days. This means that the embryo's oxygen consumption is increasing most rapidly 17 days after the egg is laid.

47. The amount of plastic depends on the area.

$A = \pi r^2 + 2\pi rh$ with $5000 = \pi r^2 h$

or $h = \frac{5000}{\pi r^2}$ and $r > 0$.

$A = \pi r^2 + 2\pi r\left(\frac{5000}{\pi r^2}\right) = \pi r^2 + \frac{10,000}{r}; \ D$

Minimize A

$\frac{dA}{dt} = 2\pi r - \frac{10,000}{r^2}$

$0 = 2\pi r - \frac{10,000}{r^2}$

$0 = 2\pi r^3 - 10,000$

$r^3 = 1592$

$r = 11.68 \quad \text{(critical point)}$

$h = \frac{5000}{\pi(11.68)^2} = 11.7$

r	1	11.7	20
A	10,000	1285	1757

Dimensions: $h = r = 11.7$ cm

345

$$-6,760,800 \quad (12 \le n \le 15)$$
$$I'(n) = 8786.4n^2 - 231,720n + 1,532,900; \quad D[12, 15]$$

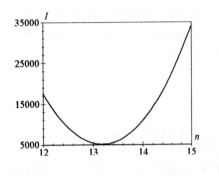

n	12	13.2	15
$I'(n)$	17,502	5137	34,040

Absolute minimum of 5137 at $n = 13.2$, absolute maximum of 34,040 at $n = 15$. The salary value per extra year of school is increasing slowly ($5137 per year) at a level of 13.2 years of schooling, and most rapidly ($34,040 per year) at a level of 15 years of schooling.

51. $p = v(1.05)^{-t}$ with $v = \dfrac{10,000}{1 + 500e^{-0.5t}}$ and $t \ge 0$

$$p = \frac{10,000(1.05)^{-t}}{1 + 500e^{-0.5t}}; \quad D[0, +\infty)$$

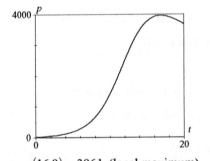

$p(16.9) = 3961$ (local maximum)
You should sell them in 17 years' time, when they will be worth approximately $3,961.

53. $C = 20,000x + 365y$ with $1000 = x^{0.4}y^{0.6}$ or
$y = \left(1000x^{-0.4}\right)^{1/0.6} = 100,000x^{-2/3}$ and $x > 0$.
$$C(x) = 20,000x + \frac{36,500,000}{x^{2/3}}; \quad D(0, +\infty)$$

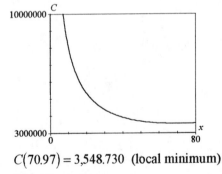

$C(70.97) = 3,548.730$ (local minimum)
Hire 71 employees

Maximize TR

$$\frac{d}{dQ}(TR) = b - 2cQ$$

$$0 = b - 2cQ$$

$$Q = b / 2c$$

Answer (d)

12.3 EXERCISES

1. $\quad y = 3x^2 - 6$

$$\frac{dy}{dx} = 6x$$

$$\frac{d^2 y}{dx^2} = 6$$

3. $\quad y = \dfrac{2}{x} = 2x^{-1}$

$$\frac{dy}{dx} = -2x^{-2}$$

$$\frac{d^2 y}{dx^2} = -4x^{-3} = -\frac{4}{x^3}$$

5. $\quad y = 4x^{0.4} - x$

$$\frac{dy}{dx} = 1.6x^{-0.6} - 1$$

$$\frac{d^2 y}{dx^2} = -0.96x^{-1.6}$$

7. $\quad y = e^{-(x-1)} - x$

$$\frac{dy}{dx} = e^{-(x-1)}(-1) - 1 = -e^{-(x-1)} - 1$$

$$\frac{d^2 y}{dx^2} = e^{-(x-1)}$$

9. $\quad y = \dfrac{1}{x} - \ln x = x^{-1} - \ln x$

$$\frac{dy}{dx} = -x^{-2} - \frac{1}{x} = -x^{-2} - x^{-1}$$

$$\frac{d^2 y}{dx^2} = 2x^{-3} + x^{-2} = \frac{2}{x^3} + \frac{1}{x^2}$$

11. $\quad s(t) = 12 + 3t - 16t^2; \ t = 2$

    (a) $\ s'(t) = 3 - 32t$

        $s''(t) = -32 \ \text{ft} / \text{s}^2$

    (b) $\ s''(2) = -32 \ \text{ft} / \text{s}^2$

13. $\quad s(t) = \dfrac{1}{t} + \dfrac{1}{t^2} = t^{-1} + t^{-2}; \ t = 1$

    (a) $\ s'(t) = -t^{-2} - 2t^{-3}$

        $s''(t) = 2t^{-3} + 6t^{-4} = \dfrac{2}{t^3} + \dfrac{6}{t^4} \ \text{ft} / \text{s}^2$

    (b) $\ s''(1) = \dfrac{2}{1^3} + \dfrac{6}{1^4} = 8 \ \text{ft} / \text{s}^2$

15. $s(t) = \sqrt{t} + t^2 = t^{1/2} + t^2$; $t = 4$

   (a) $s'(t) = \frac{1}{2} t^{-1/2} + 2t$

   $s''(t) = -\frac{1}{4} t^{-3/2} + 2 = -\frac{1}{4t^{3/2}} + 2 \text{ ft}/\text{s}^2$

   (b) $s''(4) = -\frac{1}{4(4)^{3/2}} + 2 = \frac{63}{32} \text{ ft}/\text{s}^2$

In Exercises 17 - 23, the coordinates are the approximate coordinates of the points of inflection if there are any.

17. $(1, 0)$ 19. $(1, 0)$ 21. None   23. $(-1, 0)$, $(1, 1)$

25. $f(x) = x^2 + 2x + 1$
   LOCATE CRITICAL POINTS
   $f'(x) = 2x + 2$
   $0 = 2x + 2$
   $x = -1$
   $f(-1) = (-1)^2 + 2(-1) + 1 = 0$
   Stationary points: $(-1, 0)$
   CLASSIFY CRITICAL POINTS
   $f''(x) = 2$
   $f''(-1) = 2$
   Minimum at $(-1, 0)$
   No points of inflection; $f''(x) = 2$

27. $f(x) = 2x^3 + 3x^2 - 12x + 1$
   LOCATE CRITICAL POINTS
   $f'(x) = 6x^2 + 6x - 12$
   $0 = 6x^2 + 6x - 12$
   $0 = x^2 + x - 2$
   $0 = (x + 2)(x - 1)$
   $x = -2$ or $x = 1$
   $f(-2) = 2(-2)^3 + 3(-2)^2 - 12(-2) + 1 = 21$
   $f(1) = 2(1)^3 + 3(1)^2 - 12(1) + 1 = -6$
   Stationary points: $(-2, 21)$, $(1, -6)$

27. continued

CLASSIFY CRITICAL POINTS

$f''(x) = 12x + 6$

$f''(-2) = 12(-2) + 6 = -18$ (max)

$f''(1) = 12(1) + 6 = 18$ (min)

Maximum at $(-2, 21)$, minimum at $(1, -6)$

LOCATE POINTS OF INFLECTION

$f''(x) = 0$

$0 = 12x + 6$

$x = -\dfrac{1}{2}$

$f\left(-\dfrac{1}{2}\right) = 2\left(-\dfrac{1}{2}\right)^3 + 3\left(-\dfrac{1}{2}\right)^2 - 12\left(-\dfrac{1}{2}\right) + 1 = \dfrac{15}{2}$

Point of inflection at $\left(-\dfrac{1}{2}, \dfrac{15}{2}\right)$

29. $f(x) = -4x^3 - 3x^2 + 1$

LOCATE CRITICAL POINTS

$f'(x) = -12x^2 - 6x$

$0 = -12x^2 - 6x$

$0 = -6x(2x + 1)$

$x = -\dfrac{1}{2}$ or $x = 0$

$f\left(-\dfrac{1}{2}\right) = -4\left(-\dfrac{1}{2}\right)^3 - 3\left(-\dfrac{1}{2}\right)^2 + 1 = \dfrac{3}{4}$

$f(0) = -4(0)^3 - 3(0)^2 + 1 = 1$

Stationary points: $\left(-\dfrac{1}{2}, \dfrac{3}{4}\right)$, $(0, 1)$

CLASSIFY CRITICAL POINTS

$f''(x) = -24x - 6$

$f''\left(-\dfrac{1}{2}\right) = -24\left(-\dfrac{1}{2}\right) - 6 = 6$ (min)

$f''(0) = -24(0) - 6 = -6$ (max)

Minimum at $\left(-\dfrac{1}{2}, \dfrac{3}{4}\right)$, maximum at $(0, 1)$

LOCATE POINTS OF INFLECTION

$f''(x) = 0$

$0 = -24x - 6$

$x = -\dfrac{1}{4}$

$f\left(-\dfrac{1}{4}\right) = -4\left(-\dfrac{1}{4}\right)^3 - 3\left(-\dfrac{1}{4}\right)^2 + 1 = \dfrac{7}{8}$

Point of inflection at $\left(-\dfrac{1}{4}, \dfrac{7}{8}\right)$

31. $g(x) = (x-3)\sqrt{x} = (x-3)x^{1/2}$; $D[0, +\infty)$

LOCATE CRITICAL POINTS

$$g'(x) = x^{1/2} + (x-3)\left(\frac{1}{2}\right)x^{-1/2}$$

$$= \frac{3}{2}x^{1/2} - \frac{3}{2}x^{-1/2}$$

$$0 = \frac{3}{2}x^{1/2} - \frac{3}{2}x^{-1/2}$$

$$x^{-1/2} = x^{1/2}$$

$$x = 1$$

$$g(1) = (1-3)\sqrt{1} = -2$$

Stationary points: $(1, -2)$

CLASSIFY CRITICAL POINTS

$$g''(x) = 3x^{-1/2} + 3x^{-3/2}$$

$$g''(1) = 3(1)^{-1/2} + 3(1)^{-3/2} = 6 \text{ (min)}$$

Minimum at $(1, -2)$, Maximum at $(0, 0)$

$$g(0) = 0$$

End point: $(0, 0)$

LOCATE POINTS OF INFLECTION

$$g''(x) = 0$$

$$0 \neq 3x^{-1/2} + 3x^{-3/2}$$

$g''(0)$ is not defined, but $(0, 0)$ is an end point.

No points of inflection

33. $f(x) = x - \ln x$; $D(0, +\infty)$

LOCATE CRITICAL POINTS

$$f'(x) = 1 - \frac{1}{x}$$

$$0 = 1 - \frac{1}{x}$$

$$x = 1$$

$$f(1) = 1 - \ln 1 = 1$$

Stationary points: $(1, 1)$

CLASSIFY CRITICAL POINTS

$$f''(x) = -(-1)x^{-2} = \frac{1}{x^2}$$

$$f''(1) = \frac{1}{1^2} = 1 \text{ (min)}$$

Minimum at $(1, 1)$

LOCATE POINTS OF INFLECTION

$$f''(x) = 0$$

$$0 \neq \frac{1}{x^2}$$

$f''(0)$ is not defined, but $x = 0$ is not in the domain of f.

No points of inflection

LOCATE CRITICAL POINTS

$$f'(x) = 2x + \frac{1}{x^2}(2x) = 2x + \frac{2}{x}$$

$0 \neq 2x + \frac{2}{x}$, no real value of x

$f'(0)$ is not defined, but $x = 0$ is not in the domain of f.

No critical points, no local extrema

$$f''(x) = 2 - 2x^{-2} = 2 - \frac{2}{x^2}$$

$$0 = 2 - \frac{2}{x^2}$$

$$x^2 = 1$$

$$x = \pm 1$$

$$f(-1) = (-1)^2 + \ln(-1)^2 = 1$$

$$f(1) = 1^2 + \ln 1^2 = 1$$

$f''(0)$ is not defined, but $x = 0$
is not in the domain of f.

Points of inflection at $(-1, 1)$, $(1, 1)$.

37. $f(x) = x^3 - 2.1x^2 + 4.3x$

$f'(x) = 3x^2 - 4.2x + 4.3$

$f''(x) = 6x - 4.2$

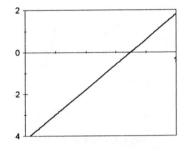

$f(0.70) = (0.70)^3 - 2.1(0.70)^2 + 4.3(0.70) = 2.324$

Point of inflection at $(0.70, 2.32)$

39. $f(x) = e^{-x^2}$

$f'(x) = e^{-x^2}(-2x)$

$f''(x) = -2e^{-x^2} + (-2x)e^{-x^2}(-2x)$

$= e^{-x^2}(4x^2 - 2)$

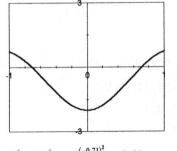

$f(-0.71) = e^{-(-0.71)^2} = 0.60$

$f(0.71) = e^{-(0.71)^2} = 0.60$

Points of inflection at $(-0.71, 0.60)$, $(0.71, 0.60)$

41. $f(x) = x^4 - 2x^3 + x^2 - 2x + 1$

$f'(x) = 4x^3 - 6x^2 + 2x - 2$

$f''(x) = 12x^2 - 12x + 2$

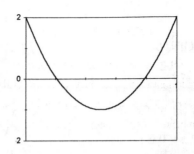

$f(0.21) = (0.21)^4 - 2(0.21)^3 + (0.21)^2 - 2(0.21) + 1 = 0.61$

$f(0.79) = (0.79)^4 - 2(0.79)^3 + (0.79)^2 - 2(0.79) + 1 = -0.55$

Points of inflection at $(0.21, 0.61)$, $(0.70, -0.55)$

43. $f(x) = x^2 - \ln x$

$f'(x) = 2x - \dfrac{1}{x}$

$f''(x) = 2 + \dfrac{1}{x^2}$

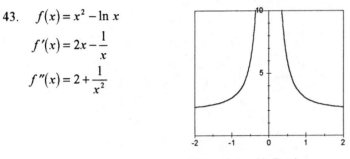

No points of inflection

45. (a) 2 years into the epidemic   (b) 2 years into the epidemic

47. (a) Maximum rate of change in 1992.   (b) Minimum rate of change in 1994   (c) Increase started in 1990

49. $N(t) = 0.028234t^3 - 1.0922t^2 + 13.029t + 146.88 \quad (0 \le t \le 39)$

$N'(t) = 0.084702t^2 - 2.1844t + 13.029$

$N''(t) = 0.169404t - 2.1844$

$0 = 0.169404t - 2.1844$

$t = 12.89$

$N(12.89) = 0.028234(12.89)^3 - 1.0922(12.89)^2 + 13.029(12.89) + 146.88$

$= 193.8$

Point of inflection at $(12.89, 193.8)$. The prison population was declining most rapidly at $t = 12.89$ (that is, in 1963), at which time the prison population was 193,800.

51. $S(n) = 904 + \dfrac{136}{(n-180)^{1.325}}$ $(192 \leq n \leq 563)$

   (a) There are no points of inflection in the graph of S.

   (b) Since the graph is concave up, the derivative of S is increasing, and so
       the rate of decrease of SAT scores with increasing numbers of prisoners
       is diminishing. In other words, the apparent effect of more prisoners is diminishing.

53. $n = -3.79 + 144.42s - 23.86s^2 + 1.457s^3$

   (a) $\dfrac{dn}{ds} = 144.42 - 47.72s + 4.371s^2$

   $\dfrac{d^2n}{ds^2} = -47.72 + 8.742s$

   $\left.\dfrac{d^2n}{ds^2}\right|_{s=3} = -47.72 + 8.742(3) = -21.494$

   Thus, the rate at which new patents are
   produced decreases with increasing firm
   size. This means that the returns (as
   measured in the number of new patents per
   increase of $1 million in sales) are
   diminishing as the firm size increases.

   (b) $\left.\dfrac{d^2n}{ds^2}\right|_{s=7} = -47.72 + 8.742(7) = 13.474$

   Thus, the rate at which new patents are produced
   increases with increasing firm size by 13.474 new
   patents per $1 million^2.

55. $n = at + bt^2$ with $n'(0) = 10$, $n''(t) = -2$

   $n' = a + 2bt$

   $n'' = 2b$

   If $n'(0) = 10$,

   $10 = a + 2b(0)$

   $a = 10$

   If $n''(t) = -2$,

   $-2 = 2b$

   $b = -1$

   $n = 10t - t^2$

353

57. $R(t) = \dfrac{a}{1+be^{-kt}}$ with $\lim_{t \to +\infty} R = 10$,

$R(0) = 0.5$, $R'(0) = 0.02375$.

(a) $R'(t) = \dfrac{-abe^{-kt}(-k)}{\left(1+be^{-kt}\right)^2} = \dfrac{abke^{-kt}}{\left(1+be^{-kt}\right)^2}$

$\lim_{t \to +\infty} R(t) = \dfrac{a}{1+b(0)} = a$

If $\lim_{t \to +\infty} R(t) = 10$

$\qquad 10 = a$

If $R(0) = 0.5$

$\qquad 0.5 = \dfrac{10}{1+be^{-k(0)}} = \dfrac{10}{1+b}$

$\qquad 0.5 + 0.5b = 10$

$\qquad\qquad b = 19$

If $R'(0) = 0.02375$

$0.02375 = \dfrac{10(19)ke^{-k(0)}}{\left(1+19e^{-k(0)}\right)^2} = \dfrac{190k}{(20)^2}$

$\qquad k = 0.05$

$a = 10,\ b = 19,\ k = 0.05$

(b) $R'(t) = \dfrac{9.5e^{-0.05t}}{\left(1+19e^{-0.05t}\right)^2}$

$R''(t) = \Big[9.5e^{-0.05t}(-0.05)\left(1+19e^{-0.05t}\right)^2$

$\qquad - 9.5e^{-0.05t}(2)\left(1+19e^{-0.05t}\right)$

$\qquad \left(19e^{-0.05t}\right)(-0.05)\Big] / \left(1+19e^{-0.05t}\right)^4$

$R''(0) = \dfrac{\left[9.5(-0.05)(1+19)^2 - 9.5(2)(1+19)(19)(-0.05)\right]}{(1+19)^4}$

$R''(0) = \$0.00106875$ million per month^2. This means that, when $t = 0$, the revenue is accelerating by $\$0.00106875$ per month each month.

---

59. $p = v(1.05)^{-t}$ with $v = 10V$, $V = \dfrac{22{,}514}{1+22{,}514t^{-2.55}}$

$p = \dfrac{225{,}140(1.05)^{-t}}{1+22{,}514t^{-2.55}}$

(a)

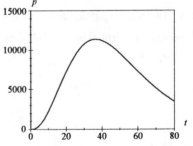

Points of inflection: 10 years and 50 years

(b) $\dfrac{dp}{dt} = \dfrac{225{,}140(1.05)^{-t}\ln(1.05)(-1)\left(1+22{,}514t^{-2.55}\right) - 225{,}140(1.05)^{-t}(-255)(22{,}514)t^{-3.55}}{\left(1+22{,}514t^{-2.55}\right)^2}$

$\qquad = \dfrac{-225{,}140\ln(1.05)(1.05)^{-t}\left(1+22{,}514t^{-2.55}\right) + 1{,}292{,}544{,}5000(1.05)^{-t}t^{-3.55}}{\left(1+22{,}514t^{-2.55}\right)^2}$

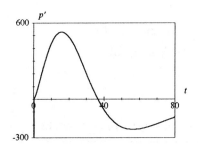

Local extrema: 9 years and 42 years

(c) $\dfrac{dp}{dt}$: $(9, 23)$

Largest rate of increase is \$23 per year after 9 years.

12.4 EXERCISES

1. $f(x) = x^2 - 4x + 1$; with domain $[0, 3]$

    $y = x^2 - 4x + 1$

    y - intercepts: $y = 0^2 - 4(0) + 1 = 1$

    x - intercepts:

    $0 = x^2 - 4x + 1$

    $x = \dfrac{-(-4) \pm \sqrt{(-4)^2 - 4(1)(1)}}{2(1)}$

    $= 2 \pm \sqrt{3}$

    $= 2 - \sqrt{3}$ only

EXTREMA:

$f'(x) = 2x - 4$

$0 = 2x - 4$

$x = 2$

$f(x) = 2^2 - 4(2) + 1 = -3$

x	0	2	3
$f(x)$	1	-3	-2

Absolute minimum at $(2, -3)$

No points of inflection

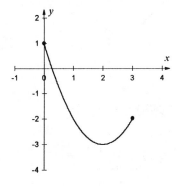

355

3. $g(x) = x^3 - 12x$; with domain $[-4, 4]$

$y = x^3 - 12x$

y - intercept: $y = 0^3 - 12(0) = 0$

x - intercepts:

$0 = x^3 - 12x$

$0 = x(x^2 - 12)$

$x = 0$ or $x^2 - 12 = 0$

$\qquad\qquad\qquad x = \pm 2\sqrt{3}$

EXTREMA:

$g'(x) = 3x^2 - 12$

$0 = 3x^2 - 12$

$x^2 = 4$

$x = \pm 2$

$g(2) = 2^3 - 12(2) = -16$

$g(-2) = (-2)^3 - 12(-2) = 16$

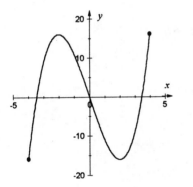

x	-4	-2	2	4
$g(x)$	-16	16	-16	16

Absolute minima at $(-4, -16)$ and $(2, -16)$

Absolute maxima at $(-2, 16)$ and $(4, 16)$

POINTS OF INFLECTION:

$g''(x) = 6x$

$0 = 6x$

$x = 0$

$g(0) = 0^3 - 12(0) = 0$

Point of inflection at $(0, 0)$

356

5. $f(t) = t^3 + t$; with domain $[-2, 2]$

$y = t^3 + t$

y - intercept: $y = 0^3 + 0 = 0$

t - intercept:

$0 = t^3 + t$

$0 = t(t^2 + 1)$

$t = 0$ or $t^2 + 1 = 0$

$\qquad\qquad\qquad t^2 \neq -1$

$t = 0$ only

EXTREMA:

$f'(t) = 3t^2 + 1$

$\quad 0 = 3t^2 + 1$

$\quad t^2 \neq -1/3$

No stationary points

t	-2	0	2
$f(t)$	-10	0	10

Absolute minimum at $(-2, -10)$

Absolute maximum at $(2, 10)$

POINTS OF INFLECTION

$f''(t) = 6t$

$\quad 0 = 6t$

$\quad t = 0$

$f(0) = 0$

Point of inflection at $(0, 0)$

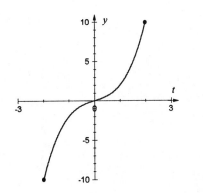

357

7. $h(t) = 2t^3 + 3t^2$; with domain $[-2, +\infty)$

$$y = 2t^3 + 3t^2$$

y - intercept: $y = 2(0)^3 + 3(0)^2 = 0$

t - intercepts:

$$0 = 2t^3 + 3t^2$$

$$0 = t^2(2t + 3)$$

$$t^2 = 0 \quad \text{or} \quad 2t + 3 = 0$$

$$t = 0 \qquad\qquad t = -3/2$$

EXTREMA:

$$h'(t) = 6t^2 + 6t$$

$$0 = 6t^2 + 6t$$

$$0 = 6t(t + 1)$$

$$6t = 0 \quad \text{or} \quad t + 1 = 0$$

$$t = 0 \qquad\qquad t = -1$$

$$h(0) = 0$$

$$h(-1) = 2(-1)^3 + 3(-1)^2 = 1$$

t	-2	0	-1	2
$h(t)$	-4	0	1	28

Absolute minimum at $(-2, -4)$

Local minimum at $(0, 0)$

Local maximum at $(-1, 1)$

POINT OF INFLECTION

$$h''(t) = 12t + 6$$

$$0 = 12t + 6$$

$$t = -1/2$$

$$h(-1/2) = 2(-1/2)^3 + 3(-1/2)^2 = 1/2$$

Point of inflection at $(-1/2, 1/2)$

358

9. $f(x) = x^4 - 4x^3$; with domain $[-1, +\infty)$

$y = x^4 - 4x^3$

y - intercept: $y = (0)^4 - 4(0)^3 = 0$

x - intercepts:

$0 = x^4 - 4x^3$

$0 = x^3(x - 4)$

$x^3 = 0$ or $x - 4 = 0$

$x = 0$ $x = 4$

EXTREMA:

$f'(x) = 4x^3 - 12x^2$

$0 = 4x^3 - 12x^2$

$0 = 4x^2(x - 3)$

$4x^2 = 0$ or $x - 3 = 0$

$x = 0$ $x = 3$

$f(0) = 0$

$f(3) = 3^4 - 4(3)^3 = -27$

x	-1	0	3	5
$f(x)$	5	0	-27	125

Local maximum at $(-1, 5)$

Absolute minimum at $(3, -27)$

POINTS OF INFLECTION

$f''(x) = 12x^2 - 24x$

$0 = 12x^2 - 24x$

$0 = 12x(x - 2)$

$12x = 0$ or $x - 2 = 0$

$x = 0$ $x = 2$

$f(0) = 0$

$f(2) = 2^4 - 4(2)^3 = -16$

Points of inflection at $(0, 0)$ and $(2, -16)$

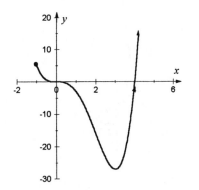

11. $g(t) = \frac{1}{4}t^4 - \frac{2}{3}t^3 + \frac{1}{2}t^2$; with domain $(-\infty, +\infty)$

$$y = \frac{1}{4}t^4 - \frac{2}{3}t^3 + \frac{1}{2}t^2$$

y - intercept: $y = \frac{1}{4}(0)^4 - \frac{2}{3}(0)^3 + \frac{1}{2}(0)^2 = 0$

t - intercepts:

$$0 = \frac{1}{4}t^4 - \frac{2}{3}t^3 + \frac{1}{2}t^2$$

$$0 = \frac{1}{12}t^2(3t^2 - 8t + 6)$$

$\frac{1}{12}t^2 = 0$ or $3t^2 - 8t + 6 = 0$

$t = 0$ 　　　　 $t = \dfrac{-(-8) \pm \sqrt{(-8)^2 - 4(3)6}}{2(3)}$

　　　　　　　　　No real values

$t = 0$ only

EXTREMA:

$g'(t) = t^3 - 2t^2 + t$

$0 = t^3 - 2t^2 + t$

$0 = t(t^2 - 2t + 1)$

$t = 0$ or $t^2 - 2t + 1 = 0$

　　　　　　　$(t-1)^2 = 0$

　　　　　　　　　$t = 1$

$g(0) = 0$

$g(1) = \frac{1}{4}(1)^4 - \frac{2}{3}(1)^3 + \frac{1}{2}(1)^2 = \frac{1}{12}$

t	-1	0	1	2
$g(t)$	$17/12$	0	$1/12$	$2/3$

Absolute minimum at $(0, 0)$

POINT OF INFLECTION

$g''(t) = 3t^2 - 4t + 1$

$0 = 3t^2 - 4t + 1$

$0 = (3t - 1)(t - 1)$

$3t - 1 = 0$ or $t - 1 = 0$

　$t = 1/3$ 　　　　 $t = 1$

$g(1/3) = \frac{1}{4}(1/3)^4 - \frac{2}{3}(1/3)^3 + \frac{1}{2}(1/3)^2 = 11/324$

$g(1) = 1/12$

Points of inflection at $\left(\frac{1}{3}, \frac{11}{324}\right)$ and $\left(1, \frac{1}{12}\right)$

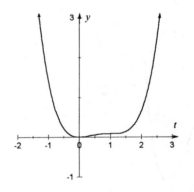

360

13. $f(t) = \dfrac{t^2+1}{t^2-1}$; with $-2 \leq t \leq 2$

If $t^2 - 1 = 0$, $t = \pm 1$ and $f(t)$ is undefined.

$\displaystyle\lim_{t \to -1^-} \frac{t^2+1}{t^2-1} = +\infty$, $\displaystyle\lim_{t \to -1^+} \frac{t^2+1}{t^2-1} = -\infty$

$\displaystyle\lim_{t \to 1^-} \frac{t^2+1}{t^2-1} = -\infty$, $\displaystyle\lim_{t \to 1^+} \frac{t^2+1}{t^2-1} = +\infty$

Vertical asymptotes at $t = \pm 1$

$y = \dfrac{t^2+1}{t^2-1}$;

y - intercept: $y = \dfrac{0^2+1}{0^2-1} = -1$

t - intercepts:

$0 = \dfrac{t^2+1}{t^2-1}$

$0 = t^2 + 1$

$t^2 \neq -1$

No t - intercepts

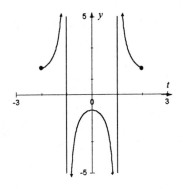

EXTREMA:

$f'(t) = \dfrac{2t(t^2-1) - 2t(t^2+1)}{(t^2-1)^2} = \dfrac{-4t}{(t^2-1)^2}$

$0 = \dfrac{-4t}{(t^2-1)^2}$

$0 = -4t$

$t = 0$

$f(0) = -1$

t	-2	-1^-	-1^+	0	1^-	1^+	2
$g(t)$	$5/3$	$+\infty$	$-\infty$	-1	$-\infty$	$+\infty$	$5/3$

Local minima at $(-2, 5/3)$ and $(2, 5/3)$

Local maximum at $(0, -1)$

No points of inflection

361

15. $f(x) = (x-1)\sqrt{x};$ with domain $[0, +\infty)$

$$y = (x-1)\sqrt{x}$$

y - intercept: $y = (0-1)\sqrt{0} = 0$

x - intercepts:

$$0 = (x-1)\sqrt{x}$$

$\sqrt{x} = 0$ or $x - 1 = 0$

$x = 0$ $x = 1$

EXTREMA:

$$f'(x) = x^{1/2} + (x-1)(1/2)(x)^{-1/2}$$

$$= \frac{3x-1}{\sqrt{x}}$$

$$0 = \frac{3x-1}{\sqrt{x}}$$

$$0 = 3x - 1$$

$$x = 1/3$$

$$f(1/3) = \left(\frac{1}{3} - 1\right)\sqrt{1/3} = -\frac{2\sqrt{3}}{9}$$

x	0	1/3	1
$f(x)$	0	$-2/3\sqrt{3}$	0

Local maximum at 0, 0

Absolute minimum at $\left(\dfrac{1}{3}, -\dfrac{2\sqrt{3}}{9}\right)$

No points of inflection

17. $g(x) = x^2 - 4\sqrt{x};$ with domain $[0, +\infty)$

$y = x^2 - 4\sqrt{x}$

y-intercept: $y = 0^2 - 4\sqrt{0} = 0$

x-intercepts:

$0 = x^2 - 4\sqrt{x}$

$0 = x^{1/2}\left(x^{3/2} - 4\right)$

$x^{1/2} = 0$ or $x^{3/2} - 4 = 0$

$x = 0$ $x = 4^{2/3}$

EXTREMA:

$g'(x) = 2x - 4\left(\dfrac{1}{2}\right)x^{-1/2}$

$ = \dfrac{2x^{3/2} - 2}{x^{1/2}}$

$0 = \dfrac{2x^{3/2} - 2}{x^{1/2}}$

$0 = 2x^{3/2} - 2$

$x^{3/2} = 1$

$x = 1$

$g(1) = 1^2 - 4\sqrt{1} = -3$

x	0	1	4
$g(x)$	0	-3	8

Local maximum at $(0, 0)$

Absolute minimum at $(1, -3)$

No points of inflection

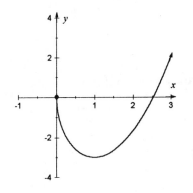

19. $g(x) = \dfrac{x^3}{x^2+3}$

$\displaystyle\lim_{x\to-\infty}\left[\dfrac{x^3}{x^2+3}\right]=-\infty$

$\displaystyle\lim_{x\to+\infty}\left[\dfrac{x^3}{x^2+3}\right]=+\infty$

$y=\dfrac{x^3}{x^2+3}$

y-intercept: $\ y=\dfrac{0^3}{0^2+3}=0$

x-intercepts:

$0=\dfrac{x^3}{x^2+3}$

$0=x^3$

$x=0$

EXTREMA:

$g'(x)=\dfrac{3x(x^2+3)-x^3(2x)}{(x^2+3)^2}=\dfrac{x^2(x^2+9)}{(x^2+3)^2}$

$0=\dfrac{x^2(x^2+9)}{(x^2+3)^2}$

$0=x^2(x^2+9)$

$x^2=0\quad$ or $\quad x^2+9=0$

$x=0\quad$ only $\quad x^2\neq-9$

$g(0)=0$

x	$-\infty$	0	$+\infty$
$g(x)$	$-\infty$	0	$+\infty$

No maxima or minima

POINTS OF INFLECTION

$g''(x)=\dfrac{\left[2x(x^2+9)+x^2(2x)\right](x^2+3)^2-x^2(x^2+9)2(x^2+3)(2x)}{(x^2+3)^4}$

$=\dfrac{-6x(x^2-9)}{(x^2+3)^3}$

$0=\dfrac{-6x(x^2-9)}{(x^2+3)^3}$

$0=-6x(x^2-9)$

$0-6x=0\quad$ or $\quad x^2-9=0$

$x=0\qquad\qquad x^2=9$

$\qquad\qquad\qquad x=\pm3$

$g(0)=0$

$g(\pm3)=\dfrac{(\pm3)^3}{3^2+3}=\pm\dfrac{9}{4}$

Points of inflection at $(0,0)$, $\left(-3,-\dfrac{9}{4}\right)$ and $\left(3,\dfrac{9}{4}\right)$

364

12.5 EXERCISES

1. Population: $P = 10,000$

   Growth rate: $\dfrac{dP}{dt} = 1000$

3. Let R be the annual revenue, and let q be the annual sales.

   $R = 7000$ and $\dfrac{dR}{dt} = 0.10(7000) = 700$

   Find $\dfrac{dq}{dt}$.

5. Let p be the price of a pair of shoes, and let q be the demand for the shoes.

   $\dfrac{dp}{dt} = 5.$ Find $\dfrac{dq}{dt}$.

7. Let T be the average global temperature, and let q be the demand for Bermuda shorts.

   $T = 60$ and $\dfrac{dT}{dt} = 0.1.$ Find $\dfrac{dq}{dt}$

9. Let A = area and r = radius

   $A = \pi r^2$

   $\dfrac{dA}{dt} = 2\pi r \dfrac{dr}{dt}$

   If $\dfrac{dA}{dt} = 12$, find $\dfrac{dr}{dt}$

   (a) When $r = 10$

   $12 = 2\pi(10)\dfrac{dr}{dt}$

   $\dfrac{dr}{dt} = \dfrac{3}{5\pi} \approx 0.1910$

   (b) When $A = 49$, $49 = \pi r^2$ or $r = \dfrac{7}{\sqrt{\pi}}$

   $12 = 2\pi \left(\dfrac{7}{\sqrt{\pi}}\right)\dfrac{dr}{dt}$

   $\dfrac{dr}{dt} = \dfrac{6}{7\sqrt{\pi}} \approx 0.4836 \text{ cm/s}$

11. Let x = the distance from wall to the base of the ladder, and y = the distance from the floor to the top of the ladder.

   $x^2 + y^2 = 50^2 = 2500$

   $2x\dfrac{dx}{dt} + 2y\dfrac{dy}{dt} = 0$

   Let $\dfrac{dx}{dt} = 10$, find $\dfrac{dy}{dt}$ when $x = 30$

   When $x = 30$, $30^2 + y^2 = 2500$ or $y = 40$

   $2(30)(10) + 2(40)\dfrac{dy}{dt} = 0$

   $\dfrac{dy}{dt} = -7\dfrac{1}{2} \text{ ft/s}$

13. $p = \dfrac{50{,}000}{q^{1.5}}$

$\dfrac{dp}{dt} = -\dfrac{75{,}000}{q^{2.5}}\dfrac{dq}{dt}$

If $q = 900$ and $\dfrac{dq}{dt} = 100$, find $\dfrac{dp}{dt}$

$\dfrac{dp}{dt} = \dfrac{-75{,}000}{(900)^{2.5}}(100) = -0.3086$

The price is decreasing at the rate of 31. cents per pound per month.

15. $p = 5 + \dfrac{100}{\sqrt{q}}$

$\dfrac{dp}{dt} = -\dfrac{50}{q^{3/2}}\dfrac{dq}{dt}$

If $\dfrac{dp}{dt} = 2$, find $\dfrac{dq}{dt}$ when $p = 15$.

When $p = 15$, $15 = 5 + \dfrac{100}{\sqrt{q}}$ or $q = 100$.

$2 = -\dfrac{50}{(100)^{3/2}}\dfrac{dq}{dt}$

$\dfrac{dq}{dt} = -40$

Monthly sales will drop by 40 T - shirts each month.

17. Let p = price per cup, q = weekly demand, and R = weekly revenue

$R = pq$

$0 = \dfrac{dp}{dt}q + p\dfrac{dq}{dt}$

If $\dfrac{dq}{dt} = -5$, $q = 50$, and $p = 30$, find $\dfrac{dp}{dt}$.

$0 = \dfrac{dp}{dt}(50) + (30)(-5)$

$\dfrac{dp}{dt} = 3$

Raise the price by 3 cents per week.

19. $P = 10x^{0.3}y^{0.7}$

$0 = 10(0.3)x^{-0.7}\dfrac{dx}{dt}y^{0.7} + 10(0.7)x^{0.3}y^{-0.3}\dfrac{dy}{dt}$

$0 = \dfrac{3y^{0.7}}{x^{0.7}}\dfrac{dx}{dt} + \dfrac{7x^{0.3}}{y^{0.3}}\dfrac{dy}{dt}$

$\dfrac{dy}{dt} = -\dfrac{3y}{7x}\dfrac{dx}{dt}$

If $\dfrac{dx}{dt} = 10$, find $\dfrac{dy}{dt}$ when $P = 1000$ and $x = 150$.

When $P = 1000$ and $x = 150$

$1000 = 10(150)^{0.3}y^{0.7}$

$y = \dfrac{100}{(150)^{0.3}}$

$y = 84.05$

$\dfrac{dy}{dt} = -\dfrac{3(84.05)}{7(150)}(10) = -2.40$

The daily operating budget is dropping by $2.40 per year.

21. $V = \dfrac{4}{3}\pi r^3$

$\dfrac{dV}{dt} = 4\pi r^2 \dfrac{dr}{dt}$

If $\dfrac{dV}{dt} = 3$, find $\dfrac{dr}{dt}$ when $r = 1$

$3 = 4\pi(1)^2 \dfrac{dr}{dt}$

$\dfrac{dr}{dt} = \dfrac{3}{4\pi} \approx 0.2387$

The radius is growing by 0.2387 ft / min.

23. $y = \dfrac{1}{x}$

$\dfrac{dy}{dt} = -\dfrac{1}{x^2}\dfrac{dx}{dt}$

If $\dfrac{dx}{dt} = 4$, find $\dfrac{dy}{dt}$ when $y = 2$

when $y = 2$, $2 = \dfrac{1}{x}$ or $x = \dfrac{1}{2}$.

$\dfrac{dy}{dt} = -\dfrac{1}{(1/2)^2}(4) = -16$

The y-coordinate is decreasing
at a rate of 16 units per second.

25. Let x = the distance east of Montauk,
y = the distance north of Montauk,
and r = the distance the ships are apart.

$x^2 + y^2 = r^2$

$2x\dfrac{dx}{dt} + 2y\dfrac{dy}{dt} = 2r\dfrac{dr}{dt}$

If $\dfrac{dx}{dt} = 30$, and $\dfrac{dy}{dt} = 20$, find $\dfrac{dr}{dt}$
when $x = 50$ and $y = 40$.

When $x = 50$ and $y = 40$,

$r^2 = 50^2 + 40^2 = 4100$

$2(50)(30) + 2(40)(20) = 2\sqrt{4100}\dfrac{dr}{dt}$

$\dfrac{dr}{dt} = \dfrac{2300}{\sqrt{4100}} \approx 35.92$

They are separating at 35.92 mph.

27. $I(n) = 2{,}928.8n^3 - 115{,}860n^2 + 1{,}532{,}900n$
$\qquad - 6{,}760{,}800 \quad (11.5 \le n \le 15.5)$

$\dfrac{dI}{dt} = 8{,}786.4n^2\dfrac{dn}{dt} - 231{,}720n\dfrac{dn}{dt} + 1{,}532{,}900\dfrac{dn}{dt}$

If $\dfrac{dn}{dt} = \dfrac{1}{3}$, find $\dfrac{dI}{dt}$ when $n = 13$

$\dfrac{dI}{dt} = 8{,}786.4(13)^2\left(\dfrac{1}{3}\right) - 231{,}720(13)\left(\dfrac{1}{3}\right) + 1{,}532{,}900\left(\dfrac{1}{3}\right)$

$\qquad = 1{,}814$

The daily operating budget is increasing by \$1,814
per year.

29. $V = 3e^2 + 5g^3$

$$0 = 6e\frac{de}{dt} + 15g^2\frac{dg}{dt}$$

If $\frac{dg}{dt} = -0.2$, find $\frac{de}{dt}$ when $g = 3.0$.

When $g = 3.0$

$$200 = 3e^2 + 5(3.0)^3$$

$$e^2 = \frac{65}{3}$$

$$e = \sqrt{65/3}$$

$$0 = 6\sqrt{65/3}\frac{de}{dt} + 15(3.0)^2(-0.2)$$

$$\frac{de}{dt} = \frac{9}{2}\sqrt{\frac{3}{65}} \approx 0.9668$$

Their prior experience must increase at a rate of 0.9668 years every year.

31. $V = \frac{1}{3}\pi r^2 h$

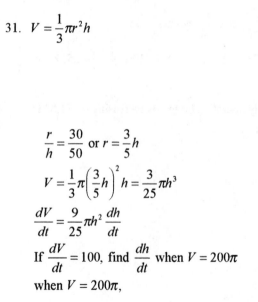

$$\frac{r}{h} = \frac{30}{50} \text{ or } r = \frac{3}{5}h$$

$$V = \frac{1}{3}\pi\left(\frac{3}{5}h\right)^2 h = \frac{3}{25}\pi h^3$$

$$\frac{dV}{dt} = \frac{9}{25}\pi h^2 \frac{dh}{dt}$$

If $\frac{dV}{dt} = 100$, find $\frac{dh}{dt}$ when $V = 200\pi$

when $V = 200\pi$,

$$200\pi = \frac{3}{25}\pi h^3$$

$$h^3 = \frac{5000}{3}$$

$$h = \left(\frac{5000}{3}\right)^{1/3}$$

$$100 = \frac{9}{25}\pi\left(\frac{5000}{3}\right)^{2/3}\frac{dh}{dt}$$

$$\frac{dh}{dt} = \frac{2500}{9\pi}\left(\frac{3}{5000}\right)^{2/3} \approx 0.6290$$

The level is rising 0.6290 m / s.

368

33. $V = \pi r^2 h$ and $V = 4t^2 - t$

so, $\pi r^2 h = 4t^2 - t$

$$h = \frac{1}{\pi r^2}\left(4t^2 - t\right)$$

$$\frac{dh}{dt} = \frac{1}{\pi r^2}(8t - 1)$$

If $h = 2$, and $r = 2$, find $\dfrac{dh}{dt}$.

When $h = 2$,

$$\pi(2^2) = 4t^2 - t$$

$$4t^2 - t - 8\pi = 0$$

$$t = \frac{-(-1) \pm \sqrt{(-1)^2 - 4(4)(-8\pi)}}{2(4)}$$

$$t = \frac{1 + \sqrt{1 + 128\pi}}{8}$$

$$\frac{dh}{dt} = \frac{1}{\pi(2)^2}\left[8\left(\frac{1 + \sqrt{1 + 128\pi}}{8}\right) - 1\right]$$

$$= \frac{\sqrt{1 + 128\pi}}{4\pi} \approx 1.598$$

The level is rising at 1.598 cm / s.

35. $S(n) = 904 + \dfrac{1326}{(n - 180)^{1.325}}$ $(192 \le n \le 563)$

$$\frac{dS}{dt} = -1.325(1326)(n - 180)^{-2.325}\frac{dn}{dt}$$

$$= \frac{-1,756.95}{(n - 180)^{2.325}}\frac{dn}{dt}$$

If $n = 475$, find $S(n)$.

$$S(475) = 904 + \frac{1326}{(475 - 180)^{1.325}} = 904.7$$

If $\dfrac{dn}{dt} = 35$, find $\dfrac{dS}{dt}$ when $n = 475$.

$$\frac{dS}{dt} = \frac{-1,756.95}{(475 - 180)^{2.325}}(35) = -0.11$$

The average SAT score was 904.70
and decreasing by 0.11 per year.

12.6 EXERCISES

Use $E = -\dfrac{dq}{dp} \cdot \dfrac{p}{q}$ for the elasticity of demand in the following exercises.

1. $q = 1000 - 20p$

$$\frac{dq}{dp} = -20$$

$$E = -(-20)\frac{p}{q} = \frac{20p}{1000 - 20p}$$

If $p = 30$,

$$E = \frac{20(30)}{1000 - 20(30)} = 1.5$$

If $E = 1$,

$$1 = \frac{20p}{1000 - 20p}$$

$$1000 - 20p = 20p$$

$$p = 25$$

If $p = 25$,

$$R = pq = 25[1000 - 20(25)] = 12,500$$

$E = 1.5$; the demand is going down 1.5% per 1% increase in price at that price level; revenue is maximized when $p = \$25$; weekly revenue at that price is $12,500.

3. $q = (100 - p)^2$

$$\frac{dq}{dp} = 2(100 - p)(-1) = -200 + 2p$$

$$E = -(-200 + 2p)\frac{p}{q} = \frac{200p - 2p^2}{(100 - p)^2}, \quad p \neq 100$$

(a) If $p = 30$, $\quad E = \dfrac{200(30) - 2(30)^2}{(100 - 30)^2} = \dfrac{6}{7}$

The demand is going down 6% per 7% increase in price at that price level; thus a price increase is in order.

(b) If $E = 1$

$$1 = \frac{200p - 2p^2}{(100 - p)^2}$$

$$10,000 - 200p + p^2 = 200p - 2p^3$$

$$3p^2 - 400p + 10,000 = 0$$

$$(3p - 100)(p - 100) = 0$$

$$p = \frac{100}{3} \text{ or } p = 100 \text{ (not allowed)}$$

Revenue is maximized when

$p = 100/3 \approx \$33.33$.

(c) If $p = \dfrac{100}{3}$,

$$q = \left(100 - \frac{100}{3}\right)^2 = \left(\frac{200}{3}\right)^2 \approx 4,444$$

Demand would be 4,444 cases per week.

5. $q = 9859.39 - 2.17p$

$\dfrac{dq}{dp} = -2.17$

$E = -(-2.17)\dfrac{p}{q} = \dfrac{2.17p}{9859.39 - 2.17p}$

(a) If $p = 2867$,

$E = \dfrac{2.17(2867)}{9859.39 - 2.17(2867)} = 1.75$

Thus the demand is elastic at the given tuition level, showing that an decrease in fees will result in an increase in revenue.

(b) If $E = 1$

$$1 = \dfrac{2.17p}{9859.39 - 2.17p}$$

$9859.39 - 2.17p = 2.17p$

$p = 2{,}271.75$

$q = 9859.39 - 2.17(2{,}271.75) = 4930$

$R = pq = 2{,}271.75(4930) = 11{,}200{,}000$

They should charge an average of $2,271.75 per student, and this will result in an enrollment of about 4930 students, giving a revenue of $11,200,000.

7. $f(p) = mp + b$

$\dfrac{d}{dp}[f(p)] = m$

(a) $E = -\dfrac{mp}{mp + b}$

(b) If $E = 1$

$1 = -\dfrac{mp}{mp + b}$

$mp + b = -mp$

$2mp = -b$

$p = -\dfrac{b}{2m}$

9. $f(p) = \dfrac{k}{p^r}$

$\dfrac{d}{dp}[f(p)] = -\dfrac{rk}{p^{r+1}}$

(a) $E = -\left(-\dfrac{rk}{p^{r+1}}\right)\dfrac{p}{k/p^r} = r$

(b) E is independent of p.

(c) If $r = 1$, then the revenue is not effected by the price. If $r > 1$, then the revenue is always elastic, while if $r < 1$, the revenue is always inelastic. This is an unrealistic model, since there should always be a price at which the revenue is a maximum.

11. $q = 100e^{-3p^2+p}$

$\dfrac{dq}{dp} = 100(-6p+1)e^{-3p^2+p}$

(a) $E = -(-6p+1)(100)e^{-3p^2+p}\dfrac{p}{100e^{-3p^2+p}}$

$\quad = 6p^2 - p$

If $p = 3$, $E = 6(3)^2 - 3 = 51$

The demand is going down 51% per 1% increase in price at that price level; thus, a large price decrease is advised.

(b) If $E = 1$

$$1 = 6p^2 - p$$

$$6p^2 - p - 1 = 0$$

$$(2p-1)(3p+1) = 0$$

$$p = \dfrac{1}{2} \quad \text{or} \quad p = -\dfrac{1}{3}\text{(not allowed)}$$

Revenue is maximized when $p =$

(c) If $p = \dfrac{1}{2}$, $q = 100e^{-3(1/2)^2+1/2}$

$\quad = 100e^{-3/4+1/2}$

$\quad \approx 77.88$

Demand would be 77.88 paint - by - number sets per month.

13. $q = mp + b$

(a) $(2, 3000)$ and $(4, 0)$

$$m = \dfrac{0-3000}{4-2} = -1500$$

$$q - 0 = -1500(p-4)$$

$$q = -1500p + 6000$$

(b) $\dfrac{dq}{dp} = -1500$

$$E = -(-1500)\dfrac{p}{-1500p+6000}$$

$$= \dfrac{1500p}{-1500p+6000}$$

If $E = 1$

$$1 = \dfrac{1500p}{-1500p+6000}$$

$$-1500p + 6000 = 1500p$$

$$p = 2$$

$$R = pq = 2\left[-1500(2) + 6000\right] = 6000$$

Maximize weekly revenue at $6,000 when the price is $2 per hamburger.

15. $f(p) = Ae^{-bp}$

(a) $407 = f(3) = Ae^{-b(3)}$

$223 = f(5) = Ae^{-b(5)}$

$\dfrac{407}{223} = \dfrac{e^{-3b}}{e^{-5b}} = e^{2b}$

$\ln \dfrac{407}{223} = 2b$

$b = \dfrac{1}{2} \ln \dfrac{407}{223} \approx 0.30$

$407 = f(3) = Ae^{-0.3(3)}$

$407 = A(0.407)$

$A = 1000$

$f(p) = 1000e^{-0.3p}$

(b) $\dfrac{df(p)}{dp} = -300e^{-0.3p}$

$E = -(-300)\dfrac{p}{1000e^{-0.3p}} = 0.3p$

$E(3) = 0.3(3) = 0.9$

$E(4) = 0.3(4) = 1.2$

$E(5) = 0.3(5) = 1.5$

(c) If $E = 1$, $1 = 0.3p$ or $p = \$3.33$ per pound

(d) Set demand at 200

$200 = 1000e^{-0.3p}$

$0.2 = e^{-0.3p}$

$\ln 0.2 = -0.3p$

$p = \$5.36$ per pound

Selling at a lower price would increase demand, but you cannot sell more than 200 pounds anyway. You should charge as much as you can and still be able to sell all 200 pounds.

17. $Q = aP^{\alpha}Y^{\beta}$

$\dfrac{dQ}{dY} = a\beta P^{\alpha}Y^{\beta-1}$

The income elasticity of demand is

$\dfrac{dQ}{dY} \cdot \dfrac{Y}{Q} = a\beta P^{\alpha}Y^{\beta-1}\left(\dfrac{Y}{aP^{\alpha}Y^{\beta}}\right) = \beta$

An increase in income of x% will result in an increase in demand of βx%. (Note that we should not take the negative here, as we expect an increase in income to product an increase in demand).

21. $f(x) = x - \ln x$; with domain $(0, +\infty)$

$y = x - \ln x$

y-intercept: none, $x \neq 0$

x-intercept:

$0 = x - \ln x$

$x \neq \ln x$

No x-intercept

EXTREMA:

$f'(x) = 1 - \dfrac{1}{x}$

$0 = 1 - \dfrac{1}{x}$

$x = 1$

$f(1) = 1 - \ln 1 = 1$

$\lim\limits_{x \to 0} (x - \ln x) = +\infty$

Vertical asymptote: $x = 0$

$\lim\limits_{x \to +\infty} (x - \ln x) = +\infty$

x	0	1	$+\infty$
$f(x)$	$+\infty$	1	$+\infty$

Absolute minimum at $(1, 1)$

No points of inflection

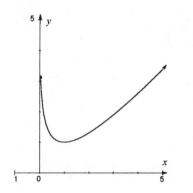

23. $g(t) = e^t - t$; with domain $[-1, 1]$

$y = e^t - t$

y-intercepts: $y = e^0 - 0 = 1$

t-intercepts:

$0 = e^t - t$

$t \neq e^t$

No t-intercepts

EXTREMA:

$g'(t) = e^t - 1$

$0 = e^t - 1$

$e^t = 1$

$t = 0$

$g(0) = 1$

t	-1	0	1
$g(t)$	$1 + e^{-1}$	1	$e - 1$

Absolute minimum at $(0, 1)$

Absolute maximum at $(1, e - 1)$

Local maximum at $(-1, 1 + e^{-1})$

No points of inflection

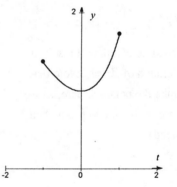

25. $f(x) = x + \dfrac{1}{x}$

$\displaystyle\lim_{x \to 0^-} \left(x + \frac{1}{x}\right) = -\infty, \quad \lim_{x \to 0^+} \left(x + \frac{1}{x}\right) = +\infty$

Vertical asymptote: $x = 0$

$\displaystyle\lim_{x \to -\infty} \left(x + \frac{1}{x}\right) = -\infty, \quad \lim_{x \to +\infty} \left(x + \frac{1}{x}\right) = +\infty$

$y = x + \dfrac{1}{x}$

y - intercept: none, $x \neq 0$

x - intercepts:

$0 \neq x + \dfrac{1}{x}$

No x - intercepts

EXTREMA:

$f'(x) = 1 - \dfrac{1}{x^2}$

$0 = 1 - \dfrac{1}{x^2}$

$x^2 = 1$

$x = \pm 1$

$f(-1) = -1 + \dfrac{1}{-1} = -2$

$f(1) = 1 + \dfrac{1}{1} = 2$

x	$-\infty$	-1	0^-	0^+	1	$+\infty$
$f(x)$	$-\infty$	-2	$-\infty$	$+\infty$	2	$+\infty$

Local minimum at $(1, 2)$

Local maximum at $(-1, -2)$

No points of inflection

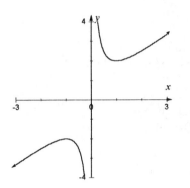

27. $p = ve^{-0.05t}$ with $v = 300{,}000 + 1000t^2$

$p(t) = (300{,}000 + 1000t^2)e^{-0.05t}$

(a) p-intercept:

$p = (300{,}000 + 0)e^0 = 300{,}000$

t-intercept: none

$\lim_{t \to +\infty} p(t) = 0$,

horizontal asymptote: $p = 0$

$p'(t) = -50(t^2 - 40t + 300)e^{-0.05t}$

$0 = t^2 - 40t + 300$

$t = 30$ or $t = 10$

$p(30) = 267{,}756$, $p(10) = 242{,}612$

Absolute maximum at $300{,}000$ at $t = 0$

Local minimum of $242{,}612$ at $t = 10$

Local maximum of $267{,}756$ at $t = 30$

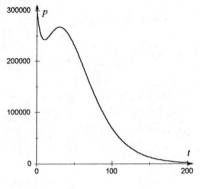

Points of inflection at $(17.6393, 252{,}955)$
and $(62.3607, 185{,}322)$
An appropriate demain is $[0, +\infty)$

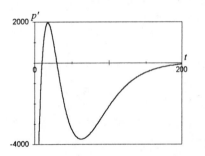

(b) Increasing most rapidly in 17.64 years, decreasing most rapidly in 0 years.

(c) The model predicts that the collection of cars will eventually become worthless in terms of discounted value. The value of the collection of classic cars, although increasing without bound, will eventually become worthless as a result of inflation. This is not reasonable if collectors continue to prize classic cars, so we may conclude that the given model is not a reliable one for long-term prediction.

29. $C(x) = 150,000 + 20x + \dfrac{x^2}{10,000}$

$\overline{C}(x) = \dfrac{150,000}{x} + 20 + \dfrac{x}{10,000}$; $D(0, +\infty)$

$\overline{C}'(x) = -\dfrac{150,000}{x^2} + \dfrac{1}{10,000}$

$0 = -\dfrac{150,000}{x^2} + \dfrac{1}{10,000}$

$x^2 = 1,500,000,000$

$x = 38,730$

$\overline{C}'(38,730) = 27.75$

$\lim\limits_{x \to 0^+} \overline{C}(x) = +\infty$

$\lim\limits_{x \to +\infty} \overline{C}(x) = +\infty$

Absolute minimum at $(38,730, 27.75)$

(b) When the production level is pushed higher and higher, the average cost per item rises without bound.

(c) The average cost per item grows without bound as the number of items produced approaches zero.

31. $P(x) = R(x) - C(x) = 500\sqrt{x} - 10,000 - 2x$

$\overline{P}(x) = \dfrac{500}{\sqrt{x}} - \dfrac{10,000}{x} - 2$; $D(0, +\infty)$

$\overline{P}'(x) = -\dfrac{250}{x^{3/2}} + \dfrac{10,000}{x^2}$

$0 = -\dfrac{250}{x^{3/2}} + \dfrac{10,000}{x^2}$

$0 = -250x^{1/2} + 10,000$

$x^{1/2} = 40$

$x = 1600$

$\overline{P}(1600) = 4.25$

$\lim\limits_{x \to 0^+} \overline{P}(x) = -\infty$; Vertical asymptote: $x = 0$

$\lim\limits_{x \to +\infty} \overline{P}(x) = 0$; Horizontal asymptote: $\overline{P} = 0$

x	0^+	1600	$+\infty$
$\overline{P}(x)$	$-\infty$	4.25	0

MARGINAL PROFIT

$P'(x) = \dfrac{250}{x^{1/2}} - 2$

$\lim\limits_{x \to 0^+} P'(x) = +\infty$; Vertical asymptote: $x = 0$

$\lim\limits_{x \to +\infty} P'(x) = 0$; Horizontal asymptote: $P' = 0$

31. continued

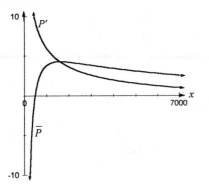

The graphs cross each other at (1,600, 4.25), which is the absolute maximum of the average profit function. When the average value is maximized, the marginal profit and the average profit are equal.

33. $A = \pi r^2 + 2\pi rh$ with $5000 = \pi r^2 h$

$$A(r) = \pi r^2 + \frac{10{,}000}{r}; \quad D(0, +\infty)$$

$$A'(r) = 2\pi r - \frac{10{,}000}{r^2} r^2$$

$$0 = 2\pi r - \frac{10{,}000}{r^2} r$$

$$r^3 = 1592$$

$$r = 11.68$$

$$A(11.68) = 1285$$

$\lim_{r \to 0^+} A(r) = +\infty$; Vertical asymptote: $r = 0$

$\lim_{r \to +\infty} A(r) = +\infty$

r	0^+	11.68	$+\infty$
$A(r)$	$+\infty$	1285	$+\infty$

33. continued

The limits at 0 and infinity are both infinite.
Since the limit at zero is infinite, the amount
of plastic needed to make very narrow buckets
(small radius) would be very large. The reason
for this is that the buckets would have to be
extremely tall in order to hold the requisite
volume. Since the limit at infinity is also
infinite, this says that a large amount of
material would be needed to make very wide
buckets (even though they would be very short).

35. $C = 20,000x + 365y$ with $1000 = x^{0.4}y^{0.6}$

$$C(x) = 20,000x + \frac{36,500,000}{x^{2/3}}$$

$$C'(x) = 20,000 - \frac{24,333,333}{x^{5/3}}$$

$$0 = 20,000 - \frac{24,333,333}{x^{5/3}}$$

$$x^{5/3} = 1217$$

$$x = 70.97$$

$$C(70.97) = 3,548,730$$

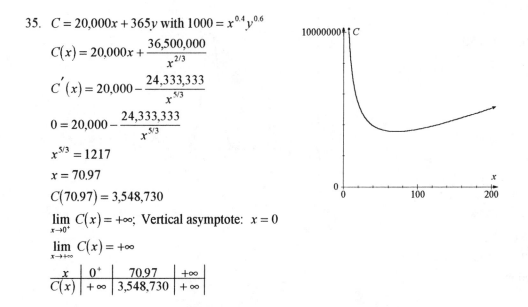

$\lim\limits_{x \to 0^+} C(x) = +\infty$; Vertical asymptote: $x = 0$

$\lim\limits_{x \to +\infty} C(x) = +\infty$

x	0^+	70.97	$+\infty$
$C(x)$	$+\infty$	3,548,730	$+\infty$

(a) Estimate the derivative at the point where $x = 50$. This is the slope of
the tangent: $-\$15,858$ per year per employee.

(b) $\lim\limits_{x \to +\infty} C'(x) = +\infty$. As the number of employees becomes large, the
additional cost per employee becomes larger without bound.

379

37. $n(s) = -3.79 + 144.42s - 23.86s^2 + 1.457s^3$ with $n \geq 0$

$n'(s) = 144.42 - 47.72s + 4.371s^2$

$0 = 144.42 - 47.72s + 4.371s^2$, no real values

$n''(s) = -47.72 + 8.742s$

$0 = -47.72 + 8.742s$

$s = 5.4587$

$n(5.4587) = 310.6$

Point of inflection at $(5.4587, 310.6)$

s	0	5.4587	10
$n(s)$	-3.79	310.6	511

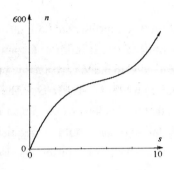

Smallest rate of increase at $s = 5.46$.

39. $p(x) = \dfrac{1}{\sigma\sqrt{2\pi}} e^{-(x-\mu)^2/2\sigma^2}$

with $\mu = 72.6$ and $\sigma = 5.2$

$p(x) = 0.07672e^{-(x-72.6)^2/54.08}$

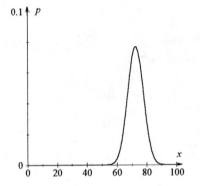

(a) Absolute max at $x = 72.6$, points of inflection at $x = 67.4$ and 77.8.

(b) $p(90) = 0.000284 = 0.03\%$

(c) $x = 72.6$ gives the maximum value of p

(d) points of inflection

41. $y = \sin x,\ -10 \le x \le 10,\ -1.0 \le y \le 1.0$

Maxima at $-4.7,\ 1.6,\ 7.9$. Minima at $-7.9,\ -1.6,\ 4.7$.

Points of inflection at $-9.4,\ -6.3,\ -3.1,\ 0,\ 3.1,\ 6.3,\ 9.4$.

The limit does not exist because the graph oscillates

between -1 and $+1$ as the x-coordinate approaches

$+\infty$ or $-\infty$.

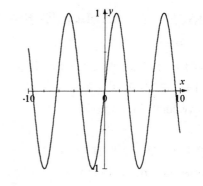

CHAPTER 12   REVIEW EXERCISES

1. $f(x) = 2x^2 - 2x - 1$; with domain $[0, 3]$

$f'(x) = 4x - 2$

Stationary points: $0 = 4x - 2$
$$x = 1/2$$

$f(1/2) = 2(1/2)^2 - 2(1/2) - 1 = -3/2$

End points: $x = 0, \; x = 3$

$f(0) = 2(0)^2 - 2(0) - 1 = -1$

$f(3) = 2(3)^2 - 2(3) - 1 = 11$

x	0	1/2	3
$f(x)$	-1	$-3/2$	11

Absolute min of $-3/2$ at $x = 1/2$, local max of -1 at $x = 0$, absolute max of 11 at $x = 3$.

3. $g(x) = 2x^3 - 6x + 1$; with domain $[-2, \infty)$

$g'(x) = 6x^2 - 6$

Stationary points: $0 = 6x^2 - 6$
$$x = \pm 1$$

$g(1) = 2(1)^3 - 6(1) + 1 = -3$

$g(-1) = 2(-1)^3 - 6(-1) + 1 = 5$

End points: $x = -2$

$g(-2) = 2(-2)^3 - 6(-2) + 1 = -3$

x	-2	-1	1	2
$g(x)$	-1	5	-3	11

Local max of 5 at $x = -1$, absolute min of -3 at $x = 1$ and $x = -2$.

5. $g(t) = \dfrac{1}{4}t^4 + t^3 + t^2$; with domain $(-\infty, \infty)$

$g'(t) = t^3 + 3t^2 + 2t = t(t^2 + 3t + 2)$

Stationary points: $g'(t) = 0$

$0 = t(t + 1)(t + 2)$

$t = 0$ or $t = -1$ or $t = -2$

$g(0) = \dfrac{1}{4}(0)^4 + (0)^3 + (0)^2 = 0$

$g(-2) = \dfrac{1}{4}(-2)^4 + (-2)^3 + (-2)^2 = 0$

$g(-1) = \dfrac{1}{4}(-1)^4 + (-1)^3 + (-1)^2 = 0$

t	-3	-2	-1	0	1
$g(t)$	2.25	0	1/4	0	2.25

Absolute min of 0 at $t = -2$ and 0, local max of $1/4$ at $t = -1$.

7. $f(t) = \dfrac{t+1}{(t-1)^2}$; with domain $[-2, 1) \cup (1, 2]$

$f'(t) = \dfrac{1(t-1)^2 - (t+1)(2)(t-1)}{(t-1)^4}$

$= \dfrac{-t^2 - 2t + 3}{(t-1)^4}$

Stationary points: $f'(t) = 0$

$0 = (-t - 3)(t - 1)$

$t = -3$ or $t = 1$

No allowed values of t, no stationary points

End points: $t = -2$ and $t = 2$

$f(-2) = \dfrac{-2+1}{(-2-1)^2} = -\dfrac{1}{9}$

$f(2) = \dfrac{2+1}{(2-1)^2} = 3$

t	-2	0	1.5	2
$f(t)$	$-1/9$	1	10	3

Absolute min of $-1/9$ at $t = -2$, absolute max of 3 at $t = 2$.

9. $f(t) = (t-1)^{2/3}$; with domain $(-\infty, \infty)$

$f'(t) = \dfrac{2}{3}(t-1)^{-1/3} = \dfrac{2}{3(t-1)^{1/3}}$

No stationary points.

Singular point: $t = 1$

$f(1) = (1-1)^{2/3} = 0$

t	0	1	2
$f(t)$	1	0	1

Absolute min of 0 at $t = 1$.

11. $g(x) = x - 3x^{1/3}$; with domain $(-\infty, \infty)$

$g'(x) = 1 - x^{-2/3} = \dfrac{x^{2/3} - 1}{x^{2/3}}$

Stationary points: $0 = \dfrac{x^{2/3} - 1}{x^{2/3}}$

$x = \pm 1$

$g(-1) = -1 - 3(-1)^{1/3} = 2$

$g(1) = 1 - 3(1)^{1/3} = -2$

No singular points

x	-8	-1	0	1	8
$g(x)$	-2	2	0	-2	2

Local max of 2 at $x = -1$, local min of -2 at $x = 1$.

13. $f(r) = \dfrac{1}{2}r^2 - \ln r$; with domain $(0, +\infty)$

$f'(r) = r - \dfrac{1}{r}$

Stationary points: $0 = r - \dfrac{1}{r}$

$r^2 = 1$

$r = \pm 1$ only

$f(1) = \dfrac{1}{2}(1)^2 - \ln 1 = \dfrac{1}{2}$

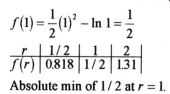

r	1/2	1	2
$f(r)$	0.818	1/2	1.31

Absolute min of $1/2$ at $r = 1$.

15. $g(t) = e^{t^2} + 1$

$g'(t) = 2te^{t^2}$

Stationary points: $0 = 2te^{t^2}$

$t = 0$

$g(0) = e^{0^2} + 1 = 2$

t	-1	0	1
$g(t)$	$1+e$	2	$1+e$

Absolute min of 2 at $t = 0$.

17. $f(x) = x^2 - x$; with domain $[0, 3]$

$f''(x) =$

No points of inflection

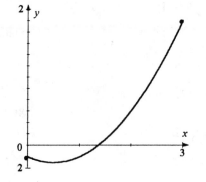

19. $g(x) = 2x^3 - 6x + 1$; with domain $[-2, \infty)$

$g'(x) = 12x$

Points of inflection $0 = 12x$

$\qquad\qquad\qquad x = 0$

$g(0) = 1$

Point of inflection at $(0, 1)$

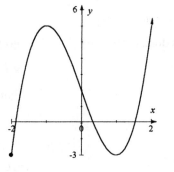

21. $g(t) = \dfrac{1}{4}t^4 + t^3 + t^2$; with domain $(-\infty, \infty)$

$g''(t) = 3t^2 + 6t + 2$

Points of inflection: $g''(t) = 0$

$0 = 3t^2 + 6t + 2$

$t = \dfrac{-6 \pm \sqrt{6^2 - 4(3)(2)}}{2(3)} = \dfrac{-3 \pm \sqrt{3}}{3}$

$t = -1.577$ or $t = -0.4227$

$g(-1.577) = 0.11$

$g(-0.4227) = 0.11$

Points of inflection at 0

$(-1.577, 0.11)$ and $(-0.4227, 0.11)$

23. $f(t) = \dfrac{t+1}{(t-1)^2}$; with domain $[-2, \infty) \cup (1, 2]$

$\lim_{t \to 1} f(t) = +8$

Vertical asymptote: $x = 1$

$f''(t) = \dfrac{2t+10}{(t-1)^4}$

Points of inflection: $f''(t) = 0$

$0 = 2t + 10$

$t = -5$, outside the domain

No point of inflection

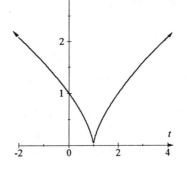

25. $f(t) = (t-1)^{2/3}$; with domain $(-\infty, \infty)$

$f''(t) = -\dfrac{2}{9(t-1)^{4/3}}$

$f''(t) \neq 0$,

No points of inflection

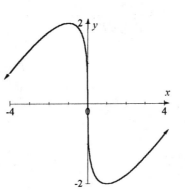

27. $g(x) = x - 3x^{1/3}$; with domain $(-\infty, \infty)$

$g''(x) = \dfrac{2}{3x^{5/3}}$

Points of inflection:

$g''(x) \neq 0$

$g''(0)$ is undefined

$g''(-1) = -\dfrac{2}{3}$

$g''(1) = \dfrac{2}{3}$

$g(0) = 0$

Points of inflection at $(0, 0)$

385

29. $f(r) = \frac{1}{2}r^2 = \ln r$; with domain $(0, +\infty)$

$f''(r) = 1 + \frac{1}{r^2}$

Points of inflection: $0 = 1 + \frac{1}{r^2}$

No real values, no points of inflection

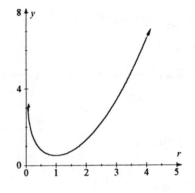

31. $g(t) = e^{t^2} + 1$

$g''(t) = 2e^{t^2} + 4t^2 e^{t^2} = 2e^{t^2}(1 + 2t^2)$

Points of inflection: $g''(t) = 0$

$0 = 2e^{t^2}(1 + 2t^2)$

$0 = 1 + 2t^2$

No real values, no points of inflection

33. $P = xy^2$ with $x^2 + y^2 = 75$, or $y^2 = 75 - x^2$

$x \geq 0$, and $y \geq 0$

If $y \geq 0$, $y^2 \geq 0$, $75 - x^2 \geq 0$ or $x \leq \sqrt{75}$

$P(x) = x(75 - x^2) = 75x - x^3$; $D[0, \sqrt{75}]$

Maximize P

$\frac{dP}{dx} = 75 - 3x^2$

$0 = 75 - 3x^2$

$x = 5$ only

$P(5) = 75(5) - 5^3 = 250$

x	0	5	$\sqrt{75}$
$P(x)$	-1	250	0

Maximum $P = 250$ at $x = 5$

35. $S = 3x + y + z$ with $xy = yz = 9$

or $x = \frac{9}{x}$ and $z = \frac{9}{y}$ and x, y, and $z > 0$.

$S(y) = 3\left(\frac{9}{y}\right) + y + \frac{9}{y} = y + \frac{36}{y}$; $D(0, \infty)$

Minimize S

$\frac{dS}{dy} = 1 - \frac{36}{y^2} = \frac{y^2 - 36}{y^2}$

$0 = y^2 - 36$

$y = 6$ only

$S(6) = 6 + \frac{36}{6} = 12$

y	1	6	36
$S(y)$	37	12	37

Minimum $S = 37$ at $y = 6$

37. $A = xy$ with $y = 1 - x^2$ and $x, y > 0$

If $y > 0$, $1 - x^2 > 0$ or $x < 1$

$A(x) = x(1 - x^2) = x - x^3$; D $(0, 1)$

Maximize A

$\dfrac{dA}{dx} = 1 - 2x^2$

$0 = 1 - 2x^2$

$x = \dfrac{1}{\sqrt{2}}$ only

$A(1/\sqrt{2}) = \dfrac{1}{\sqrt{2}} - \dfrac{1}{(\sqrt{2})^3} = \dfrac{\sqrt{2}}{4} \approx 0.35$

x	$1/4$	$1/\sqrt{2}$	$3/4$
$A(x)$	0.23	0.35	0.33

Maximum $A = \dfrac{\sqrt{2}}{4}$ at $x = \dfrac{1}{\sqrt{2}}$

39. $M(x) = \left(\dfrac{x}{1000} + \dfrac{1}{x} \right)^{-1}$

Maximize $M(x)$ when $\dfrac{dM(x)}{dx} = 0$

$\dfrac{dM(x)}{dx} = -\dfrac{\frac{1}{1000} - \frac{1}{x^2}}{\left(\frac{x}{1000} + \frac{1}{x} \right)^2}$

$0 = \dfrac{1}{1000} - \dfrac{1}{x^2}$

$x^2 = 1000$

$x = 31.62$ mph

41. $S = cwt^2$ with $w^2 + t^2 = r^2$ or

$t^2 = r^2 - w^2$ where c and r are constants.

$S = cw(r^2 - w^2) = cr^2 w - cw^3$

Maximize S when $\dfrac{dS}{dw} = 0$

$\dfrac{dS}{dw} = cr^2 - 3cw^2$

$0 = cr^2 - 3cw^2$

$w^2 = \dfrac{r^2}{3}$

$w = \dfrac{r}{\sqrt{3}}$

$t^2 = r^2 - w^2 = r^2 - \dfrac{r^2}{3} = \dfrac{2r^2}{3}$

$t = \dfrac{\sqrt{2}r}{\sqrt{3}}$

$\dfrac{t}{w} = \dfrac{\sqrt{2}r / \sqrt{3}}{r / \sqrt{3}} = \sqrt{2}$

43. $q = 1000 - 200p^2 + 20{,}000p$

$R = qp = 1000p - 200p^3 + 20{,}000p^2$

Maximize Revenue when $\dfrac{dR}{dp} = 0$

$\dfrac{dR}{dp} = 1000 - 600p^2 + 40{,}000p$

$0 = 10 - 6p^2 + 400p$

$0 = 3p^2 - 200p - 5$

$p = \dfrac{200 \pm \sqrt{(-200)^2 - 4(3)(-5)}}{2(3)}$

$= \dfrac{200 \pm \sqrt{40{,}060}}{6}$

$= 66.69$ or -0.025

387

45. $C(q) = 10,000 + 20q + \frac{1}{100}q^2$

$\overline{C}(q) = \frac{10,000}{q} + 20 + \frac{1}{100}q$

Minimize average cost when $\frac{d\overline{C}}{dq} = 0$

$\frac{d\overline{C}}{dq} = -\frac{10,000}{q^2} + \frac{1}{100}$

$0 = -\frac{10,000}{q^2} + \frac{1}{100}$

$q^2 = 1,000,000$

$q = 1000$

$\overline{C}(1000) = \frac{10,000}{1000} + 20 + \frac{1000}{100} = 40$

Minimum of \$40 per book for 1,000 books.

47. Let q = the number of occupied rooms, and p = charge per day.

$(60, 200)$ and $m = \frac{-20}{10} = -2$

$q - q_0 = m(p - p_0)$

$q - 200 = -2(p - 60)$

$q(p) = -2p + 320$

$R(p) = pq(p) = -2p^2 + 320p$

Maximize revenue when $\frac{dR}{dp} = 0$

$\frac{dR}{dp} = -4p + 320$

$0 = -4p + 320$

$p = 80$

$q(80) = -2(80) + 320 = 160$

A price of \$80 per day gives maximum revenue with a vacancy of $200 - 160 = 40$ rooms.

49. $9p^2 + 25q^2 = 22,500$

$18p\frac{dp}{dt} + 50q\frac{dq}{dt} = 0$

If $\frac{dp}{dt} = -2$, find $\frac{dq}{dt}$ when $p = 30$

when $p = 30$, $9(30)^2 + 25q^2 = 22,500$

or $q = 24$.

$18(30)(-2) + 50(24)\frac{dq}{dt} = 0$

$\frac{dq}{dt} = 0.9$

Sales are increasing at a rate of 0.9 dozen roses per week.

51. $\frac{1}{R} = \frac{1}{R_1} + \frac{1}{R_2}$

$-\frac{1}{R^2}\frac{dR}{dt} = -\frac{1}{R_1^2}\frac{dR_1}{dt} - \frac{1}{R_2^2}\frac{dR_2}{dt}$

If $\frac{dR_1}{dt} = 2$ and $\frac{dR_2}{dt} = -1$,

find $\frac{dR}{dt}$ when $R_1 = 6$ and $R_2 = 1$

when $R_1 = 6$ and $R_2 = 1$, $\frac{1}{R} = \frac{1}{6} + \frac{1}{1}$

or $R = \frac{6}{7}$.

$-\frac{1}{\left(\frac{6}{7}\right)^2}\frac{dR}{dt} = -\frac{1}{6^2}(2) - \frac{1}{1^2}(2) - \frac{1}{1^2}(-1)$

$\frac{dR}{dt} = -\frac{34}{49}$

R is decreasing at a rate of $\frac{34}{49}$ ohm / s.

53. Let x = the length of the shadow and y = the distance from the lamp post.

$$\frac{15}{6} = \frac{x+y}{x} \text{ or } x = \frac{2}{3}y$$

$$\frac{dx}{dt} = \frac{2}{3}\frac{dy}{dt}$$

If $\frac{dy}{dt} = 3$, find $\frac{dx}{dt}$

$$\frac{dx}{dt} = \frac{2}{3}(3) = 2$$

The shadow is increasing at a rate of 2 ft / s.

55. $q = 1000 - 200p^2 + 20,000p$

$$\frac{dq}{dp} = -400p + 20,000$$

$$E = -\frac{dq}{dp} \cdot \frac{p}{q}$$

If $p = 60$,

$q = 1000 - 200(60)^2 + 20,000(60) = 481,000,$

$$\frac{dq}{dp} = -400(60) + 20,000 = -4000 \text{ and}$$

$$E = -(-4000)\left(\frac{60}{481,000}\right) \approx \frac{1}{2}$$

Revenue is maximized when $E = 1$

$$E = -(-400p + 20,000)\left(\frac{p}{1000 - 200p^2 + 20,000p}\right)$$

$$\frac{400p^2 - 20,000p}{1000 - 200p^2 + 20,000p} = 1$$

$$400p^2 - 20,000p = 1000 - 200p^2 + 20,000p$$

$$600p^2 - 40,000p - 1000 = 0$$

$$3p^2 - 200p - 5 = 0$$

$$p = \frac{200 \pm \sqrt{(-200)^2 - 4(3)(-5)}}{2(3)}$$

$$= 66.69 \text{ or } -0.025$$

Revenue is maximized at $p = \$66.99$.

57. $9p^2 + 25q^2 = 22,500$

$$q^2 = \frac{22,500 - 9p^2}{25}$$

$$2q\frac{dq}{dp} = -2\left(\frac{9}{25}p\right)$$

$$\frac{dq}{dp} = -\frac{18p}{50q} = -\frac{9}{25}\frac{p}{q}$$

Revenue is maximized when $E = 1$

$$E = -\frac{dq}{dp} \cdot \frac{p}{q} = -\left(-\frac{9}{25}\frac{p}{q}\right)\left(\frac{p}{q}\right) = \frac{9p^2}{25q^2}$$

$$1 = \frac{9p^2}{25q^2}$$

$$25\left(\frac{22,500 - 9p^2}{25}\right) = 9p^2$$

$$p^2 = \frac{22,500}{18}$$

$$p = 35.36$$

Revenue is maximized when $p = \$35.36$.

CHAPTER 13
THE INTEGRAL

13.1 EXERCISES

1. $\int x^5 dx = \dfrac{x^6}{6} + C$

3. $\int 6 dx = 6x + C$

5. $\int x dx = \dfrac{x^2}{2} + C$

7. $\int (x^2 - x) dx = \dfrac{x^3}{3} - \dfrac{x^2}{2} + C$

9. $\int (1+x) dx = x + \dfrac{x^2}{2} + C$

11. $\int x^{-5} dx = -\dfrac{x^{-4}}{4} + C$

13. $\int \left(u^2 - \dfrac{1}{u}\right) du = \int (u^2 - u^{-1}) du = \dfrac{u^3}{3} - \ln|u| + C$

15. $\int (3x^4 - 2x^{-2} + x^{-5} + 4) dx = \dfrac{3x^5}{5} + 2x^{-1} - \dfrac{x^{-4}}{4} + 4x + C$

17. $\int \left(2e^x + \dfrac{5}{x}\right) dx = \int (2e^x + 5x^{-1}) dx = 2e^x + 5\ln|x| + C$

19. $\int \left(\dfrac{1}{x} + \dfrac{2}{x^2} - \dfrac{1}{x^3}\right) dx = \int (x^{-1} + 2x^{-2} - x^{-3}) dx = \ln|x| - 2x^{-1} + \dfrac{x^{-2}}{2} + C$

$\qquad\qquad = \ln|x| - \dfrac{2}{x} + \dfrac{1}{2x^2} + C$

21. $\int (3x^{0.1} - x^{4.3}) dx = \dfrac{3x^{1.1}}{1.1} - \dfrac{x^{5.3}}{5.3} + C$

23. $\int\left(\dfrac{3}{x^{0.1}} - \dfrac{4}{x^{1.1}}\right)dx = \int\left(3x^{-0.1} - 4x^{-1.1}\right)dx = \dfrac{3x^{0.9}}{0.9} + \dfrac{4x^{-0.1}}{0.1} + C = \dfrac{x^{0.9}}{0.3} + \dfrac{40}{x^{0.1}} + C$

25. $\int\dfrac{x+2}{x^3}dx = \int\left(\dfrac{1}{x^2} + \dfrac{2}{x^3}\right)dx = \int\left(x^{-2} + 2x^{-3}\right)dx = \dfrac{x^{-1}}{-1} + \dfrac{2x^{-2}}{-2} + C = -\dfrac{1}{x} - \dfrac{1}{x^2} + C$

27. $\int\sqrt{x}\,dx = \int x^{1/2}dx = \dfrac{2x^{3/2}}{3} + C$

29. $\int\left(2\sqrt[3]{x} - \dfrac{1}{2\sqrt{x}}\right)dx = \int\left(2x^{1/3} - \dfrac{x^{-1/2}}{2}\right)dx = \dfrac{2x^{4/3}}{4/3} - \dfrac{x^{1/2}}{2(1/2)} + C = \dfrac{3x^{4/3}}{2} - x^{1/2} + C$

31. $\int\dfrac{x - 2\sqrt{x}}{x^2}dx = \int\left(\dfrac{1}{x} - \dfrac{2}{x^{3/2}}\right)dx = \int\left(x^{-1} - 2x^{-3/2}\right)dx = \ln|x| - \dfrac{2x^{-1/2}}{-1/2} + C = \ln|x| + \dfrac{4}{\sqrt{x}} + C$

In Exercises 33-35 use $\int\left(\dfrac{d}{dx}f(x)\right)dx = f(x) + C$ and the slope of the tangent line at $(x, f(x)) = \dfrac{d}{dx}f(x)$.

33. If $\dfrac{d}{dx}f(x) = x$ and $f(0) = 1$, then

$\int x\,dx = f(x) + C$

$\dfrac{x^2}{2} = f(x) + C$

$\dfrac{0^2}{2} = f(0) + C$

$0 = 1 + C$

$C = -1$

Therefore

$f(x) - 1 = \dfrac{x^2}{2}$

$f(x) = \dfrac{x^2}{2} + 1$

35. If $\dfrac{d}{dx}f(x) = e^x - 1$ and $f(0) = 0$, then

$\int(e^x - 1)dx = f(x) + C$

$e^x - x = f(x) + C$

$e^0 - 0 = f(0) + C$

$1 = 0 + C$

$C = 1$

Therefore

$f(x) + 1 = e^x - x$

$f(x) = e^x - x - 1$

In Exercises 37-39 use $f(x) = \int \left(\dfrac{d}{dx} f(x) \right) dx = f(x) + K.$

37. If $\dfrac{d}{dx} C(x) = 5 - \dfrac{x}{10,000}$ and

$C(0) = 20,000$, then

$$C(x) = \int \left(5 - \frac{x}{10,000} \right) dx$$

$$= 5x - \frac{x^2}{20,000} + K$$

$$C(0) = 5(0) - \frac{0^2}{20,000} + K$$

$$20,000 = K$$

$$C(x) = 5x - \frac{x^2}{20,000} + 20,000$$

39. If $\dfrac{d}{dx} C(x) = 5 + 2x + \dfrac{1}{x}$ and

$C(1) = 1000$, then

$$C(x) = \int \left(5 + 2x + \frac{1}{x} \right) dx$$

$$= 5x + x^2 + \ln x + K$$

$$C(1) = 5(1) + (1)^2 + \ln 1 + K$$

$$1000 = 5 + 1 + 0 + K$$

$$K = 994$$

$$C(x) = 5x + x^2 + \ln x + 994$$

41. (a) If $v(t) = t^2 + 1$, then

$$s(t) = \int v(t) dt$$

$$= \int (t^2 + 1) dt$$

$$= \frac{t^3}{3} + 1 + C$$

(b) If $s = 1$ when $t = 0$, then

$$s(0) = \frac{(0)^3}{3} + 0 + C$$

$$1 = C$$

$$s(t) = \frac{t^3}{3} + t + 1$$

13.2 EXERCISES

1. $\int (3x+1)^5 dx$

   Let $u = 3x+1$

   $$\frac{du}{dx} = 3$$

   $$dx = \frac{du}{3}$$

   $$\int (3x+1)^5 dx = \int u^5 \frac{1}{3} du$$

   $$= \frac{1}{3} \int u^5 du$$

   $$= \frac{1}{3} \frac{u^6}{6} + C$$

   $$= \frac{(3x+1)^6}{18} + C$$

3. $\int (-2x+2)^{-2} dx$

   Let $u = -2x+2$

   $$\frac{du}{dx} = -2$$

   $$dx = -\frac{du}{2}$$

   $$\int (-2x+2)^{-2} dx = \int u^{-2} \left(-\frac{du}{2} \right)$$

   $$= -\frac{1}{2} \int u^{-2} du$$

   $$= -\frac{1}{2} \frac{u^{-1}}{-1} + C$$

   $$= \frac{(-2x+2)^{-1}}{2} + C$$

5. $\int x(3x^2+3)^3 dx$

   Let $u = 3x^2+3$

   $$\frac{du}{dx} = 6x$$

   $$dx = \frac{du}{6x}$$

   $$\int x(3x^2+3)^3 dx = \int xu^3 \frac{1}{6x} du$$

   $$= \frac{1}{6} \int u^3 du$$

   $$= \frac{1}{6} \frac{u^4}{4} + C$$

   $$= \frac{(3x^2+3)^4}{24} + C$$

7. $\int 2x\sqrt{3x^2-1} dx$

   Let $u = 3x^2-1$

   $$\frac{du}{dx} = 6x$$

   $$dx = \frac{du}{6x}$$

   $$\int 2x\sqrt{3x^2-1} dx = \int 2xu^{1/2} \frac{1}{6x} du$$

   $$= \frac{1}{3} \int u^{1/2} du$$

   $$= \frac{1}{3} \frac{u^{3/2}}{3/2} + C$$

   $$= \frac{2(3x^2-1)^{3/2}}{9} + C$$

9. $\int xe^{-x^2+1}dx$

Let $u = -x^2 + 1$

$$\frac{du}{dx} = -2x$$

$$dx = \frac{du}{2x}$$

$$\int xe^{-x^2+1}dx = \int xe^u\left(-\frac{1}{2x}\right)du$$

$$= -\frac{1}{2}\int e^u du$$

$$= -\frac{1}{2}e^u + C$$

$$= -\frac{e^{-x^2+1}}{2} + C$$

11. $\int (x+1)e^{-(x^2+2x)}dx$

Let $u = x^2 + 2x$

$$\frac{du}{dx} = 2x + 2$$

$$dx = \frac{du}{2(x+1)}$$

$$\int (x+1)e^{-(x^2+2x)}dx = \int (x+1)e^{-u}\left[\frac{1}{2(x+1)}\right]du$$

$$= \frac{1}{2}\int e^{-u}du$$

$$= \frac{1}{2}(-e^{-u}) + C$$

$$= \frac{e^{-(x^2+2x)}}{2} + C$$

13. $\int \frac{-2x-1}{(x^2+x+1)^3}dx$

Let $u = x^2 + x + 1$

$$\frac{du}{dx} = 2x + 1$$

$$dx = \frac{du}{2x+1}$$

$$\int \frac{-2x-1}{(x^2+x+1)^3}dx = \int \frac{-2x-1}{u^3}\left(\frac{1}{2x+1}\right)du$$

$$= -\int u^{-3}du$$

$$= -\frac{u^{-2}}{-2} + C$$

$$= \frac{(x^2+x+1)^{-2}}{2} + C$$

15. $\int \frac{x^2+x^5}{\sqrt{2x^3+x^6}}dx$

Let $u = 2x^3 + x^6 - 5$

$$\frac{du}{dx} = 6x^2 + 6x^5$$

$$dx = \frac{du}{6x^2+6x^5}$$

$$\int \frac{x^2+x^5}{\sqrt{2x^3+x^6-5}}dx = \int \frac{x^2+x^5}{u^{1/2}}\left(\frac{1}{6x^2+6x^5}\right)du$$

$$= \frac{1}{6}\int u^{-1/2}du$$

$$= \frac{1}{6}\frac{u^{1/2}}{1/2} + C$$

$$= \frac{\sqrt{2x^3+x^6-5}}{3} + C$$

17. $\int 2x\sqrt{x+1}\,dx$

Let $u = x+1 \Rightarrow x = u-1$

$\dfrac{du}{dx} = 1$

$dx = du$

$\int 2x\sqrt{x+1}\,dx = \int 2(u-1)u^{1/2}du$

$= \int \left(2u^{3/2} - 2u^{1/2}\right)du$

$= \dfrac{2u^{5/2}}{5/2} - \dfrac{2u^{3/2}}{3/2} + C$

$= \dfrac{4(x+1)^{5/2}}{5} - \dfrac{4(x+1)^{3/2}}{3} + C$

19. $\int x(x-2)^5\,dx$

Let $u = x-2 \Rightarrow x = u+2$

$\dfrac{du}{dx} = 1$

$dx = du$

$\int x(x-2)^5\,dx = \int (u+2)(u)^5 du$

$= \int \left(u^6 + 2u^5\right)du$

$= \dfrac{u^7}{7} + \dfrac{2u^6}{6} + C$

$= \dfrac{(x-2)^7}{7} + \dfrac{(x-2)^6}{3} + C$

21. $\int \dfrac{3e^{-1/x}}{x^2}\,dx$

Let $u = \dfrac{1}{x} = x^{-1}$

$\dfrac{du}{dx} = -x^{-2}$

$dx = -x^2 du$

$\int \dfrac{3e^{-1/x}}{x^2}\,dx = \int \dfrac{3e^{-u}}{x^2}\left(-x^2\right)du$

$= -3\int e^{-u}du$

$= -3\left(-e^{-u}\right) + C$

$= 3e^{-1/x} + C$

23. $\int x\left(x^2+1\right)^{1.3}\,dx$

Let $u = x^2 + 1$

$\dfrac{du}{dx} = 2x$

$dx = \dfrac{du}{2x}$

$\int x\left(x^2+1\right)^{1.3}\,dx = \int xu^{1.3}\left(\dfrac{1}{2x}\right)du$

$= \dfrac{1}{2}\int u^{1.3}du$

$= \dfrac{1}{2}\dfrac{u^{2.3}}{2.3} + C$

$= \dfrac{\left(x^2+1\right)^{2.3}}{4.6} + C$

25. $\int \left(1 + 9.3e^{3.1x-2}\right)dx$

    Let $u = 3.1x - 2$

$$\frac{du}{dx} = 3.1$$

$$dx = \frac{du}{3.1}$$

$$\int \left(1 + 9.3e^{3.1x-2}\right)dx = \int \left(1 + 9.3e^{u}\right)\left(\frac{1}{3.1}\right)du$$

$$= \frac{1}{3.1}\int \left(1 + 9.3e^{u}\right)du$$

$$= \frac{1}{3.1}\left(u + 9.3e^{u}\right) + C$$

$$= \frac{3.1x - 2}{3.1} + C$$

$$= x + 3e^{3.1x-2} + C$$

27. $\int \frac{e^{-0.05x}}{1 - e^{-0.05x}}\,dx$

    Let $u = 1 - e^{-0.05x}$

$$\frac{du}{dx} = 0.05e^{-0.05x}$$

$$dx = \frac{e^{0.05x}\,du}{0.05}$$

$$\int \frac{e^{-0.05x}}{1 - e^{-0.05x}}\,dx = \int \frac{e^{-0.05x}}{u}\left(\frac{e^{0.05x}}{0.05}\right)du$$

$$= \frac{1}{0.05}\int u^{-1}\,du$$

$$= 20 \ln |u| + C$$

$$= 20 \ln \left|1 - e^{-0.05x}\right| + C$$

29. $\int \left((2x-1)e^{2x^2-2x} + xe^{x^2}\right)dx$

    Let $u = 2x^2 - 2x$ and $v = x^2$

$$\frac{du}{dx} = 4x - 2 \qquad \frac{dv}{dx} = 2x$$

$$dx = \frac{du}{4x - 2} \qquad dx = \frac{dv}{2x}$$

$$\int \left((2x-1)e^{2x^2-2x} + xe^{x^2}\right)dx$$

$$= \int \left((2x-1)e^{2x^2-2x}\right)dx + \int xe^{x^2}\,dx$$

$$= \int (2x-1)e^{u}\left(\frac{1}{4x-2}\right)du + \int xe^{v}\left(\frac{1}{2x}\right)dv$$

$$= \frac{1}{2}\int e^{u}\,du + \frac{1}{2}\int e^{v}\,dv$$

$$= \frac{1}{2}e^{u} + \frac{1}{2}e^{v} + C$$

$$= \frac{e^{2x^2-2x} + e^{x^2}}{2} + C$$

31. $\int (ax+b)^{n}\,dx \qquad (n \neq -1)$

    Let $u = ax + b$

$$\frac{du}{dx} = a$$

$$dx = \frac{du}{a}$$

$$\int (ax+b)^{n}\,dx = \int u^{n}\left(\frac{1}{a}\right)du$$

$$= \frac{1}{a}\int u^{n}\,du$$

$$= \frac{1}{a}\frac{u^{n+1}}{n+1} + C$$

$$= \frac{(ax+b)^{n+1}}{a(n+1)} + C$$

33. $\int e^{ax+b}\,dx$

　　Let $u = ax + b$

　　$\dfrac{du}{dx} = a$

　　$dx = \dfrac{du}{a}$

　　$\int e^{ax+b}\,dx = \int e^{u}\left(\dfrac{1}{a}\right)du$

　　　　　　　$= \dfrac{1}{a}\int e^{u}\,du$

　　　　　　　$= \dfrac{1}{a}e^{u} + C$

　　　　　　　$= \dfrac{1}{a}e^{ax+b} + C$

35. $\int e^{-x}\,dx = \int e^{(-1)x}\,dx$

　　　　　　　$= -e^{-x} + C$

37. $\int e^{2x-1}\,dx = \tfrac{1}{2}e^{2x-1} + C$

39. $\int(2x+4)^{2}\,dx = \dfrac{(2x+4)^{3}}{6} + C$

41. $\int \dfrac{1}{5x-1}\,dx = \int (5x-1)^{-1}\,dx$

　　　　　　　$= \dfrac{1}{5}\ln|5x-1| + C$

43. $\int(1.5x)^{3}\,dx = \dfrac{(1.5x)^{4}}{6} + C$

In Exercises 45-47 use $\frac{d}{dx}f(x)$ = the slope of the tangent line at $(x, f(x))$ and $\int \left(\frac{d}{dx}f(x) \right)dx = f(x)+C$.

45. $\dfrac{d}{dx}f(x) = x(x^2+1)^3$ and $f(0) = 0$

Let $u = x^2 + 1$

$\dfrac{du}{dx} = 2x \Rightarrow dx = \dfrac{du}{2x}$

$f(x) + C = \displaystyle\int x(x^2+1)^3\,dx$

$\qquad = \displaystyle\int xu^3\left(\dfrac{1}{2x}\right)du$

$\qquad = \dfrac{1}{2}\displaystyle\int u^3\,du$

$\qquad = \dfrac{1}{2}\dfrac{u^4}{4}$

$f(x) = \dfrac{(x^2+1)^4}{8} + C$

$0 = \dfrac{(0^2+1)^4}{8} + C$

$C = -\dfrac{1}{8}$

$f(x) = \dfrac{(x^2+1)^4}{8} - \dfrac{1}{8}$

47. $\dfrac{d}{dx}f(x) = xe^{x^2-1}$ and $f(0) = \dfrac{1}{2e}$

Let $u = x^2 - 1$

$\dfrac{du}{dx} = 2x \Rightarrow dx = \dfrac{du}{2x}$

$f(x) + C = \displaystyle\int xe^{x^2-1}\,dx$

$\qquad = \displaystyle\int xe^u\left(\dfrac{1}{2x}\right)du$

$\qquad = \dfrac{1}{2}\displaystyle\int e^u\,du$

$\qquad = \dfrac{1}{2}e^u$

$f(x) = \dfrac{e^{x^2-1}}{2} + C$

$\dfrac{1}{2e} = \dfrac{e^{0^2-1}}{2} + C$

$C = 0$

$f(x) = \dfrac{e^{x^2-1}}{2}$

49. $v(t) = t(t^2 + 1)^4 + t$

   (a) $s(t) = \int v(t) dt$

   $= \int \left(t(t^2 + 1)^4 + t \right) dt$

   $= \int t(t^2 + 1)^4 dt + \int t dt$

   Let $u = t^2 + 1$

   $\dfrac{du}{dt} = 2t \Rightarrow dt = \dfrac{du}{2t}$

$s(t) = \int tu^4 \left(\dfrac{1}{2t} \right) du + \int t\ dt$

   $= \dfrac{1}{2} \int u^4\ du + \int t\ dt$

   $= \dfrac{1}{2} \dfrac{u^5}{5} + \dfrac{t^2}{2} + C$

   $= \dfrac{(t^2 + 1)^5}{10} + \dfrac{t^2}{2} + C$

(b) $s(0) = \dfrac{(0^2 + 1)^5}{10} + \dfrac{0^2}{2} + C$

   $1 = \dfrac{1}{10} + C$

   $C = \dfrac{9}{10}$

   $s(t) = \dfrac{(t^2 + 1)^5}{10} + \dfrac{t^2}{2} + \dfrac{9}{10}$

In Exercises 51-53 use $C(x) = \int \left(\dfrac{d}{dx} C(x) \right) dx + K$.

51. $\dfrac{d}{dx} C(x) = 5 + \sqrt{x + 1}$ and $C(0) = 20{,}000$

   $C(x) = \int \left(5 + \sqrt{x + 1} \right) dx + K$

   $= \int 5 dx + \int \sqrt{x + 1} dx + K$

   Let $u = x + 1$

   $\dfrac{du}{dx} = 1 \Rightarrow dx = du$

$C(x) = \int 5 dx + \int u^{1/2} du + K$

   $= 5x + \dfrac{u^{3/2}}{3/2} + K$

   $= 5x + \dfrac{2(x + 1)^{3/2}}{3} + K$

   $C(0) = 5(0) + \dfrac{2(0 + 1)^{3/2}}{3} + K$

   $20{,}000 = \dfrac{2}{3} + K$

   $K = 19{,}999.33$

   $C(x) = 5x + \dfrac{2(x + 1)^{3/2}}{3} + 19{,}999.33$

53. $\dfrac{d}{dx}C(x) = 5 + \dfrac{1}{x+1}$ and $C(1) = 1000$

$$C(x) = \int \left(5 + \dfrac{1}{x+1}\right)dx + K$$

$$= \int 5dx + \int \dfrac{1}{x+1}dx + K$$

Let $u = x + 1$

$$\dfrac{du}{dx} = 1 \Rightarrow dx = du$$

$$C(x) = \int 5dx + \int \dfrac{1}{u}du + K$$

$$= \int 5dx + \int u^{-1}du + K$$

$$= 5x + \ln|u| + K$$

$$= 5x + \ln|x+1| + K$$

$$C(1) = 5(1) + \ln|1+1| + K$$

$$1000 = 5 + \ln 2 + K$$

$$K = 994.31$$

$$C(x) = 5x + \ln|x+1| + 994.31$$

13.3 EXERCISES

In Exercises 1-11 use $f(x) = \int f'(x)dx$.

1. $C'(x) = 100 + 0.01x$ and $C(0) = 10,000$

$$C(x) = \int (100 + 0.01x)dx$$

$$= 100x + \dfrac{0.01x^2}{2} + K$$

$$= 100x + 0.05x^2 + K$$

$$C(0) = 100(0) + 0.005(0)^2 + K$$

$$10,000 = K$$

$$C(x) = 10,000 + 100x + 0.005x^2$$

3. $C'(x) = 100 + 20e^{-0.01x}$ and $C(0) = 10,000$

$$C(x) = \int (100 + 20e^{-0.01})dx$$

Let $u = -0.01x$, $\dfrac{du}{dx} = -0.01$, $dx = -\dfrac{du}{0.01}$

$$C(x) = \int 100dx + \int 20e^u \left(\dfrac{-1}{0.01}\right)du$$

$$= 100x - 2000e^u + K$$

$$= 100x - 2000e^{-0.01x} + K$$

$$C(0) = 100(0) - 2000e^{-0.01(0)} + K$$

$$10,000 = -2000 + K$$

$$K = 12,000$$

$$C(x) = 12,000 + 100x - 2000e^{-0.01x}$$

5. $C'(x) = 200 - 2x$ for $x \geq 2$ and $C(2) = 396$

$C(x) = \int (200 - 2x)\,dx$

$\quad = 200x - \dfrac{2x^2}{2} + K$

$\quad = 200x - x^2 + K$

$C(2) = 200(2) - 2^2 + K$

$396 = 396 + K$

$0 = K$

$C(x) = 200x - x^2$

7. $C'(x) = \dfrac{5000}{x + 10}$ and $C(0) = 0$

$C(x) = \int \dfrac{5000}{x + 10}\,dx$

Let $u = x + 10$, $\dfrac{du}{dx} = 1$, $dx = du$

$C(x) = \int 5000 u^{-1}\,du$

$\quad = 5000 \ln |u| + K$

$\quad = 5000 \ln |x + 10| + K$

$C(0) = 5000 \ln (0 + 10) + K$

$0 = 5000 \ln 10 + K$

$K = 5000 \ln 10$

$C(x) = 5000(\ln (x + 10) - \ln 10)$

$\quad = 5000 \ln \left(\dfrac{x}{10} + 1 \right)$

9. $C'(x) = 200 + 0.02x$ and $C(0) = 10{,}000$

$C(x) = \int (200 + 0.02x)\,dx$

$\quad = 200x + \dfrac{0.02x^2}{2} + K$

$\quad = 200x + 0.01x^2 + K$

$C(0) = 200(0) + 0.01(0)^2 + K$

$10{,}000 = K$

$C(x) = 10{,}000 + 200x + 0.01x^2$

$\overline{C}(x) = \dfrac{10{,}000}{x} + 200 + 0.01x$

Minimize $\overline{C}(x)$

$\overline{C}'(x) = \dfrac{-10{,}000}{x^2} + 0.01$

$0 = \dfrac{-10{,}000}{x^2} + 0.01$

$x^2 = \dfrac{10{,}000}{0.01} = 1{,}000{,}000$

$x = 1000$

$\overline{C}(1000) = \dfrac{10{,}000}{1000} + 200 + 0.01(1000)$

$\quad = 220$

The minimum average cost per box is $220 for a production level of 1000 boxes.

11. $p'(x) = 20 + 20e^{-0.5x}$ and $p(0) = 0$

$p(x) = \int (20 + 20e^{-0.5x})dx$

Let $u = -0.5x$, $\dfrac{du}{dx} = -0.5$, $dx = \dfrac{du}{-0.5}$

$p(x) = \int 20dx + \int 20e^{u}\left(-\dfrac{1}{0.5}\right)du$

$\quad = 20x - 40e^{u} + C$

$\quad = 20x - 40e^{-0.5x} + C$

$p(0) = 20(0) - 40e^{-0.5(0)} + C$

$\quad 0 = -40 + C$

$\quad C = 40$

$p(x) = 40 + 20x - 40e^{-0.5x}$

$\overline{R}(x) = \overline{p}(x) = \dfrac{40}{x} + 20 - \dfrac{40e^{-0.5x}}{x}$

$\quad = 20 - 40\dfrac{e^{-0.5x} - 1}{x}$

13. $q = 10 - 0.2t^2$ and $s(0) = 0$

(a) $s(t) = \int q\, dt$

$\quad = \int (10 - 0.2t^2)dt$

$\quad = 10t - \dfrac{0.2t^3}{3} + C$

$s(0) = 10(0) - \dfrac{0.2(0)^3}{3} + C$

$\quad 0 = C$

$s(t) = 10t - \dfrac{0.2t^3}{3}$

(b) $s(7) = 10(7) - \dfrac{0.2(7)^3}{3}$

$\quad \approx 47$ T-shirts

15. (C) $q(t) = 10 + 1.56t^2$

(b) $P(t) = \int (10 + 1.56t^2)dt$

$\quad = 10t + \dfrac{1.56t^3}{3} + C$

$P(0) = 10(0) + 0.52(0)^3 + C$

$\quad 0 = C$

$P(t) = 10t + 0.52t^3$

(c) $P(10) = 10(10) + 0.52(10)^3$

$\quad = \$620$ billion

402

17. (a) $q = \dfrac{460e^{(t-4)}}{1+e^{(t-4)}}$ and $E(0) = 0$

$$E(t) = \int \dfrac{460e^{(t-4)}}{1+e^{(t-4)}}\,dt$$

Let $u = 1 + e^{(t-4)}$

$$\dfrac{du}{dt} = e^{(t-4)},\ \ dt = \dfrac{du}{e^{(t-4)}}$$

$$E(t) = \int \dfrac{460e^{(t-4)}}{u}\left(\dfrac{1}{e^{(t-4)}}\right)du$$

$$= 460\int u^{-1}\,du$$

$$= 460\ln|u| + C$$

$$= 460\ln\left(1+e^{(t-4)}\right) + C$$

$$E(0) = 460\ln\left(1+e^{(0-4)}\right) + C$$

$$0 = 8.349 + C$$

$$C = -8.349$$

$$E(t) = 460\ln\left(1+e^{(t-4)}\right) - 8.349$$

(b) $E(15) = 460\ln\left(1+e^{(15-4)}\right) - 8.349$

$$= \$5{,}052\text{ million}$$

19. $(0, 6000)$ and $(2, 10{,}000)$ and $x = 100$

$$m = \dfrac{10{,}000 - 6000}{2 - 0} = 2000$$

$$y - 6000 = 2000(t - 0)$$

$$y = 2000t + 6000$$

(a) $P = 10x^{0.3}y^{0.7}$

$$= 10(100)^{0.3}(2000t + 6000)^{0.7}$$

$$= 39.8107(6000 + 2000t)^{0.7}$$

(b) $T(t) = \int P\,dt$

$$= \int\left(39.8107(6000 + 2000t)^{0.7}\right)dt$$

Let $u = 6000 + 2000t$

$$\dfrac{du}{dt} = 2000,\ \ dt = \dfrac{du}{2000}$$

$$T(t) = \int 39.8107u^{0.7}\left(\dfrac{1}{2000}\right)du$$

$$= 0.019905\dfrac{u^{1.7}}{1.7} + C$$

$$= 0.011709(6000 + 2000t)^{1.7} + C$$

$$T(0) = 0.011709\left[6000 + 2000(0)\right]^{1.7} + C$$

$$0 = 0.011709(6000)^{1.7} + C$$

$$C = -31{,}001$$

$$T(t) = 0.011709(6000 + 2000)^{1.7} - 31{,}001$$

$$T(2) = 0.011709\left[6000 + 2000(2)\right]^{1.7} - 31001$$

$$= 42{,}878\text{ cellular phones}$$

21. $y(x) = \dfrac{NP_0}{P_0 + (N - P_0)e^{-kx}}$

$\quad\quad = \dfrac{NP_0}{P_0 + (N - P_0)e^{-kx}} \left(\dfrac{e^{kx}}{e^{kx}} \right)$

$\quad\quad = \dfrac{NP_0 e^{kx}}{P_0 e^{kx} + (N - P_0)}$

23. $k = 2.7,\ P_0 = 0.0025,\ N = 160$ and $\int y(x)dx = 0$ at $x = 0$;

Let $S(x) = \int y(x)dx$

$S(x) = \dfrac{160}{2.7} \ln \left[0.0025 e^{2.7x} + (160 - 0.0025) \right] + C$

$\quad\quad = 59.259 \ln \left(0.0025 e^{2.7x} + 159.9975 \right) + C$

$0 = 59.259 \ln \left(0.0025 e^{2.7(0)} + 159.9975 \right) + C$

$C = -300$

$S(x) = 59.259 \ln \left(0.0025 e^{2.7x} + 159.9975 \right) - 300$

$S(10) = 59.259 \ln \left(0.0025 e^{2.7(10)} + 159.9975 \right) - 300$

$\quad\quad = \$940 \text{ million}$

25. $k = 0.5,\ P_0 = 50,\ N = 900$ and $\int y(x)dx = 0$ at $x = 0$;

Let $S(x) = \int y(x)dx$

$S(x) = \dfrac{900}{0.5} \ln \left[50 e^{0.5x} + (900 - 50) \right] + C$

$\quad\quad = 1800 \ln \left(50 e^{0.5x} + 850 \right) + C$

$0 = 1800 \ln \left(50 e^{0.5(0)} + 850 \right) + C$

$C = -12,244.31$

$S(x) = 1800 \ln \left(50 e^{0.5x} + 850 \right) - 12,244.31$

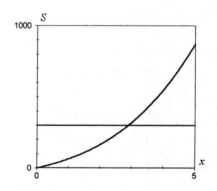

When $S(x) = 10(30) = 300$, $x \approx 3$ years

27. $v_0 = 0, \ t = 10$

$v(t) = v_0 - 32t$

$v(10) = 0 - 32(10) = -320$

$\qquad = 320 \text{ ft} / \text{s downward}$

29. At maximum height $v(t) = 0$

$v(t) = v_0 - 32t$

$0 = v_0 - 32t$

$t = \dfrac{v_0}{32}$

31. $s\left(\dfrac{v_0}{32}\right) = 20$ and $s_0 = 0$

From Exercise 30, $s\left(\dfrac{v_0}{32}\right) = \dfrac{v_0{}^2}{64}$

$\dfrac{v_0{}^2}{64} = 20$

$v_0{}^2 = 1280$

$v_0 \approx 35.78 \text{ ft} / \text{s}$

33. Maximum $s = 20$ ft and $s_0 = 0$

From Exercise 31, $v_0 = 35.78$

$s(t) = s_0 + v_0 t - 16t^2$

$s(4) = 0 + 35.78(4) - 16(4)^2$

$\qquad = -113$

It will be 113 feet below me.

35. (a) Maximum $s = 100$ and $s_0 = 0$

From Exercise 30, $s\left(\dfrac{v_0}{32}\right) = \dfrac{v_0{}^2}{64}$.

$\dfrac{v_0{}^2}{64} = 100$

$v_0{}^2 = 6400$

$v_0 = 80$

The minimum speed is 80 ft / s.

(b) $v_0 = 100, \ s = 100, \ \text{and } s_0 = 0$

$s = s_0 + v_0 t - 16t^2$

$100 = 0 + 100t - 16t^2$

$0 = -100 + 100t - 16t^2$

$0 = 25 - 25t + 4t^2$

$0 = (5 - t)(5 - 4t)$

$t = 5 \quad \text{or} \quad t = 1.25$

$v = v_0 - 32t \text{ and } t = 1.25$

$v = 100 - 32(1.25) = 60$

It will hit the ceiling at 60 ft / s.

(c) If it leaves the ceiling at

$v_0 = -60 \text{ ft} / \text{s}, \ s_0 = 100 \text{ ft and } s = 0$

$s = s_0 + v_0 t - 16t^2$

$0 = 100 - 60t - 16t^2$

$0 = 25 - 15t - 4t^2$

$0 = (5 + t)(5 - 4t)$

$t = -5 \quad \text{or} \quad t = 1.25$

It takes $1.25s$ to return.

37. From Exercise 30, $s\left(\dfrac{v_0}{32}\right) = \dfrac{v_0{}^2}{64}$

Professor Strong: $s_s = \dfrac{v_0{}^2 s}{64}$

Professor Weak: $s_w = \dfrac{v_0{}^2 w}{64}$

If $s_s = 2s_w$, then

$\dfrac{v_0{}^2 s}{64} = 2\dfrac{v_0{}^2 w}{64}$

$v_0{}^2 s = 2v_0{}^2 w$

$v_0 s = \sqrt{2} v_0 w$

Professor Strong can throw $\sqrt{2} \approx 1.414$ times as fast as Professor Weak.

39. $a = 3t^2 - 4t + 5$, $v(0) = 0$ and $s(0) = 0$

$v(t) = \int a(t)\,dt$

$\quad = \int (3t^2 - 4t + 5)\,dt$

$\quad = t^3 - 2t^2 + 5t + C$

$v(0) = (0)^3 - 2(0)^2 + 5(0) + C$

$\quad 0 = C$

$v(t) = t^3 - 2t^2 + 5t$

$s(t) = \int v(t)\,dt$

$\quad = \int (t^3 - 2t^2 + 5t)\,dt$

$\quad = \dfrac{t^4}{4} - \dfrac{2t^3}{3} + \dfrac{5t^2}{2} + C$

$s(0) = \dfrac{(0)^4}{4} - \dfrac{2(0)^3}{3} + \dfrac{5(0)^2}{2} + C$

$\quad 0 = C$

$s(t) = \dfrac{t^4}{4} - \dfrac{2t^3}{3} + \dfrac{5t^2}{2}$

$s(120) = \dfrac{(120)^4}{4} - \dfrac{2(120)^3}{3} + \dfrac{5(120)^2}{2}$

$\quad = 50{,}724$ feet

41. If $a = $ constant and $v(0) = 0$

$$v(t) = \int a\,dt$$
$$= a\int dt$$
$$= at + C$$
$$v(0) = a(0) + C$$
$$0 = C$$
$$v(t) = at$$

If $v(3) = 88$

$$88 = a(3)$$
$$a = \frac{88}{3} \approx 29.3 \text{ ft / s}^2$$

13.4 EXERCISES

1. $\int_{-1}^{1}(x^2 + 2)dx = \left[\dfrac{x^3}{3} + 2x\right]_{-1}^{1} = \left(\dfrac{1^3}{3} + 2\cdot 1\right) - \left(\dfrac{(-1)^3}{3} + 2(-1)\right) = \left(\dfrac{1}{3} + 2\right) - \left(-\dfrac{1}{3} - 2\right) = 4\dfrac{2}{3} = \dfrac{14}{3}$

3. $\int_{-2}^{2}(x^3 - 2x)dx = \left[\dfrac{x^4}{4} - x^2\right]_{-2}^{2} = \left(\dfrac{2^4}{4} - 2^2\right) - \left(\dfrac{(-2)^4}{4} - (-2)^2\right) = (4 - 4) - (4 - 4) = 0$

5. $\int_{1}^{3}\left(\dfrac{2}{x^2} + 3x\right)dx = \int_{1}^{3}(2x^{-2} + 3x)dx = \left[\dfrac{2x^{-1}}{-1} + \dfrac{3x^2}{2}\right]_{1}^{3} = \left(-\dfrac{2}{3} + \dfrac{3(3)^2}{2}\right) - \left(-\dfrac{2}{1} + \dfrac{3(1)^2}{2}\right)$

$$= \left(-\dfrac{2}{3} + \dfrac{27}{2}\right) - \left(-2 + \dfrac{3}{2}\right) = 13\dfrac{1}{3} = \dfrac{40}{3}$$

7. $\int_{0}^{1}(2.1x - 4.3x^{1.2})dx = \left[\dfrac{2.1x^2}{2} - \dfrac{4.3x^{2.2}}{2.2}\right]_{0}^{1} = \left(\dfrac{2.1(1)^2}{2} - \dfrac{4.3(1)^{2.2}}{2.2}\right) - \left(\dfrac{2.1(0)^2}{2} - \dfrac{4.3(0)^{2.2}}{2.2}\right)$

$$= \left(\dfrac{2.1}{2} - \dfrac{4.3}{2.2}\right) - (0 - 0) = -0.9045$$

9. $\int_0^1 2e^x dx = \left[2e^x\right]_0^1 = \left(2e^1\right) - \left(2e^0\right) = 2e - 2 = 2(e-1)$

11. $\int_{-1}^1 e^{2x-1} dx$

Let $u = 2x - 1$, $\dfrac{du}{dx} = 2$, $dx = \dfrac{du}{2}$. When $x = -1$, $u = -3$. When $x = 1$, $u = 1$.

$\int_{-1}^1 e^{2x-1} dx = \int_{-3}^1 \dfrac{e^u}{2} du = \left[\dfrac{e^u}{2}\right]_{-3}^1 = \left(\dfrac{e^1}{2}\right) - \left(\dfrac{e^{-3}}{2}\right) = \dfrac{1}{2}\left(e - e^{-3}\right)$

13. $\int_0^{50} e^{-0.02x-1} dx$

Let $u = -0.02x - 1$, $\dfrac{du}{dx} = -0.02$, $dx = \dfrac{du}{-0.02} = -50 du$. When $x = 0$, $u = -1$. When $x = 50$, $u = -2$.

$\int_0^{50} e^{-0.02x-1} dx = \int_{-1}^{-2} e^u du = \left[-50e^u\right]_{-1}^{-2} = \left(-50e^{-2}\right) - \left(-50e^{-1}\right) = 50\left(e^{-1} - e^{-2}\right) \approx 11.627$

15. $\int_0^1 \sqrt{x}\, dx = \int_0^1 x^{1/2} dx = \left[\dfrac{x^{3/2}}{3/2}\right]_0^1 = \left(\dfrac{2(1)^{3/2}}{3}\right) - \left(\dfrac{2(0)^{3/2}}{3}\right) = \dfrac{2}{3} - 0 = \dfrac{2}{3}$

17. $\int_{-1.1}^{1.1} e^{x+1} dx$

Let $u = x + 1$, $\dfrac{du}{dx} = 1$, $dx = du$. When $x = -1.1$, $u = -0.1$. When $x = 1.1$, $u = 2.1$.

$\int_{-1.1}^{1.1} e^{x+1} dx = \int_{-0.1}^{2.1} e^u du = \left[e^u\right]_{-0.1}^{2.1} = e^{2.1} - e^{-0.1}$

19. $\int_{-\sqrt{2}}^{\sqrt{2}} 3x\sqrt{2x^2+1}\, dx = \int_{-\sqrt{2}}^{\sqrt{2}} 3x\left(2x^2+1\right) dx$

Let $u = 2x^2 + 1$, $\dfrac{du}{dx} = 4x$, $dx = \dfrac{du}{4x}$. When $x = -\sqrt{2}$, $u = 5$. When $x = \sqrt{2}$, $u = 5$.

$\int_{-\sqrt{2}}^{\sqrt{2}} 3x\sqrt{2x^2+1}\, dx = \int_5^5 \dfrac{3xu^{1/2}}{4x} du = \int_5^5 \dfrac{3u^{1/2}}{4} du = \left[\dfrac{3u^{3/2}}{4(3/2)}\right]_5^5 = \left(\dfrac{5^{3/2}}{2}\right) - \left(\dfrac{5^{3/2}}{2}\right) = 0$

21. $\int_0^1 5xe^{x^2+2}dx$

Let $u = x^2+2$, $\dfrac{du}{dx} = 2x$, $dx = \dfrac{du}{2x}$. When $x = 0$, $u = 2$. When $x = 1$, $u = 3$.

$\int_0^1 5xe^{x^2+2}dx = \int_2^3 \dfrac{5xe^u}{2x}du = \int_2^3 \dfrac{5e^u}{2}du = \left[\dfrac{5e^u}{2}\right]_2^3 = \left(\dfrac{5e^3}{2}\right) - \left(\dfrac{5e^2}{2}\right) = \dfrac{5}{2}\left(e^3 - e^2\right)$

23. $\int_0^1 x\sqrt{2x+1}\,dx = \int_0^1 x(2x+1)^{1/2}\,dx$

Let $u = 2x+1$, $\dfrac{du}{dx} = 2$, $dx = \dfrac{du}{2}$, $x = \dfrac{u-1}{2}$. When $x = 0$, $u = 1$. When $x = 1$, $u = 3$.

$\int_0^1 x\sqrt{2x+1}\,dx = \int_1^3 \left(\dfrac{u-1}{2}\right)\dfrac{u^{1/2}}{2}du = \int_1^3\left(\dfrac{u^{3/2}}{4} - \dfrac{u^{1/2}}{4}\right)du = \left[\dfrac{u^{5/2}}{4(5/2)} - \dfrac{u^{3/2}}{4(3/2)}\right]_1^3$

$= \left(\dfrac{(3)^{5/2}}{10} - \dfrac{(3)^{3/2}}{6}\right) - \left(\dfrac{1^{5/2}}{10} - \dfrac{1^{3/2}}{6}\right) = \dfrac{3^{5/2}}{10} - \dfrac{3^{3/2}}{6} + \dfrac{1}{15}$

25. $\int_2^3 \dfrac{x^2}{\left(x^3-1\right)}dx = \int_2^3 x^2\left(x^3-1\right)^{-1}dx$

Let $u = x^3 - 1$, $\dfrac{du}{dx} = 3x^2$, $dx = \dfrac{du}{3x^2}$. When $x = 2$, $u = 7$. When $x = 3$, $u = 26$.

$\int_2^3 \dfrac{x^2}{\left(x^3-1\right)}dx = \int_7^{26}\dfrac{x^2u^{-1}}{3x^2}du = \int_7^{26}\dfrac{u^{-1}}{3}du = \left[\dfrac{\ln|u|}{3}\right]_7^{26} = \left(\dfrac{\ln 26}{3}\right) - \left(\dfrac{\ln 7}{3}\right) = \dfrac{1}{3}\left(\ln 26 - \ln 7\right)$

27. $\int_0^1 x\,dx = \left[\dfrac{x^2}{2}\right]_0^1 = \dfrac{1^2}{2} - \dfrac{0^2}{2} = \dfrac{1}{2}$

29. $\int_0^4 \sqrt{x}\,dx = \int_0^4 x^{1/2}dx = \left[\dfrac{x^{3/2}}{3/2}\right]_0^4 = \left(\dfrac{2(4)^{3/2}}{3}\right) - \left(\dfrac{2(0)^{3/2}}{3}\right) = \dfrac{16}{3}$

31. Figure 5 in the textbook shows this to be a parabola with x-intercepts of ± 1.

$$\int_0^1 \left(x^2 - 1\right)dx = \left[\frac{x^3}{3} - x\right]_0^1 = \left(\frac{1^3}{3} - 1\right) - \left(\frac{0^3}{3} - 0\right) = \frac{1}{3}$$

$$\int_1^4 \left(x^2 - 1\right)dx = \left[\frac{x^3}{3} - x\right]_1^4 = \left(\frac{4^3}{3} - 4\right) - \left(\frac{1^3}{3} - 1\right) = 18\frac{1}{3} = \frac{55}{3}$$

33. The graph of $y = xe^{x^2}$ lies above the x-axis between $x = 0$ and $x = (\ln 2)^{1/2}$

Let $u = x^2$, $\dfrac{du}{dx} = 2x$, $dx = \dfrac{du}{2x}$. When $x = 0$, $u = 0$. When $x = (\ln 2)^{1/2}$, $u = \ln 2$

$$\int_0^{(\ln 2)^{1/2}} xe^{x^2}\,dx = \int_0^{\ln 2} \frac{xe^u}{2x}\,du = \frac{1}{2}\int_0^{\ln 2} e^u\,du = \left[\frac{e^u}{2}\right]_0^{\ln 2} = \left(\frac{e^{\ln 2}}{2}\right) - \left(\frac{e^0}{2}\right) = 1 - \frac{1}{2} = \frac{1}{2}$$

35. $s = \displaystyle\int_1^6 \left(60 - e^{-t/10}\right)dt$

Let $u = -\dfrac{t}{10}$, $\dfrac{du}{dt} = -\dfrac{1}{10}$, $dt = -10\,du$. When $t = 1$, $u = -\dfrac{1}{10}$. When $t = 6$, $u = -\dfrac{6}{10}$.

$$s = \int_1^6 \left(60 - e^{-t/10}\right)dt = \int_1^6 60\,dt - \int_{-1/10}^{-6/10} -10e^u\,du = \left[60t\right]_1^6 + \left[10e^u\right]_{-1/10}^{-6/10}$$

$$= \left[(60(6)) - (60(1))\right] + \left[\left(10e^{-6/10}\right) - \left(10e^{-1/10}\right)\right] = (360 - 60) + (5.49 - 7.05) = 296 \text{ miles}$$

37. $C = \displaystyle\int_{10}^{100} \left(5 + \frac{x^2}{1000}\right)dx = \left[5x + \frac{x^3}{3000}\right]_{10}^{100} = \left(5(100) + \frac{(100)^3}{3000}\right) - \left(5(10) + \frac{10^3}{3000}\right) = \left(500 + 333\frac{1}{3}\right) - \left(50 + \frac{1}{3}\right) = \$783$

39. $R = \displaystyle\int_0^5 0.13e^{0.44t}\,dt$

Let $u = 0.44t$, $\dfrac{du}{dt} = 0.44$, $dt = \dfrac{du}{0.44}$. When $t = 0$, $u = 0$. When $t = 5$, $u = 2.2$.

$$R = \int_0^5 0.13e^{0.44t} = \int_0^{2.2} \frac{0.13e^u}{0.44}\,du = \left[\frac{0.13e^u}{0.44}\right]_0^{2.2} = \left(\frac{0.13e^{2.2}}{0.44}\right) - \left(\frac{0.13e^0}{0.44}\right) = \$2.4 \text{ billion}$$

41. (a) $(0, 9.2)$ and $(11, 0.8)$

$m = \dfrac{0.8 - 9.2}{11 - 0} = -0.7636$

$c(t) - c_0 = m(t - t_0)$

$c(t) - 9.2 = -0.7636 - (t - 0)$

$\qquad c(t) = 9.2 - 0.7636t$

(b) $C = \displaystyle\int_0^{11} (9.2 - 0.7636t)\,dt = \left[9.2t - \dfrac{0.7636t^2}{2}\right]_0^{11}$

$\qquad = \left(9.2(11) - 0.3818(11)^2\right) - \left(9.2(0) - 0.3818(0)^2\right)$

$\qquad = \$55.0022 \text{ billion}$

43. $C = \displaystyle\int_8^{10} 24\left(-0.00271t^3 + 0.137t^2 - 0.892t + 0.149\right)dt = 24\left[\dfrac{-0.00271t^4}{4} + \dfrac{0.137t^3}{3} - \dfrac{0.892t^2}{2} + 0.149t\right]_8^{10}$

$\qquad = 24\left(\dfrac{-0.00271(10)^4}{4} + \dfrac{0.137(10)^3}{3} - \dfrac{0.892(10)^2}{2} + 0.149(10)\right) - 24\left(\dfrac{-0.00271(8)^4}{4} + \dfrac{0.137(8)^3}{3} - \dfrac{0.892(8)^2}{2} + 0.149(8)\right)$

$\qquad = 24(-4.22) - 24(-6.75) = -101.28 + 162.05 \approx 60 \text{ milliliters}$

45. $E = \displaystyle\int_0^{15} \dfrac{460e^{(t-4)}}{1 + e^{(t-4)}}\,dt = \int_0^{15} 460e^{(t-4)}\left(1 + e^{(t-4)}\right)^{-1}dt$

Let $u = 1 + e^{(t-4)}$, $\dfrac{du}{dt} = e^{(t-4)}$, $dt = \dfrac{du}{e^{(t-4)}}$. When $t = 0$, $u = 1 + e^{-4}$. When $t = 15$, $u = 1 + e^{11}$.

$E = \displaystyle\int_{1+e^{-4}}^{1+e^{11}} \dfrac{460e^{(t-4)}u^{-1}}{e^{(t-4)}}\,du = \int_{1+e^{-4}}^{1+e^{11}} 460u^{-1}\,du = \left[460 \ln |u|\right]_{1+e^{-4}}^{1+e^{11}} = 460\left[\ln\left(1 + e^{11}\right) - \ln\left(1 + e^{-4}\right)\right]$

$\qquad = 460[11 - 0.01815] = \$5,052 \text{ million}$

47. $p = 10x^{0.3}y^{0.7}$, $x = 100$ and $(0, 6000)$ to $(2, 10{,}000)$ represents the linear increase in y over 2 years.

$m = \dfrac{10{,}000 - 6000}{2 - 0} = 2000 \Rightarrow y - 6000 = 2000(t - 0) \Rightarrow y = 2000t + 6000$

$P = \displaystyle\int_0^2 10(100)^{0.3}(2000t + 6000)^{0.7}\,dt = \int_0^2 39.8107(2000t + 6000)^{0.7}\,dt$

Let $u = 2000t + 6000$, $\dfrac{du}{dt} = 2000$, $dt = \dfrac{du}{2000}$. When $t = 0$, $u = 6000$. When $t = 2$, $u = 10{,}000$.

$P = \displaystyle\int_{6000}^{10{,}000} \dfrac{39.8107u^{0.7}}{2000}\,du = \left[\dfrac{0.0199054u^{1.7}}{1.7}\right]_{6000}^{10{,}000} = \left(0.011709(10{,}000)^{1.7}\right) - \left(0.011709(6000)^{1.7}\right)$

$\qquad = 73{,}879 - 31{,}001 = 42{,}878 \text{ cellular phones}$

49. Let $u = -t$, $\dfrac{du}{dt} = -1$, $dt = -du$. When $t = 0$, $u = 0$. When $t = 10$, $u = -10$

$$G = \int_0^{10}\left(1 - e^{-t}\right)dt = \int_0^{10} dt - \int_0^{-10} e^u(-du) = \left[t\right]_0^{10} + \left[e^u\right]_0^{-10}$$

$$= \left[(10) - (0)\right] + \left[\left(e^{-10}\right) - \left(e^0\right)\right] = 10 + (-1) = 9 \text{ gallons}$$

51. (a) $s(t) = Pe^{kt}$, $(0, 330)$ and $(12, 940)$

$t = 0$: $330 = Pe^{k(0)}$

$\quad p = 330$

$t = 12$: $940 = 330e^{k(12)}$

$\quad 2.8485 = e^{12k}$

$\quad \ln 2.8485 = 12k$

$\quad\quad k = 0.0872$

$s(t) = 330e^{0.0872t}$

(b) Let $u = 0.0872t$, $\dfrac{du}{dt} = 0.0872$, $dt = \dfrac{du}{0.0872}$

When $t = 0$, $u = 0$. When $t = 14$, $u = 1.2208$

$$S = \int_0^{14} 330e^{0.0872t}dt = \int_0^{1.2208} \frac{330e^u}{0.0872}\, du$$

$$= \left[3{,}784.4e^u\right]_0^{1.2208} = \left(3{,}784.4e^{1.2208}\right) - \left(3{,}784.4e^0\right)$$

$$= \$9{,}044 \text{ billion}$$

Total from chart = $8,225 billion

(c) Use the integral from part (b), but change the limits to $t = 14$ and $t = 19$.

When $t = 14$, $u = 1.2208$. When $t = 19$, $u = 1.6568$.

$$S = \left[3{,}784.4e^u\right]_{1.2208}^{1.6568} = \left(3{,}784.4e^{1.6568}\right) - \left(3{,}784.4e^{1.2208}\right)$$

$$= \$7{,}011 \text{ billion}$$

53. $s = s_0 e^{-kt}$, $k = 0.05$ and $s = 50$ when $t = 0$, therefore $s = 50e^{-0.05t}$.

Let $u = 0.05t$, $\dfrac{du}{dt} = -0.05$, $dt = \dfrac{du}{-0.05}$. When $t = 0$, $u = 0$. When $t = 52$, $u = -2.6$.

$$S = \int_0^{52} 50e^{-0.05t}\, dt = \int_0^{-2.6} \frac{50e^u}{-0.05}\, du = \left[-1000e^u\right]_0^{-2.6} = \left(1000e^{-2.6}\right) - \left(-1000e^0\right) = -74 + 1000 = 926 \text{ T-shirts}$$

55. If $p = mv$

$$W = \int_{v_0}^{v_1} v \frac{d}{dr}(p) dv = \int_{v_0}^{v_1} vm \, dv$$

$$= m \int_{v_0}^{v_1} v \, dv = \left[m \frac{v^2}{2} \right]_{v_0}^{v_1}$$

$$= \frac{1}{2} mv_1^2 - \frac{1}{2} mv_2^2$$

57. If $C(x)$ is an antiderivative of $m(x)$, then

$$\int_0^x m(t) \, dt = \left[C(t) \right]_0^x = C(x) - C(0)$$

Therfore $C(x) = C(0) + \int_0^x m(t) \, dt$

$C(0)$ is the fixed cost

13.5 EXERCISES

In Exercises 1-29 use $\Delta x = \dfrac{b-a}{n}$ and $x_k = a + k\Delta x$.

1. $\int_0^2 (4x - 1) dx$, $n = 4$

$a = 0$, $b = 2$, $\Delta x = \dfrac{2-0}{4} = \dfrac{1}{2}$

k	0	1	2	3	4
x_k	0	0.5	1	1.5	2.0
$f(x_k)$	−1	1	3	5	7

Left - hand sum $= (-1 + 1 + 3 + 5)(0.5) = 4$

Right - hand sum $= (1 + 3 + 5 + 7)(0.5) = 8$

Trapezoidal sum $= \dfrac{4+8}{2} = 6$

$\int_0^2 (4x - 1) dx = \left[2x^2 - x \right]_0^2 = 6$

Error $= |6 - 6| = 0$

3. $\int_{-2}^2 x^2 dx$, $n = 4$

$a = -2$, $b = 2$, $\Delta x = \dfrac{2 - (-2)}{4} = 1$

k	0	1	2	3	4
x_k	−2	−1	0	1	2
$f(x_k)$	4	1	0	1	4

Left - hand sum $= (4 + 1 + 0 + 1)(1) = 6$

Right - hand sum $= (1 + 0 + 1 + 4)(1) = 6$

Trapezoidal sum $= \dfrac{6+6}{2} = 6$

$\int_{-2}^2 x^2 dx = \left[\dfrac{x^3}{3} \right]_{-2}^2 = \left(\dfrac{8}{3} \right) - \left(-\dfrac{8}{3} \right) = \dfrac{16}{3} = 5.333...$

Error $= |6 - 5.333...| = 0.666... \approx \pm 0.7$

5. $\int_{-1}^{1} x^3 dx,\ n = 5$

$a = -1,\ b = 1,\ \Delta x = \dfrac{1-(-1)}{5} = 0.4$

k	0	1	2	3	4	5
x_k	−1	−0.6	−0.2	0.2	0.6	1
$f(x_k)$	−1	−0.216	−0.008	0.008	0.216	1

Left - hand sum

$= (-1 - 0.216 - 0.008 + 0.008 + 0.216)(0.4) = -0.4$

Right - hand sum

$= (-0.216 - 0.008 + 0.008 + 0.216 + 1) = (0.4)$

Trapezoidal sum $= \dfrac{-0.4 + 0.4}{2} = 0$

$\int_{-1}^{1} x^3 dx = \left[\dfrac{x^4}{4}\right]_{-1}^{1} = \left(\dfrac{1}{4}\right) - \left(\dfrac{1}{4}\right) = 0$

Error $= |0 - 0| = 0$

7. $\int_{0}^{10} e^{-x} dx,\ n = 5$

$a = 0,\ b = 10,\ \Delta x = \dfrac{10-0}{5} = 2$

k	0	1	2	3	4	5
x_k	0	2	4	6	8	10
$f(x_k)$	1	0.1353	0.0183	0.0025	0.0003	0.00005

Left - hand sum

$= (1 + 0.1353 + 0.0183 + 0.0025 + 0.0003)(2) = 2.3129$

Right - hand sum

$= (0.1353 + 0.0183 + 0.0025 + 0.0003 + 0.00005)(2) = 0.3130$

Trapezoidal sum $= \dfrac{2.3129 + 0.3130}{2} = 1.3130$

$\int_{0}^{10} e^{-x} dx = \int_{0}^{-10} e^{u}(-du) = \left[-e^{u}\right]_{0}^{10} = (-e^{-10}) - (-e^{0}) = 1.0000$

Error $= |1.3130 - 1.0000| = 0.3130 \approx \pm 0.3$

9. $\int_{0}^{1} \dfrac{1}{1+x} dx,\ n = 5$

$a = 0,\ b = 1,\ \Delta x = \dfrac{1-0}{5} = 0.2$

k	0	1	2	3	4	5
x_k	0	0.2	0.4	0.6	0.8	1
$f(x_k)$	1	0.8333	0.7143	06250	0.5556	0.5000

Left - hand sum

$= (1 + 0.8333 + 0.7143 + 0.6250 + 0.5556)(0.2) = 0.7456$

Right - hand sum

$= (0.8333 + 0.7143 + 0.6250 + 0.5556 + 0.5) = (0.2) = 0.6456$

Trapezoidal sum $= \dfrac{0.7456 + 0.6456}{2} = 0.6956$

$\int_{0}^{1} \dfrac{1}{1+x} dx = \int_{1}^{2} u^{-1} du = \ln 2 - \ln 1 \approx 0.69315$

Error $= |0.6956 - 0.69315| \approx 0.00245$

11. $\int_0^1 4\sqrt{1-x^2}\,dx$, $n=5$

$a=0$, $b=1$, $\Delta x = \dfrac{1-0}{5} = 0.2$

k	0	1	2	3	4	5
x_k	0	0.4	0.4	0.6	0.8	1.0
$f(x_k)$	4	3.9192	3.6661	3.2000	2.4000	0

Trapezoidal sum $= \dfrac{\text{left sum } + \text{ right sum}}{2}$

$= 3.0371$

13. $\int_0^{10} e^{-x^2}\,dx$, $n=10$

$a=0$, $b=10$, $\Delta x = \dfrac{10-0}{10} = 1$

k	0	1	2	3	4	$5-10$
x_k	0	1	2	3	4	$5-10$
$f(x_k)$	1	0.3679	0.0183	0.0001	1×10^{-7}	negligible

Trapezoidal sum $= \dfrac{\text{left sum } + \text{ right sum}}{2}$

$= 0.8863$

15. $\int_0^1 \ln\left(x^2+1\right)dx$, $n=10$

$a=0$, $b=1$, $\Delta x = \dfrac{1-0}{10} = 0.1$

k	0	1	2	3	4	5	6	7	8	9	10
x_k	0	0.1	0.2	0.3	0.4	0.5	0.6	0.7	0.8	0.9	1
$f(x_k)$	0	0.0100	0.0392	0.0862	0.1484	0.2231	0.3075	0.3988	0.4947	0.5933	0.6931

Trapezoidal sum $= \dfrac{\text{left sum } + \text{ right sum}}{2} = 0.2648$

17. $\int_0^1 4\sqrt{1-x^2}\,dx$

(a) $a=0$, $b=1$, $n=100$
Left - hand sum $= 3.1604$
Right - hand sum $= 3.1204$
Trapezoidal sum $= 3.1404$

(b) $a=0$, $b=1$, $n=200$
Left - hand sum $= 3.1512$
Right - hand sum $= 3.1312$
Trapezoidal sum $= 3.1412$

19. $\int_2^3 \dfrac{2x^{1.2}}{1+3.5x^{4.7}}\,dx$

(a) $a=2$, $b=3$, $n=100$
Left - hand sum $= 0.0258$
Right - hand sum $= 0.0254$
Trapezoidal sum $= 0.0256$

(b) $a=2$, $b=3$, $n=200$
Left - hand sum $= 0.0257$
Right - hand sum $= 0.0255$
Trapezoidal sum $= 0.0256$

21. $\int_0^\pi \sin(x)\,dx$

    (a) $a = 0$, $b = \pi$, $n = 100$

    Left - hand sum = 1.9998

    Right - hand sum = 1.9998

    Trapezoidal sum = 1.9998

    (b) $a = 0$, $b = \pi$, $n = 200$

    Left - hand sum = 2.0000

    Right - hand sum = 2.0000

    Trapezoidal sum = 2.0000

23. (a) $(0.5, 18.90)$ and $(4.5, 7.20)$

    $$s(t) = mt + b$$

    $$m = \frac{7.20 - 18.90}{4.5 - 0.5} = -2.925$$

    $$s(t) - 18.90 = -2.925(t - 0.5)$$

    $$s(t) = -2.925t + 20.3625$$

    (b) $\int_0^5 (-2.925t + 20.3625)\,dt$

    $$a = 0, \; b = 5, \; n = 5, \; \Delta x = \frac{5-0}{5} = 1$$

k	0	1	2	3	4	5
x_k	0	1	2	3	4	5
$f(x_k)$	20.3625	17.4375	14.5125	11.5875	8.6625	5.7375

    $$\text{Trapezoidal sum} = \frac{72.56 + 57.94}{2} = 65.25$$

    The model gives total sales of $65.25 billion.
    The actual total sales were $66.4 billion, higher
    than the sales predicted by the model. One reason
    for this is that the 1992 sales were considerably
    higher than predicted by the linear model. (If you
    join the midpoint of the top of the first column to
    that of the last, you wil find that the 1992 sales
    "stick out" above this line.

416

25. $C = \int_{11}^{100} \left(20 - \dfrac{x}{5000}\right) dx$

$n = 10, \ a = 11, \ b = 100, \ \Delta x = \dfrac{100 - 11}{10} = 8.9$

k	0	1	2	3	4	5	6	7	8	9	10
x_k	11	19.9	28.8	37.7	46.6	55.5	64.4	73.3	82.2	91.1	100
$f(x_k)$	19.998	19.996	19.994	19.992	19.991	19.989	19.987	19.985	19.984	19.982	19.980

Trapezoidal sum $= \dfrac{\text{Left sum} + \text{right sum}}{2} = \$1,779.01$

$C = \int_{11}^{100} \left(20 - \dfrac{x}{5000}\right) dx = \left[20x - \dfrac{x^2}{10,000}\right]_{11}^{100} = \left(20(100) - \dfrac{(100)^2}{10,000}\right) - \left(20(11) - \dfrac{(11)^2}{10,000}\right) = \$1,779.01$

27. $a = 0, \ b = 35, \ n = 7, \ \Delta x = \dfrac{35 - 0}{7} = 5$

k	0	1	2	3	4	5	6	7
x_k	0	5	10	15	20	25	30	35
y_k	0	15	18	8	7	16	20	0

Left sum = 420, right sum = 420

Yes, both estimates give an area

of 420 square feet.

29. $p(x) = \dfrac{1}{\sqrt{2\pi}\,\sigma} e^{-(x-\mu)^2/2\sigma^2}, \ \sigma = 5.2, \ \mu = 72.6$

(a) $\int_{60}^{100} e^{-(x-72.6)^2/54.08} dx, \ n = 40, \ a = 60, \ b = 100$

Trapezoidal sum = 0.992 = 99.2%

(b) $\int_{0}^{30} 0.07672 e^{-(x-72.6)^2/54.08} dx, \ n = 40, \ a = 0, b = 30$

Trapezoidal sum $= 1.4 \times 10^{-16} = 1.4 \times 10^{-14}\%$

13.6 EXERCISES

1. $A(x) = \int_0^x 1\, dt$

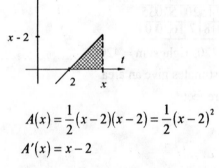

$A(x) = 1 \cdot x = x$

$A'(x) = 1$

3. $A(x) = \int_2^x (t-2)\, dt$

$A(x) = \frac{1}{2}(x-2)(x-2) = \frac{1}{2}(x-2)^2$

$A'(x) = x - 2$

5. $A(x) = \int_a^x c\, dt$

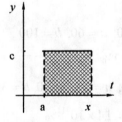

$A(x) = (x-a)(c) = cx - ac$

$A'(x) = c$

In Exercises 7 - 15 use $A(x) = \int_a^x f(t)\, dt$.

7. $A(x) = \int_0^x t\, dt = \left[\dfrac{t^2}{2}\right]_0^x = \dfrac{x^2}{2}$

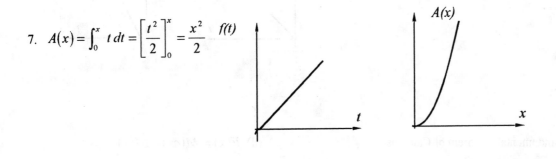

9. $A(x) = \int_0^x t^2\, dt = \left[\dfrac{t^3}{3}\right]_0^x = \dfrac{x^3}{3}$

11. $A(x) = \int_0^x e^t\, dt = \left[e^t\right]_0^x = e^x - 1$

13. $A(x) = \int_1^x \dfrac{1}{t}\, dt = \left[\ln t\right]_1^x = \ln x$

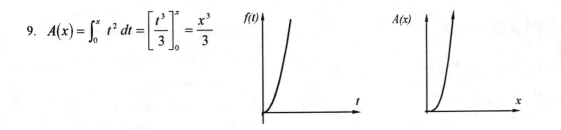

15. $A(x) = \int_0^1 t\,dt + \int_1^x 2\,dt = \left[\dfrac{t^2}{2}\right]_0^1 + [2t]_1^x = 2x - \dfrac{3}{2}$

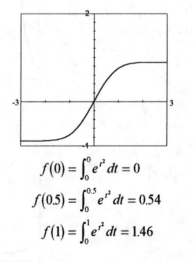

17. (a) The Fundamental Theorem of Calculus gives the antiderivative of $\dfrac{1}{x}$ as

$$\int_a^x \dfrac{1}{t}\,dt$$

(b) $M(x) = \int_1^x \dfrac{1}{t}\,dt = F(x) - F(1)$

$M(1) = \int_1^1 \dfrac{1}{t}\,dt = F(1) - F(1) = 0$

$M'(x) = \dfrac{d}{dx}\int_1^x \dfrac{1}{t}\,dt = \dfrac{1}{x}$

(c) $\dfrac{d}{dx}\big(M(ax)\big) = \dfrac{d}{du}M(u)\cdot\dfrac{d}{dx}u = \dfrac{1}{ax}\cdot a = \dfrac{1}{x}$

(d) $F(x) = M(ax) - M(x)$

$F'(x) = \dfrac{d}{dx}M(ax) - \dfrac{d}{dx}M(x)$

$= \dfrac{1}{x} - \dfrac{1}{x}$

$F(x) = \int F'(x)\,dx + cons\tan t$

$= \int 0\,dx + cons\tan t$

$= cons\tan t$

(e) $F(1) = M(a\cdot 1) - M(1)$

$= M(a) - 0$

$F(x) = F(1) = M(a)$

$M(ax) - M(x) = M(a)$

$M(ax) = M(a) + M(x)$

19. $f(x) = \int_0^x e^{t^2}\,dt \quad (-3 \le x \le 3,\ -1 \le y \le 2)$

$f(0) = \int_0^0 e^{t^2}\,dt = 0$

$f(0.5) = \int_0^{0.5} e^{t^2}\,dt = 0.54$

$f(1) = \int_0^1 e^{t^2}\,dt = 1.46$

420

21. $f(x) = \int_0^x \sqrt{1-t^2}\, dt$ with $f(0) = 0$

$(-1 \le x \le 1,\ -1 \le y \le 1)$

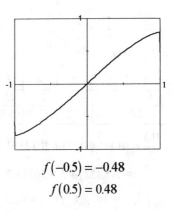

$f(-0.5) = -0.48$

$f(0.5) = 0.48$

13.7 EXERCISES

In Exercises 1 - 5 use the trapezoidal rule

$$\int_a^b f(x)dx \approx \frac{b-a}{2n}\left[f(x_0) + 2f(x_1) + 2f(x_2) + \ldots + 2f(x_{n-1}) + f(x_n)\right]$$

where $x_k = a + k\Delta x$ and $\Delta x = \dfrac{b-a}{n}$. Also use Error $\le \dfrac{(b-a)^3}{12n^2}\left|f''(M)\right|$.

1. $\int_0^2 2x\, dx$,

$a = 0,\ b = 2,\ n = 10,\ \Delta x = \dfrac{2-0}{10} = 0.2$

k	0	1	2	3	4	5	6	7	8	9	10
x_k	0	0.2	0.4	0.6	0.8	1.0	1.2	1.4	1.6	1.8	2
$f(x_k)$	0	0.4	0.8	1.2	1.6	2	2.4	2.8	3.2	3.6	4

$\int_0^2 2x\, dx \approx \dfrac{0.2}{2}\left[0 + 2(0.4) + 2(0.8) + 2(1) + 2(1.2) + 2(1.4) + 2(1.6) + 2(1.8) + 4\right] = 4$

$f(x) = 2x,\ f'(x) = 2,\ f''(x) = 0,\ f''(M) = 0,\ \left|f''(M)\right| = 0$

Error ≤ 0

Exact value $= \int_0^2 2x\, dx = \left[x^2\right]_0^2 = 4$

3. $\int_0^3 x^3\ dx,$

$a = 0,\ b = 3,\ n = 10,\ \Delta x = \dfrac{3-0}{10} = 0.3$

k	0	1	2	3	4	5	6	7	8	9	10
x_k	0	0.3	0.6	0.9	1.2	1.5	1.8	2.1	2.4	2.7	3
$f(x_k)$	0	0.027	0.216	0.729	1.728	3.375	5.832	9.261	13.824	19.683	27

$\int_0^3 x^3\ dx \approx \dfrac{0.3}{2} = \big[0 + 2(0.027) + 2(0.216) + 2(0.729) + 2(1.728) + 2(3.375)$

$\qquad\qquad + 2(5.832) + 2(9.261) + 2(13.824) + 2(19.683) + 27\big] = 20.4525$

$f(x) = x^3,\ f'(x) = 3x^2,\ f''(x) = 6x,\ f''(M) = f''(3) = 18,\ |f''(M)| = 18$

Error $\leq \dfrac{(3-0)^3}{12(10)^2}(18) = 0.405$

Exact value $= \int_0^3 x^3\ dx = \left[\dfrac{x^4}{4}\right]_0^3 = \dfrac{81}{4} = 20.25$

5. $\int_1^5 \ln x\ dx,$

$a = 1,\ b = 5,\ n = 10,\ \Delta x = \dfrac{5-1}{10} = 0.4$

k	0	1	2	3	4	5	6	7	8	9	10
x_k	1	1.4	1.8	2.2	2.6	3	3.4	3.8	4.2	4.6	5
$f(x_k)$	0	0.3365	0.5878	0.7885	0.9555	1.0986	1.2238	1.335	1.4351	1.5261	1.6094

$\int_1^5 \ln x\ dx \approx \dfrac{0.4}{2}\big[0 + 2(0.3365) + 2(0.5878) + 2(0.7885) + 2(0.9555) + 2(1.0986) + 2(1.2238)$

$\qquad\qquad + 2(1.335) + 2(1.4351) + 2(1.5261) + 1.6094\big] \approx 4.037$

$f(x) = \ln x,\ f'(x) = x^{-1},\ f''(x) = -x^{-2},\ |f''(M)| = |f''(1)| = |-1| = 1$

Error $\leq \dfrac{(5-1)^3}{12(10)^2}(1) = 0.053$

In Exercises 7-11 use Simpson's rule $\int_a^b f(x)\,dx \approx \dfrac{b-a}{3n}\big[f(a)+4f(x_1)+2f(x_2)$

$+4f(x_3)+\ldots+2f(x_{n-2})+4f(x_{n-1})+f(b)\big]$ when n is even,

$x_k = a + k\Delta x$ and $\Delta x = \dfrac{b-a}{n}$. Use Error $\leq \dfrac{(b-a)^5}{180n^4}\big|f^{(4)}(M)\big|$.

Also use the corresponding table and values of a, b, n and Δx in Exercises 1-5.

7. $\displaystyle\int_0^2 2x\,dx \approx \dfrac{0.2}{3}\big[0 + 4(0.4) + 2(0.8) + 4(1.2) + 2(1.6) + 4(2) + 2(2.4)$

$\qquad\qquad + 4(2.8) + 2(3.2) + 4(3.6) + 4\big] = 4$

$f(x) = 2x$, $f'(x) = 2$, $f''(x) = 0$, $f^{(3)}(x) = 0$, $f^{(4)}(x) = 0$, $\big|f^{(4)}(M)\big| = 0$

Error ≤ 0, Exact value $= 4$

9. $\displaystyle\int_0^3 x^3\,dx \approx \dfrac{0.3}{3}\big[0 + 4(0.027) + 2(0.216) + 4(0.729) + 2(1.728) + 4(3.375) + 2(5.832)$

$\qquad\qquad + 4(9.261) + 2(13.824) + 4(19.683) + 27\big] = 20.25$

$f(x) = x^3$, $f'(x) = 3x^2$, $f''(x) = 6x$, $f^{(3)}(x) = 6$, $f^{(4)}(x) = 0$, $\big|f^{(4)}(M)\big| = 0$

Error ≤ 0, Exact value $= 20.25$

11. $\displaystyle\int_1^5 \ln x\,dx \approx \dfrac{0.4}{3}\big[0 + 4(0.3365) + 2(0.5878) + 4(0.7885) + 2(0.9555) + 4(1.0986)$

$\qquad\qquad + 2(1.2238) + 4(1.335) + 2(1.4351) + 4(1.5261) + 1.6094\big] \approx 4.0470$

$f(x) = \ln x$, $f'(x) = x^{-1}$, $f''(x) = -x^{-2}$, $f^{(3)}(x) = 2x^{-3}$, $f^{(4)}(x) = -6x^{-4}$

$\big|f^{(4)}(M)\big| = \big|f^{(4)}(1)\big| = \big|-6\big| = 6$

Error $\leq \dfrac{(5-1)^5}{180(10)^4}(6) \approx 0.0034$

13. (a) $n = 100$, $\displaystyle\int_0^2 2x\,dx \approx 4$

    (b) $n = 200$, $\displaystyle\int_0^2 2x\,dx \approx 4$

    (c) $n = 500$, $\displaystyle\int_0^2 2x\,dx \approx 4$

15. (a) $n = 100$, $\displaystyle\int_0^3 x^3\,dx \approx 20.25202500$

    (b) $n = 200$, $\displaystyle\int_0^3 x^3\,dx \approx 20.25050625$

    (c) $n = 500$, $\displaystyle\int_0^3 x^3\,dx \approx 20.25008100$

17. (a) $n = 100$, $\int_1^5 \ln x\, dx \approx 4.0470829$

    (b) $n = 200$, $\int_1^5 \ln x\, dx \approx 4.0471629$

    (c) $n = 500$, $\int_1^5 \ln x\, dx \approx 4.0471853$

19. (a) $n = 100$, $\int_0^2 2x\, dx \approx 4$

    (b) $n = 200$, $\int_0^2 2x\, dx \approx 4$

    (c) $n = 500$, $\int_0^2 2x\, dx \approx 4$

21. (a) $n = 100$, $\int_0^3 x^3\, dx \approx 20.25$

    (b) $n = 200$, $\int_0^3 x^3\, dx \approx 20.25$

    (c) $n = 500$, $\int_0^3 x^3\, dx \approx 20.25$

23. (a) $n = 100$, $\int_1^5 \ln x\, dx \approx 4.0471895340$

    (b) $n = 200$, $\int_1^5 \ln x\, dx \approx 4.0471895604$

    (c) $n = 500$, $\int_1^5 \ln x\, dx \approx 4.0471895621$

25. $\int_2^0 x^4\, dx$, $a = 2$, $b = 0$, $f(x) = x^4$, $f'(x) = 4x^3$, $f''(x) = 12x^2$

$$\left| f''(M) \right| = \left| f''(2) \right| = \left| 12(2)^2 \right| = \left| 48 \right| = 48$$

$$\text{Max Error} = \frac{(0-2)^3}{12n^2}(48) = -\frac{32}{n^2}$$

$$-0.0001 = -\frac{32}{n^2}$$

$$n^2 = 32 \times 10^4$$

$$n = 565.7$$

$$n \geq 566$$

27. $\int_2^0 x^4\, dx$, $a = 2$, $b = 0$

$f(x) = x^4$, $f'(x) = 4x^3$, $f''(x) = 12x^2$,

$f^{(3)}(x) = 24x$, $f^{(4)}(x) = 24$

$\left| f^{(4)}(M) \right| = 24$

$$\text{Max Error} = \frac{(0-2)^5}{180n^4}(24) = -\frac{4.2667}{n^4}$$

$$-0.0001 = -\frac{4.2667}{n^4}$$

$$n^4 = 4.2667 \times 10^4$$

$$n = 14.4$$

$$n = 15$$

29. $\int_0^{10} xe^{-x}\,dx,\ a=0,\ b=10$

$f(x)=xe^{-x},\ f'(x)=e^{-x}(1-x),\ f''(x)=(-2+x)e^{-x}$

$f^{(3)}(x)=(3-x)e^{-x}=0$ when $3-x=0$ or $x=3$

x	0	3	10
$f''(x)$	-2	0.05	0.00036

$|f''(M)|=|f''(0)|=|-2|=2$

Max Error $=\dfrac{(10-0)^3}{12n^2}(2)=\dfrac{166.67}{n^2}$

$0.0001=\dfrac{166.67}{n^2}$

$n^2=166.67\times10^4$

$n=1291.0$

$n\geq1291$

31. From Exercise 29, $a=0,\ b=10$

$f^{(3)}(x)=(3-x)e^{-x},\ f^{(4)}(x)=(-4+x)e^{-x}$

$f^{(5)}(x)=(5-x)e^{-x}=0$ when $5-x=0$ or $x=5$

x	0	5	10
$f^{(4)}(x)$	-4	0.007	0.00005

$|f^{(4)}(M)|=|f^{(4)}(0)|=|-4|=4$

Max Error $=\dfrac{(10-0)^5}{180n^4}(4)=\dfrac{2222}{n^4}$

$0.0001=\dfrac{2222}{n^4}$

$n^4=2222\times10^4$

$n=68.7$

$n\geq69$

In Exercises 33 - 35 use length $=L=\int_a^b\sqrt{1+\left[f^1(x)\right]^2}\,dx$.

33. $f(x)=x^2,\ x$ in $[0,1]$

$f'(x)=2x$

$L=\int_0^1\sqrt{1+4x^2}\,dx$

(a) $a=0,\ b=1,\ n=6,\ \Delta x=\dfrac{1-0}{6}=\dfrac{1}{6}$

$L\approx1.4789405$

(b) $a=0,\ b=1,\ n=200$

$L\approx1.478942858$

35. $f(x)=e^x,\ x$ in $[0,1]$

$f'(x)=e^x$

$L=\int_0^1\sqrt{1+e^{2x}}\,dx$

(a) $a=0,\ b=1,\ n=6,\ \Delta x=\dfrac{1-0}{6}=\dfrac{1}{6}$

$L\approx2.003503$

(b) $a=0,\ b=1,\ n=200$

$L\approx2.003497112$

425

CHAPTER 13   REVIEW EXERCISES

1. $\displaystyle\int 10x^{10}\ dx = \frac{10x^{11}}{11} + C$

3. $\displaystyle\int \frac{3}{x^4}\ dx = \int 3x^{-4}\ dx = \frac{3x^{-3}}{-3} + C$

$$= -\frac{1}{x^3} + C$$

5. $\displaystyle\int \left(\sqrt{x} + \frac{1}{x}\right) dx = \int \left(x^{1/2} + x^{-1}\right) dx$

$$= \frac{x^{3/2}}{3/2} + \ln|x| + C$$

$$= \frac{2x^{3/2}}{3} + \ln|x| + C$$

7. $\displaystyle\int \left(2e^x + 3x - \frac{4}{x}\right) dx = \int \left(2e^x + 3x - 4x^{-1}\right) dx$

$$= 2e^x + \frac{3x^2}{2} - 4\ \ln|x| + C$$

9. $\displaystyle\int (x+2)^{10}\ dx$

Let $u = x+2,\ \dfrac{du}{dx} = 1,\ dx = du$

$$\int u^{10}\ du = \frac{u^{11}}{11} + C = \frac{(x+2)^{11}}{11} + C$$

11. $\displaystyle\int x\sqrt[3]{1+x^2}\ dx = \int x\left(1+x^2\right)^{1/3}\ dx$

Let $u = 1+x^2,\ \dfrac{du}{dx} = 2x,\ dx = \dfrac{du}{2x}$

$$\int xu^{1/3}\left(\frac{1}{2x}\right) du = \frac{1}{2}\frac{u^{4/3}}{4/3} + C$$

$$= \frac{3\left(1+x^2\right)^{4/3}}{8} + C$$

13. $\displaystyle\int \frac{x}{x^2+1}\ dx = \int x\left(x^2+1\right)^{-1}\ dx$

Let $u = x^2+1,\ \dfrac{du}{dx} = 2x,\ dx = \dfrac{du}{2x}$

$$\int x^2 u^{-1}\left(\frac{1}{2x}\right) du = \frac{\ln|u|}{2} + C$$

$$= \frac{\ln\left(x^2+1\right)}{2} + C$$

426

15. $\int 5e^{-2x}\,dx$

Let $u = -2x$, $\dfrac{du}{dx} = -2$, $dx = \dfrac{du}{-2}$

$\int 5e^u\left(-\dfrac{1}{2}\right)du = -\dfrac{5e^u}{2} + C$

$\qquad = -\dfrac{5e^{-2x}}{2} + C$

17. $\int\left(xe^{x^2} - 3x\right)dx = \int xe^{x^2}\,dx - \int 3x\,dx$

Let $u = x^2$, $\dfrac{du}{dx} = 2x$, $dx = \dfrac{du}{2x}$

$\int xe^u\left(\dfrac{1}{2x}\right)du - \int 3x\,dx = \dfrac{e^u}{2} - \dfrac{3x^2}{2} + C$

$\qquad = \dfrac{e^{x^2}}{2} - \dfrac{3x^2}{2} + C$

19. $\int \dfrac{4.7x^{0.2}}{x^{1.2} - 4}\,dx = \int 4.7x^{0.2}\left(x^{1.2} - 4\right)^{-1}dx$

Let $u = x^{1.2}$, $\dfrac{du}{dx} = 1.2x^{0.2}$, $dx = \dfrac{du}{1.2x^{0.2}}$

$\int 4.7x^{0.2}u^{-1}\left(\dfrac{1}{1.2x^{0.2}}\right)du = \dfrac{4.7\ \ln u}{1.2} + C$

$\qquad\qquad = \dfrac{4.7\ \ln\left|x^{1.2} - 4\right|}{1.2} + C$

21. $\int \dfrac{e^{0.3t}}{1 + 2e^{0.3t}}\,dt = \int e^{0.3t}\left(1 + 2e^{0.3t}\right)dt$

Let $u = 1 + 2e^{0.3t}$, $\dfrac{du}{dt} = 0.6e^{0.3t}$, $dt = \dfrac{du}{0.6e^{0.3t}}$

$\int e^{0.3t}u^{-1}\left(\dfrac{1}{0.6e^{0.3t}}\right)du = \dfrac{\ln|u|}{0.6} + C$

$\qquad\qquad = \dfrac{\ln\left|1 + 2e^{0.3t}\right|}{0.6} + C$

23. $\displaystyle\int_0^1\left(x^3 + 2x\right)dx = \left[\dfrac{x^4}{4} + x^2\right]_0^1 = \left(\dfrac{1}{4} + 1\right) - (0) = \dfrac{5}{4}$

25. $\displaystyle\int_1^2 \dfrac{3}{x^2}\,dx = \int_1^2 3x^{-2}\,dx = \left[\dfrac{3x^{-1}}{-1}\right]_1^2 = \left(-\dfrac{3}{2}\right) - \left(-\dfrac{3}{1}\right) = \dfrac{3}{2}$

27. $\displaystyle\int_0^1\left(e^{-x} + x\right)dx$

Let $u = -x$, $\dfrac{du}{dx} = -1$, $dx = -du$. When $x = 0$, $u = 0$. When $x = 1$, $u = -1$.

$\displaystyle\int_0^1\left(e^{-x} + x\right)dx = \int_0^{-1}e^u(-du) + \int_0^1 x\,dx = \left[-e^u\right]_0^{-1} + \left[\dfrac{x^2}{2}\right]_0^1 = \left[(-e^{-1}) - (-e^0)\right] + \left[\dfrac{1}{2} - \dfrac{0}{2}\right]$

$\qquad\qquad = \dfrac{3}{2} - \dfrac{1}{e}$

29. $\int_0^2 x^2 \sqrt{x^3+1}\,dx = \int_0^2 x^2 \left(x^3+1\right)^{1/2}\,dx$

Let $u = x^3+1$, $\dfrac{du}{dx} = 3x^2$, $dx = \dfrac{du}{3x^2}$. When $x=0$, $u=1$. When $x=2$, $u=9$.

$$\int_0^2 x^2\left(x^3+1\right)^{1/2}\,dx = \int_1^9 x^2 u^{1/2}\left(\frac{1}{3x^2}\right)du = \left[\frac{u^{3/2}}{3(3/2)}\right]_1^9 = \left(\frac{2(9)^{3/2}}{9}\right) - \left(\frac{2(1)^{3/2}}{9}\right) = 6 - \frac{2}{9} = \frac{52}{9}$$

31. $\int_0^1 \left(4 + x^{0.2}\left(2 - x^{1.2}\right)^4\right)dx$

Let $u = 2 - x^{1.2}$, $\dfrac{du}{dx} = -1.2x^{0.2}$, $dx = \dfrac{du}{-1.2x^{0.2}}$. When $x=0$, $u=2$. When $x=1$, $u=1$.

$$\int_0^1 \left(4 + x^{0.2}\left(2-x^{1.2}\right)^4\right)dx = \int_0^1 4\,dx + \int_2^1 x^{0.2}u^4\left(-\frac{1}{1.2x^{0.2}}\right)du = \left[4x\right]_0^1 + \left[-\frac{u^5}{1.2(5)}\right]_2^1$$

$$= 4(1) - 0 + \left(-\frac{(1)^5}{6}\right) - \left(-\frac{(2)^5}{6}\right) = 4 + \frac{2^5 - 1}{6}$$

33. $\int_{-1}^1 xe^{-0.3x^2}\,dx$

Let $u = -0.3x^2$, $\dfrac{du}{dx} = -0.6x$, $dx = \dfrac{du}{-0.6x}$. When $x=-1$, $u=-0.3$. When $x=1$, $u=-0.3$.

$$\int_{-1}^1 xe^{-0.3x^2}\,dx = \int_{-0.3}^{-0.3} xe^u\left(-\frac{1}{0.6x}\right)du = \left[-\frac{e^u}{0.6}\right]_{-0.3}^{-0.3} = 0$$

35. $\int_{-1}^1 \left(1-x^2\right)dx = \left[x - \frac{x^3}{3}\right]_{-1}^1 = \left(1 - \frac{(1)^3}{3}\right) - \left(-1 - \frac{(-1)^3}{3}\right) = \frac{2}{3} + \frac{2}{3} = \frac{4}{3}$

37. $\int_1^{10} \frac{1}{x}\,dx = \int_1^{10} x^{-1}\,dx = \left[\ln|x|\right]_1^{10} = \ln 10 - \ln 1 = \ln 10$

39. $\int_0^5 xe^{-x^2}\,dx$. The graph of $y = xe^{-x^2}$ lies above the x axis between $x=0$ and $x=5$.

Let $u = -x^2$, $\dfrac{du}{dx} = -2x$, $dx = \dfrac{du}{-2x}$. When $x=0$, $u=0$. When $x=5$, $u=-25$.

$$\int_0^5 xe^{-x^2}\,dx = \int_0^{-25} xe^u\left(-\frac{1}{2x}\right)du = \left[-\frac{e^u}{2}\right]_0^{-25} = \left(\frac{-e^u}{2}\right) - \left(\frac{-e^0}{2}\right) = \frac{1 - e^{-25}}{2}$$

41. $y = 1 - 2x^2$ crosses the x - axis at $y = 0$, so $1 - 2x^2 = 0$ or $x = \pm \dfrac{1}{\sqrt{2}}$

$$\int_{-1/\sqrt{2}}^{1/\sqrt{2}} \left(1 - 2x^2\right) dx = \left[x - \frac{2x^3}{3} \right]_{-1/\sqrt{2}}^{1/\sqrt{2}} = \left(\frac{1}{\sqrt{2}} - \frac{2\left(1/\sqrt{2}\right)^3}{3} \right) - \left(-\frac{1}{\sqrt{2}} - \frac{2\left(-1/\sqrt{2}\right)^3}{3} \right) = \frac{2\sqrt{2}}{3}$$

43.

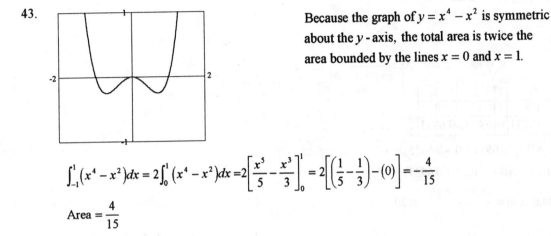

Because the graph of $y = x^4 - x^2$ is symmetric about the y - axis, the total area is twice the area bounded by the lines $x = 0$ and $x = 1$.

$$\int_{-1}^{1} \left(x^4 - x^2\right) dx = 2\int_{0}^{1} \left(x^4 - x^2\right) dx = 2\left[\frac{x^5}{5} - \frac{x^3}{3} \right]_{0}^{1} = 2\left[\left(\frac{1}{5} - \frac{1}{3}\right) - (0) \right] = -\frac{4}{15}$$

Area $= \dfrac{4}{15}$

45.

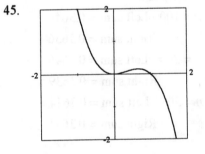

The graph of $y = x^2 - x^3$ is zero at $x = 0$ and $x = 1$ and lies above the x - axis between $x = -1$ and $x = 0$ as well as between $x = 0$ and $x = 1$.

$$\int_{-1}^{1} \left(x^2 - x^3\right) dx = \left[\frac{x^3}{3} - \frac{x^4}{4} \right]_{-1}^{1} = \left(\frac{1}{3} - \frac{1}{4} \right) - \left(\frac{(-1)^3}{3} - \frac{(-1)^4}{4} \right) = \frac{2}{3}$$

47. $\int_{-1}^{1} e^{-x^2}$, $n = 4$, $a = -1$, $b = 1$, $\Delta x = \dfrac{1-(-1)}{4} = \dfrac{1}{2}$

k	0	1	2	3	4
x_k	-1	-0.5	0	0.5	1
$f(x_k)$	0.3679	0.7788	1	0.7788	0.3679

Left sum $= 0.5(0.3679 + 0.7788 + 1 + 0.7788) = 1.46$

Right sum $= 0.5(0.7788 + 1 + 0.7788 + 0.3679) = 1.46$

Trapezoidal sum $= \dfrac{1.46 + 1.46}{2} = 1.46$

49. $\int_{0.5}^{2} \ln x \, dx$, $n = 3$, $a = 0.5$, $b = 2$, $\Delta x = \dfrac{2-0.5}{3} = 0.5$

k	0	1	2	3
x_k	0.5	1	1.5	2
$f(x_k)$	-0.6931	0	0.4055	0.6931

Left sum $= 0.5(-0.6931 + 0 + 0.4055) = -0.14$

Right sum $= 0.5(0 + 0.4055 + 0.6931) = 0.55$

Trapezoidal sum $= \dfrac{-0.14 + 0.55}{2} = 0.20$

51. $\int_{-1}^{1} e^{x^2} \, dx$, $a = -1$, $b = 1$

$n = 100$: Left sum $= 2.9257$

Right sum $= 2.9257$

$n = 200$: Left sum $= 2.9254$

Right sum $= 2.9254$

$n = 500$: Left sum $= 2.9253$

Right sum $= 2.9253$

53. $\int_{0.5}^{2} \dfrac{dx}{4 + \ln x}$, $a = 0.5$, $b = 2$

$n = 100$: Left sum $= 0.3649$

Right sum $= 0.3636$

$n = 200$: Left sum $= 0.3646$

Right sum $= 0.3639$

$n = 500$: Left sum $= 0.3644$

Right sum $= 0.3641$

55. $\int_{-1}^{1} e^{-x^2} \, dx$, $a = -1$, $b = 1$

$n = 50$: 1.49364831

$n = 100$: 1.493648268

$n = 500$: 1.493648266

57. $\int_{0.5}^{2} \ln x \, dx$, $a = 0.5$, $b = 2$

$n = 50$: 0.232867881

$n = 100$: 0.232867947

$n = 500$: 0.232867951

59. $\int_{-1}^{1} e^{x^2}\, dx, \ a = -1, \ b = 1$

    $n = 50$:   2.925305

    $n = 100$:   2.925304

    $n = 500$:   2.925303

61. $\int_{0.5}^{2} \dfrac{dx}{4 + \ln x}, \ a = 0.5, \ b = 2$

    $n = 50$:   0.364234050

    $n = 100$:   0.364234037

    $n = 500$:   0.364234036

63. $\int e^{-x^2}\, dx, \ a = -1, \ b = 1$

    $f(x) = e^{-x^2}, \ f'(x) = -2x e^{-x^2}$

    $f''(x) = \left(4x^2 - 2\right) e^{-x^2}, \ f^{(3)}(x) = \left(12x - 8x^3\right) e^{-x^2}$

    $12x - 8x^3 = 4x\left(2 - 2x^2\right) = 0 \ \text{at} \ x = 0$

    $\text{or} \ x = \pm\sqrt{3/2}$

x	-1	$-\sqrt{3/2}$	0	$\sqrt{3/2}$	1
$f''(x)$	0.74	0.89	-2	0.89	0.74

    $\left|f''(M)\right| = \left|f''(0)\right| = |-2| = 2$

    $\text{Max Error} = \dfrac{\left[1 - (-1)\right]^3}{12n^2}(2) = \dfrac{1.333}{n^2}$

    $0.001 = \dfrac{1.333}{n^2}$

    $n^2 = 1333$

    $n = 36.5$

    $n \geq 37$

65. $\int_{0.5}^{2} \ln x\, dx, \ a = 0.5, \ b = 2$

    $f(x) = \ln x, \ f'(x) = \dfrac{1}{x}, \ f''(x) = -\dfrac{1}{x^2}$

    $f^{(3)}(x) = -\dfrac{6}{x^4} \neq 0$

x	0.5	2
$f''(x)$	-4	$-1/4$

    $\left|f''(M)\right| = \left|f''(0.5)\right| = |-4| = 4$

    $\text{Max Error} = \dfrac{(2 - 0.5)^3}{12n^2}(4) = \dfrac{1.125}{n^2}$

    $0.001 = \dfrac{1.125}{n^2}$

    $n^2 = 1125$

    $n = 33.5$

    $n \geq 34$

67. $\int_{-1}^{1} e^{x^2}\, dx,\ a = -1,\ b = 1$

$f(x) = e^{x^2},\ f'(x) = 2xe^{x^2}$

$f''(x) = \left(4x^2 + 2\right)e^{x^2},\ f^{(3)}(x) = \left(8x^3 + 12x\right)e^{x^2}$

$8x^3 + 12x = 4x\left(2x^2 + 3\right) = 0$ at $x = 0$

or $x = \pm i\sqrt{3/2}$.

x	-1	0	1
$f''(x)$	16.3	2	16.3

$\left|f''(M)\right| = \left|f''(\pm 1)\right| = 16.3$

Max Error $= \dfrac{\left[1 - (-1)\right]^3}{12n^2}(16.3) = \dfrac{10.867}{n^2}$

$0.001 = \dfrac{10.867}{n^2}$

$n^2 = 10{,}867$

$n = 104.2$

$n \geq 105$

69. $\int_{0.5}^{2} \dfrac{1}{4 + \ln x}\, dx,\ a = 0.5,\ b = 2$

$f(x) = \dfrac{1}{4 + \ln x},\ f'(x) = -\dfrac{1}{\left(4 + \ln x\right)^2 x}$

$f''(x) = \dfrac{2}{\left(4 + \ln x\right)^3 x^2} + \dfrac{1}{\left(4 + \ln x\right)^2 x^2}$

The graph of $f''(x)$ has no local max or min.

x	0.5	2
$f''(x)$	0.587	0.016

$\left|f''(M)\right| = \left|f''(0.5)\right| = 0.587$

Max Error $= \dfrac{\left(2 - 0.5\right)^3}{12n^2}(0.587) = \dfrac{0.165}{n^2}$

$0.001 = \dfrac{0.165}{n^2}$

$n^2 = 165$

$n = 12.8$

$n \geq 13$

In Exercises 71 - 77 assume the results of the derivatives from Exercises 63 - 69 respectively.

Use Maximum error $= \dfrac{(b - a)^5}{180n^4}\left|f^{(4)}(M)\right|$.

71. $\int_{-1}^{1} e^{-x^2}\, dx,\ a = -1,\ b = 1$

$f^{(4)}(x) = \left(16x^4 - 48x^2 + 12\right)e^{-x^2}$

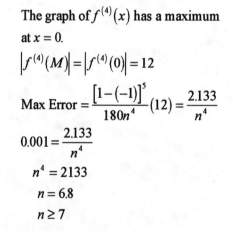

The graph of $f^{(4)}(x)$ has a maximum at $x = 0$.

$\left|f^{(4)}(M)\right| = \left|f^{(4)}(0)\right| = 12$

Max Error $= \dfrac{\left[1 - (-1)\right]^5}{180n^4}(12) = \dfrac{2.133}{n^4}$

$0.001 = \dfrac{2.133}{n^4}$

$n^4 = 2133$

$n = 6.8$

$n \geq 7$

73. $\int_{0.5}^{2} \ln x \, dx, \ a = 0.5, \ b = 2$

$f^{(4)}(x) = -\dfrac{6}{x^4}$ which has no local max or min.

x	0.5	2
$f^{(4)}(x)$	-96	-0.4

$\left| f^{(4)}(M) \right| = \left| f^{(4)}(0.5) \right| = |-96| = 96$

Max Error $= \dfrac{(2-0.5)^5}{180n^4}(96) = \dfrac{4.05}{n^4}$

$0.001 = \dfrac{4.05}{n^4}$

$n^4 = 4050$

$n = 7.98$

$n \geq 8$

75. $\int_{-1}^{1} e^{x^2} \, dx, \ a = -1, \ b = 1$

$f^{(4)}(x) = \left(16x^4 + 48x^2 + 12\right)e^{x^2}$

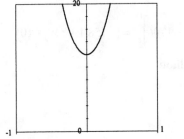

The graph of $f^{(4)}(x)$ has a local minimum at $x = 0$.

x	-1	0	1
$f^{(4)}(x)$	207	12	207

$\left| f^{(4)}(M) \right| = \left| f^{(4)}(\pm 1) \right| = 207$

Max Error $= \dfrac{[1-(-1)]^5}{180n^4}(207) = \dfrac{36.8}{n^4}$

$0.001 = \dfrac{36.8}{n^4}$

$n^4 = 36{,}800$

$n = 13.9$

$n \geq 14$

77. $\int_{0.5}^{2} \dfrac{dx}{4 + \ln x}$, $a = 0.5$, $b = 2$

The graph of $f^{(4)}(x)$ has no local max or min.

x	0.5	2
$f^{(4)}(x)$	24.3	0.04

$\left| f^{(4)}(M) \right| = \left| f^{(4)}(0.5) \right| = 24.3$

$\text{Max Error} = \dfrac{(2 - 0.5)^5}{180 n^4}(24.3) = \dfrac{1.025}{n^4}$

$0.001 = \dfrac{1.025}{n^4}$

$\quad n^4 = 1025$

$\quad n = 5.7$

$\quad n \geq 6$

79. (a) $s(t) = mt + s_0$, $t = $ years since 1982

$(0, 70)$ and $(10, 48)$

$m = \dfrac{48 - 70}{10 - 0} = -2.2$

$s(t) - 70 = -2.2(t - 0)$

$\quad\quad s(t) = -2.2t + 70$

(b) Actual total $= 70 + 68 + 66 + 64 + 62 + 60$
$\quad\quad\quad\quad\quad + 58 + 56 + 54 + 51 + 48 + 47$
$\quad\quad\quad = 704$ million

$S \approx \int_{0}^{12}(-2.2t + 70)\,dt$

$\quad = \left[\dfrac{-2.2t^2}{2} + 70t \right]_{0}^{12} = (-158.4 + 840) - (0)$

$\quad = 682$ million

(c) $S \approx \int_{-2}^{8}(-2.2t + 70)\,dt$

$\quad = \left[\dfrac{-2.2t^2}{2} + 70t \right]_{-2}^{8}$

$\quad = (-70.4 + 560) - (-4.4 + 140)$

$\quad = 634$ million

81. At its maximum height, $v(t) = 0$

$$v(t) = v_0 - 5.3t$$
$$0 = 100 - 5.3t$$
$$t = 18.87$$

$$s(t) = s_0 + v_0 t - \frac{5.3}{2}t^2$$

$$s(18.87) = 0 + 100(18.87) - \frac{5.3}{2}(18.87)^2 = 943$$

Maximum height = 943 feet

83. $a = 320 - 320e^{-t/60}$

$$v(t) = \int a(t)\, dt$$
$$= \int \left(320 - 320e^{-t/60}\right) dt$$
$$= 320t - 320(-60)e^{-t/60} + C$$

If $v(0) = 0$, then

$$0 = 320(0) + 19{,}200e^{-0/60} + C$$
$$C = -19{,}200$$
$$v(t) = -19{,}200 + 320t + 19{,}200e^{-t/60}$$

$$s(t) = \int v(t)\, dt$$
$$= \int \left(-19{,}200 + 320t + 19{,}200e^{-t/60}\right) dt$$
$$= -19{,}200t + 160t^2 + 19{,}200(-60)e^{-t/60} + C$$

If $s(0) = 0$, then

$$0 = -19{,}200(0) + 160(0)^2 - 1{,}152{,}000e^{-0/60} + C$$
$$C = 1{,}152{,}000$$
$$s(t) = 1{,}152{,}000 - 19{,}200t + 160t^2 - 1{,}152{,}000e^{-t/60}$$
$$s(60) = 1{,}152{,}000 - 19{,}200(60) + 160(60)^2 - 1{,}152{,}000e^{-60/60}$$
$$= 152{,}202 \text{ feet}$$

85. $C'(x) = 50 + 40/(x + 40)$ and $C(0) = 50{,}000$

$$C(x) = \int \left(50 + 40(x + 40)^{-1}\right) dx$$
$$= 50x + 40 \ln (x + 40) + K$$
$$C(0) = 50(0) + 40 \ln (0 + 40) + K$$
$$50{,}000 = 147 + K$$
$$K = 49{,}853$$
$$C(x) = 49{,}853 + 50x + 40 \ln (x + 40)$$

87. $R'(x) = 75 + 25e^{-x}$ and $R(0) = 0$

$$R(x) = \int \left(75 + 25e^{-x}\right) dx$$
$$= 75x - 25e^{-x} + K$$
$$R(0) = 75(0) - 25e^{-0} + K$$
$$0 = -25 + K$$
$$K = 25$$
$$R(x) = 75x + 25\left(1 - e^{-x}\right)$$

89. $q(p) = \dfrac{268.227}{p^{1.1990}}$

(a) If $q = 100$, then

$$100 = \dfrac{268.227}{p^{1.1990}}$$

$$p^{1.1990} = 2.68227$$

$$p = 2.28$$

$$\approx \$2 \text{ per watt of capacity}$$

(b) Assume $p = 2e^{-kt}$

If $p = \dfrac{1}{2}p_0$ when $t = 4.342$, then

$$1 = 2e^{-k(4.342)}$$

$$\ln\left(\dfrac{1}{2}\right) = -4.342K$$

$$K = 0.1596$$

$$p = 2e^{-0.1596t}$$

(c) $q = \dfrac{268.227}{\left(2e^{-0.1596t}\right)^{1.1990}}$

$$= 116.8338e^{0.19136t}$$

$$S = \int_0^3 116.8338e^{0.19136t}\, dt$$

$$= \left[116.8338e^{0.19136t}\right]_0^3$$

$$= 610.544\left[e^{0.19136(3)} - e^0\right]$$

$$= 473.5 \text{ megawatts}$$

CHAPTER 14
FURTHER INTEGRATION TECHNIQUES AND APPLICATIONS OF THE INTEGRAL

14.1 EXERCISES

1. $\displaystyle\int 2xe^x\,dx \quad \begin{vmatrix} D & I \\ 2x & e^x \\ -2 & e^x \\ 0 & e^x \end{vmatrix}$

$= 2xe^x - 2e^x + C$

$= 2e^x(x-1) + C$

3. $\displaystyle\int (3x-1)e^{-x}\,dx \quad \begin{vmatrix} D & I \\ 3x-1 & e^{-x} \\ -3 & -e^{-x} \\ 0 & e^{-x} \end{vmatrix}$

$= (3x-1)(-e^{-x}) - 3e^{-x} + C$

$= -e^{-x}(2+3x) + C$

5. $\displaystyle\int (x^2-1)e^{2x}\,dx \quad \begin{vmatrix} D & I \\ x^2-1 & e^{2x} \\ -2x & e^{2x}/2 \\ 2 & e^{2x}/4 \\ 0 & e^{2x}/8 \end{vmatrix}$

$= (x^2-1)\left(\dfrac{e^{2x}}{2}\right) - 2x\left(\dfrac{e^{2x}}{4}\right) + 2\left(\dfrac{e^{2x}}{8}\right) + C$

$= \dfrac{e^{2x}}{4}(2x^2 - 2x - 1) + C$

7. $\displaystyle\int (x^2+1)e^{-2x+4}\,dx \quad \begin{vmatrix} D & I \\ x^2-1 & e^{2x} \\ -2x & e^{2x}/2 \\ 2 & e^{2x}/4 \\ 0 & e^{2x}/8 \end{vmatrix}$

$= (x^2+1)\left(-\dfrac{e^{-2x+4}}{2}\right) - 2x\left(\dfrac{e^{-2x+4}}{4}\right) + 2\left(-\dfrac{e^{-2x+4}}{8}\right) + C$

$= -\dfrac{e^{-2x+4}}{4}(2x^2 + 2x + 3) + C$

9. $\displaystyle\int \dfrac{x^2-x}{e^x}\,dx \quad \begin{vmatrix} D & I \\ x^2-x & e^{-x} \\ -(2x-1) & -e^{-x} \\ 2 & e^{-x} \\ 0 & -e^{-x} \end{vmatrix}$

$= (x^2-x)(-e^{-x}) - (2x-1)e^{-x} + 2(-e^{-x}) + C$

$= -e^{-x}(x^2 + x + 1) + C$

11. $\displaystyle\int x^3 \ln x\,dx \quad \begin{vmatrix} D & I \\ \ln x & x^3 \\ -1/x & x^4/4 \end{vmatrix}$

$= \dfrac{x^4}{4}\ln x - \int \dfrac{x^3}{4}\,dx$

$= \dfrac{x^4}{4}\ln x - \dfrac{x^4}{16} + C$

$= \dfrac{x^4}{16}4(\ln x - 1) + C$

13. $\displaystyle\int (t^2+1)\ln(2t)\,dt$ $\quad\begin{vmatrix} D & I \\ \ln(2t) & t^2+1 \\ -1/t & t^3/3+t \end{vmatrix}$

$\displaystyle= \left(\frac{t^3}{3}+t\right)\ln(2t) - \int\left(\frac{t^2}{3}+1\right)dt$

$\displaystyle= \left(\frac{t^3}{3}+t\right)\ln(2t) - \frac{t^3}{9} - t + C$

15. $\displaystyle\int t^{1/3}\ln t\,dt$ $\quad\begin{vmatrix} D & I \\ \ln t & t^{1/3} \\ -1/t & 3t^{4/3}/4 \end{vmatrix}$

$\displaystyle= \frac{3t^{4/3}\ln t}{4} - \int \frac{3t^{1/3}}{4}\,dt$

$\displaystyle= \frac{3t^{4/3}\ln t}{4} - \frac{9t^{4/3}}{16} + C$

$\displaystyle= \frac{3t^{4/3}}{4}\left(\ln t - \frac{3}{4}\right) + C$

17. $\displaystyle\int_0^1 (x+1)e^x\,dx$ $\quad\begin{vmatrix} D & I \\ x+1 & e^x \\ -1 & e^x \\ 0 & e^x \end{vmatrix}$

$\displaystyle= \left[(x+1)e^x - e^x\right]_0^1$

$\displaystyle= \left[xe^x\right]_0^1 = 1e^1 - 0e^0 = e$

19. $\displaystyle\int_0^1 x^2(x+1)^{10}\,dx$ $\quad\begin{vmatrix} D & I \\ x^2 & (x+1)^{10} \\ -2x & (x+1)^{11}/11 \\ 2 & (x+1)^{12}/11\cdot 12 \\ 0 & (x+1)^{13}/11\cdot 12\cdot 13 \end{vmatrix}$

$\displaystyle= \left[\frac{x^2(x+1)^{11}}{11} - \frac{2x(x+1)^{12}}{11\cdot 12} + \frac{2(x+1)^{13}}{11\cdot 12\cdot 13}\right]_0^1$

$\displaystyle= \left[\frac{1^2(1+1)^{11}}{11} - \frac{2(1+1)^{12}}{11\cdot 12} + \frac{2(1+1)^{13}}{11\cdot 12\cdot 13}\right] - \left[0 - 0 + \frac{2(1)^{13}}{11\cdot 12\cdot 13}\right]$

$\displaystyle= \frac{6\cdot 13\cdot 2^{11} - 13\cdot 2^{12} + 2^{13} - 1}{11\cdot 6\cdot 13} = \frac{38{,}229}{286}$

21. $\displaystyle\int_1^2 x\ln(2x)\,dx$ $\quad\begin{vmatrix} D & I \\ \ln(2x) & x \\ -1/x & x^2/2 \end{vmatrix}$

$\displaystyle= \left[\frac{x^2\ln(2x)}{2}\right]_1^2 - \int_1^2 \frac{x}{2}\,dx$

$\displaystyle= \left[\frac{x^2\ln(2x)}{2} - \frac{x^2}{4}\right]_1^2$

$\displaystyle= \left(\frac{2^2\ln(2\cdot 2)}{2} - \frac{2^2}{4}\right) - \left(\frac{1^2\ln(2\cdot 1)}{2} - \frac{1^2}{4}\right)$

$\displaystyle= 2\ln 2^2 - 1 - \frac{\ln 2}{2} + \frac{1}{4}$

$\displaystyle= \frac{7}{2}\ln 2 - \frac{3}{4}$

23. $\displaystyle\int_0^1 x \ln(x+1)\,dx$ $\quad\begin{vmatrix} D & I \\ \ln(x+1) & x \\ -1/x+1 & x^2/2 \end{vmatrix}$

$\displaystyle = \left[\frac{x^2 \ln(x+1)}{2}\right]_0^1 - \int_0^1 \frac{x^2}{2(x+1)}\,dx$

$\displaystyle = \left(\frac{1^2 \ln(1+1)}{2}\right) - 0 - \int_1^2 \frac{(u-1)^2}{u}\,du$

$\displaystyle = \frac{\ln 2}{2} - \int_1^2 \frac{u^2 - 2u + 1}{2u}\,du$

$\displaystyle = \frac{\ln 2}{2} - \frac{1}{2}\int_1^2 \left(u - 2 + \frac{1}{u}\right)du$

$\displaystyle = \frac{\ln 2}{2} - \frac{1}{2}\left[\frac{u^2}{2} - 2u + \ln u\right]_1^2$

$\displaystyle = \frac{\ln 2}{2} - \left(\frac{2^2}{4} - 2 + \frac{\ln 2}{2}\right) + \left(\frac{1^2}{4} - 1 + \ln 1\right)$

$\displaystyle = \frac{1}{4}$

25. $\displaystyle\int_0^{10} xe^{-x}\,dx$ $\quad\begin{vmatrix} D & I \\ x & e^{-x} \\ -1 & -e^{-x} \\ 0 & e^{-x} \end{vmatrix}$

$\displaystyle = \left[x(-e^{-x}) - e^{-x}\right]_0^{10} = \left[-e^{-x}(x+1)\right]_0^{10}$

$\displaystyle = \left(-e^{-10}(10+1)\right) - \left(-e^{-0}(0+1)\right)$

$\displaystyle = 1 - 11e^{-10}$

27. $\displaystyle\int_1^2 (x+1)\ln x\,dx$ $\quad\begin{vmatrix} D & I \\ \ln x & x+1 \\ -1/x & x^2/2 + x \end{vmatrix}$

$\displaystyle = \left[\left(\frac{x^2}{2} + x\right)\ln x\right]_1^2 - \int_1^2 \left(\frac{x}{2} + 1\right)dx$

$\displaystyle = \left(\frac{2^2}{2} + 2\right)\ln 2 - \left(\frac{1^2}{2} + 1\right)\ln 1 - \left[\frac{x^2}{4} + x\right]_1^2$

$\displaystyle = 4\ln 2 - 0 - \left(\frac{2^2}{4} + 2\right) + \left(\frac{1^2}{4} + 1\right)$

$\displaystyle = 4\ln 2 - \frac{7}{4}$

29. $s = \int v\,dt = \int_0^{120} 2000te^{-t/120}\,dt$

$$\begin{array}{c|c} D & I \\ 2000t & e^{-t/120} \\ -2000 & -120e^{-t/120} \\ 0 & 14{,}400e^{-t/120} \end{array}$$

$s = \left[-240{,}000te^{-t/120} - 28{,}800{,}000e^{-t/120} \right]_0^{120}$

$\quad = \left((-240{,}000)(120)e^{-120/120} - 28{,}800{,}000e^{-120/120} \right) - \left(0 - 28{,}800{,}000e^{-0/120} \right)$

$\quad = 28{,}800{,}000\left(1 - 2e^{-1} \right)$ feet

31. $C = \int_0^x \left(10 + \dfrac{\ln(t+1)}{(t+1)^2} \right) dt + 5000 = \int_0^x 10\,dt + \int_0^x \dfrac{\ln(t+1)}{(t+1)^2}\,dt + 5000$
$$\begin{array}{c|c} D & I \\ \ln(t+1) & 1/(t+1)^2 \\ -1/t+1 & -1/t+1 \end{array}$$

$\quad = 5000 + \left[10t\right]_0^x + \left[\dfrac{-\ln(t+1)}{t+1} \right]_0^x - \int_0^x -\dfrac{1}{(t+1)^2}\,dt$

$\quad = 5000 + [10x - 0] + \left(\dfrac{-\ln(t+1)}{x+1} \right) - \left(\dfrac{-\ln(0+1)}{0+1} \right) - \left[\dfrac{1}{t+1} \right]_0^x$

$\quad = 5000 + 10x - \dfrac{\ln(x+1)}{x+1} - 0 - \dfrac{1}{x+1} + \dfrac{1}{0+1}$

$\quad = 5001 + 10x - \dfrac{1}{x+1} - \dfrac{\ln(x+1)}{x+1}$

33. $R = \int_0^{52} (10 + 0.5x)(50)e^{-0.02x}\,dx$
$$\begin{array}{c|c} D & I \\ 10+0.5x & e^{-0.02x} \\ -0.5 & -50e^{-0.02x} \\ 0 & 2500e^{-0.02x} \end{array}$$

$\quad = 50\left[-50(10+0.5x)e^{-0.02x} - 1250e^{-0.02x} \right]_0^{52}$

$\quad = 50\left[\left(-50(10+0.5(52))e^{-0.02(52)} - 1250e^{-0.02(52)} \right) - \left(-50(10+0)e^0 - 1250e^0 \right) \right]$

$\quad = \$33{,}598$

35. $R = \int_0^{10} (70.5 - 2.5t)(2)e^{0.05t}\,dt$
$$\begin{array}{c|c} D & I \\ 70.5-2.5t & e^{0.05t} \\ 2.5 & 20e^{0.05t} \\ 0 & 400e^{0.05t} \end{array}$$

$\quad = 2\left[(70.5 - 2.5t)(20)e^{0.05t} + 1000e^{0.05t} \right]_0^{10}$

$\quad = 2\left[\left((70.5 - 2.5(10))(20)e^{0.05(10)} + 1000e^{0.05(10)} \right) - \left((70.5 - 0)(20)e^0 + 1000e^0 \right) \right]$

$\quad = \$1{,}478$ million

14.2 EXERCISES

1. $\displaystyle\int \frac{x}{1+x}\, dx;\ \int \frac{u}{a+bu}\, du,\ a=1,\ b=1,\ u=x,\ \frac{du}{dx}=1,\ dx=du$

$\displaystyle = \int \frac{u}{1+u}\, du = \frac{1}{1^2}\Big[1\cdot u - 1\cdot \ln|1+1\cdot u|\Big] + C = x - \ln|1+x| + C$

3. $\displaystyle\int x\sqrt{1+2x}\, dx;\ \int u\sqrt{a+bu}\, du,\ a=1,\ b=2,\ u=x,\ \frac{du}{dx}=1,\ dx=du$

$\displaystyle = \int u\sqrt{1+2u}\, du = \frac{2}{15\cdot 2^2}(3\cdot 2u - 2\cdot 1)(1+2u)^{3/2} + C = \frac{1}{30}(6x-2)(1+2x)^{3/2} + C$

5. $\displaystyle\int \sqrt{x^2+4}\, dx;\ \int \sqrt{u^2 \pm a^2}\, du,\ a=2,\ u=x,\ \frac{du}{dx}=1,\ dx=du$

$\displaystyle = \int \sqrt{u^2 + 2^2}\, du = \frac{u}{2}\sqrt{u^2+2^2} + \frac{2^2}{2}\ln\left|u+\sqrt{u^2+2^2}\right| + C$

$\displaystyle = \frac{x}{2}\sqrt{x^2+4} + 2\ln\left|x+\sqrt{x^2+4}\right| + C$

7. $\displaystyle\int \frac{dx}{1+x^2};\ \int \frac{du}{a^2+u^2},\ a=1,\ u=x,\ \frac{du}{dx}=1,\ dx=du$

$\displaystyle = \int \frac{du}{1^2+u^2} = \frac{1}{1}\tan^{-1}\frac{u}{1} + C = \tan^{-1}x + C$

9. $\displaystyle\int \frac{dx}{\sqrt{3+x^2}};\ \int \frac{du}{\sqrt{u^2 \pm a^2}},\ a=\sqrt{3},\ u=x,\ \frac{du}{dx}=1,\ dx=du$

$\displaystyle = \int \frac{du}{\sqrt{u^2 + \left(\sqrt{3}\right)^2}} = \ln\left|u + \sqrt{u^2 + \left(\sqrt{3}\right)^2}\right| + C = \ln\left|x + \sqrt{3+x^2}\right| + C$

11. $\displaystyle\int \frac{dx}{1-4x^2};\ \int \frac{du}{a^2-u^2},\ a=1,\ u=2x,\ \frac{du}{dx}=2,\ dx=\frac{du}{2}$

$\displaystyle = \frac{1}{2}\int \frac{du}{1^2-u^2} = \frac{1}{2}\left(\frac{1}{2\cdot 1}\right)\ln\left|\frac{1+u}{1-u}\right| + C = \frac{1}{4}\ln\left|\frac{1+2x}{1-2x}\right| + C$

441

13. $\int 2x\sqrt{2x^2-1}\ dx;\ \int u\sqrt{u^2\pm a^2}\ du,\ a=1,\ u=\sqrt{2}x,\ \dfrac{du}{dx}=\sqrt{2},\ dx=\dfrac{du}{\sqrt{2}}$

$=\dfrac{1}{\sqrt{2}}\int 2\left(\dfrac{u}{\sqrt{2}}\right)\sqrt{u^2-1^2}\ du=\dfrac{1}{3}\left(u^2-1^2\right)^{3/2}+C=\dfrac{1}{3}\left(2x^2-1\right)^{3/2}+C$

15. $\int\dfrac{dx}{3x^2-1}=-\int\dfrac{dx}{1-3x^2};\ \int\dfrac{du}{a^2-u^2},\ a=1,\ u=\sqrt{3}x,\ \dfrac{du}{dx}=\sqrt{3},\ dx=\dfrac{du}{\sqrt{3}}$

$=-\dfrac{1}{\sqrt{3}}\int\dfrac{du}{1^2-u^2}=-\dfrac{1}{\sqrt{3}}\left(\dfrac{1}{2\cdot 1}\right)\ln\left|\dfrac{1+u}{1-u}\right|+C$

$=-\dfrac{1}{2\sqrt{3}}\ln\left|\dfrac{1+\sqrt{3}x}{1-\sqrt{3}x}\right|+C$

$=\dfrac{1}{2\sqrt{3}}\ln\left|\dfrac{1-\sqrt{3}x}{1+\sqrt{3}x}\right|+C$

17. $\int\dfrac{dx}{x^2+6x+10}=\int\dfrac{dx}{x^2+6x+3^2-3^2+10}=\int\dfrac{dx}{(x+3)^2+1};\ \int\dfrac{du}{a^2+u^2},\ a=1,\ u=x+3,\ dx=du$

$=\int\dfrac{du}{1^2+u^2}=\dfrac{1}{1}\tan^{-1}\dfrac{u}{1}+C=\tan^{-1}(x+3)+C$

19. $\int\sqrt{x^2-x+3}\ dx=\int\sqrt{x^2-x+\left(-\dfrac{1}{2}\right)^2-\left(-\dfrac{1}{2}\right)^2+3}\ dx=\int\sqrt{\left(x-\dfrac{1}{2}\right)^2+\dfrac{11}{4}}\ dx;$

$\int\sqrt{u^2\pm a^2}\ du,\ a=\sqrt{11}/2,\ u=x-\dfrac{1}{2},\ \dfrac{du}{dx}=1,\ dx=du$

$\int\sqrt{\left(x-\dfrac{1}{2}\right)^2+\dfrac{11}{4}}\ dx=\int\sqrt{u^2+\left(\sqrt{11}/2\right)^2}\ du=\dfrac{u}{2}\sqrt{u^2+\left(\sqrt{11}/2\right)^2}+\dfrac{\left(\sqrt{11}/2\right)^2}{2}\ln\left|u+\sqrt{u^2+\left(\sqrt{11}/2\right)^2}\right|+C$

$=\dfrac{2x-1}{4}\sqrt{x^2-x+3}+\dfrac{11}{8}\ln\left|x-\dfrac{1}{2}+\sqrt{x^2-x+3}\right|+C$

21. $\int\dfrac{dx}{3x^2(2+4x)};\ \int\dfrac{du}{u^2(a+bu)},\ a=2,\ b=4,\ u=x,\ \dfrac{du}{dx}=1,\ dx=du$

$=\dfrac{1}{3}\int\dfrac{du}{u^2(2+4u)}=\dfrac{1}{3}\left(-\dfrac{1}{2u}\right)+\dfrac{1}{3}\left(\dfrac{4}{2^2}\right)\ln\left|\dfrac{2+4u}{u}\right|+C=-\dfrac{1}{6x}+\dfrac{1}{3}\ln\left|\dfrac{2+4x}{x}\right|+C$

23. $\int \dfrac{e^x}{9-e^{2x}}\,dx,\ \int \dfrac{du}{a^2-u^2},\ a=3,\ u=e^x,\ \dfrac{du}{dx}=e^x,\ dx=\dfrac{du}{e^x}$

$= \int \dfrac{e^x}{3^2-u^2}\dfrac{du}{e^x} = \int \dfrac{du}{3^2-u^2} = \dfrac{1}{2(3)}\ln\left|\dfrac{3+u}{3-u}\right|+C = \dfrac{1}{6}\ln\left|\dfrac{3+e^x}{3-e^x}\right|+C$

25. $\int \dfrac{1}{x\left(1+(\ln x)^2\right)}\,dx;\ \int \dfrac{du}{a^2+u^2},\ a=1,\ u=\ln x,\ \dfrac{du}{dx}=\dfrac{1}{x},\ dx=x\,du$

$= \int \dfrac{x\,du}{x\left(1^2+u^2\right)} = \int \dfrac{du}{1^2+u^2} = \dfrac{1}{1}\tan^{-1}\dfrac{u}{1}+C = \tan^{-1}(\ln x)+C$

27. $C = 10{,}000 + \int_0^x 105x^2\sqrt{1+x}\,dx = 10{,}000 + 105\int_0^u u^2\sqrt{1+1\cdot u}\,du$

$= 10{,}000 + 105\left[\left(\dfrac{2}{105(1)^3}\right)\left(15(1)^2u^2 - 12(1)(1)u + 8(1)^2(1+1\cdot u)^{3/2}\right)\right]_0^u$

$= 10{,}000 + \left(2(15u^2-12u+8)(1+u)^{3/2}\right) - \left(2(15\cdot0 - 12\cdot0 + 8)(1+0)^{3/2}\right)$

$= 2(15x^2 - 12x + 8)(1+x)^{3/2}$

29. $C = 10{,}000 + \int_0^x \dfrac{30}{\sqrt{x^2+9}}\,dx = 10{,}000 + 30\int_0^u \dfrac{du}{\sqrt{u^2+3^2}} = 10{,}000 + 30\left[\ln\left|u+\sqrt{u^2+3^2}\right|\right]_0^u$

$= 10{,}000 + 30\ln\left|u+\sqrt{u^2+3^2}\right| - 30\ln\left|0+\sqrt{0^2+3^2}\right|$

$= 30\ln\left|x+\sqrt{x^2+9}\right| + 10{,}000 - 30\ln 3$

31. $P = \int_0^x \dfrac{5000}{\sqrt{x^2+1}}\,dx = 5000\int_0^u \dfrac{du}{\sqrt{u^2+1^2}} = 5000\left[\ln\left|u+\sqrt{u^2+1^2}\right|\right]_0^u$

$= 5000\ln\left|u+\sqrt{u^2+1^2}\right| - 5000\ln\left|0+\sqrt{0^2+1^2}\right|$

$= \$5000\ln\left|x+\sqrt{x^2+1}\right|$

33. $C = 10,000 + \int_0^x \dfrac{10}{\sqrt{4+x^2}}\,dx = 10,000 + 10\int_0^u \dfrac{du}{\sqrt{u^2+2^2}} = 10,000 + 10\left[\ln\left|u+\sqrt{u^2+2^2}\right|\right]_0^u$

$\qquad = 10,000 + 10\ln\left|u+\sqrt{u^2+4}\right| - 10\ln\left|0+\sqrt{0^2+4}\right|$

$\qquad = 10\ln\left|x+\sqrt{x^2+4}\right| + 10,000 - 10\ln 2$

$\qquad = 10\ln\left|\dfrac{x+\sqrt{x^2+4}}{2}\right| + 10,000$

$\qquad \overline{C} = \dfrac{10}{x}\ln\left|\dfrac{x+\sqrt{x^2+4}}{2}\right| + \dfrac{10,000}{x}$

35. $v(t) = \int \dfrac{1000}{\left(1+t^2\right)^{3/2}}\,dt = 1000\int \dfrac{du}{\left(1^2+u^2\right)^{3/2}} = \dfrac{1000u}{1^2\sqrt{u^2+1^2}} + C = \dfrac{1000t}{\sqrt{t^2+1}} + C$

When $t = 0$, $v = 0$, $\quad 0 = \dfrac{1000(0)}{\sqrt{0^2+1}} + C \Rightarrow C = 0$

$s(t) = \int \dfrac{1000t}{\sqrt{t^2+1}}\,dt = 1000\int \dfrac{u}{\sqrt{u^2+1^2}}\,du = 1000\sqrt{u^2+1^2} + C = 1000\sqrt{t^2+1} + C$

When $t = 0$, $s = 0$, $\quad 0 = 1000\sqrt{0^2+1} + C \Rightarrow C = -1000$

$s(t) = 1000\sqrt{t^2+1} - 1000$

$s(120) = 1000\sqrt{(120)^2+1} - 1000 = 119{,}004 \text{ feet}$

14.3 EXERCISES

1. $A = \int_{-1}^{1}(x^2 - (-1))\,dx = \int_{-1}^{1}(x^2 + 1)\,dx = \left[\dfrac{x^3}{3} + x\right]_{-1}^{1} = \left(\dfrac{1^3}{3} + 1\right) - \left(\dfrac{(-1)^3}{3} + (-1)\right) = \dfrac{8}{3}$

3. $A = \int_{0}^{2}(-x - x)\,dx = \int_{0}^{2}(-2x)\,dx = \left[-x^2\right]_{0}^{2} = (-2^2) - (-0^2) = -4, \ A = 4$

5. The area is the sum of the piece from $x = -1$ to $x = 0$
   and the piece from $x = 0$ to $x = 1$.

$A = \int_{-1}^{0}(x^2 - x)\,dx + \int_{0}^{1}(x - x^2)\,dx = \left[\dfrac{x^3}{3} - \dfrac{x^2}{2}\right]_{-1}^{0} + \left(\dfrac{x^2}{2} - \dfrac{x^3}{3}\right)_{0}^{1}$

$= (0 - 0) - \left(\dfrac{(-1)^3}{3} - \dfrac{(-1)^2}{2}\right) + \left(\dfrac{1^2}{2} - \dfrac{1^3}{3}\right) - (0 - 0) = 1$

7. $A = \int_{0}^{1}(e^x - x)\,dx = \left[e^x - \dfrac{x^2}{2}\right]_{0}^{1} = \left(e^1 - \dfrac{1^2}{2}\right) - (e^0 - 0) = e - \dfrac{3}{2}$

9. $A = \int_{0}^{1}\left((x-1)^2 - \left[-(x-1)^2\right]\right)dx = \int_{0}^{1}2(x-1)^2\,dx = 2\int_{-1}^{0}u^2\,du = \left[2\dfrac{u^3}{3}\right]_{-1}^{0} = (0) - \left(\dfrac{2(-1)^3}{3}\right) = \dfrac{2}{3}$

11. $x^4 = x$ when $x^4 - x = 0$, $x(x^3 - 1) = 0$, $x = 0$, or $x = 1$.

$A = \int_{0}^{1}(x - x^4)\,dx = \left[\dfrac{x^2}{2} - \dfrac{x^5}{5}\right]_{0}^{1} = \left(\dfrac{1^2}{2} - \dfrac{1^5}{5}\right) - (0 - 0) = \dfrac{3}{10}$

13. $x^4 = x^3$ when $x^4 - x^3 = 0$, $x^3(x - 1) = 0$, $x = 0$, or $x = 1$.

$A = \int_{0}^{1}(x^3 - x^4)\,dx = \left[\dfrac{x^4}{4} - \dfrac{x^5}{5}\right]_{0}^{1} = \left(\dfrac{1^4}{4} - \dfrac{1^5}{5}\right) - (0 - 0) = \dfrac{1}{20}$

15. $x^4 = x^2$ when $x^4 - x^2 = 0$, $x^2(x^2 - 1) = 0$, $x = 0$, or $x = \pm 1$.

$$A = \int_{-1}^{1}(x^2 - x^4)\,dx = \left[\frac{x^3}{3} - \frac{x^5}{5}\right]_{-1}^{1} = \left(\frac{1^3}{3} - \frac{1^5}{5}\right) - \left(\frac{(-1)^3}{3} - \frac{(-1)^5}{5}\right) = \frac{4}{15}$$

17. $e^x = 2$ when $x = \ln 2$.

$$A = \int_{0}^{\ln 2}(2 - e^x)\,dx = \left[2x - e^x\right]_{0}^{\ln 2} = (2\ln 2 - e^{\ln 2}) - (0 - e^0) = 2\ln 2 - 1$$

19. $\ln x = 2 - \ln x$ when $2\ln x = 2$, $\ln x = 1$, $x = e$.

$$A = \int_{e}^{4}(\ln x - (2 - \ln x))\,dx = 2\int_{e}^{4}(\ln x - 1)\,dx = 2\int_{e}^{4}\ln x\,dx - 2\int_{e}^{4}dx$$

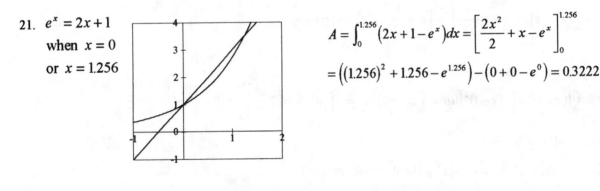

$$2\int_{e}^{4}\ln x\,dx \Rightarrow \begin{vmatrix} D & I \\ \ln x & 1 \\ -1/x & x \end{vmatrix} \Rightarrow [2\,x\ln x]_{e}^{4} - 2\int_{e}^{4}dx = 2(4)\ln 4 - 2e\ln e - 2[x]_{e}^{4}$$

For future Exercises use $\int \ln x\,dx = x\ln x - x$

$$A = 8\ln 4 - 2e - 2[x]_{e}^{4} - 2[x]_{e}^{4} = 8\ln 4 - 2e - 4(4) + 4(e) = 8\ln 4 + 2e - 16$$

21. $e^x = 2x + 1$
when $x = 0$
or $x = 1.256$

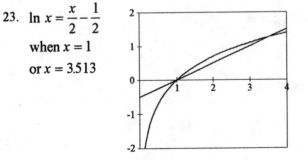

$$A = \int_{0}^{1.256}(2x + 1 - e^x)\,dx = \left[\frac{2x^2}{2} + x - e^x\right]_{0}^{1.256}$$

$$= ((1.256)^2 + 1.256 - e^{1.256}) - (0 + 0 - e^0) = 0.3222$$

23. $\ln x = \dfrac{x}{2} - \dfrac{1}{2}$
when $x = 1$
or $x = 3.513$

See Exercise 19 for $\int \ln x\,dx = x\ln x - x$

$$A = \int_1^{3.513} \left(\ln x - \left(\frac{x}{2} - \frac{1}{2} \right) \right) dx = \int_1^{3.513} \ln x \, dx - \frac{1}{2} \int_1^{3.513} x \, dx + \frac{1}{2} \int_1^{3.513} dx = [x \ln x]_1^{3.513} - \int_1^{3.513} dx - \frac{1}{2} \int_1^{3.513} x \, dx + \frac{1}{2} \int_1^{3.513} dx$$

$$= 3.513 \ln 3.513 - 1 \ln 1 - \frac{1}{2}(3.513) + \frac{1}{2}(1) - \frac{(3.513)^2}{4} + \frac{1^2}{4} = 0.3222$$

In Exercises 25-35 use $CS = \int_0^{\bar{q}} \left(f(q) - \bar{p} \right) dq$ where $p = f(q)$ and $f(\bar{q}) = \bar{p}$.

25. $p = 10 - 2q, \ \bar{p} = 5 \Rightarrow 5 = 10 - 2\bar{q} \Rightarrow \bar{q} = 2.5$

$$CS = \int_0^{2.5} (10 - 2q - 5) dq = \left[5q - \frac{2q^2}{2} \right]_0^{2.5} = \left(5(2.5) - (2.5)^2 \right) - (0 - 0) = \$6.25$$

27. $p = 100 - 3\sqrt{q}, \ \bar{p} = 76 \Rightarrow 76 = 100 - 3\sqrt{q} \Rightarrow \bar{q} = 64$

$$CS = \int_0^{64} \left(100 - 3\sqrt{q} - 76 \right) dq = \left[24q - 2q^{3/2} \right]_0^{64} = \left(24(64) - 2(64)^{3/2} \right) - (0 - 0) = \$512$$

29. $p = 500e^{-2q}, \ \bar{p} = 100 \Rightarrow 100 = 500e^{-2q} \Rightarrow 0.2 = e^{-2q} \Rightarrow \bar{q} = -\dfrac{\ln 0.2}{2}$

$$CS = \int_0^{-\ln 0.2/2} \left(500e^{-2q} - 100 \right) dq = \left[-250e^{-2q} - 100q \right]_0^{-\ln 0.2/2}$$

$$= \left(-250e^{-2\left(-\frac{\ln 0.2}{2} \right)} - 100 \left(-\frac{\ln 0.2}{2} \right) \right) - \left(-250e^{-2(0)} - 0 \right) = \$159.76$$

31. $\bar{p} = 20, \ q = 100 - 2p \Rightarrow p = 50 - \dfrac{8}{2} \Rightarrow 20 = 50 - \dfrac{\bar{q}}{2} \Rightarrow \bar{q} = 60$

$$CS = \int_0^{60} \left(50 - \frac{q}{2} - 20 \right) dq = \left[30q - \frac{q^2}{4} \right]_0^{60} = \left(30(60) - \frac{(60)^2}{4} \right) - (0 - 0) = \$900$$

33. $\bar{p} = 10, \ q = 100 - 0.25p^2 \Rightarrow p = \sqrt{400 - 4q} \Rightarrow 10 = \sqrt{400 - 4\bar{q}} \Rightarrow \bar{q} = 75$

$$CS = \int_0^{75} \left(\sqrt{400 - 4q} - 10 \right) dq = \int_{400}^{100} u^{1/2} \left(-\frac{du}{4} \right) - \int_0^{75} 10 \, dq = \left[-\frac{1}{6} u^{3/2} \right]_{400}^{100} - [10q]_0^{75}$$

$$= -\frac{1}{6} \left((100)^{3/2} - (400)^{3/2} \right) - (10(75) - 0) = \$416.67$$

447

35. $\bar{p}=1,\ q=500e^{-0.5p}-50 \Rightarrow e^{-0.5p}=\dfrac{9+50}{500} \Rightarrow p=-2\ln\left(\dfrac{q+50}{500}\right)$

$1=-2\ln\dfrac{\bar{q}+50}{500} \Rightarrow e^{-1/2}=\dfrac{\bar{q}+50}{500} \Rightarrow \bar{q}=500e^{-1/2}-50$

$CS=\displaystyle\int_0^{500e^{-1/2}-50}\left(-2\ln\left(\dfrac{q+50}{500}\right)-1\right)dq=-2\int_{1/10}^{e^{-1/2}}\ln u(500\,du)-\int_0^{500e^{-1/2}-50}dq$

$\left(\text{See Exercise 19 for }\displaystyle\int\ln x\,dx=x\ln x-x\right)$

$=-1000\big[u\ln u-u\big]_{1/10}^{e^{-1/2}}-\big[q\big]_0^{500e^{-1/2}-50}$

$=-1000\left[\left(e^{-1/2}\ln e^{-1/2}-e^{-1/2}\right)-\left(\dfrac{1}{10}\ln\dfrac{1}{10}-\dfrac{1}{10}\right)\right]-\big[(500e^{-1/2}-50)-0\big]=\$326.27$

In Exercises 37-47 use $PS=\displaystyle\int_0^{\bar{q}}\left(\bar{p}-f(q)\right)dq$ where $p=f(q)$ and $f(\bar{q})=\bar{p}$.

37. $p=10+2q,\ \bar{p}=20 \Rightarrow 20=10+2\bar{q} \Rightarrow \bar{q}=5$

$PS=\displaystyle\int_0^5\left(20-(10+2q)\right)dq=\left[10q-q^2\right]_0^5=\left(10(5)-5^2\right)-(0-0)=\$25$

39. $p=10+2q^{1/3},\ \bar{p}=12 \Rightarrow 12=10+2\left(\bar{q}\right)^{1/3} \Rightarrow \bar{q}=1$

$PS=\displaystyle\int_0^1\left(12-(10+2q^{1/3})\right)dq=\left[2q-\dfrac{3}{2}q^{4/3}\right]_0^1=\left(2(1)-\dfrac{3}{2}(1)^{4/3}\right)-(0-0)=\$0.50$

41. $p=500e^{0.5q},\ \bar{p}=1000 \Rightarrow 1000=500e^{0.5\bar{q}} \Rightarrow e^{0.5\bar{q}}=2 \Rightarrow \bar{q}=2\ln 2$

$PS=\displaystyle\int_0^{2\ln 2}\left(1000-500e^{0.5q}\right)dq=\big[1000q\big]_0^{2\ln 2}-500\big[2e^{0.5q}\big]_0^{2\ln 2}$

$=1000(2\ln 2-0)-1000\left(e^{0.5(2\ln 2)}-e^0\right)=\$386.29$

43. $\bar{p}=40,\ q=2p-50 \Rightarrow p=\dfrac{q}{2}+25 \Rightarrow 40=\dfrac{\bar{q}}{2}+25 \Rightarrow \bar{q}=30$

$PS=\displaystyle\int_0^{30}\left(40-\left(\dfrac{q}{2}+25\right)\right)dq=\left[15q-\dfrac{q^2}{4}\right]_0^{30}=\left(15(30)-\dfrac{(30)^2}{4}\right)-(0-0)=\$225$

45. $\bar{p}=10,\ q=0.25p^2-10\Rightarrow p=\sqrt{4q+40}\Rightarrow 10=\sqrt{4q+40}\Rightarrow \bar{q}=15$

$$PS=\int_0^{15}\left(10-\sqrt{4q+40}\right)dq=\int_0^{15}10\,dq-\int_{40}^{100}u^{1/2}\left(\frac{du}{4}\right)=\left[10q\right]_0^{15}-\left[\frac{1}{6}u^{3/2}\right]_{40}^{100}$$

$$=\left(10(15)-0\right)-\left(\frac{(100)^{3/2}}{6}-\frac{(40)^{3/2}}{6}\right)=\$25.50$$

47. $\bar{p}=10,\ q=500e^{0.05P}-50\Rightarrow e^{0.05P}=\dfrac{q+50}{500}\Rightarrow p=20\ln\left(\dfrac{q+50}{500}\right)$

$\bar{q}=500e^{0.05(10)}-50=500e^{0.5}-50$

$$PS=\int_0^{500e^{0.5}-50}\left(10-20\ln\left(\frac{q+50}{500}\right)\right)dq=\int_0^{500e^{0.5}-50}10\,dq-20\int_{0.1}^{e^{0.5}}\ln u(500\,du)$$

$\left(\text{See Exercise 19 for }\int \ln x\ dx = x\ln x - x\right)$

$$=\left[10q\right]_0^{500e^{0.5}-50}-10{,}000\left[u\ln u-u\right]_{0.1}^{e^{0.5}}$$

$$=\left[10(500e^{0.5}-50)-0\right]-10{,}000\left[\left(e^{0.5}\ln e^{0.5}-e^{0.5}\right)-\left(0.1\ln 0.1-0.1\right)\right]=\$12{,}684.63$$

49. Demand: $q=9859.39-2.17p\Rightarrow p=\dfrac{9859.39-q}{2.17}$

Supply: $q=100+0.5p\Rightarrow p=2q-200$

Equilibrium exists when supply = demand.

$9859.39-2.17\bar{p}=100+0.5\bar{p}\Rightarrow \bar{p}=3{,}655.20\Rightarrow \bar{q}=100+0.5(3{,}655.20)=1{,}927.60$

$$CS=\int_0^{1{,}927.60}\left(\frac{9859.39}{2.17}-\frac{q}{2.17}-3{,}655.20\right)dq=\left[888.30q-\frac{q^2}{4.34}\right]_0^{1{,}927.60}$$

$$=\left(888.30(1{,}927.60)-\frac{(1{,}927.60)^2}{4.34}\right)-(0-0)=\$856{,}139$$

$$PS=\int_0^{1{,}927.60}\left(3{,}655.20-(2q-200)\right)dq=\left[3855.20q-q^2\right]_0^{1{,}927.60}$$

$$=\left((3855.20)(1927.60)-(1927.60)^2\right)-(0-0)=\$3{,}715{,}642$$

The university would earn $3,715,642 less than if it charged the equilibrium price of $3655.

51. You save more because the producers surplus is greater because there is more area between the equilibrium price line of $\bar{p}\approx 12$ and the supply curve than between the demand curve and $\bar{p}\approx 12$.

53. $q = -mp + b \Rightarrow p = \dfrac{b-q}{m}$, $\bar{q} = -m\bar{p} + b$

$$CS = \int_0^{-m\bar{p}+b} \left(\left(\frac{b-q}{m} \right) - \bar{p} \right) dq = \left[\left(\frac{b}{m} - \bar{p} \right) q - \frac{q^2}{2m} \right]_0^{-m\bar{p}+b}$$

$$= \left(\left(\frac{b-m\bar{p}}{m} \right)(b-m\bar{p}) - \frac{(b-m\bar{p})^2}{2m} \right) - (0-0) = \frac{1}{2m}(b-m\bar{p})^2$$

55. (a) $C = \int(R-P)\,dt = \int_0^4 \left(16.15e^{0.87t} - 3.93e^t \right) dt = \left[\dfrac{16.15}{0.87} e^{0.87t} - 3.93e^t \right]_0^4$

$= \left(18.5632e^{0.87(4)} - 3.93e^4 \right) - \left(18.5632e^0 - 3.93e^0 \right) = \$373.35$ million

(b) This is the area of the region between the graphs of $R(t)$ and $P(t)$ for $0 \le t \le 4$.

(c) Since the exponent for P is larger, this tells us that the ratio of profit to revenue was increasing; that is, cost accounted for a decreasing proportion of the revenues.

57. (a) $C'(8) = 1{,}000{,}000\left(41.25(8)^2 - 697.5(8) + 3210 \right) = 270{,}000{,}000$

$C'(10) = 1{,}000{,}000\left(41.25(10)^2 - 697.5(10) + 3210 \right) = 360{,}000{,}000$

$C'(12) = 1{,}000{,}000\left(41.25(12)^2 - 697.5(12) + 3210 \right) = 780{,}000{,}000$

(b) $C = \int_0^{12} 1{,}000{,}000\left(41.25q^2 - 697.5q + 3210 \right) dq = 1{,}000{,}000 \left[\dfrac{41.25}{3} q^3 - \dfrac{697.5}{2} q^2 + 3210q \right]_0^{12}$

$= 1{,}000{,}000\left[\left(\dfrac{41.25}{3} \right)(12)^3 - \dfrac{697.5}{2}(12)^2 + 3210(12) - (0-0+0) \right] = 12{,}060$ million

$C_{net} = 12{,}060 - 400(12) = \$7{,}260$ million

(c) The area between $y = 1{,}000{,}000\left(41.25q^2 - 697.5q + 3210 \right)$ and $y = 400{,}000{,}000$ for $0 \le q \le 12$.

14.4 EXERCISES

In Exercises 1-3 use $\bar{y} = \dfrac{y_1 + y_2 + \cdots + y_n}{n}$.

1. $\bar{y} = \dfrac{6648 + 7684 + 11{,}356 + 17{,}938 + 28{,}389}{5}$

   $= \$14{,}403$ million

3. $\bar{y} = \dfrac{210 + 205 + 340 + 285 + 175 + 140 + 155}{7}$

   $= \$215.714$ billion

In Exercises 5-9 use $\bar{f} = \dfrac{1}{b-a}\int_a^b f(x)\,dx$.

5. $f(x) = x^3$ over $[0, 2]$

   $\bar{f} = \dfrac{1}{2-0}\int_0^2 x^3\,dx = \dfrac{1}{2}\left[\dfrac{x^4}{4}\right]_0^2 = \dfrac{1}{2}\left[\dfrac{2^4}{4} - \dfrac{0^4}{4}\right] = 2$

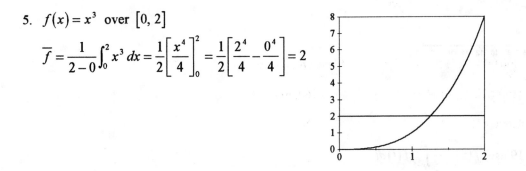

7. $f(x) = x^3 - x$ over $[0, 2]$

   $\bar{f} = \dfrac{1}{2-0}\int_0^2 (x^3 - x)\,dx = \dfrac{1}{2}\left[\dfrac{x^4}{4} - \dfrac{x^2}{2}\right]_0^2 = \dfrac{1}{2}\left(\dfrac{2^4}{4} - \dfrac{2^2}{2}\right) - \dfrac{1}{2}(0 - 0) = 1$

451

9. $f(x) = e^{-x}$ over $[0, 2]$

$$\overline{f} = \frac{1}{2-0}\int_0^2 e^{-x}\,dx = \frac{1}{2}\left[-e^{-x}\right]_0^2 = \frac{1}{2}\left[-e^{-2}\right] - \frac{1}{2}\left(-e^0\right) = \frac{1-e^{-2}}{2}$$

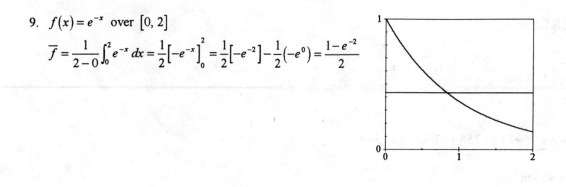

11.

y	1	2	3	4	3	2	3	4	5	6	5	6	7	8	9	8	7	8	9	10
Moving Average					2.6	2.8	3.0	3.2	3.4	4.0	4.6	5.2	5.8	6.4	7.0	7.6	7.8	8.0	8.2	8.4

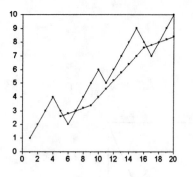

In Exercises 13 - 19 use $\overline{f}(x) = \dfrac{1}{5}\displaystyle\int_{x-5}^x f(t)\,dt.$

13. $f(x) = x^3$

$$\overline{f}(x) = \frac{1}{5}\int_{x-5}^x t^3\,dt = \frac{1}{5}\left[\frac{t^4}{4}\right]_{x-5}^x$$

$$= \frac{1}{20}\left[x^4 - (x-5)^4\right]$$

$$= \frac{1}{20}\left[x^4 - \left(x^4 - 20x^3 + 150x^2 - 500x + 625\right)\right]$$

$$= x^3 - \frac{15}{2}x^2 + 25x - \frac{125}{4}$$

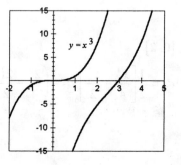

452

15. $f(x) = x^{2/3}$

$\overline{f}(x) = \dfrac{1}{5}\displaystyle\int_{x-5}^{x} t^{2/3}\, d = \dfrac{3}{25}\Big[t^{5/3}\Big]_{x-5}^{x}$

$\qquad = \dfrac{3}{25}\Big[x^{5/3} - (x-5)^{5/3}\Big]$

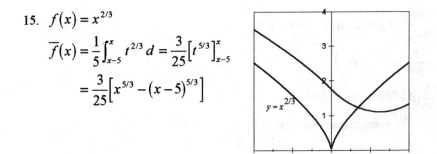

17. $f(x) = e^{0.5x}$

$\overline{f}(x) = \dfrac{1}{5}\displaystyle\int_{x-5}^{x} e^{0.5t}\, dt = \dfrac{2}{5}\Big[e^{0.5t}\Big]_{x-5}^{x}$

$\qquad = \dfrac{2}{5}\Big(e^{0.5x} - e^{0.5(x-5)}\Big)$

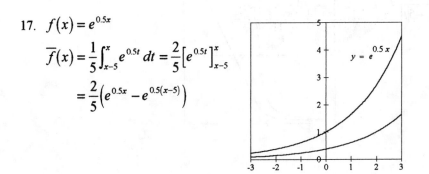

19. $f(x) = \sqrt{x}$

$\overline{f}(x) = \dfrac{1}{5}\displaystyle\int_{x-5}^{x} \sqrt{t}\, dt = \dfrac{2}{15}\Big[t^{3/2}\Big]_{x-5}^{x}$

$\qquad = \dfrac{2}{15}\Big(x^{3/2} - (x-5)^{3/2}\Big)$

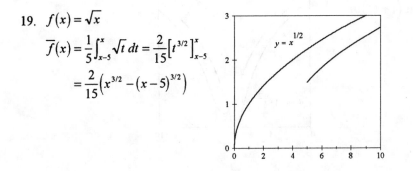

21. $A = \dfrac{1}{1-0}\displaystyle\int_{0}^{1} 10{,}000 e^{0.08t}\, dt = 10{,}000\left[\dfrac{e^{0.08t}}{0.08}\right]_{0}^{1} = \dfrac{10{,}000}{0.08}\Big(e^{0.08(1)} - e^{0}\Big) = \$10{,}410.88$

453

23. $(0, 3000)$ and $(30, 0) \Rightarrow m = \dfrac{0 - 3000}{30 - 0} = -100, \ A(t) = 3000 - 100t$

$$\overline{A} = \frac{1}{30 - 0} \int_0^{30} (3000 - 100t)\, dt = \frac{1}{30} \Big[3000t - 50t^2 \Big]_0^{30}$$

$$= \frac{1}{30}\big(3000(30) - 50(30)^2\big) - \frac{1}{30}(0 - 0) = \$1500$$

This process repeats each month so the average over any number of months will be \$1500.

25. $f(x) = \dfrac{10x}{1 + 5|x|}, \quad \overline{f}(x) = \dfrac{1}{3} \displaystyle\int_{x-3}^{x} \dfrac{10t}{1 + 5|t|}\, dt$

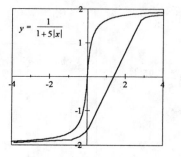

27. $f(x) = \ln\left(1 + x^2\right), \quad \overline{f}(x) = \dfrac{1}{3} \displaystyle\int_{x-3}^{x} \ln\left(1 + t^2\right) dt$

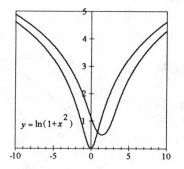

454

29. **(a)** $(2, 7)$ and $(12, 10) \Rightarrow m = \dfrac{10-7}{12-2} = 0.3, \ n(t) - 7 = 0.3(t-2)$

$$n(t) = 0.3t + 6.4$$

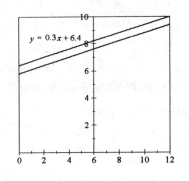

where t is the number of years since 1983.

(b) $\overline{n}(t) = \dfrac{1}{4}\displaystyle\int_{t-4}^{t}(0.3x + 6.4)\,dx = \dfrac{1}{4}\left[\dfrac{0.3x^2}{2} + 6.4x\right]_{t-4}^{t}$

$$= \dfrac{1}{4}\left(\dfrac{0.3t^2}{2} + 6.4t\right) - \dfrac{1}{4}\left(\dfrac{0.3}{2}(t-4)^2 + 6.4(t-4)\right)$$

$$= \dfrac{0.3t^2}{8} + 1.6t - \dfrac{0.3t^2}{8} + 0.3t - 0.6 - 1.6t + 6.4$$

$$= 0.3t + 5.8$$

(c) The slope of the moving average is the same as the slope of the original function.

(d)

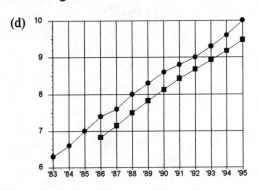

31. $f(x) = mx + b$

$$\overline{f}(x) = \dfrac{1}{a}\displaystyle\int_{x-a}^{x}(mt + b)\,dt = \dfrac{1}{a}\left[\dfrac{mt^2}{2} + bt\right]_{x-a}^{x}$$

$$= \dfrac{1}{a}\left(\dfrac{m}{2}x^2 + bx\right) - \dfrac{1}{a}\left(\dfrac{m}{2}(x-a)^2 + b(x-a)\right)$$

$$= \dfrac{m}{2a}x^2 + \dfrac{b}{a}x - \dfrac{m}{2a}x^2 + mx - \dfrac{ma}{2} - \dfrac{b}{a}x + b$$

$$= mx + b - \dfrac{ma}{2}$$

33. (a) Let $p(t) = at^2 + bt + c$ for $(0, 12)$, $(10, 18)$, and $(20, 16)$

$12 = a(0)^2 + b(0) + c$ $\qquad$ $18 = a(10)^2 + b(10) + 12$ $\qquad$ $16 = a(20)^2 + b(20) + 12$

$c = 12$ $\qquad$ (1) $\quad 3 = 50a + 5b$ $\qquad$ (2) $\quad 1 = 100a + 5b$

Add -1 times equation (2) to equation (1) to get $a = -0.04$

Substitute -0.04 for a in equation (2) to get $b = 1$

$p(t) = -0.04t^2 + t + 12$

(b) $\overline{p} = \dfrac{1}{20 - 0} \int_0^{20} \left(-0.04t^2 + t + 12\right) dt = \dfrac{1}{20}\left[\dfrac{-0.04}{3}t^3 + \dfrac{1}{2}t^2 + 12t\right]_0^{20}$

$\qquad = \dfrac{1}{20}\left(-\dfrac{0.04}{3}(20)^3 + \dfrac{1}{2}(20)^2 + 12(20)\right) - \dfrac{1}{20}(-0 + 0 + 0) = 16.67\%$

(c) $\overline{p} = \dfrac{1}{28 - 25} \int_{25}^{28} \left(-0.04t^2 + t + 12\right) dt = \dfrac{1}{3}\left[\dfrac{-0.04}{3}t^3 + \dfrac{1}{2}t^2 + 12t\right]_{25}^{28}$

$\qquad = \dfrac{1}{3}\left(\dfrac{-0.04}{3}(28)^3 + \dfrac{1}{2}(28)^2 + 12(28)\right) - \dfrac{1}{3}\left(\dfrac{-0.04}{3}(25)^3 + \dfrac{1}{2}(25)^2 + 12(25)\right) = 10.38\%$

35. (a)

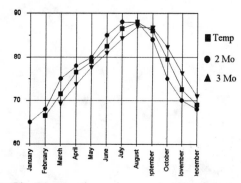

(b) The 24 - month moving average is constant and equal to the year - long average of approximately $77°$.

(c) A quadratic model could not be used to predict temperatures beyond the given 12 - month period, since temperature patterns are periodic, whereas parabolas are not.

37. (a)

Year	86	87	88	89	90	91	92	93
% Inc.	9	7	7	9	11	7	6	4
%	109	116.6	124.8	136	151	161.6	171.3	178.1

Total Increase $= 78.1\%$

(b) $A = Pe^{rt}$

$1.781 = 1e^{r(8)}$

$\ln 1.781 = 8r$

$r = 0.0721 = 7.21\%$

This is slightly less than the answer to Exercise 4. (This is expected because of compounding over eight years.)

39. $y = 5 + 4\cos(x - 0.1)$

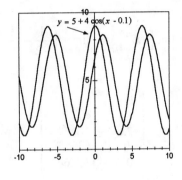

41. $V(t) = 165\cos(120\pi t)$

(a) $V = 0$ for $\left[0, \dfrac{1}{6}\right]$

Number of cycles per second

$= \dfrac{120}{2\pi}$

$= 60$

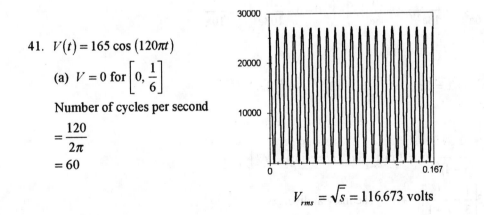

$V_{rms} = \sqrt{s} = 116.673$ volts

14.5 EXERCISES

1. $\int_1^{+\infty} x\,dx = \lim_{M\to+\infty} \int_1^M x\,dx = \lim_{M\to+\infty}\left[\dfrac{x^2}{2}\right]_1^M = \lim_{M\to+\infty}\left(\dfrac{M^2}{2} - \dfrac{1}{2}\right) \to +\infty$

   The integral diverges

3. $\int_{-2}^{+\infty} e^{-0.5x}\,dx = \lim_{M\to+\infty}\int_{-2}^M e^{-0.5x}\,dx = \lim_{M\to+\infty}\left[-2e^{-0.5x}\right]_{-2}^M = \lim_{M\to+\infty} -2\left(e^{-0.5M} - e^{0.5(-2)}\right) = 2e$

5. $\int_{-\infty}^2 e^x\,dx = \lim_{M\to-\infty}\int_{M1}^2 e^x\,dx = \lim_{M\to-\infty}\left[e^x\right]_M^2 = \lim_{M\to-\infty}\left(e^2 - e^M\right) = e^2$

7. $\int_{-\infty}^{-2} \dfrac{1}{x^2}\,dx = \lim_{M\to-\infty}\int_M^{-2} x^{-2}\,dx = \lim_{M\to-\infty}\left[-x^{-1}\right]_M^{-2} = \lim_{M\to-\infty} -\left(\dfrac{1}{-2} - \dfrac{1}{M}\right) = \dfrac{1}{2}$

9. $\int_0^{+\infty} x^2 e^{-6x}\,dx = \lim_{M\to+\infty}\int_0^M x^2 e^{-6x}\,dx$

   $\begin{array}{c|c} D & I \\ x^2 & e^{-6x} \\ -2x & -1/6e^{-6x} \\ 2 & 1/36e^{-6x} \\ 0 & -1/216e^{-6x} \end{array}$

   $= \lim_{M\to+\infty}\left[-\dfrac{x^2 e^{-6x}}{6} - \dfrac{2xe^{-6x}}{36} - \dfrac{2e^{-6x}}{216}\right]_0^M$

   $= \lim_{M\to+\infty}\left[\left(\dfrac{-M^2}{6e^{6M}} - \dfrac{M}{18e^{6M}} - \dfrac{1}{108e^{6M}}\right) - \left(-0 - 0 - \dfrac{1}{108e^0}\right)\right] = \dfrac{1}{108}$

11. $\int_0^5 \dfrac{2}{x^{1/3}}\,dx = \lim_{r\to0^+}\int_r^5 2x^{-1/3}\,dx = \lim_{r\to0^+}\left[3x^{2/3}\right]_r^5 = \lim_{r\to0^+} 3\left(5^{2/3} - r^{2/3}\right) = 3\cdot 5^{2/3}$

13. $\int_{-1}^2 \dfrac{3}{(x+1)^2}\,dx = \lim_{r\to-1^+}\int_r^2 \dfrac{3}{(x+1)^2}\,dx = \lim_{r\to-1^+}\int_{r+1}^3 3u^{-2}\,du = \lim_{r\to-1^+}\left[-3u^{-1}\right]_{r+1}^3$

   $= \lim_{r\to-1^+} -3\left[\dfrac{1}{3} - \dfrac{1}{r+1}\right] = +\infty$ The integral diverges

15. $\displaystyle\int_{-1}^{2}\frac{3x}{x^2-1}\,dx=\int_{-1}^{0}\frac{3x}{x^2-1}dx+\int_{0}^{1}\frac{3x}{x^2-1}dx+\int_{1}^{2}\frac{3x}{x^2-1}dx$

$\displaystyle=\lim_{r\to-1^{+}}\int_{r}^{0}\frac{3x}{x^2-1}dx+\lim_{r\to1^{-}}\int_{0}^{r}\frac{3x}{x^2-1}dx+\lim_{r\to1^{+}}\int_{r}^{2}\frac{3x}{x^2-1}dx$

$\displaystyle=\lim_{r\to-1^{+}}\int_{r^2-1}^{-1}\frac{3}{2u}du+\lim_{r\to1^{-}}\int_{-1}^{r^2-1}\frac{3}{2u}du+\lim_{r\to1^{+}}\int_{r^2-1}^{3}\frac{3}{2u}du$

$\displaystyle=\lim_{r\to-1^{+}}\frac{3}{2}\Big[\ln|u|\Big]_{r^2-1}^{-1}+\lim_{r\to1^{-}}\frac{3}{2}\Big[\ln|u|\Big]_{-1}^{r^2-1}+\lim_{r\to1^{+}}\frac{3}{2}\Big[\ln|u|\Big]_{r^2-1}^{3}$

$\displaystyle=\lim_{r\to-1^{+}}\frac{3}{2}\Big(\ln|-1|-\ln|r^2-1|\Big)+\lim_{r\to1^{-}}\frac{3}{2}\Big(\ln|r^2-1|-\ln|-1|\Big)$

$\displaystyle\qquad+\lim_{r\to1^{+}}\frac{3}{2}\Big(\ln|3|-\ln|r^2-1|\Big)=+\infty-\infty+\infty$

The integral diverges

17. $\displaystyle\int_{-2}^{2}\frac{1}{(x+1)^{1/5}}\,dx=\int_{-2}^{-1}(x+1)^{-1/5}\,dx+\int_{-1}^{2}(x+1)^{-1/5}\,dx=\lim_{r\to-1^{-}}\int_{-2}^{r}(x+1)^{-1/5}\,dx+\lim_{r\to-1^{+}}\int_{r}^{2}(x+1)^{-1/5}\,dx$

$\displaystyle=\lim_{r\to-1^{-}}\int_{-1}^{r+1}u^{-1/5}\,du+\lim_{r\to-1^{+}}\int_{r+1}^{3}u^{-1/5}\,du=\lim_{r\to-1^{-}}\left[\frac{5u^{4/5}}{4}\right]_{-1}^{r+1}+\lim_{r\to-1^{+}}\left[\frac{5u^{4/5}}{4}\right]_{R+1}^{3}$

$\displaystyle=\lim_{r\to-1^{-}}\frac{5}{4}\Big((r+1)^{4/5}-(-1)^{4/5}\Big)+\lim_{r\to-1^{+}}\frac{5}{4}\Big((3)^{4/5}-(r+1)^{4/5}\Big)=\frac{5}{4}\Big(3^{4/5}-1\Big)$

19. $\displaystyle\int_{-1}^{1}\frac{2x}{x^2-1}\,dx=\int_{-1}^{0}\frac{2x}{x^2-1}dx+\int_{0}^{1}\frac{2x}{x^2-1}dx=\lim_{r\to-1^{+}}\int_{r}^{0}\frac{2x}{x^2-1}dx+\lim_{r\to1^{-}}\int_{0}^{r}\frac{2x}{x^2-1}dx$

$\displaystyle=\lim_{r\to-1^{+}}\int_{r^2-1}^{-1}\frac{2x}{2xu}\,du+\lim_{r\to1^{-}}\int_{-1}^{r^2-1}\frac{2x}{2xu}\,du=\lim_{r\to-1^{+}}\Big[\ln|u|\Big]_{r^2-1}^{-1}+\lim_{r\to1^{-}}\Big[\ln|u|\Big]_{-1}^{r^2-1}$

$\displaystyle=\lim_{r\to-1^{+}}\Big(\ln|-1|-\ln|r^2-1|\Big)+\lim_{r\to1^{-}}\Big(\ln|r^2-1|-\ln|-1|\Big)=+\infty-\infty$

The integral diverges

21. $\displaystyle\int_{-\infty}^{+\infty}xe^{-x^2}\,dx=\int_{-\infty}^{0}xe^{-x^2}dx+\int_{0}^{+\infty}xe^{-x^2}dx=\lim_{M\to-\infty}\int_{M}^{0}xe^{-x^2}dx+\lim_{M\to+\infty}\int_{0}^{M}xe^{-x^2}dx$

$\displaystyle=\lim_{M\to-\infty}\left(-\frac{1}{2}\right)\int_{-M^2}^{0}e^{u}\,du+\lim_{M\to+\infty}\left(-\frac{1}{2}\right)\int_{0}^{-M^2}e^{u}\,du$

$\displaystyle=\lim_{M\to-\infty}\left(-\frac{1}{2}\right)\Big[e^{u}\Big]_{-M^2}^{0}+\lim_{M\to+\infty}\left(-\frac{1}{2}\right)\Big[e^{u}\Big]_{0}^{-M^2}$

$\displaystyle=\lim_{M\to-\infty}\left(-\frac{1}{2}\right)\Big(e^{0}-e^{-M^2}\Big)+\lim_{M\to+\infty}\left(-\frac{1}{2}\right)\Big(e^{-M^2}-e^{0}\Big)=-\frac{1}{2}+\frac{1}{2}=0$

23. $\displaystyle\int_0^{+\infty} \frac{1}{x\ln x}\,dx = \int_0^{1/2}\frac{1}{x\ln x}\,dx + \int_{1/2}^1\frac{1}{x\ln x}\,dx + \int_1^2\frac{1}{x\ln x}\,dx + \int_2^{+\infty}\frac{1}{x\ln x}\,dx$

$\displaystyle = \lim_{r\to 0^+}\int_r^{1/2}\frac{1}{x\ln x}\,dx + \lim_{r\to 1^-}\int_{1/2}^r\frac{1}{x\ln x}\,dx + \lim_{r\to 1^+}\int_r^2\frac{1}{x\ln x}\,dx + \lim_{M\to +\infty}\int_2^M\frac{1}{x\ln x}\,dx$

Let $u = \ln x$, then $\displaystyle\int_a^b\frac{1}{x\ln x}\,dx = \int_{\ln a}^{\ln b}\frac{du}{u} = \Big[\ln|u|\Big]_{\ln a}^{\ln b} = \ln|\ln b| - \ln|\ln a|$

$\displaystyle\int_0^{+\infty}\frac{1}{x\ln x}\,dx = \lim_{r\to 0^+}\big(\ln|\ln 1/2| - \ln|\ln r|\big) + \lim_{r\to 1^-}\big(\ln|\ln r| - \ln|\ln 1/2|\big)$

$\displaystyle \quad + \lim_{r\to 1^+}\big(\ln|\ln 2| - \ln|\ln r|\big) + \lim_{M\to +\infty}\big(\ln|\ln M| - \ln|\ln 2|\big) = +\infty - \infty + \infty + \infty$

The integral diverges

25. $\displaystyle\int_0^{+\infty}\frac{2x}{x^2-1}\,dx = \int_0^1\frac{2x}{x^2-1}\,dx + \int_1^2\frac{2x}{x^2-1}\,dx + \int_2^{+\infty}\frac{2x}{x^2-1}\,dx$

$\displaystyle = \lim_{r\to 1^-}\int_0^r\frac{2x}{x^2-1}\,dx + \lim_{r\to 1^+}\int_r^2\frac{2x}{x^2-1}\,dx + \lim_{M\to+\infty}\int_2^M\frac{2x}{x^2-1}\,dx$

See Exercise 19 for the integration and the conversions

$\displaystyle = \lim_{r\to 1^-}\Big[\ln|u|\Big]_{-1}^{r^2-1} + \lim_{r\to 1^+}\Big[\ln|u|\Big]_{r^2-1}^3 + \lim_{M\to+\infty}\Big[\ln|u|\Big]_3^{M^2-1}$

$\displaystyle = \lim_{r\to 1^-}\big(\ln|r^2-1| - \ln|-1|\big) + \lim_{r\to 1^+}\big(\ln|3| - \ln|r^2-1|\big) + \lim_{M\to+\infty}\big(\ln|M^2-1| - \ln|3|\big)$

The integral diverges

27. $\displaystyle q = \int_0^{+\infty} 200e^{-0.1t} = \lim_{M\to+\infty}\int_0^M 200e^{-0.1t} = \lim_{M\to+\infty}\int_0^{-0.1M} -2000e^u\,du$

$\displaystyle = \lim_{M\to+\infty}\Big[-2000e^u\Big]_0^{-0.1M} = \lim_{M\to+\infty} -2000\big[e^{-0.1M} - e^0\big]$

$= 2000$ No, you will not sell more than 2,000 T-shirts.

29. $S = S_0 e^{-kt}$, $(0, 600)$ and $(3, 300)$

$300 = 600e^{-k(3)}$

$\dfrac{1}{2} = e^{-3k}$

$-3k = \ln\dfrac{1}{2}$

$k = \dfrac{\ln(1/2)}{3}$

$\displaystyle S = \int_0^{+\infty} 600e^{\frac{\ln(1/2)}{3}t}$

$\displaystyle = \lim_{M\to+\infty}\int_0^M 600e^{\frac{\ln(1/2)}{3}t}\,dt$

$\displaystyle = \lim_{M\to+\infty}\int_0^{\frac{\ln(1/2)}{3}M} \frac{600\cdot 3}{\ln(1/2)}e^u\,du$

$\displaystyle = \lim_{M\to+\infty}\frac{1800}{\ln(1/2)}\Big[e^u\Big]_0^{\frac{\ln(1/2)}{3}M}$

$\displaystyle = \lim_{M\to+\infty}\frac{1800}{\ln(1/2)}\left(e^{\frac{\ln(1/2)}{3}M} - e^0\right)$

$= \$2,596.85$ million

460

31. $p = 85e^{0.4t}$, $q = 500e^{-1t}$, $R = \int_0^{+\infty} 85e^{0.4t} 500e^{-t} \, dt$

$$R = \lim_{M \to +\infty} \int_0^M 42{,}500e^{-0.6t} \, dt = \lim_{M \to +\infty} \int_0^{-0.6M} \frac{42{,}500}{-0.6} e^u \, du$$

$$= \lim_{M \to +\infty} \left(-\frac{42{,}500}{0.6} \right) \left(e^{-0.6M} - e^0 \right) = \$70{,}833$$

33. $Q = \int_0^{+\infty} \left(10 + 1.56t^2 \right) e^{-0.05t} \, dt = \lim_{M \to +\infty} \int_0^M \left(10 + 1.56t^2 \right) e^{-0.05t} \, dt$

$$= \lim_{M \to +\infty} \left[\left(10 + 1.56t^2 \right)\left(-20e^{-0.05t} \right) - 3.12t\left(400e^{-0.05t} \right) + 3.12\left(-8000e^{-0.05t} \right) \right]_0^M$$

$$= \lim_{M \to +\infty} \left[\left(-25{,}160 - 1{,}248(M) - 31.2(M)^2 \right)e^{-0.05M} - \left(-25{,}160 - 0 - 0 \right)e^{-0.05(0)} \right]$$

$$= \$25{,}160 \text{ billion}$$

35. $\int_0^{+\infty} \dfrac{460e^{t-4}}{1+e^{t-4}} \, dt = \lim_{M \to +\infty} \int_0^M \dfrac{460e^{t-4}}{1+e^{t-4}} \, dt = \lim_{M \to +\infty} 460 \int_{1+e^{-4}}^{1+e^{M-4}} \dfrac{(u-1)\, du}{(u-1)u}$ (where $u = 1 + e^{t-4}$)

$$= \lim_{M \to +\infty} 460 [\ln u]_{1+e^{-4}}^{1+e^{M-4}} = \lim_{M \to +\infty} 460 \left[\ln\left(1 + e^{M-4} \right) - \ln\left(1 + e^{-4} \right) \right]$$

$$= +\infty$$

$\int_{-\infty}^0 \dfrac{460e^{t-4}}{1+e^{t-4}} \, dt = $ (from above with the limits changed) $= \lim_{M \to -\infty} [\ln u]_{1+e^{M-4}}^{1+e^{-4}}$

$$= \lim_{M \to -\infty} 460 \left[\ln\left(1 + e^{-4} \right) - \ln\left(1 + e^{M-4} \right) \right]$$

$$= 460 \left[\ln\left(1 + e^{-4} \right) - \ln\left(1 + 0 \right) \right] \approx 8.3490$$

$\int_0^{+\infty} q(t) \, dt$ diverges, indicating that there is no bound to the expected future exports of pork. $\int_{-\infty}^0 q(t) \, dt$ converges to approximately 8.3490, indicating that total exports of pork prior to 1985 amounted to $\approx \$8.3490$ million.

37. (a) $\displaystyle\int_{0.2}^{+\infty} \frac{1}{5.6997k^{1.081}}\,dk = \lim_{M\to+\infty}\int_{0.2}^{M} \frac{1}{5.6997k^{1.081}}\,dk = \lim_{M\to+\infty}\left[\frac{k^{-0.081}}{-0.081}\right]_{0.2}^{M}$

$$= \lim_{M\to+\infty}\frac{1}{5.6997}\left[\frac{M^{-0.081}}{-0.081} - \frac{(0.2)^{-0.081}}{-0.081}\right]$$

$$= 2.468$$

(b) $\displaystyle\int_{0}^{1} \frac{1}{5.6997k^{1.081}}\,dk = \lim_{r\to 0}\int_{r}^{1} \frac{1}{5.6997k^{1.081}}\,dk = \lim_{r\to 0}\frac{1}{5.6997}\left[\frac{k^{-0.081}}{-0.081}\right]_{r}^{1}$

$$= \lim_{r\to 0}\frac{1}{5.6997}\left[\frac{1^{-0.081}}{-0.081} - \frac{r^{-0.081}}{-0.081}\right] = +\infty$$

The integral diverges. We can interpret this as saying that the number of impacts by meteors smaller than 1 megaton is very large. (This makes sense because, for example, this number includes meteors no larger than a grain of dust.)

39. $\displaystyle\Gamma(x) = \int_{0}^{+\infty} t^{x-1}e^{-t}\,dt$

(a) $\displaystyle\Gamma(1) = \int_{0}^{+\infty} e^{-t}\,dt = \lim_{M\to+\infty}\int_{0}^{M} e^{-t}\,dt = \lim_{M\to+\infty}\left[-e^{-t}\right]_{0}^{M}$

$$= \lim_{M\to+\infty} -\left(e^{-M} - e^{-0}\right) = 1$$

$\displaystyle\Gamma(2) = \int_{0}^{+\infty} te^{-t}\,dt = \lim_{M\to+\infty}\int_{0}^{M} te^{-t}\,dt$

$$= \lim_{M\to+\infty}\left[t(-e^{-t}) - 1(e^{-t})\right]_{0}^{M} = \lim_{M\to+\infty}\left[-\left(Me^{-M} + e^{-M}\right) + \left(0e^{-0} + e^{-0}\right)\right]$$

$$= 1$$

(b) $\displaystyle\Gamma(n+1) = \int_{0}^{+\infty} t^{n+1-1}e^{-t}\,dt = \int_{0}^{+\infty} t^{n}e^{-t}\,dt$

$$= \lim_{M\to+\infty}\left[t^{n}(-e^{-t})\right]_{0}^{M} - \int_{0}^{+\infty} nt^{n-1}(-e^{-t})\,dt$$

$$= \lim_{M\to+\infty}\left[M^{n}(-e^{-M}) - 0^{n}(-e^{-0})\right] + n\int_{0}^{+\infty} t^{n-1}e^{-t}\,dt$$

$$= n\int_{0}^{+\infty} t^{n-1}e^{-t}\,dt$$

$$= n\Gamma(n)$$

39. (c) Deduce $\Gamma(n) = (n-1)!$

$\Gamma(1) = (1-1)! = 1$

True for $n = 1$

Assume $\Gamma(k+1) = (k+1-1)! = k!$

From part (c), $\Gamma(k+1) = k\Gamma(k)$

$$k! = k\Gamma(k)$$

$$\frac{k!}{k} = \Gamma(k)$$

$$(k-1)! = \Gamma(k)$$

Therefore $\Gamma(n) = (n-1)!$

41. $\displaystyle\int_{-\infty}^{+\infty} \frac{1}{\sqrt{2\pi}(4)} e^{-(x-1)^2/32}\, dx = 1$

43. $\displaystyle\int_{1}^{+\infty} \frac{1}{\sqrt{2\pi}} e^{-x^2/2}\, dx = 0.1587$

14.6 EXERCISES

1. $\dfrac{dy}{dx} = x^2 + \sqrt{x}$

$\quad y = \int \left(x^2 + \sqrt{x} \right) dx$

$\quad\quad = \dfrac{x^3}{3} + \dfrac{2x^{3/2}}{3} + C$

3. $\dfrac{dy}{dx} = \dfrac{x}{y}$

$\quad ydy = xdx$

$\quad\quad \int ydy = \int xdx$

$\quad \dfrac{y^2}{2} = \dfrac{x^2}{2} + C$

$\quad y^2 = x^2 + C$

$\quad\quad y = \pm\sqrt{x^2 + C}$

5. $\dfrac{dy}{dx} = xy$

$\quad \dfrac{1}{y} dy = x\, dx$

$\quad \int \dfrac{1}{y} dy = \int x\, dx$

$\quad \ln|y| = \dfrac{x^2}{2} + C$

$\quad |y| = e^{\frac{x^2}{2} + C}$

$\quad y = \pm e^C e^{x^2/2}$

$\quad\quad = \pm A e^{x^2/2}, \text{ where } A = e^C$

7. $\dfrac{dy}{dx} = (x+1)y^2$

$\quad \dfrac{1}{y^2} dy = (x+1)dx$

$\quad \int y^{-2}\, dy = \int (x+1)dx$

$\quad -y^{-1} = \dfrac{(x+1)^2}{2} + C_1$

$\quad y = -\dfrac{2}{(x+1)^2} + C$

464

9. $x\dfrac{dy}{dx} = \dfrac{1}{y}\ln x$

$y\,dy = \dfrac{\ln x}{x}dx$

$\displaystyle\int y\,dy = \int \dfrac{\ln x}{x}dx$

$\dfrac{y^2}{2} = \displaystyle\int u\,du,\ \text{where } u = \ln x$

$\dfrac{y^2}{2} = \dfrac{(\ln x)^2}{2} + C_1$

$y^2 = (\ln x)^2 + C$

$y = \left((\ln x)^2 + C\right)^{1/2}$

11. $\dfrac{dy}{dx} = x^3 - 2x$

$dy = (x^3 - 2x)dx$

$\displaystyle\int dy = \int (x^3 - 2x)dx$

$y = \dfrac{x^4}{4} - x^2 + C$

$y = 1$ when $x = 0$

$1 = \dfrac{0^4}{4} - 0^2 + C$

$C = 1$

$y = \dfrac{x^4}{4} - x^2 + 1$

13. $\dfrac{dy}{dx} = \dfrac{x^2}{y^2}$

$y^2\,dy = x^2\,dx$

$\displaystyle\int y^2\,dy = \int x^2\,dx$

$\dfrac{y^3}{3} = \dfrac{x^3}{3} + C_1$

$y^3 = x^3 + C$

$y = 2$ when $x = 0$

$2^3 = 0^3 + C$

$C = 8$

$y^3 = x^3 + 8$

$y = \left(x^3 + 8\right)^{1/3}$

15. $x\dfrac{dy}{dx} = y$

$\dfrac{dy}{y} = \dfrac{dx}{x}$

$\displaystyle\int \dfrac{1}{y}dy = \int \dfrac{1}{x}dx$

$\ln|y| = \ln|x| + C_1$

$|y| = e^{\ln|x|+C}$

$|y| = e^C|x|$

If $y(1) = 2$, then $2 = e^C(1)$

$|y| = 2|x|$

$y = \pm 2x$

17. $$\frac{dy}{dx} = x(y+1)$$

$$\frac{dy}{y+1} = x\,dx$$

$$\int \frac{1}{y+1}\,dy = \int x\,dx$$

$$\ln|y+1| = \frac{x^2}{2} + C$$

$$|y+1| = e^{\frac{x^2}{2}+C}$$

If $y(0) = 0$, then $1 = e^C$

$$|y+1| = e^C e^{\frac{x^2}{2}}$$

$$y+1 = \pm e^{\frac{x^2}{2}}$$

$$y = \pm e^{\frac{x^2}{2}} - 1$$

19. $$\frac{1}{x}\frac{dy}{dx} = \frac{y^2}{x^2+1}$$

$$\frac{dy}{y^2} = \frac{x\,dx}{x^2+1}$$

$$\int \frac{1}{y^2}\,dy = \int \frac{x}{x^2+1}\,dx$$

$$-\frac{1}{y} = \frac{1}{2}\ln|x^2+1| + C$$

If $y(0) = -1$, then $\frac{-1}{-1} = \frac{1}{2}\ln 1 + C \Rightarrow C = 1$

$$-\frac{1}{y} = \frac{1}{2}\ln(x^2+1) + 1$$

$$y = -\frac{2}{\ln(x^2+1)+2}$$

21. Let $s(t) = $ monthly sales after t months

$$\frac{ds}{dt} = -0.05s; \quad s = 1000 \text{ when } t = 0.$$

$$\int \frac{1}{s}\,ds = \int -0.05\,dt$$

$$\ln s = -0.05t + C$$

$$s = e^{-0.05t+c}$$

$$s = e^C e^{-0.05t}$$

If $s = 1000$ when $t = 0$

$$1000 = e^C$$

$$s = 1000e^{-0.05t}$$

23. Let $S(t) = $ total sales after t months

$$\frac{dS}{dt} = 0.1(100{,}000 - S); \quad S(0) = 0$$

$$\int \frac{1}{S-100{,}000}\,dS = \int -0.1\,dt$$

$$\ln|S-100{,}000| = -0.1t + C$$

$$|S-100{,}000| = e^{-0.1t+C}$$

$$100{,}000 - S = e^C e^{-0.1t}$$

If $S(0) = 0$, then $100{,}000 = e^C$

$$100{,}000 - S = 100{,}000e^{-0.1t}$$

$$S = 100{,}000(1 - e^{-0.1t})$$

25. $\dfrac{dy}{dt} = ay(L - y), \quad y = \dfrac{CL}{e^{-aLt} + C}$

$$\frac{d}{dt}\left(\frac{CL}{e^{-aLt} + C}\right) = a\left(\frac{CL}{e^{-aLt} + C}\right)\left(L - \frac{CL}{e^{-aLt} + C}\right)$$

$$\frac{0 - CL\left(-aLe^{-aLt}\right)}{\left(e^{-aLt} + C\right)^2} = \frac{acL^2}{e^{-aLt} + C} - \frac{ac^2 L^2}{\left(e^{-aLt} + C\right)^2}$$

$$\frac{aCL^2 e^{-aLt}}{\left(e^{-aLt} + C\right)^2} = \frac{aCL^2 e^{-aLt}}{\left(e^{-aLt} + C\right)^2}$$

27. $\dfrac{dS}{dt} = \dfrac{1}{4}S(2 - S)$

From Exercise 25

$$S = \frac{C(2)}{e^{-\left(\frac{1}{4}\right)2t} + C}$$

$$S = \frac{2C}{e^{-t/2} + C}$$

If $S = 0.001$ when $t = 0$

$$0.001 = \frac{2C}{1 + C} \Rightarrow C = \frac{1}{1999}$$

$$S = \frac{2/1999}{e^{-0.5t} + 1/1999}$$

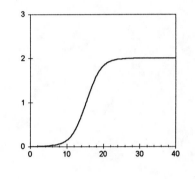

It will take about 27 months to saturate the market.

29. (a) $\dfrac{dy}{dt} = -ay \ln\left(\dfrac{y}{b}\right)$

$$\frac{dy}{y \ln\left(\frac{y}{b}\right)} = -a\,dt$$

$$\int \frac{1}{y \ln\left(\frac{y}{b}\right)}\,dy = -\int a\,dt$$

$$\int \frac{1}{yu}(y\,du) = -\int a\,dt \quad \text{where} \quad u = \ln\left(\frac{y}{b}\right)$$

$$\ln|u| = -at + C_1$$

$$|u| = e^{-at + C_1}$$

$$= e^{C_1} e^{-at}$$

$$u = Ae^{-at}$$

$$\ln\left(\frac{y}{b}\right) = Ae^{-at}$$

$$\frac{y}{b} = e^{Ae^{-at}}$$

$$y = be^{Ae^{-at}}$$

467

29. (b) If $a = 1$, $b = 10$ and $y(0) = 5$

$$5 = 10e^{Ae^{-1(0)}}$$
$$0.5 = e^{A}$$
$$\ln 0.5 = A$$
$$A = -0.69315$$
$$y = 10e^{-0.69315e^{-t}}$$

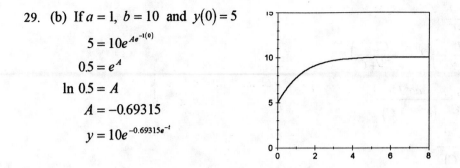

CHAPTER 14   REVIEW EXERCISES

1. $\displaystyle\int_0^1 (x+1)e^x\,dx$ $\begin{array}{c|cc} & D & I \\ & x+1 & e^x \\ & -1 & e^x \\ & 0 & e^x \end{array}$

$= \left[(x+1)e^x - e^x\right]_0^1$

$= \left[xe^x\right]_0^1 = 1e^1 - 0e^0 = e$

3. $\displaystyle\int x^2 \ln 2x\,dx$ $\begin{array}{c|cc} & D & I \\ & \ln 2x & x^2 \\ & -1/x & x^3/3 \end{array}$

$= \dfrac{x^3}{3}\ln 2x - \displaystyle\int \dfrac{x^2}{3}\,dx + C$

$= \dfrac{x^3}{3}\ln 2x - \dfrac{x^3}{9} + C$

5. $\displaystyle\int_0^1 x^2 e^x\,dx$ $\begin{array}{c|cc} & D & I \\ & x^2 & e^x \\ & -2x & e^x \\ & 2 & e^x \\ & 0 & e^x \end{array}$

$= \left[x^2 e^x - 2xe^x + 2e^x\right]_0^1$

$= \left[(x^2 - 2x + 2)e^x\right]_0^1$

$= \left[(1^2 - 2\cdot 1 + 2)e^1\right] - \left[(0 - 0 + 2)e^0\right]$

$= e - 2$

7. $\displaystyle\int_1^e x^2 \ln x\,dx$ $\begin{array}{c|cc} & D & I \\ & \ln x & x^2 \\ & -1/x & x^3/3 \end{array}$

$= \left[\dfrac{x^3}{3}\ln x\right]_1^e - \displaystyle\int_1^e \dfrac{1}{x}\left(\dfrac{x^3}{3}\right)dx$

$= \left[\dfrac{x^3}{3}\ln x\right]_1^e - \left[\dfrac{x^3}{9}\right]_1^e$

$= \left(\dfrac{e^3 \ln e}{3} - \dfrac{e^3}{9}\right) - \left(\dfrac{1^3 \ln 1}{3} - \dfrac{1^3}{9}\right)$

$= \dfrac{2e^3 + 1}{9}$

In Exercises 9 - 29 use the integral tables in Appendix D.

9. $\displaystyle\int \dfrac{dx}{9 + 4x^2} = \dfrac{1}{4}\int \dfrac{dx}{\frac{9}{4} + x^2} = \left(\dfrac{1}{4}\right)\left(\dfrac{2}{3}\right)\tan^{-1}\dfrac{x}{3/2} + C = \dfrac{1}{6}\tan^{-1}\dfrac{2x}{3} + C$

11. $\displaystyle\int \dfrac{dx}{\sqrt{9 + 4x^2}} = \dfrac{1}{2}\int \dfrac{d(2x)}{\sqrt{3^2 + (2x)^2}} = \dfrac{1}{2}\ln\left|2x + \sqrt{9 + 4x^2}\right| + C$

469

13. $\displaystyle\int \sqrt{x^2 + 2x + 2}\ dx = \int \sqrt{x^2 + 2x + \left(\frac{2}{2}\right)^2 - \left(\frac{2}{2}\right)^2 + 2}\ dx = \int \sqrt{(x+1)^2 + 1^2}\ d(x+1)$

$\displaystyle = \frac{x+1}{2}\sqrt{(x+1)^2 + 1} + \frac{1}{2}\ln\left|x+1 + \sqrt{(x+1)^2}\right| + C$

$\displaystyle = \frac{x+1}{2}\sqrt{x^2 + 2x + 2} + \frac{1}{2}\ln\left|x+1 + \sqrt{x^2 + 2x + 2}\right| + C$

15. $\displaystyle\int \frac{e^{2x}}{1+e^{4x}}\ dx = \frac{1}{2}\int \frac{d(e^{2x})}{1^2 + \left(e^{2x}\right)^2} = \frac{1}{2}\left(\frac{1}{1}\right)\tan^{-1}\left(\frac{e^{2x}}{1}\right) + C = \frac{1}{2}\tan^{-1}\left(e^{2x}\right) + C$

17. $\displaystyle\int_0^1 \frac{e^{2x}}{\sqrt{9+4e^{4x}}}\ dx = \frac{1}{4}\int_0^1 \frac{d(2e^{2x})}{\sqrt{3^2 + \left(2e^{2x}\right)^2}} = \frac{1}{4}\left[\ln\left|2e^{2x} + \sqrt{4e^{4x}+9}\right|\right]_0^1$

$\displaystyle = \frac{1}{4}\left(\ln\left|2e^{2(1)} + \sqrt{4e^{4(1)+9}}\right|\right) - \frac{1}{4}\left(\ln\left|2e^0 + \sqrt{4e^0 + 9}\right|\right)$

$= 0.41817$

19. $\displaystyle\int_1^e \frac{\sqrt{(\ln x)^2 + 1}}{x}\ dx = \int_1^e \sqrt{(\ln x)^2 + 1^2}\ d(\ln x)$

$\displaystyle = \left[\frac{\ln x}{2}\sqrt{(\ln x)^2 + 1^2} + \frac{1^2}{2}\ln\left|\ln x + \sqrt{(\ln x)^2 + 1^2}\right|\right]_1^e$

$\displaystyle = \left(\frac{\ln e}{2}\sqrt{(\ln e)^2 + 1} + \frac{1}{2}\ln\left|\ln e + \sqrt{(\ln e)^2 + 1}\right|\right)$

$\displaystyle \quad - \left(\frac{\ln 1}{2}\sqrt{(\ln 1)^2 + 1} + \frac{1}{2}\ln\left|\ln 1 + \sqrt{(\ln 1)^2 + 1}\right|\right)$

$\displaystyle = \left(\frac{1}{2}\sqrt{2} + \frac{1}{2}\ln\left|1 + \sqrt{2}\right|\right) - \left(\frac{0}{2}\sqrt{0+1} + \frac{1}{2}\ln\left|0 + \sqrt{0+1}\right|\right)$

$= 1.14779$

21. $\displaystyle\int_1^\infty \frac{1}{x^5}\ dx = \lim_{M\to+\infty}\int_1^M x^{-5}\ dx = \lim_{M\to+\infty}\left[-\frac{1}{4x^4}\right]_1^M$

$\displaystyle = \lim_{M\to+\infty}\left[\left(-\frac{1}{4M^4}\right) - \left(-\frac{1}{4(1)^4}\right)\right] = \frac{1}{4}$

470

23. $\displaystyle\int_{-\infty}^{1} e^{x/2}\,dx = \lim_{M\to+\infty}\int_M^1 e^{x/2}\,dx = \lim_{M\to+\infty}\left[2e^{x/2}\right]_M^1 = \lim_{M\to+\infty}\left[2e^{1/2}-2e^{M/2}\right]=2\sqrt{e}$

25. $\displaystyle\int_{-\infty}^{\infty}\frac{1}{x^2}\,dx = \int_{-\infty}^{-1}\frac{1}{x^2}\,dx+\int_{-1}^{0}\frac{1}{x^2}\,dx+\int_{0}^{1}\frac{1}{x^2}\,dx+\int_{1}^{\infty}\frac{1}{x^2}\,dx$

$\displaystyle = \lim_{M\to-\infty}\int_M^{-1}x^{-2}\,dx+\lim_{r\to0}\int_{-1}^{r}x^{-2}\,dx+\lim_{r\to0}\int_{r}^{1}x^{-2}\,dx+\lim_{M\to+\infty}\int_1^{M}x^{-2}\,dx$

$\displaystyle = \lim_{M\to-\infty}\left[-\frac{1}{x}\right]_M^{-1}+\lim_{r\to0}\left[-\frac{1}{x}\right]_{-1}^{r}+\lim_{r\to0}\left[-\frac{1}{x}\right]_r^{1}+\lim_{M\to+\infty}\left[-\frac{1}{x}\right]_1^{M}$

$\displaystyle = \lim_{M\to-\infty}\left[\left(-\frac{1}{-1}\right)-\left(-\frac{1}{M}\right)\right]+\lim_{r\to0}\left[\left(-\frac{1}{r}\right)-\left(-\frac{1}{-1}\right)\right]+\lim_{r\to0}\left[\left(-\frac{1}{1}\right)-\left(-\frac{1}{r}\right)\right]+\lim_{M\to+\infty}\left[\left(-\frac{1}{M}\right)-\left(-\frac{1}{1}\right)\right]$

The integral diverges

27. $\displaystyle\int_0^1\frac{1}{(x-1)^2}\,dx = \lim_{r\to1}\int_0^r\frac{1}{(x-1)^2}\,dx = \lim_{r\to1}\left[-\frac{1}{x-1}\right]_0^r = \lim_{r\to1}\left[\left(-\frac{1}{r-1}\right)-\left(-\frac{1}{0-1}\right)\right]$

The integral diverges

29. $\displaystyle\int_0^1\frac{1}{\sqrt{1-x}}\,dx = \lim_{r\to1}\int_0^r(1-x)^{-1/2}\,dx = \lim_{r\to1}\left[-2(1-x)^{1/2}\right]_0^r$

$\displaystyle = \lim_{r\to1}\left[\left(-2(1-r)^{1/2}\right)-\left(-2(1-0)^{1/2}\right)\right]=2$

31. $\displaystyle A = \int_0^1\left((1-x^3)-x^3\right)dx = \int_0^1(1-2x^3)\,dx = \left[x-\frac{x^4}{2}\right]_0^1 = \left(1-\frac{1}{2}\,^4\right)-(0-0)=\frac{1}{2}$

33. The area is the sum of two pieces because the graphs of e^x and $2e^{-x}$

cross where $e^x = 2e^{-x}$ or $x = \dfrac{\ln 2}{2} = \ln\sqrt{2}$.

$\displaystyle A = \int_0^{\ln\sqrt{2}}\left(2e^{-x}-e^x\right)dx + \int_{\ln\sqrt{2}}^{2}\left(e^{-x}-2e^{-x}\right)dx = \left[-2e^{-x}-e^x\right]_0^{\ln\sqrt{2}} + \left[e^x+2e^{-x}\right]_{\ln\sqrt{2}}^{2}$

$\displaystyle = \left[\left(-2e^{-\ln\sqrt{2}}-e^{\ln\sqrt{2}}\right)-\left(-2e^{-0}-e^0\right)\right]+\left[\left(e^2+2e^{-2}\right)-\left(e^{\ln\sqrt{2}}+2e^{-\ln\sqrt{2}}\right)\right]$

$\displaystyle = -\frac{2}{\sqrt{2}}-\sqrt{2}+3+e^2+\frac{2}{e^2}-\sqrt{2}-\frac{2}{\sqrt{2}}=3-4\sqrt{2}+e^2+\frac{2}{e^2}$

35. The graphs intersect where $x^2 = 1 - x^2 \Rightarrow x^2 = \dfrac{1}{2} \Rightarrow x = \pm \dfrac{\sqrt{2}}{2}$.

$$A = \int_{-\sqrt{2}/2}^{\sqrt{2}/2} \left((1 - x^2) - x^2 \right) dx = \int_{-\sqrt{2}/2}^{\sqrt{2}/2} (1 - 2x^2)\, dx = \left[x - \frac{2x^3}{3} \right]_{-\sqrt{2}/2}^{\sqrt{2}/2} = \left(\frac{\sqrt{2}}{2} - \frac{2\left(\frac{\sqrt{2}}{2}\right)^3}{3} \right) - \left(\frac{-\sqrt{2}}{2} - \frac{2\left(-\frac{\sqrt{2}}{2}\right)^3}{3} \right) = \frac{2\sqrt{2}}{3}$$

37. The graphs intersect where $x^3 - x = x - x^3 \Rightarrow 2x^3 - 2x = 0 \Rightarrow x(x^2 - 1) = 0 \Rightarrow x = \pm 1,\ 0$.
The area is the sum of two pieces.

$$A = \int_{-1}^{0} \left((x - x^3) - (x^3 - x) \right) dx + \int_{0}^{1} \left((x^3 - x) - (x - x^3) \right) dx = \int_{-1}^{0} (2x - 2x^3)\, dx + \int_{0}^{1} (2x^3 - 2x)\, dx$$

$$= \left[x - \frac{x^4}{2} \right]_{-1}^{0} + \left[\frac{x^4}{2} - x \right]_{0}^{1} = \left[(0 - 0) - \left(-1 - \frac{(-1)^4}{2} \right) \right] + \left[\left(\frac{1^4}{2} - 1 \right) - (0 - 0) \right] = 1$$

39. $A = \displaystyle\int_{0}^{1/2} \left(\frac{1}{x^2 + 1} - \frac{1}{x^2 - 1} \right) dx = \int_{0}^{1/2} \frac{1}{x^2 + 1}\, dx - \int_{0}^{1/2} \frac{1}{x^2 - 1}\, dx = \int_{0}^{1/2} \frac{1}{x^2 + 1^2}\, dx + \int_{0}^{1/2} \frac{1}{1^2 - x^2}\, dx$

$$= \left[\frac{1}{1} \tan^{-1} \frac{x}{1} \right]_{0}^{1/2} + \left[\frac{1}{2(1)} \ln \left| \frac{1 + x}{1 - x} \right| \right]_{0}^{1/2}$$

$$= \left[\tan^{-1}(1/2) - \tan^{-1}(0) \right] + \left[\left(\frac{1}{2} \ln \left| \frac{1 + 1/2}{1 - 1/2} \right| \right) - \left(\frac{1}{2} \ln \left| \frac{1 + 0}{1 - 0} \right| \right) \right] = 1.01295$$

In Exercises 41-43 use the results of Exercises 5-7.

41. $\overline{f} = \dfrac{1}{1 - 0} \displaystyle\int_{0}^{1} x^2 e^x\, dx = e - 2 = 0.71828$

43. $\overline{f} = \dfrac{1}{e - 1} \displaystyle\int_{1}^{e} x^2 \ln x\, dx = \dfrac{2e^3 + 1}{(e - 1)(9)} = 2.66229$

In Exercises 45-47 use the results of Exercises 17-19.

45. $\overline{f} = \dfrac{1}{1 - 0} \displaystyle\int \dfrac{e^{2x}}{\sqrt{9 + 4e^{4x}}}\, dx = 0.418171$

47. $\overline{f} = \dfrac{1}{e - 1} \displaystyle\int \dfrac{\sqrt{(\ln x)^2 + 1}}{x}\, dx = \dfrac{1.14779}{e - 1} = 0.66799$

49. $\overline{f}(x) = \dfrac{1}{2}\displaystyle\int_{x-2}^{x} t^{2/3}\,dt = \dfrac{1}{2}\left[\dfrac{t^{5/3}}{5/3}\right]_{x-2}^{x}$

$\phantom{\overline{f}(x)} = \dfrac{3}{10}\left[x^{5/3} - (x-2)^{5/3}\right]$

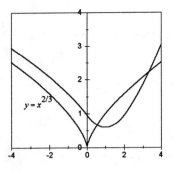

51. $\overline{f}(x) = \dfrac{1}{2}\displaystyle\int_{x-2}^{x}\ln t\,dt = \text{From Exercise 19, Section 7.3} = \dfrac{1}{2}\left[t\ln t - t\right]_{x-2}^{x}$

$\phantom{\overline{f}(x)} = \dfrac{1}{2}(x\ln x - x) - \dfrac{1}{2}\big((x-2)\ln(x-2) - (x-2)\big)$

$\phantom{\overline{f}(x)} = \dfrac{1}{2}\big[x\ln x - (x-2)\ln(x-2) - 2\big]$

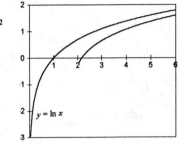

53. $\dfrac{dy}{dx} = x^2 y^2$

$\dfrac{dy}{y^2} = x^2\,dx$

$\displaystyle\int y^{-2}\,dy = \int x^2\,dx$

$-\dfrac{1}{y} = \dfrac{x^3}{3} + C_1$

$y = -\dfrac{3}{x^3} + C$

55. $\dfrac{dy}{dx} = xy + x + y + 1 = (x+1)(y+1)$

$\dfrac{dy}{y+1} = (x+1)\,dx$

$\displaystyle\int\dfrac{1}{y+1} = \int(x+1)\,dx$

$\ln|y+1| = \dfrac{x^2}{2} + x + C$

$|y+1| = e^{\frac{x^2}{2}+x+C}$

$ = e^{C}e^{\frac{1}{2}(x^2+2x)}$

$ = e^{\frac{1}{2}(x^2+2x+1)-\frac{1}{2}}$

$y = Ae^{(x+1)2/2} - 1$

57. $\dfrac{dy}{dx} = \dfrac{1}{y}$

$y\,dy = dx$

$\displaystyle\int y\,dy = \int dx$

$\dfrac{y^2}{2} = x + C_1$

$y = \sqrt{2x + C}$

59. $xy\dfrac{dy}{dx} = 1$

$y\,dy = \dfrac{dx}{x}$

$\displaystyle\int y\,dy = \int \dfrac{dx}{x}$

$\dfrac{y^2}{2} = \ln|x| + C_1$

$y = \sqrt{2\ln|x| + C}$

If $y(1) = 1$, then $1 = \sqrt{2\ln 1 + C} \Rightarrow C = 1$

$y = \sqrt{2\ln|x| + 1}$

61. $y(x^2 + 1)\dfrac{dy}{dx} = xy^2$

$\dfrac{dy}{y} = \dfrac{x}{x^2 + 1}dx$

$\displaystyle\int \dfrac{dy}{y} = \int \dfrac{x}{x^2 + 1}dx$

$\ln|y| = \dfrac{1}{2}\ln|x^2 + 1| + C_1$

$2\ln|y| = \ln C|x^2 + 1|$

$|y|^2 = C|x^2 + 1|$

$y = \sqrt{C(x^2 + 1)}$

If $y(0) = 2$, then $2 = \sqrt{C(0 + 1)} \Rightarrow C$

$y = 2\sqrt{x^2 + 1}$

63. (a) Approximating a linear function with $x = 0$ in 1973, $(0, 56)$ and $(20, 44)$ we get the area to be $\dfrac{1}{2}(56 + 44)(20) = 1000$.

There were approximately 1,000,000 deaths resulting from motor vehicles in the period 1973 - 1992.

(b) Approximating a linear function with $x = 0$ in 1973 $(0, 32)$ and $(20, 40)$ we get the area of the lower graph to be

$\dfrac{1}{2}(32 + 40)(20) = 720$. The area between the graphs is $1000 - 720 = 280$.

There were approximately 280,000 more deaths resulting from motor vehicles than guns in the period 1973 - 1992.

65. $C = \int_0^{12} pq\,dt = \int_0^{12} \left(2e^{0.03t}\right)(100t + 1000)\,dt = 200\int_0^{12} e^{0.03t}(t + 10)\,dt$

$= \left[200(t + 10)\left(\frac{100e^{0.03t}}{3}\right) - 200\left(\frac{10,000e^{0.03t}}{9}\right)\right]_0^{12}$

$= \frac{20,000}{3}\left[e^{0.03t}\left(t + 10 - \frac{100}{3}\right)\right]_0^{12} = \frac{20,000}{3}\left[\left(12 + 10 - \frac{100}{3}\right)e^{0.03(12)} - \left(0 + 10 - \frac{100}{3}\right)e^0\right]$

$= \$47,259.56$

In Exercises 67 $CS = \int_0^{\bar{q}}\left(f(q) - \bar{p}\right)dq$ where p is the demand curve and

$PS = \int_0^{\bar{q}}\left(\bar{p} - f(q)\right)dq$ where p is the supply curve and for both $p = f(q)$ and $f(\bar{q}) = \bar{p}$.

67. Supply: $p = (q - 20)^{0.5} \Rightarrow p^2 = q - 20$ or $q = p^2 + 20$, $q \geq 20$

Demand: $p = (70 - q)^{0.5} \Rightarrow p^2 = 70 - q$ or $q = 70 - p^2$, $q \leq 70$

Equilibrium: $\bar{p}^2 + 20 = 70 - \bar{p}^2 \Rightarrow 2\bar{p}^2 = 50 \Rightarrow \bar{p} = \$5$, $\bar{q} = 70 - \bar{p}^2 = 70 - 25 = 45$

$CS = \int_0^{45}\left((70 - q)^{0.5} - 5\right)dq = \left[\frac{-(70 - q)^{1.5}}{1.5} - 5q\right]_0^{45}$

$= \left(\frac{-(70 - 45)^{1.5}}{1.5} - 5(45)\right) - \left(\frac{-(70 - 0)^{1.5}}{1.5} - 5(0)\right) = \$82.11$

$PS = \int_{20}^{45}\left(5 - (q - 20)^{0.5}\right)dq = \left[5q - \frac{(q - 20)^{1.5}}{1.5}\right]_{20}^{45}$

$= \left(5(45) - \frac{(45 - 20)^{1.5}}{1.5}\right) - \left(5(20) - \frac{(20 - 20)^{1.5}}{1.5}\right) = \$41.67$

69. (a) $\int_0^{10}(320 - 32t)\,dt = \left[320t - \frac{32t^2}{2}\right]_0^{10} = \left(320(10) - 16(10)^2\right) - (0 - 0) = 1600$

This is the total distance the cannonball has traveled.

(b) When the ball returns to the ground $V(T) = -320$

$-320 = 320 - 32T$

$T = 20$ sec.

The area will be the total distance traveled, 1600 ft up $+$ 1600 ft down $= 3200$ ft.

71. (a) Average $= \dfrac{300+310+340+285+320+350+375+365+300+275}{10} = 322$ pedestrians per year

(b)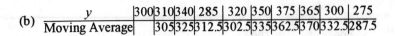

y	300	310	340	285	320	350	375	365	300	275
Moving Average		305	325	312.5	302.5	335	362.5	370	332.5	287.5

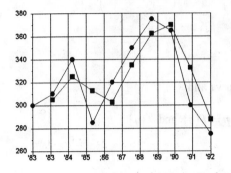

73. $\overline{f} = \dfrac{1}{1-0}\int_0^1 10{,}000e^{-0.06t}\,dt = \left[10{,}000\left(-\dfrac{1}{0.06}\right)e^{-0.06t}\right]_0^1$

$= -\dfrac{10{,}000}{0.06}\left[e^{-0.06t}\right]_0^1 = -\dfrac{10{,}000}{0.06}\left[e^{-0.06} - e^0\right] = \$9{,}705.91$

75. $\dfrac{dy}{dt} = 0.1\sqrt{y}$

$\dfrac{dy}{y^{1/2}} = 0.1\,dt$

$\int y^{-1/2}\,dy = \int 0.1\,dt$

$2y^{1/2} = 0.1t + C_1$

$y = (0.05t + C)^2$

If $y(0) = 10{,}000$, then

$10{,}000 = (0 + C)^2$

$C = 100$

$y = (0.05t + 100)^2$

CHAPTER 15
FUNCTIONS OF SEVERAL VARIABLES

15.1 EXERCISES

1. $f(x, y) = x^2 + y^2 - x + 1$

   (a) $f(0, 0) = 0^2 + 0^2 - 0 + 1 = 1$

   (b) $f(1, 0) = 1^2 + 0^2 - 1 + 1 = 1$

   (c) $f(0, -1) = 0^2 + (-1)^2 - 0 + 1 = 2$

   (d) $f(a, 2) = a^2 + 2^2 - a + 1 = a^2 - a + 5$

   (e) $f(y, x) = y^2 + x^2 - y + 1$

   (f) $f(x + h, y + k) = (x + h)^2 + (y + k)^2 - (x + h) + 1$

3. $f(x, y) = \sqrt{(x - 1)^2 + (y - 2)^2}$

   (a) $f(0, 0) = \sqrt{(0 - 1)^2 + (0 - 2)^2} = \sqrt{5}$

   (b) $f(1, 0) = \sqrt{(1 - 1)^2 + (0 - 2)^2} = 2$

   (c) $f(0, -1) = \sqrt{(0 - 1)^2 + (-1 - 2)^2} = \sqrt{10}$

   (d) $f(a, 2) = \sqrt{(a - 1)^2 + (2 - 2)^2} = |a - 1|$

   (e) $f(y, x) = \sqrt{(y - 1)^2 + (x - 2)^2}$

   (f) $f(x + h, y + k)$
   $= \sqrt{(x + h - 1)^2 + (y + k - 2)^2}$

5. $g(x, y, z) = e^{x + y + z}$

   (a) $g(0, 0, 0) = e^{0 + 0 + 0} = 1$

   (b) $g(1, 0, 0) = e^{1 + 0 + 0} = e$

   (c) $g(0, 1, 0) = e^{0 + 1 + 0} = e$

   (d) $g(z, x, y) = e^{z + x + y}$

   (e) $g(x + h, y + k, z + l) = e^{x + h + y + k + z + l}$

7. $g(x, y, z) = \dfrac{xyz}{x^2 + y^2 + z^2}$

   (a) $g(0, 0, 0) = \dfrac{(0)(0)(0)}{0^2 + 0^2 + 0^2};$

   does not exist

   (b) $g(1, 0, 0) = \dfrac{1(0)(0)}{1^2 + 0^2 + 0^2} = 0$

   (c) $g(0, 1, 0) = \dfrac{(0)(1)(0)}{0^2 + 1^2 + 0^2} = 0$

   (d) $g(z, x, y) = \dfrac{zxy}{z^2 + x^2 + y^2} = \dfrac{xyz}{x^2 + y^2 + z^2}$

   (e) $g(x + h, y + k, z + l)$
   $= \dfrac{(x + h)(y + k)(z + l)}{(x + h)^2 + (y + k)^2 + (z + l)^2}$

In Exercises 9-13 use $d = \sqrt{(x_2 - x_1)^2 + (y_2 - y_1)^2}$ for the distance between the two points (x_1, y_1) and (x_2, y_2).

9. $(1, -1)$ and $(2, -2)$

$d = \sqrt{(2-1)^2 + [-2-(-1)]^2} = \sqrt{2}$

11. $(a, 0)$ and $(0, b)$

$d = \sqrt{(0-a)^2 + (b-0)^2} = \sqrt{a^2 + b^2}$

13. For $(1, k)$ and $(0, 0)$

$d = \sqrt{(0-1)^2 + (0-k)^2} = \sqrt{1+k^2}$

For $(1, k)$ and $(2, 1)$

$d = \sqrt{(2-1)^2 + (1-k)^2} = \sqrt{1+(1-k)^2}$

If they are equal distances

$\sqrt{1+k^2} = \sqrt{1+(1-k)^2}$

$1+k^2 = 1+(1-k)^2$

$1+k^2 = 1+1-2k+k^2$

$2k = 1$

$k = \dfrac{1}{2}$

15. $(x-2)^2 + (y+1)^2 = 9$

The set of points (x, y) is a circle with center $(2, -1)$ and radius 3.

17. $x^2 + 6x + y^2 + 4y + 7 = 0$

$x^2 + 6x + (3)^2 + y^2 + 4y + (2)^2 = -7 + 3^2 + 2^2$

$(x+3)^2 + (y+2)^2 = 6$

The set of points (x, y) is a circle with center $(-3, -2)$ and radius $\sqrt{6}$.

In Exercises 19-21 use the Cobb-Douglas production function $P(x, y) = Kx^a y^{1-a}$.

19. $P(100, 500{,}000) = 1{,}000(100)^{0.5}(500{,}000)^{1-0.5} = 7{,}071{,}068$ to the nearest item

21. (a) Two years ago:

$$100 = K(1{,}000)^a (1{,}000{,}000)^{1-a}$$

Last year

$$10 = K(1{,}000)^a (10{,}000)^{1-a}$$

(b) Two years ago:

$$100 = 10^6 K(10^3)^a (10^6)^{-a}$$

$$K = 10^{3a-4}$$

$$\log K = 3a - 4 \qquad (1)$$

Last year:

$$10 = 10^4 K(10^3)^a (10^4)^{-a}$$

$$K = 10^{a-3}$$

$$\log K = a - 3 \qquad (2)$$

(c) Right hand side (Rhs) of (1) = Rhs (2)

$$3a - 4 = a - 3$$

$$a = 0.5$$

So $K = 10^{0.5-3} = 0.003162$

(d) $p = 0.003162(500)^{0.5}(1{,}000{,}000)^{1-0.5}$

    $= 71$ pianos (to the nearest piano)

23. $s(i, t) = ai + bt + c$

$$4 = a(5.05) + b(0) + c \qquad (1)$$
$$4.05 = a(5.05) + b(1) + c \qquad (2)$$
$$4.15 = a(5.1) + b(2) + c \qquad (3)$$

Add -2 times (2) to (3) to eliminate b

$$-8.1 = -10.1a - 2b - 2c$$
$$\underline{4.15 = 5.1a + 2b + c}$$
$$-3.95 = -5a - c \qquad (4)$$

Add (1) to (4) to eliminate c

$$4 = 5.05a + c$$
$$\underline{-3.95 = -5a - c}$$
$$.05 = .05a$$
$$a = 1$$

Substitute 1 for a in (1)

$$4 = 1(5.05) + c$$
$$c = -1.05$$

Substitute 1 for a and -1.05 for c in (3)

$$4.15 = 1(5.1) + 2b - 1.05$$
$$b = 0.05$$
$$s(i, t) = i + 0.05t - 1.05$$

25. $Q(y, p, r) = Ky^{-0.023}p^{-1.040}r^{0.939}$

(a) The demand function increases with increasing values of r.

(b) $Q(2 \times 10^8, 0.5, 500)$

$$= 200(2 \times 10^8)^{-0.023}(0.5)^{-1.040}(500)^{0.939} = 90,680$$

This means that, if the total real income in Great Britain is 2×10^8 units of currency, and the average retail price of beer is 0.5 units of currency per unit of beer, and if the average retail price of all other commodities is 500 units of currency, then 90,680 units of beer will be sold per year.

27. $U(x, y) = 6x^{0.8}y^{0.2} + x$

$U(10, 10) = 6(10)^{0.8}(10)^{0.2} + 10 = 70$

$U(11, 10) = 6(11)^{0.8}(10)^{0.2} + 11 \approx 75.75$

$U(11, 10) - U(10, 10) \approx 5.75$

This means that, if your company now has 10 copies of Macro Publish and 10 copies of Turbo Publish, then the purchase of one additional copy of Macro Publish will result in a productivity increase of approximately 5.75 pages per day.

29. $V(a, b, c) = \frac{4}{3}\pi abc$

(a) If $V(a, b, c) = 1$, then $abc = \frac{3}{4\pi}$

$(a, b, c) = (3, 1/4, 1/\pi);\ (a, b, c) = (1/\pi, 3, 1/4)$

(b) $V(a, a, a) = \frac{4}{3}\pi a^3$

$$1 = \frac{4}{3}\pi a^3$$

$$a^3 = \frac{3}{4\pi}$$

$$a = \left(\frac{3}{4\pi}\right)^{1/3}$$

31.

x	y	$f(x, y) = x^2\sqrt{1 + xy}$
-1	-1	$\sqrt{2}$
1	12	$\sqrt{13}$
0.3	0.5	0.096514
41	42	$1681\sqrt{1723}$

33.

x	y	$f(x, y) = x\ln(x^2 + y^2)$
3	1	$3\ln 10$
1.4	-1	1.5193
e	0	$2e$
0	e	0

35. $f(x, y) = y^2 - x^2$

(a) $0 = y^2 - x^2$

$1 = y^2 - x^2$

$2 = y^2 - x^2$

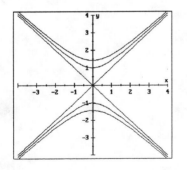

481

(b) $3 = y^2 - x^2$

(c) $1 = x^2 - y^2$; $2 = x^2 - y^2$

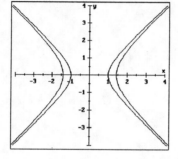

37. $B(x, n) = \dfrac{xn}{A}$

(a) $B(1.5 \times 10^{14},\ 0.014) = \dfrac{(1.5 \times 10^{14})(0.014)}{5.1 \times 10^{14}}$
$$= 4.118 \times 10^{-3} \ g/m^2$$

(b) The total weight of sulfates in the earth's atmosphere.

39. $N(R,\ f_p,\ n_e,\ f_l,\ f_i,\ f_c,\ L) = Rf_p n_e f_l f_i f_c L$

(a) The value of N would be doubled.

(b) $N(R,\ f_p,\ n_e,\ f_{l,\ f_i},\ L) = Rf_p n_e f_l f_i L$, where here L is the average lifetime of an intelligent civilization.

(c) The function is not linear, since it involves a product of variables.

482

41. $c(x, y, z) = 72.3 - 0.8x - 0.2y - 0.7z$

    (a) $38.8 = 72.3 - 0.8x - 0.2(20.1) - 0.7(32.9)$

        $x \approx 8.06$

(a) The model predicts 8.06% (The actual figure was 8.2%, showing the accuracy of the model).

(b) Foreign manufacturers, since each 1% gain of the market by foreign manufacturers decreases Chrysler's share by 0.8%, the largest of the three.

EXERCISES 15.2

1. Vertices: $(0, 0, 0)$, $(1, 0, 0)$, $(0, 1, 0)$, $(0, 0, 1)$, $(1, 1, 0)$, $(1, 0, 1)$, $(0, 1, 1)$, $(1, 1, 1)$

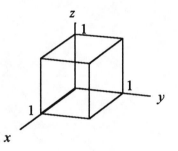

3. Vertices: $(1, 1, 0)$, $(1, -1, 0)$, $(-1, 1, 0)$, $(-1, -1, 0)$, $(0, 0, 2)$

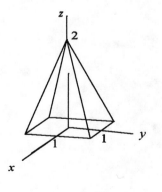

5. $z = -2$

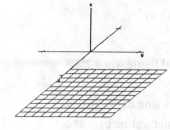

7. $y = 2$

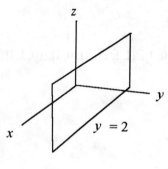

9. $x = -3$

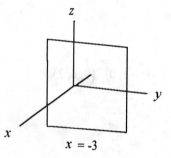

11. H   13. A   15. F   17. C

19. $f(x,y) = 1 - x - y$

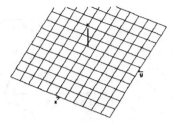

21. $g(x,y) = 2x + y - 2$

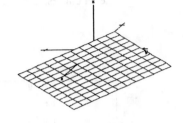

23. $h(x,y) = x + 2$

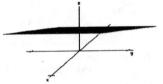

25. $r(x,y) = x + y$

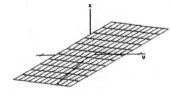

27. $s(x,y) = 2x^2 + 2y^2$

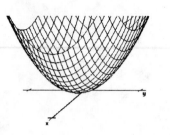

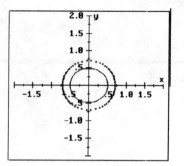

Level curves: $z = 1, z = 2$

29. $t(x,y) = x^2 + 2y^2$

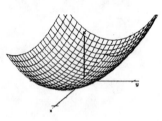

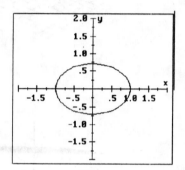

Level curve: $z = 1$

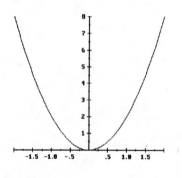

Cross section: $x = 0$

486

31. $f(x,y) = 2 + \sqrt{x^2 + y^2}$

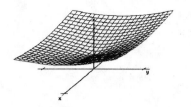

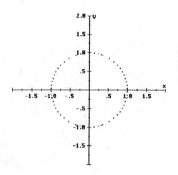

Level curve: $z = 3$

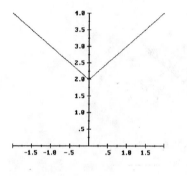

Cross section: $y = 0$

33. $f(x,y) = -2\sqrt{x^2 + y^2}$

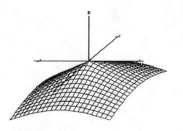

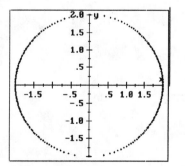

Level curve: $z = -4$

487

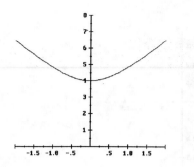

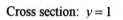

Cross section: $y = 1$

35. $f(x,y) = y^2$

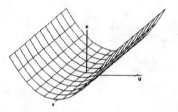

37. $h(x,y) = \frac{1}{y}$

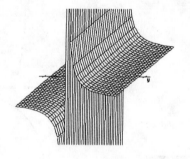

39. $f(x, y) = e^{-(x^2+y^2)}$

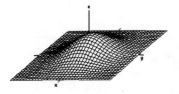

41. $P(x, y) = Kx^a y^{1-a}, \quad K = 1, \quad a = 0.5$
$P(x, y) = x^{0.5} y^{0.5}$

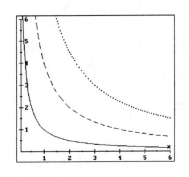

43. $U(x, y) = 6x^{0.8} y^{0.2} + x$

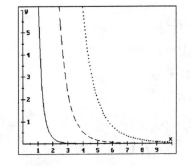

45. $C(k, e) = 15,000 + 50k^2 + 60e^2$

47. $A(h, w) = hw$

The following figure shows several lines
of the form $h + w = c$.

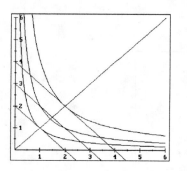

15.3 EXERCISES

1. $f(x, y) = 3x^2 - y^3 + x - 1$

   $f_x = 6x + 1; \ f_x(1, -1) = 7$

   $f_y = -3y^2; \ f_y(1, -1) = -3$

   $f_{xx} = 6; \ f_{xx}(1, -1) = 6$

   $f_{yy} = -6y; \ f_{yy}(1, -1) = 6$

   $f_{xy} = f_{yx} = 0; \ f_{xy}(1, -1) = f_{yx}(1, -1) = 0$

3. $f(x, y) = 3x^2 y$

   $f_x = 6xy; \ f_x(1, -1) = -6$

   $f_y = 3x^2; \ f_y(1, -1) = 3$

   $f_{xx} = 6y; \ f_{xx}(1, -1) = -6$

   $f_{yy} = 0; \ f_{yy}(1, -1) = 0$

   $f_{xy} = f_{yx} = 6x; \ f_{xy}(1, -1) = f_{yx}(1, -1) = 6$

5. $f(x, y) = x^2 y^3 - x^3 y^2 - xy$

   $f_x(x, y) = 2xy^3 - 3x^2 y^2 - y; \ f_x(1, -1) = -4$

   $f_y(x, y) = 3x^2 y^2 - 2x^3 y - x; \ f_y(1, -1) = 4$

   $f_{xx}(x, y) = 2y^3 - 6xy^2; \ f_{xx}(1, -1) = -8$

   $f_{yy}(x, y) = 6x^2 y - 2x^3; \ f_{yy}(1, -1) = -8$

   $f_{xy}(x, y) = f_{yx}(x, y) = 6xy^2 - 6x^2 y - 1; \ f_{xy}(1, -1) = f_{yx}(1, -1) = 11$

490

7. $f(x, y) = (2xy + 1)^3$

$f_x(x, y) = 6y(2xy + 1)^2$; $f_x(1, -1) = -6$

$f_y(x, y) = 6x(2xy + 1)^2$; $f_y(1, -1) = 6$

$f_{xx}(x, y) = 24y^2(2xy + 1)$; $f_{xx}(1, -1) = -24$

$f_{yy}(x, y) = 24x^2(2xy + 1)$; $f_{yy}(1, -1) = -24$

$f_{xy}(x, y) = f_{yx}(x, y) = 6(2xy + 1)^2 + 24xy(2xy + 1) = 6(6xy + 1)(2xy + 1)$;

$f_{xy}(1, -1) = f_{yx}(1, -1) = 30$

9. $f(x, y) = e^{x+y}$

$f_x(x, y) = e^{x+y}$; $f_x(1, -1) = 1$

$f_y(x, y) = e^{x+y}$; $f_y(1, -1) = 1$

$f_{xx}(x, y) = e^{x+y}$; $f_{xx}(1, -1) = 1$

$f_{yy}(x, y) = e^{x+y}$; $f_{yy}(1, -1) = 1$

$f_{xy}(x, y) = f_{yx}(x, y) = e^{x+y}$; $f_{xy}(1, -1) = f_{yx}(1, -1) = 1$

11. $f(x, y) = 5x^{0.6}y^{0.4}$

$f_x(x, y) = 3x^{-0.4}y^{0.4}$; $f_x(1, -1)$ not defined

$f_y(x, y) = 2x^{0.6}y^{-0.6}$; $f_y(1, -1)$ not defined

$f_{xx}(x, y) = -1.2x^{-1.4}y^{0.4}$; $f_{xx}(1, -1)$ not defined

$f_{yy}(x, y) = -1.2x^{0.6}y^{-1.6}$; $f_{yy}(1, -1)$ not defined

$f_{xy}(x, y) = f_{yx}(x, y) = 1.2x^{-0.4}y^{-0.6}$; $f_{xy}(1, -1)$ and $f_{yx}(1, -1)$ not defined

13. $f(x, y) = 4.1x^{1.2}e^{-0.2y}$

$f_x(x, y) = 4.92x^{0.2}e^{-0.2y}$; $f_x(1, -1) = 4.92e^{0.2}$

$f_y(x, y) = -0.82x^{1.2}e^{-0.2y}$; $f_y(1, -1) = -0.82e^{0.2}$

$f_{xx}(x, y) = -0.164x^{-0.8}e^{-0.2y}$; $f_{xx}(1, -1) = -0.164e^{0.2}$

$f_{yy}(x, y) = 0.164x^{1.2}e^{-0.2y}$; $f_{yy}(1, -1) = 0.164e^{0.2}$

$f_{xy}(x, y) = f_{yx}(x, y) = -0.984x^{0.2}e^{-0.2y}$; $f_{xy}(1, -1) = f_{yx}(1, -1) = -0.984e^{0.2}$

15. $f(x, y) = e^{0.2xy}$

$f_x(x, y) = 0.2y e^{0.2xy} f_x(1, -1) = -0.2 e^{-0.2}$

$f_y(x, y) = 0.2x e^{0.2xy} f_y(1, -1) = 0.2 e^{-0.2}$

$f_{xx}(x, y) = 0.04 y^2 e^{0.2xy} f_{xx}(1, -1) = 0.04 e^{-0.2}$

$f_{yy}(x, y) = 0.04 x^2 e^{0.2xy} f_{yy}(1, -1) = 0.04 e^{-0.2}$

$f_{xy}(x, y) = f_{yx}(x, y) = 0.2(1 + 0.2xy) e^{0.2xy}; \; f_{xy}(1, -1) = f_{yx}(1, -1) = 0.16 e^{-0.2}$

17. $f(x, y, z) = xyz$

$f_x(x, y, z) = yz; \; f_x(0, -1, 1) = -1$

$f_y(x, y, z) = xz; \; f_y(0, -1, 1) = 0$

$f_z(x, y, z) = xy; \; f_z(0, -1, 1) = 0$

19. $f(x, y, z) = -\dfrac{4}{x + y + z^2}$

$f_x(x, y) = -4 / (x + y + z^2)^2; \; f_x(0, -1, 1) \; \text{undefined}$

$f_y(x, y, z) = -4 / (x + y + z^2)^2; \; f_y(0, -1, 1) \; \text{undefined}$

$f_z(x, y, z) = -8z(x + y + z)^2; \; f_z(0, -1, 1) \; \text{undefined}$

21. $f(x, y, z) = xe^{yz} + ye^{xz}$

$f_x(x, y, z) = e^{yz} + yze^{xz}; \; f_x(0, -1, 1) = e^{-1} - 1$

$f_y(x, y, z) = xze^{yz} + e^{xz}; \; f_y(0, -1, 1) = 1$

$f_z(x, y, z) = xy(e^{yz} + e^{xz}); \; f_z(0, -1, 1) = 0$

23. $f(x, y, z) = x^{0.1} y^{0.4} z^{0.5}$

$f_x(x, y, z) = 0.1 \dfrac{y^{0.4} z^{0.5}}{x^{0.9}}; \; f_y(x, y, z) = 0.4 \dfrac{x^{0.1} z^{0.5}}{y^{0.6}}; \; f_x(x, y, z) = 0.5 \dfrac{x^{0.1} y^{0.4}}{z^{0.5}}$

$f_x(0, -1, 1)$ is not defined, $f_y(0, -1, 1)$ is not defined, $f_z(0, -1, 1)$ is not defined.

25. $f(x, y, z) = e^{xyz}$

$f_x(x, y, z) = yze^{xyz}, \; f_y(x, y, z) = xze^{xyz}, \; f_z(x, y, z) = xye^{xyz}$

$f_x(0, -1, 1) = -1, \; f_y(0, -1, 1) = f_z(0, -1, 1) = 0$

27. $f(x, y, z) = \dfrac{2000z}{1 + y^{0.3}}$

$f_x(x, y, z) = 0,\ f_y(x, y, z) = -\dfrac{600z}{y^{0.7}\left(1 + y^{0.3}\right)^2},\ f_z(x, y, z) = \dfrac{2000}{1 + y^{0.3}}$

$f_x(0, -1, 1) = 0,\ f_y(0, -1, 1)$ is not defined, $f_z(0, -1, 1)$ is not defined.

29. $c(x, y, z) = 72.3 - 0.8x - 0.2y - 0.7z$

$\dfrac{\partial c}{\partial x} = -0.8$, Showing that Chrysler's percentage of the market decreases by 0.8% for every

1% rise in foreign manufacturers' share. $\dfrac{\partial c}{\partial y} = -0.2$, showing that Chrysler's percentage

of the market decreases by 0.2% for every 1% rise in Ford's share. $\dfrac{\partial c}{\partial z} = -0.7$, showing

that Chrysler's percentage of the market decreases by 0.7% for every 1% rise in G.M.'s share.

31. $C(x, y) = 200{,}000 + 6000x + 4000y - 100{,}000e^{-0.01(x+y)}$

The marginal cost of cars $= \$6{,}000 + 1000e^{-0.01(x+y)}$ per car. The marginal cost of trucks

$= \$4000 + 1000e^{-0.01(x+y)}$ per truck. Both marginal costs decrease as production rises.

33. $\overline{C}(x, y) = \dfrac{C(x, y)}{x + y} = \dfrac{200{,}000 + 6000x + 4000y - 100{,}000e^{-0.01(x+y)}}{x + y}$

$\overline{C}_x(50, 50) = \$42.36$ per car. This means that, at a production level of 50 cars
and 50 trucks per week, the average cost per car is increasing by $42.36 for each
additional car manufactured. $\overline{C}_y(50, 50) = \$22.36$ per truck. This means that,
at a production level of 50 cars and 50 trucks per week, the average cost per
truck is increasing by $22.36 for each additional truck manufactured.

35. $C(x, y) = 15{,}000x + 10{,}000y - 5000\sqrt{x + y}$

No, the dealer's marginal revenue from the sale of cars is $C_x = \$15{,}000 - \dfrac{2500}{\sqrt{x + y}}$ per car

and $C_y = \$10{,}000 - \dfrac{2500}{\sqrt{x + y}}$ per truck from the sale of trucks. These increase with increasing

x and y. In other words, the dealer will earn more revenue per vehicle with increasing sales,
and so the rental company will pay more for each additional vehicle it buys.

37. $P(x, y, z) = 0.04x^{0.4}y^{0.2}z^{0.4}$; $P_z = 0.016x^{0.4}y^{0.2}z^{-0.6}$

$P_z(10, 100,000, 1,000,000) \approx .0001010$ papers/\$

39. $U(x, y) = 6x^{0.8}y^{0.2} + x$; $U_x = 4.8x^{-0.2}y^{0.2} + 1$; $U_y = 1.2x^{0.8}y^{-0.8}$

(a) $U_x(10, 5) = 5.18$; $U_y(10, 5) = 2.09$. This means that, if 10 copies of Macro Publish and 5 copies of Turbo Publish are purchased, the company's daily productivity is increasing at a rate of 5.18 pages per day for each additional copy of Macro purchased, and by 2.09 pages per day for each additional coy of Turbo purchased.

(b) $\dfrac{U_x(10, 5)}{U_y(10, 5)} \approx 2.48$ is the ratio of the usefulness of one additional copy of Macro to one of Turbo. Thus, with 10 copies of Macro and 5 copies of Turbo, the company can expect approximately 2.48 times the productivity per additional copy of Macro than Turbo.

41. $F(x, y, z) = K\dfrac{Qq}{(x-a)^2 + (y-b)^2 + (z-c)^2} = K\dfrac{50}{x^2 + y^2 + z^2}$

$F_y = -K\dfrac{100y}{\left[x^2 + y^2 + z^2\right]^2}$

$F_y(2, 3, 3) = \dfrac{-9 \times 10^9 (100)(3)}{(4+9+9)^2} \approx -6 \times 10^9 \, n/s$

43. $A(P, r, t) = P(1+r)^t$

(a) $A_p = (1+r)^t$; $A_r = Pt(1+r)^{t-1}$; $A_t = P(1+r)^t \ln(1+r)$

$A_p(100, 0.10, 10) = (1 + 0.10)^{10} = 2.59$

$A_r = (100, 0.10, 10) = 100(10)(1 + 0.10)^{10-1} = 2,357.95$

$A_t = (100, 0.10, 10) = 100(1 + 0.10)^{10} \ln(1 + 0.10) = 24.72$

Thus, for a \$100 investment at 10% interest invested for 10 years, the accumulated amount is increasing at a rate of \$2.59 per \$1 of principal, at a rate of \$2,357.95 per increase of 1 in r (note that this would correspond to an increase in the interest rate of 100%), and at a rate of \$24.72 per year.

(b) $A_p(100, 0.1, t)$ tells you the rate at which the accumulated amount in an account bearing 10% interest with a principal of \$100 is growing per \$1 increase in the principal, t years after the investment.

45. $P(x, y) = Kx^a y^b$ $(a + b = 1)$

(a) $P_x = Kax^{a-1}y^b = Kax^{-b}y^b = Ka\left(\dfrac{y}{x}\right)^b$

$P_y = Kbx^a y^{b-1} = Kbx^a y^{-a} = Kb\left(\dfrac{x}{y}\right)^a$

If $P_x = P_y$, then $Ka\left(\dfrac{y}{x}\right)^b = Kb\left(\dfrac{x}{y}\right)^a \Rightarrow \dfrac{a}{b} = \left(\dfrac{x}{y}\right)^b \left(\dfrac{x}{y}\right)^a = \left(\dfrac{x}{y}\right)^{a+b} = \left(\dfrac{x}{y}\right)$

Therefore $\dfrac{a}{b} = \dfrac{x}{y}$

(b) Given $P_x(100, 200) = P_y(100, 200)$, then $\dfrac{a}{b} = \dfrac{x}{y} = \dfrac{100}{200} = \dfrac{1}{2}$

But $b = 1 - a$, so $\dfrac{a}{1-a} = \dfrac{1}{2} \Rightarrow 2a = 1 - a \Rightarrow a = \dfrac{1}{3}$ and $b = \dfrac{2}{3}$

47. $u(r, t) = \dfrac{1}{4\pi Dt}e^{-r^2/4Dt}$, $D = 1$

$u(r, t) = \dfrac{1}{4\pi t}e^{-r^2/4t}$

$u_t = \dfrac{e^{-r^2/4t}\left(\frac{r^2}{4t^2}\right)(4\pi t) - e^{-r^2/4t}(4\pi)}{16\pi^2 t^2} = \dfrac{e^{-r^2/4t}}{4\pi t^2}\left(\dfrac{r^2}{4t} - 1\right)$

$u_t(1, 3) = \dfrac{e^{-r^2}(4)(3)}{4\pi(3)^2}\left(\dfrac{1^2}{4(3)} - 1\right) = -0.007457$

Decreasing at a rate of 0.007457 cc / s.

15.4 EXERCISES

1. Local max at P and R, saddle point at Q.

3. Critical point (neither saddle point nor local extremum) at Q, non-critical point at P and R.

5. Non-critical point at P and Q, saddle point at R.

7.

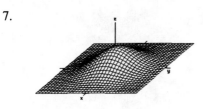

In Exercises 9 - 25 use $H = f_{xx}(a, b)f_{yy}(a, b) - \left[f_{xy}(a, b)\right]^2$.

9. $f(x, y) = x^2 + y^2 + 1$
   $f_x = 2x;\ 2x = 0 \Rightarrow x = 0$
   $f_y = 2y;\ 2y = 0 \Rightarrow y = 0$
   Critical point: $(0, 0)$
   $f_{xx} = 2,\ f_{yy} = 2,\ f_{xy} = 0$
   $H = 2(2) - (0)^2 = 4 > 0$ and $f_{xx} > 0$
   $f(x, y)$ has a local minimum of 1 at $(0, 0)$

11. $g(x, y) = 1 - x^2 - x - y^2 + y$
    $g_x = -2x - 1;\ -2x - 1 = 0 \Rightarrow x = -\dfrac{1}{2}$
    $g_y = -2y + 1;\ -2y + 1 = 0 \Rightarrow y = \dfrac{1}{2}$
    Critical point: $\left(-\dfrac{1}{2}, \dfrac{1}{2}\right)$
    $g_{xx} = -2,\ g_{yy} = -2,\ g_{xy} = 0$
    $H = (-2)(-2) - (0)^2 = 4 > 0$ and $g_{xx} < 0$
    $g\left(-\dfrac{1}{2}, \dfrac{1}{2}\right) = 1 - \left(-\dfrac{1}{2}\right)^2 - \left(-\dfrac{1}{2}\right)^2 + \dfrac{1}{2} = \dfrac{3}{2}$
    $g(x, y)$ has a maximum of $\dfrac{3}{2}$ at $\left(-\dfrac{1}{2}, \dfrac{1}{2}\right)$

496

13. $h(x, y) = x^2 y - 2x^2 - 4y^2$

$h_x = 2xy - 4x; \ 2xy - 4x = 0$

$\qquad\qquad 2x(y - 2) = 0$

$\qquad\qquad x = 0 \ \text{ or } \ y = 2$

$h_y = x^2 - 8y; \ x^2 - 8y = 0$

If $x = 0, \ y = 0$ or if $y = 2, \ x = \pm 4$

Critical points: $(0, 0), \ (-4, 2), \ (4, 2)$

$h_{xx} = 2y - 4, \ h_{yy} = -8, \ h_{xy} = 2x$

Point $(0, 0)$: $h_{xx} = -4, \ h_{yy} = -8$

$h_{xy} = 0, \ H = (-4)(-8) - (0)^2 = 32 > 0$ and $h_{xx} < 0$

$h(x, y)$ has a maximum of 0 at $(0, 0)$

Point $(-4, 2)$: $h = -16, \ h_{xx} = 0$

$h_{yy} = -8, \ h_{xy}(-4, 2) = -8$

$H = 0(-8) - (-8)^2 = -64 < 0$

$h(x, y)$ has a saddle point at $(-4, 2, -16)$

Point $(4, 2)$: $h(4, 2) = -16, \ h_{xx}(4, 2) = 0$

$h_{yy}(4, 2) = -8, \ h_{xy}(4, 2) = 8$

$H = 0(8) - (8)^2 = -64 < 0$

$h(x, y)$ has a saddle point at $(4, 2, -16)$

15. $s(x, y) = e^{x^2 + y^2}$

$s_x = 2xe^{x^2 + y^2}; \ 2xe^{x^2 + y^2} = 0$

$\qquad\qquad\qquad\qquad x = 0$

$s_y = 2ye^{x^2 + y^2}; \ 2ye^{x^2 + y^2} = 0$

$\qquad\qquad\qquad\qquad y = 0$

Critical point: $(0, 0)$

$s_{xx} = (4x^2 + 2)e^{x^2 + y^2}$

$s_{yy} = (4y^2 + 2)e^{x^2 + y^2}$

$s_{xy} = 4xye^{x^2 + y^2}$

Point $(0, 0)$: $s = 1, \ s_{xx} = 2, \ s_{yy} = 2, \ s_{xy} = 0$

$H = 2(2) - (0)^2 = 4 > 0$ and $s_{xx} > 0$

$s(x, y)$ has a minimum of 1 at $(0, 0)$

17. $t(x, y) = x^4 + 8xy^2 + 2y^4$

$t_x = 4x^3 + 8y^2$, $4x^3 + 8y^2 = 0$

$$y^2 = \frac{-x^3}{2}$$

$y_y = 16xy + 8y^3$, $16xy + 8y^3 = 0$

$8y(2x + y^2) = 0$

$y = 0$ or $y^2 = -2x$

$$\frac{x^3}{2} = -2x$$

$x^3 - 4x = 0$

$x(x^2 - 4) = 0$

$x = 0$ or $x = \pm 2$

If $x = 0$, $y = 0$ or if $x = -2$, $y = \pm 2$

Critical points: $(0, 0)$, $(-2, 2)$, $(-2, -2)$

$t_{xx} = 12x^2$, $t_{yy} = 16x + 24y^2$, $t_{xy} = 16y$

Point $(0, 0)$: $t = 0$, $t_{xx} = 0$, $t_{yy} = 0$, $t_{xy} = 0$

$H = (48)(64) - (-32) = 2048 > 0$, $t_{xx} > 0$

$t(x, y)$ has a local minimum of -16 at $(-2, 2)$

Point $(-2, -2)$, $t = -16$, $t_{xx} = 48$, $t_{yy} = 64$

$t_{xy} = -32$

$H = (48)(64) - (-32)^2 = 2048 > 0$, $t_{xx} > 0$

$t(x, y)$ has a local minimum of -16 at $(-2, -2)$

19. $f(x, y) = x^2 + y - e^y$

$f_x = 2x$, $2x = 0$

$x = 0$

$f_y = 1 - e^y$, $1 - e^y = 0$

$e^y = 1$

$y = 0$

Critical Points: $(0, 0)$

$f_{xx} = 2$, $f_{yy} = -e^y$, $f_{xy} = 0$

Point $(0, 0)$: $f = -1$, $f_{xx} = 2$, $f_{yy} = -1$,

$f_{xy} = 0$

$H = (-1)(2) - (0)^2 = -2 < 0$

$f(x, y)$ has a saddle point at $(0, 0, -1)$

21. $f(x, y) = e^{-\left(x^2+y^2+2x\right)}$

$\quad f_x = (-2x-2)e^{-\left(x^2+y^2+2x\right)}, \ (-2x-2)e^{-\left(x^2+y^2+2x\right)} = 0, \ -2x-2 = 0, \ x = -1$

$\quad f_y = -2ye^{-\left(x^2+y^2+2x\right)}, \ (-2y)e^{-\left(x^2+y^2+2x\right)} = 0, \ -2y = 0, \ y = 0$

Critical Point: $(-1, 0)$

$\quad f_{xx} = (-2x-2)(-2x-2)e^{-\left(x^2+y^2+2x\right)} + (-2)e^{-\left(x^2+y^2+2x\right)} = \left(4x^2+8x+2\right)e^{-\left(x^2+y^2+2x\right)}$

$\quad f_{yy} = (-2y)(-2y)e^{-\left(x^2+y^2+2x\right)} + (-2)e^{-\left(x^2+y^2+2x\right)} = \left(4y^2-2\right)e^{-\left(x^2+y^2+2x\right)}$

$\quad f_{xy} = (-2x-2)(-2y)e^{-\left(x^2+y^2+2x\right)} = 4y(x+1)e^{-\left(x^2+y^2+2x\right)}$

Point $(-1, 0)$: $f = e, \ f_{xx} = -2e, \ f_{yy} = -2e, \ f_{xy} = 0$

$H = (-2e)(-2e) - (0)^2 = 4e^2 > 0, \ f_{xx} < 0$

$f(x, y)$ has a maximum of e at $(-1, 0)$

23. $f(x, y) = xy + \dfrac{2}{x} + \dfrac{2}{y}$

$\quad f_x = y - 2x^{-2}, \ y - \dfrac{2}{x^2} = 0, \ y = \dfrac{2}{x^2}$

$\quad f_y = x - 2y^{-2}, \ x - \dfrac{2}{y^2} = 0, \ x = \dfrac{2}{y^2} \Rightarrow x^2 = \dfrac{4}{y^4}$ so, $y = \dfrac{2}{4/y^4} = \dfrac{y^4}{2}$ or $y^3 = 2, \ y = 2^{1/3}$

Critical Point: $\left(2^{1/3}, 2^{1/3}\right) \ f_{xx} = 4x^{-3}, \ f_{yy} = 4y^{-3}, \ f_{xy} = 1$

Point $\left(2^{1/3}, 2^{1/3}\right)$: $f = 3(2)^{2/3}, \ f_{xx} = 2, \ f_{yy} = 2, \ f_{xy} = 1$

$H = (2)(2) - (1)^2 = 3 > 0, \ f_{xx} > 0$

$f(x, y)$ has a minimum of $3(2)^{2/3}$ at $\left(2^{1/3}, 2^{1/3}\right)$

25. $g(x, y) = x^2 + y^2 + \dfrac{2}{xy}$

$g_x = 2x - \dfrac{2}{x^2 y}, \quad 2x - \dfrac{2}{x^2 y} = 0, \quad x^3 = \dfrac{1}{y} \Rightarrow y = \dfrac{1}{x^3}$

$g_y = 2y - \dfrac{2}{xy^2}, \quad 2y - \dfrac{2}{xy^2} = 0, \quad y^3 = \dfrac{1}{x} \Rightarrow x = \dfrac{1}{y^3} \Rightarrow x^3 = \dfrac{1}{y^9}$

so $y = \dfrac{1}{1/y^9} = \dfrac{y^9}{1}$ or $y^9 - y = 0 \Rightarrow y(y^8 - 1) = 0, \quad y = 0, \quad y = \pm 1$

If $y = 0$, x is undefined; if $y = 1$, $x = 1$ or if $y = -1$, $x = -1$.

Critical Points: $(1, 1)$ and $(-1, -1)$

$g_{xx} = 2 + \dfrac{4}{x^3 y}, \quad g_{yy} = 2 + \dfrac{4}{x^3 y}, \quad g_{xy} = \dfrac{2}{x^2 y^2}$

Point $(1, 1)$ or $(-1, -1)$: $g = 4, \quad g_{xx} = 6, \quad g_{yy} = 6, \quad g_{xy} = 2$

$H = (6)(6) - (2)^2 = 32 > 0, \quad g_{xx} > 0$

$g(x, y)$ has a minimum of 4 at $(1, 1)$ and $(-1, -1)$

27. $\overline{C}(x, y) = \dfrac{10{,}000 + 50x + 70y + 0.0125xy}{x + y}$

$\overline{C}_x = \dfrac{(50 + 0.0125y)(x + y) - (10{,}000 + 50x + 70y + 0.0125xy)}{(x + y)^2}$

$\overline{C}_x = \dfrac{0.0125y^2 - 20y - 10{,}000}{(x + y)^2} = 0 \Rightarrow y = 2000 \text{ or } -400 \text{ (not allowed)}$

$\overline{C}_y = \dfrac{(70 + 0.0125x)(x + y) - (10{,}000 + 50x + 70y + 0.0125xy)}{(x + y)^2}$

$\overline{C}_y = \dfrac{0.0125x^2 + 20x - 10{,}000}{(x + y)^2} = 0 \Rightarrow x = 400 \text{ or } -2000 \text{ (not allowed)}$

Critical Point: $(400, 2000)$

$\overline{C}_{xx} = (0.0125y^2 - 20y - 10{,}000)(-2)(x + y)^{-3} \qquad \overline{C}_{yy} = (0.0125x^2 + 20x - 10{,}000)(-2)(x + y)^{-3}$

$\overline{C}_{xy} = (0.025y - 20)(x + y)^2 - 2(0.0125y^2 - 20y - 10{,}000)(x + y)^{-3}$

Point $(400, 2000)$: $\overline{C} = 75, \quad \overline{C}_{xx} = 0, \quad \overline{C}_{yy} = 0, \quad \overline{C}_{xy} = 0.0125$

$H = (0)(0) - (0.0125)^2 < 0$

Product four hundred 5-speeds and two thousand 10-speeds.

29. Let $C(x, y)$ be any cost function, then $\overline{C} = \dfrac{C}{x+y}$.

$\overline{C}_x = \dfrac{\partial}{\partial x}\left(\dfrac{C}{x+y}\right) = \dfrac{(x+y)C_x - C}{(x+y)^2}$. This is zero when $(x+y)C_x = C$, or $C_x = \dfrac{C}{x+y} = \overline{C}$.

Similarly, $\overline{C}_y = 0$ when $C_y = \overline{C}$. This is reasonable because, if the average cost is decreasing with increasing x, then the average cost is greater than the marginal cost C_x. Similarly, if the average cost is increasing with increasing x, the average cost is less than the marginal cost C_x. Thus, if the average cost is stationary with increasing x, then the average cost equals the marginal cost C_x. (Similarly for the case of increasing y).

31. $R = p_1 q_1 + p_2 q_2 = 100{,}000 p_1 - 100 p_1^2 + 10 p_1 p_2 + 150{,}000 p_2 + 10 p_1 p_2 - 100 p_2^2$

$R_{p_1} = 100{,}000 - 200 p_1 + 20 p_2 = 0 \Rightarrow p_2 = 10 p_1 - 5000$

$R_{p_2} = 20 p_1 + 150{,}000 - 200 p_2 = 0 \Rightarrow p_1 = 10 p_2 - 7500$

Therefore $p_2 = 10(10 p_2 - 7500) - 5000 \Rightarrow p_2 = 808.08$ and $p_1 = 580.81$.

Critical Point: $(580.81, 808.08)$

$R_{p_1 p_1} = -200,\ R_{p_2 p_2} = -200,\ R_{p_1 p_2} = 20$

$H = (-200)(-200) - (20)^2 > 0,\ R_{p_1 p_1} < 0$

Revenue is a maximum if they charge \$580.81 for the Ultra Mini and \$808.08 for the Big Stack.

15.5 EXERCISES

1. $f(x, y) = x^2 + y^2$, $0 \le x \le 2$, $0 \le y \le 2$

   Interior of domain bounded by the square with vertices $(0, 0)$, $(0, 2)$, $(2, 2)$, and $(2, 0)$

   $f_x = 2x = 0 \Rightarrow x = 0$, $f_y = 2y = 0 \Rightarrow y = 0$; critical point $(0, 0)$

   The critical point $(0, 0)$ lies on the boundary of the domain, so there are none in the interior.

   Side 1: $y = 0$, $0 \le x \le 2$; The endpoints are $(0, 0, 0)$ and $(2, 0, 4)$

   Side 2: $x = 0$, $0 \le y \le 2$; The endpoints are $(0, 0, 0)$ and $(0, 2, 4)$

   Side 3: $y = 2$, $0 \le x \le 2$; The endpoints are $(0, 2, 4)$ and $(2, 2, 8)$

   Side 4: $x = 2$, $0 \le y \le 2$; The endpoints are $(2, 0, 4)$ and $(2, 2, 8)$

   Absolute maximum of 8 at $(2, 2)$ and absolute minimum of 0 at $(0, 0)$.

3. $h(x, y) = (x - 1)^2 + y^2$, $x^2 + y^2 \le 4$

   Interior of domain bounded by a circle of radius 2 centered at the origin.

   $h_x = 2x - 2 = 0 \Rightarrow x = 1$, $h_y = 2y = 0 \Rightarrow y = 0$. Critical point $(1, 0)$

   $h_{xx} = 2$, $h_{yy} = 2$, $h_{xy} = 0$, $h(1, 0) = 0$. $H = (2)(2) - 0 > 0$, $h_{xx} > 0$.

   Local minimum of 0 at $(1, 0)$

   On the boundary $x^2 + y^2 = 4$ or $y^2 = 4 - x^2$, the curve becomes

   $z = (x - 1)^2 + 4 - x^2 = -2x + 5$, $z' = -2$, so there is no critical point.

   But $-2 \le x \le 2$ so consider the endpoints of $z = -2x + 5$, $(-2, 0, 9)$ and $(2, 0, 1)$

   Maximum value of 9 at $(-2, 0)$ and minimum value of 0 at $(1, 0)$

5. $f(x, y) = e^{x^2+y^2}$, $4x^2 + y^2 \leq 4$

Interior of domain bounded by the ellipse of semi - major axis 2 and semi - minor axis 1 centered at the origin.

$f_x = 2xe^{x^2+y^2} = 0 \Rightarrow x = 0$, $f_y = 2ye^{x^2+y^2} = 0 \Rightarrow y = 0$. Critical point $(0, 0)$

$f_{xx} = (4x^2 + 2)e^{x^2+y^2}$, $f_{xx}(0, 0) = 2$, $f_{yy} = (4y^2 + 2)e^{x^2+y^2}$, $f_{yy}(0, 0) = 2$

$f_{xy} = 4xye^{x^2+y^2}$, $f_{xy}(0, 0) = 0$, $f(0, 0) = 1$. $H = (2)(2) - 0 > 0$, $f_{xx} > 0$.

Local minimum of 1 at $(0, 0)$

On the boundary $4x^2 + y^2 = 4$ or $y^2 = 4 - x^2$, the curve becomes

$z = e^{x^2+4-4x^2} = e^{4-3x^2}$, $z' = -6xe^{4-3x^2} = 0 \Rightarrow x = 0$ and $y^2 = 4 - 4(0)^2 = 4 \Rightarrow y = \pm 2$.

Critical points $(0, \pm 2)$. $z''(0, \pm 2) = (36(0)^2 - 6)e^{4-3(0)^2} < 0$.

Maximum of e^4 at $(0, \pm 2)$ and minimum of 1 at $(0, 0)$.

7. $h(x, y) = e^{4x^2+y^2}$, $x^2 + y^2 \leq 1$

Interior of domain bounded by the circle of radius 1 centered at the origin.

$h_x = 8xe^{4x^2+y^2} = 0 \Rightarrow x = 0$, $h_y = 2ye^{4x^2+y^2} = 0 \Rightarrow y = 0$. Critical point $(0, 0)$

$h_{xx} = (64x^2 + 8)e^{4x^2+y^2}$, $h_{xx}(0, 0) = 8$, $h_{yy} = (4y^2 + 2)e^{4x^2+y^2}$, $h_{yy}(0, 0) = 2$

$h_{xy} = 16xye^{4x^2+y^2}$, $h_{xy}(0, 0) = 0$, $h(0, 0) = 1$. $H = 8(2) - 0^2 > 0$, $h_{xx} > 0$.

Local minimum of 1 at $(0, 0)$

On the boundary $x^2 + y^2 = 1$ or $x^2 = 1 - y^2$, the curve becomes

$z = e^{4(1-y^2)+y^2} = e^{4-3y^2}$, $z' = -6ye^{4-3y^2} = 0 \Rightarrow y = 0$ and

$x^2 = 1 - 0^2 = 1 \Rightarrow x = \pm 1$. Critical points $(\pm 1, 0)$

$z''(\pm 1, 0) = (36(0)^2 - 6)e^{4-3(0)^2} < 0$

Maximum value of e^4 at $(\pm 1, 0)$ and minimum value of 1 at $(0, 0)$

9. $f(x, y) = x + y + \dfrac{1}{xy}$, $x \geq \dfrac{1}{2}$, $y \geq \dfrac{1}{2}$, $x + y \leq 3$

Interior of domain bounded by the triangle with vertices $\left(\dfrac{1}{2}, \dfrac{1}{2}\right)$, $\left(\dfrac{1}{2}, \dfrac{5}{2}\right)$ and $\left(\dfrac{5}{2}, \dfrac{1}{2}\right)$.

$f_x = 1 - \dfrac{1}{x^2 y} = 0 \Rightarrow y = \dfrac{1}{x^2}$, $f_y = 1 - \dfrac{1}{xy^2} = 0 \Rightarrow x = \dfrac{1}{y^2}$ so $y = y^4$ or

$y(y^3 - 1) = 0 \Rightarrow y = 0$ (not allowed) or $y = 1$ and $x = 1$. Critical point $(1, 1)$

$f_{xx} = \dfrac{2}{x^3 y}$, $f_{xx}(1, 1) = 2$, $f_{yy} = \dfrac{2}{xy^3}$, $f_{yy}(1, 1) = 2$

$f_{xy} = \dfrac{1}{x^2 y^2}$, $f_{xy}(1, 1) = 1$, $f(1, 1) = 3$. $H = 2(2) - 1^2 > 0$, $f_{xx} > 0$

Local minimum of 3 at $(1, 1)$

Side 1: $y = \dfrac{1}{2}$, $\dfrac{1}{2} \leq x \leq \dfrac{5}{2}$. The endpoints are $\left(\dfrac{1}{2}, \dfrac{1}{2}, 5\right)$ and $\left(\dfrac{5}{2}, \dfrac{1}{2}, \dfrac{19}{5}\right)$

Side 2: $x = \dfrac{1}{2}$, $\dfrac{1}{2} \leq y \leq \dfrac{5}{2}$. The endpoints are $\left(\dfrac{1}{2}, \dfrac{1}{2}\right)$ and $\left(\dfrac{1}{2}, \dfrac{5}{2}, \dfrac{19}{5}\right)$

Maximum value of 5 at $\left(\dfrac{1}{2}, \dfrac{1}{2}\right)$ and minimum value of 3 at $(1, 1)$

11. $h(x, y) = xy + \dfrac{8}{xy} + \dfrac{8}{y}$, $x \geq 1$, $y \geq 1$, $xy \leq 9$

Interior of domain bounded by the horizontal line $y = 1$, the vertical line $x = 1$, and the hyperbola $xy = 9$ which have the vertices $(1, 1)$, $(1, 9)$ and $(9, 1)$.

$h_x = y - \dfrac{8}{x^2} = 0 \Rightarrow y = \dfrac{8}{x^2}$, $h_y = x - \dfrac{8}{y^2} = 0 \Rightarrow x = \dfrac{8}{y^2}$ so $y = \dfrac{y^4}{8}$ or

$y(y^3 - 8) = 0 \Rightarrow y = 0$ (not allowed) or $y = 2$ and $x = 2$. Critical point $(2, 2)$

$h_{xx} = \dfrac{24}{x^3}$, $h_{xx}(2, 2) = 3$, $h_{yy} = \dfrac{24}{y^3}$, $h_{yy}(2, 2) = 3$, $h_{xy} = 1$, $h(2, 2) = 12$

$H = 3(3) - 1^2 > 0$, $h_{xx} > 0$

Local minimum of 12 at $(2, 2)$

Side 1: $y = 1$, $1 \le x \le 9$. The endpoints are $(1, 1, 17)$ and $\left(9, 1, \dfrac{161}{9}\right)$

Side 2: $x = 1$, $1 \le y \le 9$. The endpoints are $(1, 1, 17)$ and $\left(1, 9, \dfrac{161}{9}\right)$

Side 3: $xy = 9$ or $x = \dfrac{9}{y}$ so the curve becomes

$z = 9 + \dfrac{8y}{9} + \dfrac{8}{y}$, $z' = \dfrac{8}{9} - \dfrac{8}{y^2} = 0 \Rightarrow y = -3$ (outside the domain) or

$y = 3$ and $x = 3$. Critical point $(3, 3)$ $z''(3, 3) = \dfrac{16}{3^3} > 0$

Local minimum of $14\dfrac{1}{3}$ at $(3, 3)$

13. $f(x, y) = x^2 + 2x + y^2$, bounded by the square with vertices $(-2, 0)$, $(0, 2)$, $(2, 0)$, and $(0, -2)$.
Interior of domain. $f_x = 2x + 2 = 0 \Rightarrow x = -1$, $f_y = 2y = 0 \Rightarrow y = 0$
Critical point $(-1, 0)$ $f_{xx} = 2$, $f_{yy} = 2$, $f_{xy} = 0$, $f(-1, 0) = -1$
$H = 2(2) - 0^2 > 0$, $f_{xx} > 0$. Local minimum of -1 at $(-1, 0)$
Side 1: $y = x + 2$, $-2 \le x \le 0$; The endpoints are $(-2, 0, 0)$ and $(0, 2, 4)$
Side 2: $y = -x + 2$, $0 \le x \le 2$; The endpoints are $(0, 2, 4)$ and $(2, 0, 8)$
Side 3: $y = x - 2$, $0 \le x \le 2$; The endpoints are $(2, 0, 8)$ and $(0, -2, 4)$
Maximum of 8 at $(2, 0)$ and minimum of -1 at $(-1, 0)$.

15. $h(x, y) = x^3 + y^3$, bounded by the triangle with vertices $(0, 0)$, $(1, 1)$ and $(2, 0)$.
Interior of domain: $h_x = 3x^2 = 0 \Rightarrow x = 0$, $h_y = 3y^2 = 0 \Rightarrow y = 0$
Critical point $(0, 0)$ is on the boundary. No extremum in the interior.
Side 1: $y = 0$, $0 \le x \le 2$; The endpoints are $(0, 0, 0)$ and $(2, 0, 8)$
Side 2: $y = x$, $0 \le x \le 1$; The endpoints are $(0, 0, 0)$ and $(1, 1, 2)$
Maximum of 8 at $(2, 0)$ and minimum of 0 at $(0, 0)$

17. Maximize $f(x, y, z) = xyz$ subject to $g(x, y, z) = x^2 + y^2 + z^2 - 1 = 0$

$f_x = yz$, $f_y = xz$, $f_z = xy$, $g_x = 2x$, $g_y = 2y$, $g_z = 2z$

$yz = \lambda 2x$, $xz = \lambda 2y$, $xy = \lambda 2z$. $y = \dfrac{\lambda 2x}{z}$, $y = \dfrac{\lambda 2z}{x}$

so $z^2 = x^2$. $z = \dfrac{\lambda 2x}{y}$, $z = \dfrac{\lambda 2y}{x}$ so $y^2 = x^2$.

$x^2 + x^2 + x^2 - 1 = 0 \Rightarrow x^2 = \dfrac{1}{3}$ or $x = \pm\dfrac{1}{\sqrt{3}}$, $y = \pm\dfrac{1}{\sqrt{3}}$, $z = \pm\dfrac{1}{\sqrt{3}}$.

To maximize xyz, at least two coordinates must be positive.

$\left(\pm\dfrac{1}{\sqrt{3}}, \pm\dfrac{1}{\sqrt{3}}, \pm\dfrac{1}{\sqrt{3}}\right) \Rightarrow xyz = \dfrac{1}{3\sqrt{3}}$. The maximum occurs at the four points

$\left(\dfrac{1}{\sqrt{3}}, \dfrac{1}{\sqrt{3}}, \dfrac{1}{\sqrt{3}}\right)$, $\left(-\dfrac{1}{\sqrt{3}}, -\dfrac{1}{\sqrt{3}}, \dfrac{1}{\sqrt{3}}\right)$, $\left(-\dfrac{1}{\sqrt{3}}, \dfrac{1}{\sqrt{3}}, -\dfrac{1}{\sqrt{3}}\right)$ and $\left(\dfrac{1}{\sqrt{3}}, -\dfrac{1}{\sqrt{3}}, -\dfrac{1}{\sqrt{3}}\right)$

19. $c(x, y) = 10{,}000 + 50x + 70y - 0.5xy$, $100 \le x \le 150$, $80 \le y \le 120$

Interior of domain: $c_x = 50 - 0.5y = 0 \Rightarrow y = 100$, $c_y = 70 - 0.5x = 0 \Rightarrow x = 140$

Critical point $(140, 100)$. $c_{xx} = 0$, $c_{yy} = 0$, $c_{xy} = -0.5$

$H = 0(0) - (-0.5)^2 < 0$. No extremum in the interior.

Side 1: $y = 80$, $100 \le x \le 150$; The endpoints are $(100, 80, 16{,}600)$ and $(150, 80, 17{,}100)$

Side 2: $x = 100$, $80 \le y \le 120$; The endpoints are $(100, 80, 16{,}600)$ and $(100, 120, 17{,}400)$

Side 3: $y = 120$, $100 \le x \le 150$; The endpoints are $(100, 120, 17{,}400)$ and $(150, 120, 16{,}900)$

For minimum cost make one hundred 5-speeds and eighty 10-speeds. For maximum cost make one hundred 5-speeds and one hundred twenty 10-speeds.

21. $P(x, y) = 20x + 40y - 0.1(x^2 + y^2)$, $0 \le x \le 200$, $0 \le y \le 200$

Interior:

$P_x = 20 - 0.2x = 0 \Rightarrow x = 100$, $P_y = 40 - 0.2y = 0 \Rightarrow y = 200$

Critical point $(100, 200)$ is outside the domain.

Boundary: Triangle with vertices $(0, 0)$, $(0, 200)$ and $(200, 0)$

Side 1: $y = 0$, $0 \le x \le 200$; The endpoints are $(0, 0, 0)$ and $(200, 0, 0)$

Side 2: $x = 0$, $0 \le y \le 200$. The endpoints are $(0, 0, 0)$ and $(0, 200, 4000)$

Side 3: $x + y = 200$ or $x = 200 - y$. The curve becomes

$z = 20(200 - y) + 40y - 0.1\left[(200 - y)^2 + y^2\right] = 60y - 0.2y^2$

$z' = 60 - 0.4y = 0 \Rightarrow y = 150$ and $x = 50$. $z'' = -0.4 < 0$

Critical point $(50, 150)$ is a maximum of 4500.

For a maximum profit of \$4,500, sell 50 copies of Walls and 150 copies of Doors.

23. $T(x, y) = x^2 + 2y^2$ bounded by the square with vertices $(0, 0)$, $(0, 1)$, $(1, 0)$ and $(1, 1)$.

Interior:

$T_x = 2x = 0 \Rightarrow x = 0$, $T_y = 4y = 0 \Rightarrow y = 0$

Critical point $(0, 0)$ is on the boundary. No extremum in the interior.

Boundary:

Side 1: $y = 0$, $0 \le x \le 1$; The endpoints are $(0, 0, 0)$ and $(1, 0, 1)$

Side 2: $x = 0$, $0 \le y \le 1$. The endpoints are $(0, 0, 0)$ and $(0, 200, 4000)$

Side 3: $y = 1$, $0 \le x \le 1$. The endpoints are $(0, 1, 2)$ and $(1, 1, 3)$

Maximum of 3 at $(1, 1)$ and minimum of 0 at $(0, 0)$

Hottest point $(1, 1)$. Coldest point $(0, 0)$.

507

25. $T(x, y) = x^2 + 2y^2 - x$, $x^2 + y^2 \leq 1$

Interior of domain bounded by the circle of radius 1 centered at the origin.

$T_x = 2x - 1 = 0 \Rightarrow x = \dfrac{1}{2}$, $T_y = 4y = 0 \Rightarrow y = 0$. Critical point $\left(\dfrac{1}{2}, 0\right)$

$T_{xx} = 2$, $T_{yy} = 4$, $T_{xy} = 0$, $T\left(\dfrac{1}{2}, 0\right) = -\dfrac{1}{4}$. $H = 2(4) - 0^2 > 0$, $T_{xx} > 0$.

Local minimum of $-\dfrac{1}{4}$ at $\left(\dfrac{1}{2}, 0\right)$

On the boundary $x^2 + y^2 = 1$ or $y^2 = 1 - x^2$, the curve becomes

$z = x^2 + 2(1 - x^2) - x = 2 - x - x^2$, $z' = -1 - 2x = 0 \Rightarrow x = -\dfrac{1}{2}$ and $y = \pm\dfrac{\sqrt{3}}{2}$

$z'' = -2 < 0$. Maximum value of $\dfrac{9}{4}$ at $\left(-\dfrac{1}{2}, \pm\dfrac{\sqrt{3}}{2}\right)$

Hottest points $\left(-\dfrac{1}{2}, \pm\dfrac{\sqrt{3}}{2}\right)$. Coldest point $\left(\dfrac{1}{2}, 0\right)$.

27. Minimize $f(x, y, z) = (x + 1)^2 + (y - 1)^2 + (z - 3)^2$ subject to

$g(x, y, z) = -2x + 2y + z - 5 = 0$. $f_x = 2(x + 1)$, $f_y = 2(y - 1)$,

$f_z = 2(z - 3)$, $g_x = -2$, $g_y = 2$, $g_z = 1$.

$x \& y$: $2(x + 1) = -\lambda 2 \Rightarrow x + 1 = -\lambda$, $2(y - 1) = 2\lambda \Rightarrow y - 1 = \lambda$ so

$\quad\quad y - 1 = -(x + 1) \Rightarrow y = -x$

$x \& z$: $2(x + 1) = -\lambda 2 \Rightarrow x + 1 = -\lambda$, $2(z - 3) = \lambda$ so

$\quad\quad 2(z - 3) = -(x + 1) \Rightarrow z = -\dfrac{1}{2}x + \dfrac{5}{2}$

g: $-2x + 2(-x) + \left(-\dfrac{1}{2}x + \dfrac{5}{2}\right) - 5 = 0 \Rightarrow x = -\dfrac{5}{9}$, $y = \dfrac{5}{9}$, $z = \dfrac{25}{9}$

$\quad\quad \left(-\dfrac{5}{9}, \dfrac{5}{9}, \dfrac{25}{9}\right)$

15.6 Exercises

Use $\left(\sum p_i^2\right)m + \left(\sum p_i\right)b = \sum p_i q_i$ and $\left(\sum p_i\right)m + nb = \sum q_i$ in Exercises 1 - 3.

1. $(1, 1),\ (2, 2),\ (3, 4)$

$\sum p_i^2 = 1^2 + 2^2 + 3^2 = 14$

$\sum p_i = 1 + 2 + 3 = 6$

$\sum q_i = 1 + 2 + 4 = 7$

$\sum p_i q_i = (1 \cdot 1) + (2 \cdot 2) + (3 \cdot 4) = 17$

$n = 3$

$14m + 6b = 17 \qquad (1)$

$6m + 3b = 7 \qquad (2)$

Add -2 times equation (2) to equation (1)

$2m = 3$

$m = \dfrac{3}{2}$

Substitute $\dfrac{3}{2}$ for m in equation (2)

$6\left(\dfrac{3}{2}\right) + 3b = 7$

$b = -\dfrac{2}{3}$

$y = \dfrac{3}{2}x - \dfrac{2}{3}$

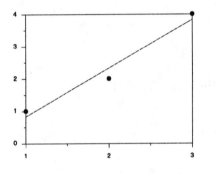

3. $(0, -1),\ (1, 3),\ (4, 6),\ (5, 0)$

$\sum p_i^2 = 0^2 + 1^2 + 4^2 + 5^2 = 42$

$\sum p_i = 0 + 1 + 4 + 5 = 10$

$\sum q_i = -1 + 3 + 6 + 0 = 8$

$\sum p_i q_i = 0(-1) + (1 \cdot 3) + (4 \cdot 6) + (5 \cdot 0) = 27$

$n = 4$

$42m + 10b = 27 \qquad (1)$

$10m + 4b = 8 \qquad (2)$

Add 2 times equation (1) to -5 times equation (2)

$34m = 14$

$m = \dfrac{7}{17} = 0.4118$

Substitute $\dfrac{7}{17}$ for m in equation (2)

$10\left(\dfrac{7}{17}\right) + 4b = 8$

$b = \dfrac{33}{34} = 0.9706$

$y = 0.4118x + 0.9706$

3.

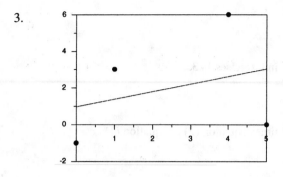

5. $(0, 9.3), (5, 4.8), (6, 4.3), (9, 3.3), (12, 4.3)$

   $y = -0.4347x + 7.9823$

   1985: $x = 10$

      $y = -0.4347(10) + 7.9923$

      $y = 3.6$

   3.6%

7. $(12, 21.2), (14, 29.4), (22, 41.6), (30, 49.2)$

   $y = 1.477x + 6.542$

   $x = 10$

   $y = 1.477(10) + 6.542$

   $y = 21.3$

   21.3 billion barrels

9. $(0, 54.1), (10, 59.7), (20, 62.9), (30, 68.2), (40, 69.7), (50, 70.8), (60, 73.7), (70, 75.4)$

   $y = 0.291x + 56.6$

   Life expectancy was increasing by 0.291 years each year.

11. $(4, 25), (5, 175), (6, 225), (7, 380), (8, 525), (9, 700), (10, 825), (11, 630), (12, 820)$

   $p(t) = 101.1t - 330.3$

   1993: $t = 13$

      $p(13) = 101.1(13) - 330.3$

      profit = \$984 million

13. $(0, 515)$, $(10, 1016)$, $(15, 1598)$, $(20, 2732)$, $(29, 5201)$, $(30, 5465)$

$y = 176.5x - 304.9$

2000: $x = 40$

$\quad y = 176.5(40) - 304.9$

$\quad y = 6755$

$6755 billion

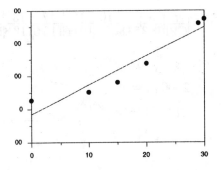

15.

P	0	10	20	30	40	50	60	70	80	90	100
A	3929	5308	7240	9638	12861	17063	23192	31443	38558	50189	62980

110	120	130	140	150	160	170	180	190	200
76212	92228	106022	123203	132165	151326	179323	203302	226542	248710

$A = 5835.54e^{0.020769t}$

2000: $t = 210$

$A(210) = 5835.54e^{0.020769(210)}$

$\quad = 457{,}353$ thousand

$\quad = 457{,}353{,}000$

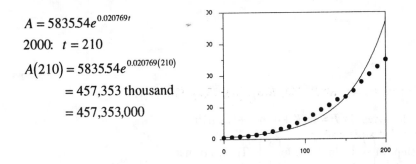

17.

t	0	1	2	3	4	5	6	7	8	9	10
y	1000	1024	1071	1287	1287	1553	1956	2722	2184	2791	3000

Linear function: $y = 217.69t + 718.36$

2000: $t = 20$

$\quad y = 217.69(20) + 718.36$

$\quad\quad = 5072$, High

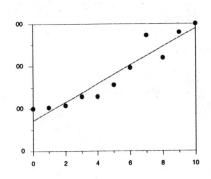

19.

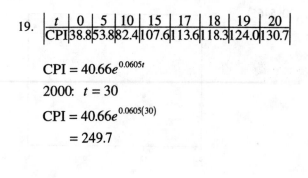

t	0	5	10	15	17	18	19	20
CPI	38.8	53.8	82.4	107.6	113.6	118.3	124.0	130.7

$\text{CPI} = 40.66e^{0.0605t}$

2000: $t = 30$

$\text{CPI} = 40.66e^{0.0605(30)}$

$\qquad = 249.7$

21.

p	10	15	20	25	30
q	30	15	10	5	3

$q = 3803.02\,p^{-2.058}$

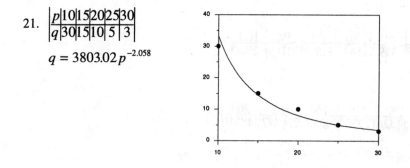

23. Taking the natural log of both sides gives $\ln P = \ln K + a \ln x + (1-a)\ln y$. This leads to the linear relationship $\ln\left(\dfrac{P}{y}\right) = \ln K + a \ln\left(\dfrac{x}{y}\right)$. Thus, given a number of data points $(x_i,\, y_i,\, P_i)$, we can use them to calculate the best fit linear relationship between $\ln\left(\dfrac{P}{y}\right)$ and $\ln\left(\dfrac{x}{y}\right)$. The slope is then a and the intercept is $\ln K$.

15.7 Exercises

1. $\int_0^1 \int_0^1 (x-2y)\,dx\,dy = \int_0^1 \left[\frac{x^2}{2} - 2xy\right]_0^1 dy = \int_0^1 \left[\left(\frac{1^2}{2} - 2y\right) - (0-0)\right] dy$

$= \int_0^1 \left(\frac{1}{2} - 2y\right) dy = \left[\frac{y}{2} - y^2\right]_0^1 = \left(\frac{1}{2} - 1^2\right) - (0-0) = -\frac{1}{2}$

3. $\int_0^1 \int_0^2 (ye^x - x - y)\,dx\,dy = \int_0^1 \left[ye^x - \frac{x^2}{2} - xy\right]_0^2 dy = \int_0^1 \left[\left(ye^2 - \frac{2^2}{2} - 2y\right) - (ye^0 - 0 - 0)\right] dy$

$= \int_0^1 \left[(e^2 - 3)y - 2\right] dy = \left[\frac{(e^2-3)y^2}{2} - 2y\right]_0^1 = \left(\frac{(e^2-3)(1)^2}{2} - 2(1)\right) - (0-0) = \frac{e^2-7}{2}$

5. $\int_0^3 \int_0^2 e^{x+y}\,dx\,dy = \int_0^3 \left[e^{x+y}\right]_0^2 dy = \int_0^3 \left(e^{2+y} - e^{0+y}\right) dy = \int_0^3 \left(e^{y+2} - e^y\right) dy$

$= \left[e^{y+2} - e^y\right]_0^3 = \left(e^{3+2} - e^3\right) - \left(e^{0+2} - e^0\right) = e^5 - e^3 - e^2 + 1 = (e^3 - 1)(e^2 - 1)$

7. $\int_0^1 \int_0^{2-y} x\,dx\,dy = \int_0^1 \left[\frac{x^2}{2}\right]_0^{2-y} dy = \int_0^1 \left[\frac{(2-y)^2}{2} - 0\right] dy = \begin{bmatrix} \text{Let } 2-y = u \\ -dy = du \end{bmatrix}$

$= \int_2^1 -\frac{u^2}{2}\,du = \left[-\frac{u^3}{6}\right]_2^1 = \left(-\frac{1^3}{6}\right) - \left(-\frac{2^3}{6}\right) = \frac{7}{6}$

9. $\int_{-1}^1 \int_{y-1}^{y+1} e^{x+y}\,dx\,dy = \int_{-1}^1 \left[e^{x+y}\right]_{y-1}^{y+1} dy = \int_{-1}^1 \left(e^{y+1+y} - e^{y-1+y}\right) dy = \int_{-1}^1 \left(e^{2y+1} - e^{2y-1}\right) dy$

$= \left[\frac{e^{2y+1}}{2} - \frac{e^{2y-1}}{2}\right]_{-1}^1 = \left(\frac{e^{2\cdot1+1}}{2} - \frac{e^{2\cdot1-1}}{2}\right) - \left(\frac{e^{2(-1)+1}}{2} - \frac{e^{2(-1)-1}}{2}\right) = \frac{1}{2}\left(e^3 - e - e^{-1} + e^{-3}\right)$

11. $\int_0^1 \int_{-x^2}^{x^2} x\,dx\,dy = \int_0^1 \int_{-x^2}^{x^2} x\,dy\,dx = \int_0^1 \left[xy\right]_{-x^2}^{x^2} dx = \int_0^1 \left[x^3 - (-x^3)\right] dx$

$= \int_0^1 2x^3\,dx = \left[\frac{2x^4}{4}\right]_0^1 = \frac{1^4}{2} - 0 = \frac{1}{2}$

13. $\displaystyle\int_0^1 \int_0^x e^{x^2}\, dy\, dx = \int_0^1 \left[y e^{x^2} \right]_0^x\, dx = \int_0^1 \left(x e^{x^2} - 0 \right) dx = \left[\begin{array}{l} \text{Let } u = x^2 \\ du = 2x\, dx \end{array} \right]$

$\displaystyle = \int_0^1 \frac{1}{2} e^u\, du = \left[\frac{1}{2} e^u \right]_0^1 = \frac{1}{2}\left(e^1 - e^0 \right) = \frac{1}{2}(e-1)$

15. $\displaystyle\int_1^3 \int_{1-x}^{8-x} \sqrt[3]{x+y}\, dy\, dx = \int_1^3 \left[\frac{(x+y)^{4/3}}{4/3} \right]_{1-x}^{8-x} dx = \frac{3}{4} \int_1^3 \left[(x+8-x)^{4/3} - (x+1-x)^{4/3} \right] dx$

$\displaystyle = \frac{3}{4} \int_1^3 (16-1)\, dx = \frac{45}{4} \int_1^3 dx = \frac{45}{4}[x]_1^3 = \frac{45}{4}(3-1) = \frac{45}{2}$

17. $\displaystyle\int_{-1}^1 \int_0^{1-x^2} 2\, dy\, dx = \int_{-1}^1 \left[2y \right]_0^{1-x^2} dx = \int_{-1}^1 \left[2(1-x^2) - 0 \right] dx = \int_{-1}^1 \left(2 - 2x^2 \right) dx$

$\displaystyle = \left[2x - \frac{2x^3}{3} \right]_{-1}^1 = \left(2(1) - \frac{2(1)^3}{3} \right) - \left(2(-1) - \frac{2(-1)^3}{3} \right) = 4 - \frac{4}{3} = \frac{8}{3}$

19. $\displaystyle\int_{-1}^1 \int_0^{1-y^2} (1+y)\, dx\, dy = \int_{-1}^1 \left[x(1+y) \right]_0^{1-y^2} dy = \int_{-1}^1 \left[(1-y^2)(1+y) - 0 \right] dy$

$\displaystyle = \int_{-1}^1 \left(1 + y - y^2 - y^3 \right) dy = \left[y + \frac{y^2}{2} - \frac{y^3}{3} - \frac{y^4}{4} \right]_{-1}^1 = \left(1 + \frac{1^2}{2} - \frac{1^3}{3} - \frac{1^4}{4} \right)$

$\displaystyle - \left((-1) + \frac{(-1)^2}{2} - \frac{(-1)^3}{3} - \frac{(-1)^4}{4} \right) = \frac{11}{12} + \frac{5}{12} = \frac{16}{12} = \frac{4}{3}$

21. $\displaystyle\int_{-1}^1 \int_{-\sqrt{1-y^2}}^{\sqrt{1-y^2}} xy^2\, dx\, dy = \int_{-1}^1 \left[\frac{x^2 y^2}{2} \right]_{-\sqrt{1-y^2}}^{\sqrt{1-y^2}} dy = \int_{-1}^1 \left[\frac{(1-y^2)y^2}{2} - \frac{(1-y^2)y^2}{2} \right] dy$

$\displaystyle = \int_{-1}^1 0\, dy = 0$

23. $\int_{-1}^{0}\int_{-x-1}^{x+1}\left(x^{2}+y^{2}\right)dy\,dx+\int_{0}^{1}\int_{x-1}^{-x+1}\left(x^{2}+y^{2}\right)dy\,dx=\int_{-1}^{0}\left[x^{2}y+\frac{y^{3}}{3}\right]_{-x-1}^{x+1}dx+\int_{0}^{1}\left[x^{2}y+\frac{y^{3}}{3}\right]_{x-1}^{-x+1}dx$

$=\int_{-1}^{0}\left[\left(x^{2}(x+1)+\frac{(x+1)^{3}}{3}\right)-\left(x^{2}(-x-1)+\frac{(-x-1)^{3}}{3}\right)\right]dx+\int_{0}^{1}\left[\left(x^{2}(-x+1)+\frac{(-x+1)^{3}}{3}\right)-\left(x^{2}(x-1)+\frac{(x-1)^{3}}{3}\right)\right]dx$

$=\int_{-1}^{0}\left(\frac{8}{3}x^{3}+4x^{2}+2x+\frac{2}{3}\right)dx+\int_{0}^{1}\left(-\frac{8}{3}x^{2}+4x^{2}-2x+\frac{2}{3}\right)dx$

$=\left[\frac{2x^{4}}{3}+\frac{4x^{3}}{3}+x^{2}+\frac{2x}{3}\right]_{-1}^{0}+\left[-\frac{2x^{4}}{3}+\frac{4x^{3}}{3}-x^{2}+\frac{2x}{3}\right]_{0}^{1}$

$=0-\left[\frac{2(-1)^{4}}{3}+\frac{4(-1)^{3}}{3}+(-1)^{2}+\frac{2(-1)}{3}\right]+\left[\frac{-2(1)^{4}}{3}+\frac{4(1)^{3}}{3}-(1)^{2}+\frac{2(1)}{3}\right]-0$

$=-\frac{2}{3}+\frac{4}{3}-1+\frac{2}{3}-\frac{2}{3}+\frac{4}{3}-1+\frac{2}{3}=\frac{2}{3}$

In Exercises 25 - 31 use $\overline{f}=\frac{1}{A}\int_{R}\int f(x,\,y)\,dx\,dy.$

25. $A=\int_{-1}^{1}\int_{0}^{1-x^{2}}dy\,dx=\int_{-1}^{1}\left[y\right]_{0}^{1-x^{2}}dx=\int_{-1}^{1}\left(1-x^{2}\right)dx=\left[x-\frac{x^{3}}{3}\right]_{-1}^{1}=\left(1-\frac{1}{3}\right)-\left(-1-\frac{(-1)^{3}}{3}\right)=\frac{4}{3}$

Use the result of Exercise 17. $\overline{f}=\frac{8/3}{4/3}=2.$

27. $A=\int_{-1}^{1}\int_{0}^{1-y^{2}}dx\,dy=$ (same A as Exercise 25) $=\frac{4}{3}$

Use the result of Exercise 19. $\overline{f}=\frac{4/3}{4/3}=1$

29. $A=\int_{-1}^{1}\int_{-\sqrt{1-y^{2}}}^{\sqrt{1-y^{2}}}dx\,dy=\pi(1)^{2}=\pi$

Use the result of Exercise 21. $\overline{f}=\frac{0}{\pi}=0.$

31. $A = \int_{-1}^{0} \int_{-x-1}^{x+1} dy \, dx + \int_{0}^{1} \int_{x-1}^{-x+1} dy \, dx = \left(\sqrt{2}\right)^2 = 2$

Use the result of Exercise 23. $\overline{f} = \dfrac{2/3}{2} = \dfrac{1}{3}$.

33.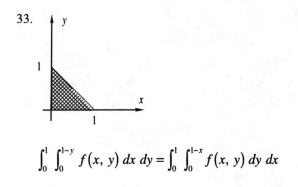

$$\int_{0}^{1} \int_{0}^{1-y} f(x, y) \, dx \, dy = \int_{0}^{1} \int_{0}^{1-x} f(x, y) \, dy \, dx$$

35.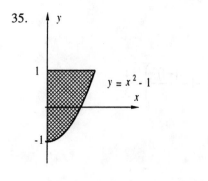

$$\int_{0}^{1} \int_{0}^{\sqrt{1+y}} f(x, y) \, dx \, dy = \int_{0}^{1} \int_{x^2-1}^{1} f(x, y) \, dy \, dx$$

37.

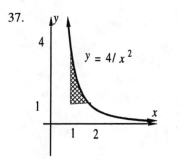

$$\int_1^2 \int_1^{4/x^2} f(x, y) \, dy \, dx = \int_1^2 \int_1^{2/\sqrt{y}} f(x, y) \, dx \, dy$$

39.

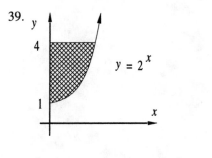

$$\int_0^2 \int_{2^x}^4 f(x, y) \, dy \, dx = \int_1^4 \int_0^{\log_2 y} f(x, y) \, dx \, dy$$

41. $V = \int_0^2 \int_0^1 \left(1 - x^2\right) dx \, dy = \int_0^2 \left[x - \dfrac{x^3}{3} \right] \, dy = \int_0^2 \left[\left(1 - \dfrac{1^3}{3} \right) - (0 - 0) \right] dy = \int_0^2 \dfrac{2}{3} \, dy$

$= \left[\dfrac{2}{3} y \right]_0^2 = \dfrac{2}{3}(2 - 0) = \dfrac{4}{3}$

43. The region R in the triangle in the $x - y$ plane bounded by the x and y axes and the line $x + y = 1$. The volume is bounded above by the plane $x + y + z = 1$.

$$V = \int_0^1 \int_0^{-x+1} (1 - x - y)\, dy\, dx = \int_0^1 \left[(1-x)y - \frac{y^2}{2} \right]_0^{1-x} dx = \int_0^1 \left[\left((1-x)(1-x) - \frac{(1-x)^2}{2} \right) - 0 \right] dx$$

$$= \int_0^1 \frac{(1-x)^2}{2}\, dx = \frac{1}{2} \int_0^1 (1 - 2x + x^2)\, dx = \frac{1}{2} \left[x - x^2 + \frac{x^3}{3} \right]_0^1 = \frac{1}{2} \left[\left(1 - 1^2 + \frac{1^3}{3} \right) - 0 \right] = \frac{1}{6}$$

45. $\displaystyle \int_R \int P\, dx\, dy = \int_{45}^{55} \int_8^{12} 10{,}000 x^{0.3} y^{0.7}\, dy\, dx = 10{,}000 \int_{45}^{55} \left[\frac{x^{0.3} y^{1.7}}{1.7} \right]_8^{12} dx$

$$= \int_{45}^{55} 10{,}000 \frac{x^{0.3}}{1.7} \left[(12)^{1.7} - (8)^{1.7} \right] dx = \int_{45}^{55} 200{,}192.5 x^{0.3}\, dx$$

$$= \left[200{,}192.5 \frac{x^{1.3}}{1.3} \right]_{45}^{55} = 153{,}994.2 \left[(55)^{1.3} - (45)^{1.3} \right] = 6{,}471{,}223$$

$$\overline{P} = \frac{6{,}471{,}223}{(55 - 45)(12 - 8)} = 161{,}781$$

47. Max: $R = pq = 50 \left[10{,}000 - (50)^2 \right] = \$375{,}000$

Min: $R = pq = 40 \left[8000 - (40)^2 \right] = \$256{,}000$

$$A = \int_{40}^{50} \int_{8000-p^2}^{10{,}000-p^2} dq\, dp = \int_{40}^{50} [q]_{8000-p^2}^{10{,}000-p^2} dp = \int_{40}^{50} (10{,}000 - p^2) - (8000 - p^2)\, dp$$

$$= \int_{40}^{50} 2000\, dp = 2000[p]_{40}^{50} = 2000[50 - 40] = 20{,}000$$

$$= \int_{40}^{50} \int_{8000-p^2}^{10{,}000-p^2} pq\, dq\, dp = \int_{40}^{50} \left[\frac{pq^2}{2} \right]_{8000-p^2}^{10{,}000-p^2} dp = \frac{1}{2} \int_{40}^{50} \left[p(10{,}000 - p^2)^2 - p(8000 - p^2)^2 \right] dp$$

$$= \frac{1}{2} \int_{40}^{50} (36{,}000{,}000 - 4000p^2)p\, dq = \left[\frac{36{,}000{,}000}{4} p^2 - \frac{4000}{8} p^4 \right]_{40}^{50}$$

$$= \left(9{,}000{,}000(50)^2 - 500(50)^4 \right) - \left(9{,}000{,}000(40)^2 - 500(40)^4 \right) = 6{,}255{,}000{,}000$$

$$\overline{R} = \frac{6{,}255{,}000{,}000}{20{,}000} = \$312{,}750$$

49. Max: $R = pq = (20{,}000/q)(q) = \$20{,}000$

Min: $R = pq = (15{,}000/q)(q) = \$15{,}000$

$$\int_{500}^{1000} \int_{15{,}000/q}^{20{,}000/q} pq \, dp \, dq = \int_{500}^{1000} \left[q\frac{p^2}{2} \right]_{15{,}000/q}^{20{,}000/q} dp = \frac{1}{2}\int_{500}^{1000} q\left[\left(\frac{20{,}000}{q}\right)^2 - \left(\frac{15{,}000}{q}\right)^2 \right] dq$$

$$= \frac{1}{2}\int_{500}^{1000} \frac{175{,}000{,}000}{q} \, dq = 87{,}500{,}000[\ln q]_{500}^{1000} = 87{,}500{,}000[\ln 1000 - \ln 500]$$

$$= 60{,}650{,}378$$

$$A = \int_{500}^{1000} \int_{15{,}000/q}^{20{,}000/q} dp \, dq = \int_{500}^{1000} [p]_{15{,}000/q}^{20{,}000/q} \, dq = \int_{500}^{1000} \left[\frac{20{,}000}{q} - \frac{15{,}000}{q} \right] dq$$

$$= \int_{500}^{1000} \frac{5000}{q} \, dq = 5000[\ln q]_{500}^{1000} = 5000(\ln 1000 - \ln 500) = 3{,}465.7$$

$$\overline{R} = \frac{60{,}650{,}378}{3{,}465.7} = \$17{,}500$$

51. $$\int_{0}^{20} \int_{0}^{30} e^{-0.1(x+y)} \, dy \, dx = \int_{0}^{20} \frac{1}{-0.1}\left[e^{-0.1(x+y)} \right]_{0}^{30} dx = -10\int_{0}^{20}\left[e^{-0.1(x+30)} - e^{-0.1x} \right] dx$$

$$= -10\int_{0}^{20}\left(e^{-3} - 1\right)e^{-0.1x} \, dx = -10\left(-\frac{1}{0.1}\right)\left(e^{-3} - 1\right)\left[e^{-0.1x} \right]_{0}^{20}$$

$$= 100\left(e^{-3} - 1\right)\left(e^{-0.1(20)} - e^0\right) = 100\left(e^{-3} - 1\right)\left(e^{-2} - 1\right) = 82.16$$

Total population $= 8216$

53. $A = 1^2 = 1$

$$\int_{0}^{1} \int_{0}^{1} \left(x^2 + 2y^2\right) dy \, dx = \int_{0}^{1}\left[x^2 y + \frac{2y^3}{3} \right]_{0}^{1} dx = \int_{0}^{1}\left[\left(x^2(1) + \frac{2(1)^3}{3} \right) - (0+0) \right] dx$$

$$= \int_{0}^{1}\left(x^2 + \frac{2}{3} \right) dx = \left[\frac{x^3}{3} + \frac{2x}{3} \right]_{0}^{1} = \left(\frac{1^3}{3} + \frac{2(1)}{3} \right) - (0+0) = 1$$

$$\overline{T} = \frac{1}{1} = 1 \text{ degree}$$

519

Chapter 15 Review

1. $g(x, y, z) = xy(x + y - z) + x^2$

$g(0, 0, 0) = 0$

$g(1, 0, 0) = 1(0)(0 + 0 - 0) + 1^2 = 1$

$g(0, 1, 0) = 0(1)(0 + 1 - 0) + 0^2 = 0$

$g(x, x, x) = xx(x + x - x) + x^2 = x^3 + x^2$

$g(x, y + k, z) = x(y + k)(x + y + k - z) + x^2$

3. $g(x, y, z) = xe^{xy+z}$

$g(0, 0, 0) = 0 \cdot e^{0 \cdot 0 + 0} = 0$

$g(1, 0, 0) = 1 \cdot e^{1 \cdot 0 + 0} = 1$

$g(0, 1, 0) = 0 \cdot e^{0 \cdot 1 + 0} = 0$

$g(x, x, x) = xe^{x^2 + x}$

$g(x, y + k, z) = xe^{x(y+k)+z}$

Use $d = \sqrt{(x_2 - x_1)^2 + (y_2 - y_1)^2}$ in Exercises 5-7.

5. For $(1, 2)$ and $(3, 0)$

$d = \sqrt{(3 - 1)^2 + (0 - 2)^2} = 2\sqrt{2}$

7. For (a, b) and (a, c)

$d = \sqrt{(a - a)^2 + (c - b)^2}$

$= \sqrt{(b - c)^2} = |b - c|$

Use $P(x, y) = K\, x^9 y^{1-a}$ in Exercise 9.

9. $P(1000, 100{,}000) = 100(1000)^{0.3}(100{,}000)^{1-0.3} = 2{,}511{,}886$

11. $f(x, y, z) = x^2 + xyz$

$f_x = 2x + yz$

$f_y = xz$

$f_z = xy$

$f_{xy} = z$

$f_{xz} = y$

$f_{zz} = 0$

13. $f(x, y, z) = \dfrac{x}{\left(x^2 + y^2 + z^2\right)}$

$f_x = \dfrac{\left(x^2 + y^2 + z^2\right) - x(2x)}{\left(x^2 + y^2 + z^2\right)^2}$

$\quad = \dfrac{-x^2 + y^2 + z^2}{\left(x^2 + y^2 + z^2\right)^2}$

$f_y = \dfrac{0 - x(2y)}{\left(x^2 + y^2 + z^2\right)^2}$

$\quad = \dfrac{-2xy}{\left(x^2 + y^2 + z^2\right)^2}$

$f_z = \dfrac{0 - x(2z)}{\left(x^2 + y^2 + z^2\right)^2}$

$\quad = \dfrac{-2xz}{\left(x^2 + y^2 + z^2\right)^2}$

$f_x(0, 1, 0) = \dfrac{1^2}{\left(1^2\right)^2} = 1$

15. $f(x, y, z) = \left(x^2 + y^2 + z^2\right)^{-\frac{1}{2}}$

$f_x = -\dfrac{1}{2}\left(x^2 + y^2 + z^2\right)^{-\frac{3}{2}}(2x) = -x\left(x^2 + y^2 + z^2\right)^{-\frac{3}{2}}$

$f_{xx} = -1\left(x^2 + y^2 + z^2\right)^{-\frac{3}{2}} + (-x)\left(-\dfrac{3}{2}\right)\left(x^2 + y^2 + z^2\right)^{-\frac{5}{2}}(2x) = \left(x^2 + y^2 + z^2\right)^{-\frac{5}{2}}\left(-x^2 - y^2 - z^2 + 3x^2\right)$

$f_y = -\dfrac{1}{2}\left(x^2 + y^2 + z^2\right)^{-\frac{3}{2}}2y = -2y\left(x^2 + y^2 + z^2\right)^{-\frac{3}{2}}$

$f_{yy} = -1\left(x^2 + y^2 + z^2\right)^{-\frac{3}{2}} + (-y)\left(-\dfrac{3}{2}\right)\left(x^2 + y^2 + z^2\right)^{-\frac{5}{2}}(2y) = \left(x^2 + y^2 + z^2\right)^{-\frac{5}{2}}\left(-x^2 - y^2 - z^2 + 3y^2\right)$

$f_z = -\dfrac{1}{2}\left(x^2 + y^2 + z^2\right)^{-\frac{3}{2}}(2z) = -z\left(x^2 + y^2 + z^2\right)^{-\frac{3}{2}}$

$f_{zz} = -1\left(x^2 + y^2 + z^2\right)^{-\frac{3}{2}} + (-z)\left(-\dfrac{3}{2}\right)\left(x^2 + y^2 + z^2\right)^{-\frac{5}{2}}(2z) = \left(x^2 + y^2 + z^2\right)^{-\frac{5}{2}}\left(-x^2 - y^2 - z^2 + 3z^2\right)$

$f_{xx} + f_{yy} + f_{zz} = \left(x^2 + y^2 + z^2\right)^{-\frac{5}{2}}\left(-3x^2 - 3y^2 - 3z^2 + 3x^2 + 3y^2 + 3z^2\right) = 0$

17. $F(x,y,z) = G\dfrac{Mm}{(x-a)^2 + (y-b)^2 + (z-c)^2} = G\dfrac{(3000)(10)}{(x-1)^2 + (y-1)^2 + (z-1)^2}$

$F_z = \dfrac{-G(30{,}000)(2)(z-1)}{\left[(x-1)^2 + (y-1)^2 + (z-1)^2\right]^2}$

$F_z(10,100,100) = \dfrac{-6.67 \times 10^{-11}(60{,}000)(100-1)}{\left[(10-1)^2 + (100-1)^2 + (100-1)^2\right]^2} \approx -1.02266 \times 10^{-12} \dfrac{\text{newtons}}{\text{second}}$

Use $H = f_{xx}(a,b)f_{yy}(a,b) - \left[f_{xy}(a,b)\right]^2$ in Exercises 19 - 31.

19. $f(x,y) = (x-1)^2 + (2y-3)^2$

$f_x = 2(x-1) = 0 \Rightarrow x = 1$

$f_y = 2(2y-3) = 0 \Rightarrow y = \dfrac{3}{2}$

Critical point: $\left(1, \dfrac{3}{2}\right)$

$f_{xx} = 2, f_{yy} = 4, f_{xy} = 0$

$H = (2)(4) - (0)^2 > 0, f_{xx} > 0$

Absolute minimum at $\left(1, \dfrac{3}{2}\right)$.

21. $g(x,y) = (x-1)^2 - 3y^2 + 9$

$g_x = 2(x-1) = 0 \Rightarrow x = 1$

$g_y = -6y = 0 \Rightarrow y = 0$

Critical point: $(1, 0)$

$g_{xx} = 2, g_{yy} = -6, g_{xy} = 0$

$H = (2)(-6) - (0)^2 < 0$

Saddle point at $(1, 0)$.

23. $h(x,y) = e^{xy}$

$h_x = ye^{xy} = 0 \Rightarrow y = 0$

$h_y = xe^{xy} = 0 \Rightarrow x = 0$

Critical point: $(0, 0)$

$h_{xx} = y^2 e^{xy}, h_{xx}(0, 0) = 0$

$h_{yy} = y^2 e^{xy}, h_{yy}(0, 0) = 0$

$h_{xy} = e^{xy} + y^2 e^{xy}, h_{xy}(0, 0) = 1$

$H = (0)(0) - (1)^2 < 0$

Saddle point at $(0, 0)$.

25. $f(x,y) = xy + x^2$

$f_x = y + 2x = 0 \Rightarrow y = -2x$

$f_y = x = 0 \Rightarrow y = 0$

Critical point: $(0, 0)$

$f_{xx} = 2, f_{yy} = 0, f_{xy} = 1$

$H = (2)(0) - (1)^2 < 0$

Saddle point at $(0, 0)$.

27. $f(x, y) = \ln(x^2 + y^2) - (x^2 + y^2)$

$f_x = \dfrac{2x}{(x^2 + y^2)} - 2x = 0 \Rightarrow x^2 + y^2 = 1, \; f_y = \dfrac{2y}{(x^2 + y^2)} - 2y = 0 \Rightarrow x^2 + y^2 = 1$

Critical points: Each point on the circle $x^2 + y^2 = 1$

$f_{xx} = \dfrac{2(x^2 + y^2) - 2x(2x)}{(x^2 + y^2)^2} - 2 = \dfrac{-2x^2 + y^2}{(x^2 + y^2)^2}, \; f_{yy} = \dfrac{2(x^2 + y^2) - 2y(2y)}{(x^2 + y^2)^2} - 2 = \dfrac{-2y^2 + x^2}{(x^2 + y^2)^2}, \; f_{xy} = \dfrac{-2x(2y)}{(x^2 + y^2)^2}$

At critical points, $f_{xx} = \dfrac{2 - 2x^2}{1^2} - 2 = -2x^2, \; f_{yy} = \dfrac{2 - 2y^2}{1^2} - 2 = -2y^2, \; f_{xy} = \dfrac{-4xy}{1^2} = -4xy$

$H = (-2x^2)(-2y^2) - (-4xy)^2 > 0, \; f_{xx} < 0, \;$ Absolute maximum at each point where $x^2 + y^2 = 1$

29. Minimize $f(x, y, z) = x^2 + y^2 + z^2$ subject to $z^2 = 2(y - 3)^2 + x^2$

On the boundary, $z^2 = 2(y - 3)^2 + x^2$ the function $f(x, y, z)$ becomes, $f = x^2 + y^2 + 2(y - 3)^2 + x^2$

$= 2x^2 + y^2 + 2(y - 3)^2, \; f_x = 4x = 0 \Rightarrow x = 0, \; f_y = 2y + 4(y - 3) = 0 \Rightarrow y = 2.$ Critical point $(0, 2)$.

$f_{xx} = 4, \; f_{yy} = 6, \; f_{xy} = 0, \; f(0, 2) = \sqrt{2}. \; H = 4(6) - 0^2 > 0, \; f_{xx} > 0.$ Minimum at $(0, 2, \sqrt{20})$

31. Minimize $f(x, y, z) = x^2 + y^2 + z^2$ subject to $z = \dfrac{1}{x}$

On the boundary, $z = \dfrac{1}{x}$, the function $f(x, y, z)$ becomes, $f(x, y) = x^2 + y^2 + \dfrac{1}{x^2}$

$f_x = 2x - \dfrac{2}{x^3} = 0 \Rightarrow x^4 = 1 \Rightarrow x = \pm 1 \Rightarrow z = \pm 1, \; f_y = 2y = 0 \Rightarrow y = 0.$ Critical point $(\pm 1, 0)$.

$f_{xx} = 2 + \dfrac{6}{x^4} > 0, \; f_{yy} = 2, \; f_{xy} = 0, \; f(\pm 1, 2) = \pm 1. \; H = \left(2 + \dfrac{6}{x^4}\right)(2) - 0^2 > 0, \; f_{xx} > 0.$

Minimum distance at $(\pm 1, 0, \pm 1)$

33. $T(x, y) = (x-1)^2 + 2y^2$ bounded by the triangle with vertices $(0, 1)$, $(3, 0)$ and $(0, 0)$.

Interior:

$T_x = 2(x-1) = 0 \Rightarrow x = 1$, $T_y = 2y = 0 \Rightarrow y = 0$

Critical point $(1, 0)$ is on the boundary. No extremum in the interior.

Boundary:

Side 1: $y = 0$, $0 \le x \le 3$; The endpoints are $(0, 0, 1)$ and $(3, 0, 4)$

   On the side $y = 0$, the function becomes $z = (x-1)^2$, $z' = 2(x-1) = 0 \Rightarrow x = 1$,

   $z'' = 2$. The critical point $(1, 0)$ is a minimum of 0.

Side 2: $x = 0$, $0 \le y \le 1$. The endpoints are $(0, 0, 1)$ and $(0, 1, 2)$

Maximum of 4 at $(3, 0)$ and minimum of 0 at $(1, 0)$

Hottest point $(3, 0)$. Coldest point $(1, 0)$.

35. $C(x, y) = 0.001x^2 + 0.002y^2 - 2x - 4.8y + 10,880$, $000 \le x \le 1100$, $1000 \le y \le 1500$

Interior:

$C_x = 0.002x - 2 = 0 \Rightarrow x = 1000$, $C_y = 0.004y - 4.8 = 0 \Rightarrow y = 1200$

Critical point $(1000, 1200)$. $C_{xx} = 0.002$, $C_{yy} = 0.004$, $C_{xy} = 0$

$H = 0.002(0.004) - 0^2 > 0$, $C_{xx} > 0$. Minimum at $(1000, 1200, 7000)$

Boundary:

Side 1: $y = 1000$, $900 \le x \le 1100$. The endpoints are $(900, 1000, 7090)$ and $(1100, 1000, 7090)$

Side 2: $x = 900$, $1000 \le y \le 1500$; The endpoints are $(900, 1000, 7090)$ and $(900, 1500, 7190)$

Side 3: $y = 1500$, $900 \le x \le 1100$; The endpoints are $(900, 1500, 7190)$ and $(1100, 1500, 7190)$

For minimum cost make one thousand pencils and twelve hundred pens. For maximum

cost make either nine hundred or eleven hundred pencils and fifteen hundred pens.

37. $(1, 2)$, $(3, -4)$, $(5, 0)$

$\sum p_i^2 = 1^2 + 3^2 + 5^2 = 35$

$\sum p_i = 1 + 3 + 5 = 9$

$\sum q_i = 2 + -4 + 0 = -2$

$\sum p_i q_i = (1 \cdot 2) + 3(-4) + (5 \cdot 0) = 10$

$n = 3$

$35m + 9b = -10$ (1)

$9m + 3b = -2$ (2)

Solve (1) and (2) to get

$m = -\dfrac{1}{2}$

$b = \dfrac{5}{6}$

$y = -\dfrac{x}{2} + \dfrac{5}{6}$

524

39. $(0, 1),\ (1, 2),\ (4, 2),\ (5, 0)$

$$\sum p_i^2 = 0^2 + 1^2 + 4^2 + 5^2 = 42$$

$$\sum p_i = 0 + 1 + 4 + 5 = 10$$

$$\sum q_i = -1 + 2 + 2 + 0 = 5$$

$$\sum p_i q_i = (0 \cdot 1) + (1 \cdot 2) + (4 \cdot 2) + (5 \cdot 0) = 10$$

$$n = 4$$

$$42m + 10b = 10 \qquad (1)$$

$$10m + 4b = 5 \qquad (2)$$

Solve (1) and (2) to get

$m = -0.1471$

$b = 1.6176$

$y = -0.1471x + 1.6176$

41.

Year	1	1650	1850	1930	1975	1991
Population	0.2	0.5	1	2	4	5.384

43. $\displaystyle\int_1^2 \int_1^3 \sqrt{x+y}\ dx\ dy = \int_1^2 \left[\frac{2}{3}(x+y)^{3/2}\right]_1^3 dy = \int_1^2 \left[\frac{2}{3}(3+y)^{3/2} - \frac{2}{3}(1+y)^{3/2}\right] dy$

$$= \left[\frac{4}{15}(3+y)^{5/2}\right]_1^2 - \left[\frac{4}{15}(1+y)^{5/2}\right]_1^2 = \frac{4}{15}\left[(3+2)^{5/2} - (3+1)^{5/2}\right] - \frac{4}{15}\left[(1+2)^{5/2} - (1+1)^{5/2}\right]$$

$$= \frac{4}{15}\left(5^{5/2} - 4^{5/2} - 3^{5/2} + 2^{5/2}\right)$$

45. $\displaystyle\int_0^2 \int_x^{2x} \frac{1}{x^2+1}\ dy\ dx = \int_0^2 \left[\frac{y}{x^2+1}\right]_x^{2x} dx = \int_0^2 \left(\frac{2x}{x^2+1} - \frac{x}{x^2+1}\right) dx = \int_0^2 \frac{x}{x^2+1}\ dx$

$$= \frac{1}{2}\left[\ln\left(x^2+1\right)\right]_0^2 = \frac{1}{2}\left[\ln\left(2^2+1\right) - \ln\left(0^2+1\right)\right] = \frac{1}{2}\ln 5$$

47. $\displaystyle\int_0^2 \int_0^{\sqrt{4-x^2}} xy\ dy\ dx = \int_0^2 \left[\frac{y^2 x}{2}\right]_0^{\sqrt{4-x^2}} dx = \int_0^2 \left(\frac{(4-x^2)x}{2} - 0\right) dx$

$$= \int_0^2 \left(2x - \frac{x^3}{2}\right) dx = \left[x^2 - \frac{x^4}{8}\right]_0^2 = \left(2^2 - \frac{2^4}{8}\right) - (0-0) = 2$$

In Exercises 49 - 53, use the results of 43 - 47 and $\overline{f} = \dfrac{\int_R \int f(x, y)\, dx\, dy}{A}$.

49. $A = \int_1^2 \int_1^3 dx\, dy = \int_1^2 [x]_1^3\, dy = \int_1^2 (3-1)\, dy = [2y]_1^2 = 2(2-1) = 2$

$\overline{f} = \dfrac{\dfrac{4}{15}\left(5^{5/2} - 4^{5/2} - 3^{5/2} + 2^{5/2}\right)}{2} = \dfrac{2}{15}\left(5^{5/2} - 4^{5/2} - 3^{5/2} + 2^{5/2}\right)$

51. $A = \int_0^2 \int_x^{2x} dy\, dx = \int_0^2 [y]_x^{2x}\, dx = \int_0^2 (2x - x)\, dx = \int_0^2 x\, dx = \left[\dfrac{x^2}{2}\right]_0^2 = \dfrac{1}{2}(2^2 - 0) = 2$

$\overline{f} = \dfrac{\dfrac{1}{2}\ln 5}{2} = \dfrac{1}{4}\ln 5$

53. $A = \int_0^2 \int_0^{\sqrt{4-x^2}} dy\, dx = \dfrac{1}{4}\pi(2)^2 = \pi$

$\overline{f} = \dfrac{2}{\pi}$

55. $V = \int_{-1}^1 \int_{-1}^1 (4 - x^2 - y^2)\, dx\, dy = \int_{-1}^1 \left[4x - \dfrac{x^3}{3} - xy^2\right]_{-1}^1 dy = \int_{-1}^1 \left[\left(1(4 - y^2) - \dfrac{1^3}{3}\right) - \left((-1)(4 - y^2) - \dfrac{(-1)^3}{3}\right)\right] dy$

$= \int_{-1}^1 \left(\dfrac{22}{3} - 2y^2\right) dy = \left[\dfrac{22}{3}y - \dfrac{2y^3}{3}\right]_{-1}^1 = \left(\dfrac{22}{3}(1) - \dfrac{2(1)^3}{3}\right) - \left(\dfrac{22}{3}(-1) - \dfrac{2(-1)^3}{3}\right) = \dfrac{40}{3}$

57. $\int_{1800}^{2000} \int_{1200}^{1500} (0.03x + 0.10y)\, dx\, dy = \int_{1800}^{2000} \left[0.015x^2 + 0.10xy\right]_{1200}^{1500} dy$

$= \int_{1800}^{2000} \left[(0.015)(1500)^2 + 0.10(1500)y - \left(0.015(1200)^2 + 0.10(1200)y\right)\right] dy$

$= \int_{1800}^{2000} (12{,}150 + 30y)\, dy = \left[12{,}150y + 30y^2\right]_{1800}^{2000}$

$= \left(12{,}150(2000) + 15(2000)^2\right) - \left(12{,}150(1800) + 15(1800)^2\right) = 13{,}830{,}000$

$A = \int_{1800}^{2000} \int_{1200}^{1500} dx\, dy = (200)(300) = 60{,}000$

$\overline{P} = \dfrac{13{,}830{,}000}{60{,}000} = \$230.50$